Applied Mathematical Sciences

Volume 196

More information about this series at http://www.springer.com/series/34

Zhongqiang Zhang • George Em Karniadakis

Numerical Methods for Stochastic Partial Differential Equations with White Noise

Zhongqiang Zhang
Department of Mathematical Sciences
Worcester Polytechnic Institute
Worcester, Massachusetts, USA

George Em Karniadakis
Division of Applied Mathematics
Brown University
Providence, Rhode Island, USA

ISSN 0066-5452 ISSN 2196-968X (electronic)
Applied Mathematical Sciences
ISBN 978-3-319-57510-0 ISBN 978-3-319-57511-7 (eBook)
DOI 10.1007/978-3-319-57511-7

Library of Congress Control Number: 2017941192

Mathematics Subject Classification (2010): 65Cxx, 65Qxx, 65Mxx

Printed on acid-free paper

This Springer imprint is published by Springer Nature
The registered company is Springer International Publishing AG
The registered company address is: Gewerbestrasse 11, 6330 Cham, Switzerland

Preface

In his forward-looking paper [374] at the conference "Mathematics Towards the Third Millennium," our esteemed colleague at Brown University Prof. David Mumford argued that "...stochastic models and statistical reasoning are more relevant to i) the world, ii) to science and many parts of mathematics and iii) particularly to understanding the computations in our mind, than exact models and logical reasoning." Deterministic modeling and corresponding simulations are computationally much more manageable than stochastic simulations, but they also offer much less, i.e., a single point in the "design space" instead of a "sphere" of possible solutions that reflect the various random (or not) perturbations of the physical or biological problem we study.

In the last twenty years, three-dimensional simulations of physical and biological phenomena have gone from an exception to the rule, and they have been widely adopted in Computational Science and Engineering (CSE). This more realistic approach to simulating physical phenomena together with the continuing fast increase of computer speeds has also led to the desire in performing more ambitious and even more realistic simulations with multiscale physics. However, not all scales can be modeled directly, and stochastic modeling has been used to account for the un-modeled physics. In addition, there is a fundamental need to quantify the uncertainty of large-scale simulations, and this has led to a new emerging field in CSE, namely that of Uncertainty Quantification or UQ. Hence, computational scientists and engineers are interested in endowing their simulations with "error bars" that reflect not only numerical discretization errors but also uncertainties associated with unknown precise values of material properties, ill-defined or even unknown boundary conditions, or uncertain constitutive laws in the governing equations. However, performing UQ taxes the computational resources greatly, and hence the selection of the proper numerical method for stochastic simulations is of paramount importance, in some sense much more important than selecting a numerical method for deterministic modeling.

The most popular simulation method for stochastic modeling is the Monte Carlo method and its various extensions, but it requires a lot of computational effort to compute thousands and often millions of sample paths required to obtain certain statistics of the quantity of interest. Specifically, Monte Carlo methods are quite general, but they suffer from slow convergence so they are usually employed in conjunction with some variance reduction techniques to produce satisfactory accuracy in practice. More recently and for applications that employ stochastic models with *color* noise, deterministic integration methods in random space have been used with great success as they lead to high accuracy, especially for a modest number of uncertain parameters. However, they are not directly applicable to stochastic partial differential equations (SPDEs) with temporal white noise, since their solutions are usually non-smooth and would require a very large number of random variables to obtain acceptable accuracy.

Methodology. For linear SPDEs, we can still apply deterministic integration methods by exploiting the linear property of these equations for a long-time numerical integration. This observation has been made in [315] for Zakai-type equations, where a recursive Wiener chaos method was developed. In this book, we adopt this idea and we further formulate a recursive strategy to solve linear SPDEs of general types using Wiener chaos methods and stochastic collocation methods in conjunction with sparse grids for efficiency. In order to apply these deterministic integration methods, we first truncate the stochastic processes (e.g., Brownian motion) represented by orthogonal expansions. In this book, we show that the orthogonal expansions can lead to higher-order schemes with proper time discretization when Wiener chaos expansion methods are employed. However, we usually keep only a small number of truncation terms in the orthogonal expansion of Brownian motion to efficiently use deterministic integration methods for *temporal* noise. For *spatial* noise, the orthogonal expansion of Brownian motions leads to higher-order methods in both random space and physical space when the solutions to the underlying SPDEs are smooth.

The framework we use is the Wong-Zakai approximation and Wiener chaos expansion. Wiener chaos expansion is associated with the Ito-Wick product which was used intensively in [223]. The methodology and the proofs are introduced in [315] and in some subsequent papers by Rozovsky and his collaborators. In this framework, we are led to systems of deterministic partial differential equations with unknowns being the Wiener chaos expansion coefficients, and it is important to understand the special structure of these linear systems. Following another framework, such as viewing the SPDEs as infinite-dimensional stochastic differential equations (SDEs), admits a ready application of numerical methods for SDEs. We emphasize that there are multiple points of view for treating SPDEs and accordingly there are many views on what are proper numerical methods for SPDEs. However, irrespective of such views, the common difficulty remains: the solutions are typically very rough and do not have first-order derivatives either in time or in space. Hence, no high-order (higher than first-order) methods are known except in very special cases.

Since the well-known monographs on numerical SDEs by Kloeden & Platen (1992) [259], Milstein (1995) [354], and Milstein & Tretyakov (2004) [358], numerical SPDEs with white noise have gained popularity, and there have been some new books on numerical SPDEs available, specifically:

- The book by Jentzen & Kloeden (2011) [251] on the development of stochastic Taylor's expansion for mild solutions to stochastic parabolic equations and their application to numerical methods.
- The book by Grigoriu (2012) [174] on the application of stochastic Galerkin and collocation methods as well as Monte Carlo methods to partial differential equations with random data, especially elliptic equations. Numerical methods for SODEs with random coefficients are discussed as well.
- The book by Kruse (2014) [277] on numerical methods in space and time for semi-linear parabolic equations driven by space-time noise addressing strong (L^p or mean-square) and weak (moments) sense of convergence.
- The book by Lord, Powell, & Shardlow (2014) [308] on the introduction of numerical methods for stochastic elliptic equations with color noise and stochastic semi-linear equations with space-time noise.

For numerical methods of stochastic differential equations with *color* noise, we refer the readers to [294, 485]. On the theory of SPDEs, there are also some new developments, and we refer the interested readers to the book [36] covering amplitude equations for nonlinear SPDEs and to the book [119] on homogenization techniques for effective dynamics of SPDEs.

How to use this book. This book can serve as a reference/textbook for graduate students or other researchers who would like to understand the state-of-the-art of numerical methods for SPDEs with white noise.

Reading this book requires some basic knowledge of probability theory and stochastic calculus, which are presented in Chapter 2 and Appendix A. Readers are also required to be familiar with numerical methods for partial differential equations and SDEs in Chapter 3 before further reading. The reader can also refer to Chapter 3 for MATLAB implementations of test problems. More MATLAB codes for examples in this book are available upon request. For those who want to take a glance of numerical methods for stochastic partial differential equation, they are encouraged to read a review of these methods presented in Chapter 3. Exercises with hints are provided in most chapters to nurture the reader's understanding of the presented materials.

Part I. Numerical stochastic ordinary differential equations. We start with numerical methods for SDEs with delay using the Wong-Zakai approximation and finite difference in time in Chapter 4. The framework of Wong-Zakai approximation is used throughout the book. If the delay time is zero, we then recover the standard SDEs. We then discuss how to deal with strong nonlinearity and stiffness in SDEs in Chapter 5.

Part II. Temporal white noise. In Chapters 6–8, we consider SPDEs as PDEs driven by white noise, where discretization of white noise (Brownian motion) leads to PDEs with smooth noise, which can then be treated by numerical methods for PDEs. In this part, recursive algorithms based on Wiener chaos expansion and stochastic collocation methods are presented for linear stochastic advection-diffusion-reaction equations. Stochastic Euler equations in Chapter 9 are exploited as an application of stochastic collocation methods, where a numerical comparison with other integration methods in random space is made.

Part III. Spatial white noise. We discuss in Chapter 10 numerical methods for nonlinear elliptic equations as well as other equations with additive noise. Numerical methods for SPDEs with multiplicative noise are discussed using the Wiener chaos expansion method in Chapter 11. Some SPDEs driven by non-Gaussian white noise are discussed, where some model reduction methods are presented for generalized polynomial chaos expansion methods.

We have attempted to make the book self-contained. Necessary background knowledge is presented in the appendices. Basic knowledge of probability theory and stochastic calculus is presented in Appendix A. In Appendix B, we present some semi-analytical methods for SPDEs. In Appendix C, we provide a brief introduction to Gauss quadrature. In Appendix D, we list all the conclusions we need for proofs. In Appendix E, we present a method to compute convergence rate empirically.

MATLAB codes accompanying this book are available at the following website:

https://github.com/springer-math/Numerical-Methods-for-Stochastic-Partial-Differential-Equations-with-White-Noise

Acknowledgments. This book is dedicated to Dr. Fariba Fahroo who with her quiet and steady leadership at AFOSR and DARPA and with her generous advocacy and support has established the area of uncertainty quantification and stochastic modeling as an independent field, indispensable in all areas of computational simulation of physical and biological problems.

This book is based on our research through collaboration with Professor Boris L. Rozovsky at Brown University and Professor Michael V. Tretyakov at the University of Nottingham. We are indebted to them for their valuable advice. Specifically, Chapters 6, 7, and 8 are based on collaborative research with them. Chapter 5 is also from a collaboration with Professor Michael V. Tretyakov. We would also like to thank Professor Wanrong Cao at Southeast University for providing us numerical results in Chapter 4 and Professor Guang Lin at Purdue University and Dr. Xiu Yang at Pacific Northwest National Laboratory for providing their code for Chapter 9.

The research of this book was supported by OSD/MURI grant FA9550-09-1-0613, NSF/DMS grant DMS-0915077, NSF/DMS grant DMS-1216437, and ARO and NIH grants and also by the Collaboratory on Mathematics for Mesoscopic Modeling of Materials (CM4) which is sponsored by DOE. The first author was also supported by a start-up funding at Worcester Polytechnic Institute during the writing of the book.

Worcester, MA, USA Zhongqiang Zhang
Providence, RI, USA George Em Karniadakis

Contents

Appendices

1

Prologue

Stochastic mathematical models have received increasing attention for their ability of representing intrinsic uncertainty in complex systems, e.g., representing various scales in particle simulations at molecular and mesoscopic scales, as well as extrinsic uncertainty, e.g., stochastic external forces, stochastic initial conditions, or stochastic boundary conditions. One important class of stochastic mathematical models is stochastic partial differential equations (SPDEs), which can be seen as deterministic partial differential equations (PDEs) with finite or infinite dimensional stochastic processes – either with color noise or white noise. Though white noise is a purely mathematical construction, it can be a good model for rapid random fluctuations, and it is also the limit of color noise when the correlation length goes to zero.

In the following text, we first discuss why random variables/processes are used using a simple model and introduce the color of noise. Then we present some typical models using PDEs with random (stochastic) processes. At last, we preview the topics covered in this book and the methodology we use.

1.1 Why random and Brownian motion (white noise)?

Consider the following simple population growth model

$$dy(t) = k(t)y(t)\,dt, \quad y(0) = y_0. \tag{1.1.1}$$

Here $y(t)$ is the size of population and $k(t)$ is the relative growth rate. In practice, $k(t)$ is not completely known and is disturbed around some known quantity $\bar{k}(t)$:

$$k(t) = \bar{k}(t) + \text{`perturbation (noise)'}.$$

Here $\bar{k}(t)$ is deterministic and is usually known while no exact behavior exists of the perturbation (noise) term. The uncertainty (lack of information) about

Z. Zhang, G.E. Karniadakis, *Numerical Methods for Stochastic Partial Differential Equations with White Noise*,
Applied Mathematical Sciences 196, DOI 10.1007/978-3-319-57511-7_1

$k(t)$ (the perturbation term) is naturally represented as a stochastic quantity, denoted as $v(t, \omega)$. Here ω represents the randomness.

To address the dependence on ω, we then write the ordinary differential equation as

$$dy(t, \omega) = k(t, \omega) y(t, \omega)\, dt,^{1} \quad y(0) = y_0. \tag{1.1.2}$$

Here y_0 can be a random variable but we take $y_0 = 1$ (deterministic) for simplicity.

What color is the noise?

In many situations, it is assumed that a stochastic process $v(t)$ satisfies the following properties

1. The expectation of $v(t)$ is zero for all t, i.e., $\mathbb{E}[v(t)] = 0$.
2. The covariance (two-point correlation) function of $v(t)$ is more or less known. That is,

$$\mathbb{C}\mathrm{ov}[(v(t), v(s))] = \mathbb{E}[(v(t) - \mathbb{E}[v(t)])(v(s) - \mathbb{E}[v(s)])].$$

When the covariance function is proportional to the Dirac function $\delta(t - s)$, the process $v(t)$ is called uncorrelated, which is usually referred to as *white noise*. Otherwise, it is correlated and is referred to as *color noise*. The white noise can be intuitively described as a stochastic process, which has independent values at each time instance and has an infinite variance.

Due to the simplicity of Gaussian processes, $v(t, \omega)$ is modeled with a Gaussian process. One important Gaussian process is Brownian motion in physics, which describes a random motion of particles in a fluid with constantly varying fluctuations. It is fully characterized by its expectation (usually taken as zero) and its covariance function, see Chapter 2.2.

Another way to represent the Gaussian noise is through Fourier series. A real-valued Gaussian process can be represented as

$$v(t) = \sum_{k=-\infty}^{\infty} e^{-ikt} a_k \xi_k,$$

where ξ_k's are i.i.d. standard Gaussian random variables (i.e., mean zero and variance 1) and i is the imaginary unit ($i^2 = -1$). When a_k's are the same constant, the process is called white noise. When $|a_k|^2$ is proportional to $1/k^\alpha$, it is called $1/f^\alpha$ noise. When $\alpha = 2$, the process is called red noise (Brownian noise). When $\alpha = 1$, the process is called pink noise. It is called blue noise when $\alpha = -1$. However, white noise ($\alpha = 0$) is not observed in experiments and nature, but $1/f^\alpha$ noise, $0 < \alpha \leq 2$ was first observed back in

[1] We assume that $k(t, \omega)$ is sufficiently smooth in t. It will be clear why some smoothness is needed when we introduce the stochastic product and calculus in Chapter 2.3.

1910s, e.g., $\alpha = 1$ [144]. When α is closer to 0, the process $v(t)$ becomes less correlated and when $\alpha = 0$, white noise can be treated as a limit of processes with extremely small correlation. This is illustrated in Figure 1.1, which is generated by the Matlab code in [429] using Mersenne Twister pseudorandom number generator with seed 100. The smaller α is, the closer is the noise to white noise.

Fig. 1.1. Sample paths of different Gaussian $1/f^{\alpha}$ processes.

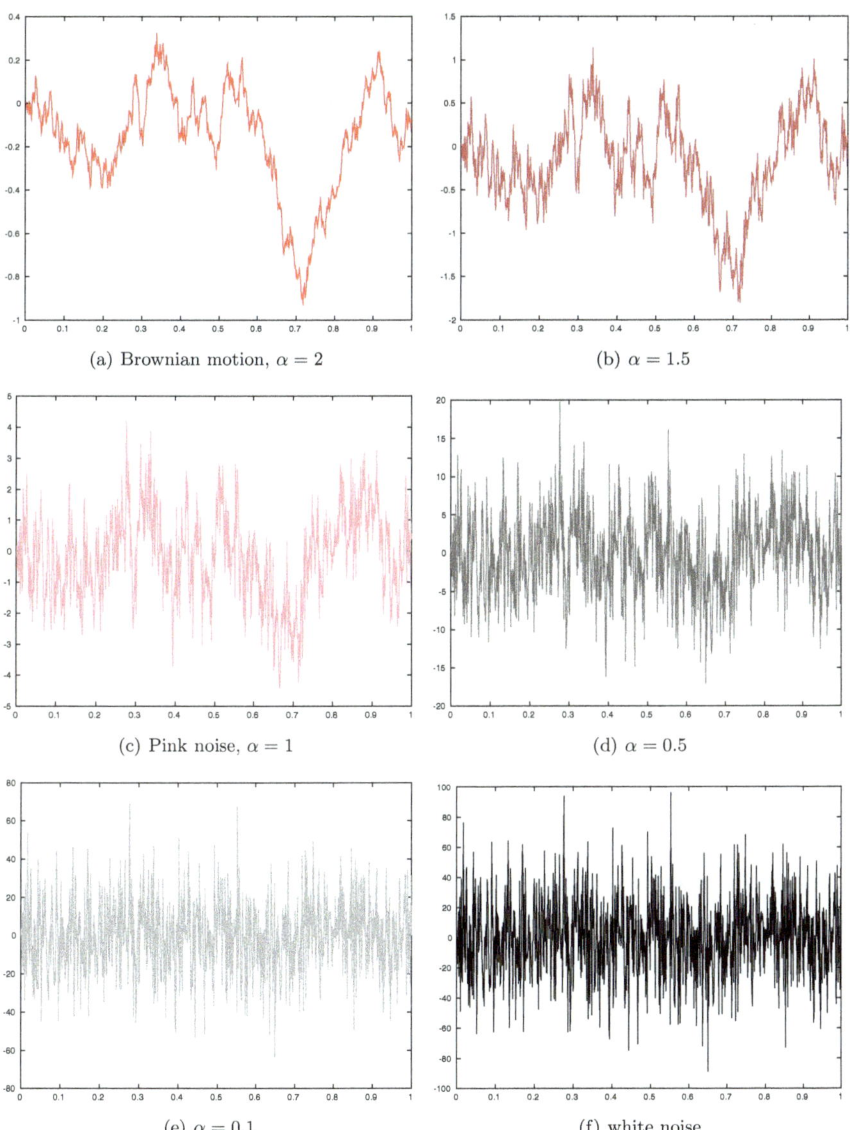

(a) Brownian motion, $\alpha = 2$

(b) $\alpha = 1.5$

(c) Pink noise, $\alpha = 1$

(d) $\alpha = 0.5$

(e) $\alpha = 0.1$

(f) white noise

Solutions to (1.1.2)

When $\bar{k}(t) = 0$, $k(t,\omega) = v(t,\omega)$ may take the following form:

- $v(t,\omega) = \xi \sim \mathcal{N}(0,1)$ is a standard Gaussian random variable. The covariance (in this case variance) of $v(t,\omega)$ is 1.
- $v(t,\omega)$ where the covariance function of $V(t)$ is $\exp(-\frac{|t-s|}{A})$, with A the correlation time.
- $v(t,\omega) = W(t,\omega)$ is a standard Brownian motion where the covariance function is $\min(t,s)$.
- $v(t,\omega) = \dot{W}(t,\omega)$ is the white noise and the covariance function is $\delta(t-s)$.

When $k(t,\omega) =: \dot{W}(t,\omega)$, Equation (1.1.2) is understood in the Stratonovich sense

$$dy = y \circ dW(t), \quad y(0) = y_0 \tag{1.1.3}$$

so that one can apply the classical chain rule. We discuss what the circle means in Chapter 2.3, but roughly speaking the circle denotes Stratonovich product, which can be understood in some limit of Riemann sum using the midpoint rule.

The exact solution to Equation (1.1.2) is $y = y_0 \exp(K(t))$, where $K(t) = \int_0^t k(s)\, ds$ is again Gaussian with mean zero. It can be readily checked by the definition of moments that

- $K(t) \sim \mathcal{N}(0,t^2)$ when $k(t,\omega) =: \xi \sim \mathcal{N}(0,1)$;
- $K(t) \sim \mathcal{N}\big(0, 2At + 2A^2(\exp(-\frac{t}{A}) - 1)\big)$ when $k(t,\omega)$ has the two-point correlation function of $\exp(-\frac{|t_1-t_2|}{A})$.
- $K(t) \sim \mathcal{N}(0, \frac{t^3}{3})$ when $k(t,\omega) = W(t,\omega)$ is a standard Brownian motion.
- $K(t) \sim \mathcal{N}(0,t)$ is the Brownian motion when $k(t,\omega) = \dot{W}(t,\omega)$ is the white noise.

Then the moments of the solution y are, for $m = 1, 2, \cdots$,

$$\mathbb{E}[y^m(t)] = y_0^m \exp(\frac{m^2}{2}\sigma^2), \tag{1.1.4}$$

where $\sigma^2 = t^2,\ 2At + 2A^2(\exp(-\frac{t}{A}) - 1),\ \frac{t^3}{3},\ t$ for the listed processes, respectively. These results are useful and can be used in checking the accuracy of different numerical schemes applied to Equation (1.1.2).

1.2 Modeling with SPDEs

SPDEs with white noise have been formulated for various applications, such as nonlinear filtering (see, e.g., [499]), turbulent flows (see, e.g., [43, 348]), fluid flows in random media (see, e.g., [223]), particle systems (see, e.g., [268]), population biology (see, e.g., [98]), neuroscience (see, e.g., [463]), etc. Since analytic solutions to SPDEs can rarely be obtained, numerical methods have to be developed to solve SPDEs. One of the motivations for numerical SPDEs in the early literature was to solve the Zakai equation of nonlinear filtering, see, e.g., [31, 78, 130, 150–152]. In the next section, we review some numerical methods for a semilinear equation (3.4.1), the advection-diffusion-reaction equation of nonlinear filtering (3.4.13), the stochastic Burgers equation (3.4.20), and the stochastic Navier-Stokes equation (1.2.4).

Let $W(t)$ be a r-dimensional Brownian motion, i.e., $W(t) = (W_1(t), \ldots, W_r(t))^\top$ where $W_i(t)$'s are mutually independent Brownian motions – Gaussian processes with the covariance function $\min(t, s)$. For a rigorous definition of Brownian motion, see Chapter 2.2.

Example 1.2.1 (Zakai equation of nonlinear filtering) *Let $y(t)$ be a r-dimensional observation of a signal $x(t)$ and $y(t)$ given by*

$$y(t) = y_0 + \int_0^t h(x(s))\, ds + W(t),$$

where $h = (h_1, h_2, \ldots, h_r)^\top$ is a $\mathbb{R}^r$-vector-valued function defined on $\mathbb{R}^d$ and the observational signal $x(t)$ satisfies the following stochastic differential equation

$$dx(t) = b(x(t))\, dt + \sum_{k=1}^{q} \sigma_k(x(t))\, dB_k, \quad x(0) = x_0,$$

where b and σ_k's are d-dimensional vector functions on $\mathbb{R}^d$, and $B(t) = (B_1(t), B_2(t), \ldots, B_q(t))^\top$ is a q-dimensional Brownian motion on $(\Omega, \mathcal{F}, \mathbb{P})$ and is independent of $W(t)$.

The conditional probability density function (un-normalized) of the signal $x(t)$ given $y(t)$ satisfies, see, e.g., [499],

$$\begin{aligned} du(t,x) = \frac{1}{2} \sum_{i,j=1}^{d} D_i D_j [(\sigma\sigma^\top)_{ij} u(t,x)]\, dt - \sum_{i}^{d} D_i (b_i u(t,x))\, dt \\ + \sum_{l=1}^{r} h_l u(t,x)\, dy_l(t), \quad x \in \mathbb{R}^d. \end{aligned} \tag{1.2.1}$$

Here $D_i := \partial_{x_i}$ is the partial derivative in the x_i-th direction and $\sigma^\top$ is the transpose of σ.

Equation (1.2.1) provides an analytical solution to the aforementioned filtering problem. Equation (1.2.1) and its variants have been a major motivation for the development of theory for SPDEs (see, e.g., [280, 408]) and corresponding numerical methods; see, e.g., [31, 78, 130, 150–152] and for a comprehensive review on numerical methods see [52].

Example 1.2.2 (Pressure equation) *The following model was introduced as an example of the pressure of a fluid in porous and heterogeneous (but isotropic) media at the point x over the physical domain $\mathcal{D}$ in $\mathbb{R}^d$ ($d \leq 3$):*

$$-\operatorname{div}(K(x)\nabla p) = f(x), \quad p|_{\partial\mathcal{D}} = 0, \tag{1.2.2}$$

where $K(x) > 0$ is the permeability of media at the point x and $f(x)$ is a mass source. In a typical porous media, $K(x)$ is fluctuating in an unpredictable and irregular way and can be modeled with a stochastic process.

In [140, 141], $K(x) = \exp(\int_{\mathcal{D}} \phi(x-y)\, dW(y) - \frac{1}{2}\left\|\phi\right\|^2)$ is represented as a lognormal process, see also [223] for a formulation of the lognormal process using the Ito-Wick product (see its definition in Chapter 2.3). Here ϕ is a continuous function in $\mathcal{D}$ and $\|\phi\| = (\int_{\mathcal{D}} \phi^2\, dx)^{1/2} < \infty$.

We define infinite dimensional Gaussian processes, see, e.g., [94, 408], as follows

$$W^Q(t,x) = \sum_{i\in\mathbb{N}^d} \sqrt{q_i} e_i(x) W_i(t), \tag{1.2.3}$$

where $W_i(t)$'s are mutually independent Brownian motions. Here $q_i \geq 0$, $i \in \mathbb{N}^d$ and $\{e_i(x)\}$ is an orthonormal basis in $L^2(\mathcal{D})$. The following noise is usually considered in literature: $\dot{W}^Q(t,x) = \sum_{i\in\mathbb{N}^d} \sqrt{q_i} e_i(x) \dot{W}_i(t)$. When $q_i = 1$ for all i's, we have a space-time *white noise*. When $\sum_{i=1}^{\infty} q_i < \infty$, we call it the space-time **color noise**. Sometimes it is called a Q-Wiener process. We call the noise finite-dimensional when $q_i = 0$ for all sufficient large i.

Example 1.2.3 (Turbulence model) *The stochastic incompressible Navier-Stokes equation reads, see, e.g., [108, 348]*

$$\partial_t \mathbf{u} + \mathbf{u}\cdot\nabla\mathbf{u} - \nu\Delta\mathbf{u} + \nabla p = \sigma(\mathbf{u})\dot{W}^Q, \quad \operatorname{div}\mathbf{u} = 0, \tag{1.2.4}$$

where σ is Lipschitz continuous over the physical domain in $\mathbb{R}^d$ ($d = 2, 3$). Here $\mathbb{E}[W^Q(x,t)W^Q(y,s)] = q(x,y)\min(s,t)$ and $q(x,x)$ is square-integrable over the physical domain.

Example 1.2.4 (Reaction-diffusion equation)

$$du = \nu\Delta u + f(u) + \sigma(u)\dot{W}^Q(t,x).$$

This may represent a wide class of SPDEs:

- *In materials, the stochastic Allen-Cahn equation, where $f(u) = u(1-u)(1+u)$, $\sigma(u)$ is a constant and $\dot{W}^Q(t,x)$ is space-time white noise, see, e.g., in [204, 302].*
- *In population genetics, this equation has been used to model changes in the structure of population in time and space, where $\dot{W}^Q(t,x)$ is a Gaussian process white in time but color in space. For example, $f(u) = 0$, $\sigma(u) = \gamma\sqrt{\max(u,0)}$, where γ is a constant and u is the mass distribution of the population, see, e.g., [97]. Also, $f(u) = \alpha u - \beta$, $\sigma(u) = \gamma\sqrt{\max(u(1-u),0)}$, where α, β, γ are constants, see, e.g., [129].*

- *When $\sigma(u)$ is a small constant, the equation can be treated as random perturbation of deterministic equations ($\sigma(u) = 0$), see, e.g., [136].*

In the first and third cases, we say that the equation has an additive noise *as the coefficients of noise are independent of the solution. In the second case, we say that the equation has a* multiplicative noise *since the coefficient of noise depends on the solution itself.*

For more details on modeling with SPDEs, we refer to the books [223, 268]. For well-posedness (existence, uniqueness, and stability) of SPDEs, we refer to the books [94, 145, 170, 223, 399, 408]. For studies of the dynamics of SPDEs, we refer to [145] for asymptotic behavior of solutions to SPDEs; also to [92] for Kolmogrov equations for SPDEs, to [36] for amplitude equations of nonlinear SPDEs, and to [119] for homogenization of multiscale SPDEs. In this book, we present numerical methods for SPDEs. Specifically, we focus on *forward problems*, i.e., predicting the quantities of interest from known SPDEs with known inputs, especially those SPDEs driven by white noise.

1.3 Specific topics of this book

In this book, we focus on two issues related to numerical methods for SPDEs with white noise: one is deterministic integration methods in random space while the other one is the effect of truncation of the Brownian motion using the spectral approximation.[2] In Figure 1.2, we provide an overview on how we organize this book. In addition, we present some necessary details in the appendices to make the book self-contained.

In Chapter 2, we review some numerical methods for SPDEs with white noise addressing primarily their discretization in random space. We then explore the effect of truncation of Brownian motion using its spectral expansion (Wong-Zakai approximation) for stochastic differential equations with time delay in Chapter 4 and show that the Wong-Zakai approximation can facilitate the derivation of various numerical schemes.

For deterministic integration methods of SPDEs in random space, we aim at performing accurate long-time numerical integration of time-dependent equations, especially of linear stochastic advection-reaction-diffusion equations. We study Wiener chaos expansion methods (WCE) and stochastic collocation methods (SCM), compare their performance and prove their convergence order. To achieve longer time integration, we adopt the recursive WCE proposed in [315] for the Zakai equation for nonlinear filtering and develop algorithms for the first two moments of solutions. Numerical results show that when high accuracy is required, WCE is superior to

---

[2] There are different numerical methods for SPDEs using different approximation of Brownian motion and integration methods in random space, see Chapter 2 for a review, where discretization in time and space is also presented.

Monte Carlo methods while WCE is not as efficient if only low accuracy is sufficient, see Chapter 6. We show that the recursive approach for SCM for linear advection-reaction-diffusion equations is efficient for long-time integration in Chapter 7. We first analyze the error of SCM (sparse grid collocation of Smolyak type) with an Euler scheme in time for linear SODEs, and show that the error is small only when the noise magnitude is small and/or the integration time is relatively short. We compare WCE and SCM using the recursive procedure in Chapter 8, where we derive error estimates of WCE and SCM for linear advection-reaction-diffusion equations and show that WCE and SCM are competitive in practice by presenting careful numerical comparisons, even though WCE can be of one order higher than SCM.

Among almost all approximations for WCE and SCM, we use the Wong-Zakai approximation with spectral approximation of Brownian motion. The convergence order with respect to the number of truncation modes is half order, see Chapter 8. However, WCE can be of higher convergence order since it can preserve the orthogonality over the Wiener space (infinite dimensional) while SCM cannot as the orthogonality is only valid on discrete spaces (finite dimensional), see Chapter 8. In Chapter 9, we test the Wong-Zakai approximation in conjunction with the stochastic collocation method for the stochastic Euler equations modeling a stochastic piston problem and show the effectiveness of this approximation in a practical situation. To further investigate the effect of truncation of Brownian motions, we study the elliptic equation with additive white noise in Chapter 10. We show that the convergence of numerical solutions with truncation of Brownian motion depends on the smoothing effects of the resolvent of the elliptic operator. We also show similar convergence when finite element methods are used.

As shown in Figure 1.2, we focus on deterministic integration methods in random space, i.e., Wiener chaos and stochastic collocation, in Chapters 6–9 and compare their performance with Monte Carlo methods and/or quasi-Monte Carlo methods. In Chapter 6, we compare WCE and Monte Carlo methods and show that WCE is superior to Monte Carlo methods if high accuracy is needed. In Chapters 7 and 9, we show theoretically and numerically the efficiency of SCM for short time integration and for small magnitudes of noises. In Chapter 8, we compare WCE and SCM in conjunction with a recursive multistage procedure and show that they are comparable in performance. In Chapter 11, we consider WCE for elliptic equations with multiplicative noise. We use the Wick product for the interaction of two random variables as well as the Wick-Malliavin approximation to reduce the computational cost. We use Monte Carlo methods in Chapters 4 and 10 as the dimensionality in random space is beyond deterministic integration methods.

In all chapters except Chapters 5 and 7, we apply the Wong-Zakai approximation with the Brownian motion approximated by its spectral truncation. It is shown that the convergence of numerical schemes based on the Wong-Zakai approximation is determined by further discretization in space (Chapter 10) or in time (Chapter 4).

Fig. 1.2. Conceptual and chapter overview of this book.

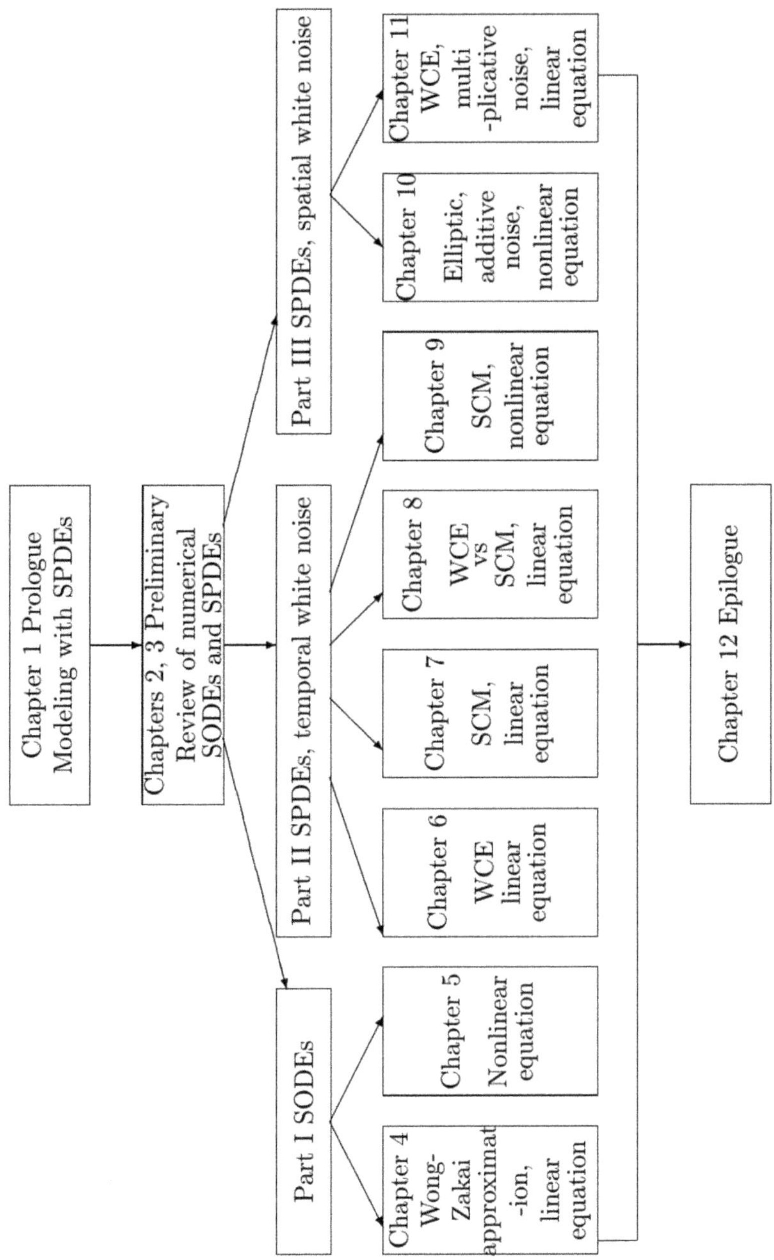

2

Brownian motion and stochastic calculus

In this chapter, we review some basic concepts for stochastic processes and stochastic calculus as well as numerical integration methods in random space for obtaining statistics of stochastic processes.

We start from Gaussian processes and their representations in Chapter 2.1 and then introduce Brownian motion and its properties and approximations in Chapter 2.2. We discuss basic concepts in stochastic calculus: Ito integral in Chapter 2.3 and Ito formula in Chapter 2.4. We then focus on numerical integration methods in random space such as Monte Carlo methods, quasi-Monte Carlo methods, Wiener chaos method, and stochastic collocation method (sparse grid collocation method) in Chapter 2.5. Examples of applying these methods to a simple equation are provided in Chapter 2.5.5 with Matlab code. In Chapter 2.6 of bibliographic notes, we present a review on different types of approximation of Brownian motion and a brief review on *pros* and *cons* of numerical integration methods in random space. Various exercises are provided for readers to familiarize themselves with basic concepts presented in this chapter.

2.1 Gaussian processes and their representations

On a given probability space $(\Omega, \mathcal{F}, \mathbb{P})$ $(\Omega = \mathbb{R})$, if a cumulative distribution function of a random variable X is normal, i.e.,

$$\mathbb{P}(X < x) = \int_{-\infty}^{x} \frac{1}{\sqrt{2\pi}\,\sigma} e^{-\frac{(y-\mu)^2}{2\sigma^2}}\, dy, \quad \sigma > 0.$$

Z. Zhang, G.E. Karniadakis, *Numerical Methods for Stochastic Partial Differential Equations with White Noise*,
Applied Mathematical Sciences 196, DOI 10.1007/978-3-319-57511-7_2

then the random variable X is called a Gaussian (normal) random variable on the probability space $(\Omega, \mathcal{F}, \mathbb{P})$. Here X is completely characterized by its mean μ and its standard deviation σ. We denote $X \sim \mathcal{N}(\mu, \sigma^2)$. The probability density function of X is

$$p(x) = \frac{1}{\sqrt{2\pi}\,\sigma} e^{-\frac{(x-\mu)^2}{2\sigma^2}}.$$

When $\mu = 0$ and $\sigma = 1$, we call X a standard Gaussian (normal) random variable. If $X \sim \mathcal{N}(\mu, \sigma^2)$, then $Z = \frac{X-\mu}{\sigma} \sim \mathcal{N}(0,1)$, i.e., Z is a standard Gaussian (normal) random variable.

Example 2.1.1 *If X_i are mutually independent Gaussian random variables, then $\sum_{j=1}^{N} a_i X_i$ is a Gaussian random variable for any $a_i \in \mathbb{R}$. In particular, if $X_1 \sim \mathcal{N}(\mu_1, \sigma_1^2)$ and $X_2 \sim \mathcal{N}(\mu_2, \sigma_2^2)$ are independent, then*

$$\alpha X_1 + \beta X_2 \sim \mathcal{N}(\mu_1 + \mu_2, \alpha^2\sigma_1^2 + \beta^2\sigma_2^2).$$

Definition 2.1.2 (Gaussian random vector) *A $\mathbb{R}^n$-valued random vector $X = (X_1, X_2, \ldots, X_n)^\top$ has an n-variate Gaussian distribution with mean $\boldsymbol{\mu}$ and covariance matrix Σ if $X = \boldsymbol{\mu} + AZ$ where the matrix A is of size $n \times n$, $\Sigma = AA^\top$, and $Z = (Z_1, Z_2, \ldots, Z_n)^\top$ is a vector with independent standard Gaussian (normal) components.*

When $n = 1$, X is a (univariate) Gaussian random variable. The probability density of X is

$$p(\boldsymbol{x}) = \frac{1}{\sqrt{(2\pi)^n}\,|\Sigma|^{1/2}} e^{-\frac{(\boldsymbol{x}-\boldsymbol{\mu})^\top \Sigma^{-1}(\boldsymbol{x}-\boldsymbol{\mu})}{2}}.$$

Example 2.1.3 *A set of random variables $\{X_i\}_{i=1}^n$ are called* jointly Gaussian *if $\sum_{i=1}^n a_i X_i$ is a Gaussian random variable for any $a_i \in \mathbb{R}$. Then $X = (X_1, X_2, \ldots, X_n)^\top$ is a Gaussian random vector.*

The correlation of two random variables (vectors) is a normalized version of the covariance, with values ranging from -1 to 1:

$$\mathrm{Corr}(X,Y) = \frac{\mathrm{Cov}[(X,Y)]}{\sqrt{\mathbb{V}\mathrm{ar}[X]\mathbb{V}\mathrm{ar}[Y]}}, \quad \mathrm{Cov}[(X,Y)] = \mathbb{E}[(X - \mathbb{E}[X])(Y - \mathbb{E}[Y])^\top].$$

When $\mathrm{Corr}(X,Y) = 0$, we say X and Y are uncorrelated.

Definition 2.1.4 (Gaussian process) *A collection of random variables is called a Gaussian process, if the joint distribution of any finite number of its members is Gaussian. In other words, a Gaussian process is a $\mathbb{R}^d$-valued stochastic process with continuous time (or with index) t such that $(X(t_0), X(t_1), \cdots, X(t_n))^\top$ is a $n+1$-dimensional Gaussian random vector for any $0 \le t_0 < t_1 < \cdots < t_n$.*

The Gaussian process is denoted as $X = \{X(t)\}_{t\in T}$ where T is a set of indexes. Here $T = [0, \infty)$.

The consistency theorem of Kolmogorov [255, Theorem 2.2] implies that the finite dimensional distribution of a Gaussian stochastic process $X(t)$ is uniquely characterized by two functions: the mean function $\mu_t = \mathbb{E}[X(t)]$ and the covariance function $C(t, s) = \mathbb{C}\text{ov}[X(t), X(s)]$.

A Gaussian process $\{X(t)\}_{t\in T}$ is called a centered Gaussian process if the mean function $\mu(t) = \mathbb{E}[X(t)] = 0$ for all $t \in T$.

Given a function $\mu(t) : T \to \mathbb{R}$ and a nonnegative definite function $C(t, s) : T\times T \to \mathbb{R}$, there exists a Gaussian process $\{X(t)\}_{t\in T}$ with the mean function $\mu(t)$ and the covariance function $C(t, s)$.

To find such a Gaussian process, we can use the following expansion.

Theorem 2.1.5 (Karhunen-Loève expansion) *Let $X(t)$ be a Gaussian stochastic process defined over a probability space $(\Omega, \mathcal{F}, \mathbb{P})$ and $t \in [a, b]$, $-\infty < a < b < \infty$. Suppose that $X(t)$ has a continuous covariance function $C(t, s) = \mathbb{C}ov[(X(t), X(s))] = \mathbb{E}[(X(t) - \mathbb{E}[X(t)])(X(s) - \mathbb{E}[X(s)])]$. Then $X(t)$ admits the following representation*

$$X(t) = \mathbb{E}[X(t)] + \sum_{k=1}^{\infty} Z_k e_k(t),$$

where the convergence is in L^2, uniform in t (i.e., $\lim_{n\to\infty} \max_{t\in[a,b]} \mathbb{E}[(X(t) - \mathbb{E}[X(t)] - \sum_{k=1}^{n} Z_k e_k(t))^2] = 0$) and

$$Z_k = \int_a^b (X(t) - \mathbb{E}[X(t)]) e_k(t)\, dt.$$

Here the eigenfunctions e_k's of C_X with respective eigenvalues λ_k's form an orthonormal basis of $L^2([a, b])$ and

$$\int_a^b C(t, s) e_k(t)\, dt = \lambda_k e_k(s), \quad k \geq 1.$$

Furthermore, the random variables Z_k's have zero-mean, are uncorrelated, and have variance λ_k

$$\mathbb{E}[Z_k] = 0, \quad \text{for all } k \geq 1 \qquad \text{and} \qquad \mathbb{E}[Z_i Z_j] = \delta_{ij}\lambda_j, \quad \text{for all } i, j \geq 1.$$

This is a direct application of Mercer's theorem [497] on a representation of a symmetric positive-definite function as a sum of a convergent sequence of product functions. The stochastic process $X(t)$ can be non-Gaussian.

The covariance function $C(t, s)$ can be represented as $C(t, s) = \sum_{k=1}^{\infty} \lambda_k e_k (t) e_k(s)$. The variance of $X(t)$ is the sum of the variances of the individual components of the sum:

$$\mathbb{V}\mathrm{ar}[X(t)] = \mathbb{E}[(X(t) - \mathbb{E}[X(t)])^2] = \sum_{k=0}^{\infty} e_k^2(t)\mathbb{V}\mathrm{ar}[Z_k] = \sum_{k=1}^{\infty} \lambda_k e_k^2(t).$$

Here Z_k are uncorrelated random variables.

The domain where the process is defined can be extended to domains in $\mathbb{R}^d$. In Table 2.1 we present a list of covariance functions commonly used in practice. Here the constant l is called correlation length, K_ν is the modified

Table 2.1. A list of covariance functions.

Wiener process	$\min(x, y),\ x, y \geq 0$
White Noise	$\sigma^2\delta(x - y), \quad x, y \in \mathbb{R}^d$
Gaussian	$\exp\left(-\frac{\lvert x-y\rvert^2}{2l^2}\right), \quad x, y \in \mathbb{R}^d$
Exponential	$\exp\left(-\frac{\lvert x-y\rvert}{l}\right), \quad x, y \in \mathbb{R}^d$
Matérn kernel	$\frac{2^{1-\nu}}{\Gamma(\nu)}\left(\frac{\sqrt{2\nu}\,\lvert x-y\rvert}{l}\right)^{\nu} K_\nu\left(\frac{\sqrt{2\nu}\,\lvert x-y\rvert}{l}\right), \quad x, y \in \mathbb{R}^d$
Rational quadratic	$(1 + \lvert x - y\rvert^2)^{-\alpha}, \quad x, y \in \mathbb{R}^d, \quad \alpha \geq 0$

Bessel function of order ν, and $\Gamma(\cdot)$ is the gamma function:

$$\Gamma(t) = \int_0^{\infty} x^{t-1}e^{-x}\,dx,\ t > 0.$$

Here are some examples of Karhunen-Loève expansion for Gaussian processes.

Example 2.1.6 (Brownian motion) *When* $C(t, s) = \min(t, s)$, $t \in [0, 1]$, *then the Gaussian process* $X(t)$ *can be written as*

$$X(t) = \sqrt{2}\sum_{k=1}^{\infty} \xi_k \frac{\sin\left(\left(k - \frac{1}{2}\right)\pi t\right)}{\left(k - \frac{1}{2}\right)\pi}.$$

Here ξ_k*'s are mutually independent standard Gaussian random variables. One can show that for* $t, s \in [0, 1]$, *the eigenvectors of the covariance function* $\min(t, s)$ *are*

$$e_k(t) = \sqrt{2}\sin\left(\left(k - \tfrac{1}{2}\right)\pi t\right),$$

and the corresponding eigenvalues are

$$\lambda_k = \frac{1}{(k - \frac{1}{2})^2\pi^2}.$$

In the next section, we know that the process in Example 2.1.6 is actually a *Brownian motion.*

Example 2.1.7 (Brownian Bridge) *Let $X(t)$, $0 \leq t \leq 1$, be the Gaussian process in Example 2.1.6. Then $Y(t) = X(t) - tX(1)$, $0 \leq t \leq 1$, is also a Gaussian process and admits the following Karhunen-Loève expansion:*

$$Y(t) = \sum_{k=1}^{\infty} \eta_k \frac{\sqrt{2}\sin(k\pi t)}{k\pi}.$$

Here η_k's are mutually independent standard Gaussian random variables.

Example 2.1.8 (Ornstein-Uhlenbeck process) *Consider a centered one-dimensional Gaussian process with an exponential covariance function $\exp(-\frac{|t-s|}{l})$. The Karhunen-Loève expansion of such a Gaussian process over $[-a, a]$ is*

$$O(t) = \sum_{k=1}^{\infty} \xi_k \sqrt{\lambda_k} e_k(t),$$

where $\lambda_k = \frac{2l}{l^2\theta_k^2+1}$ and the corresponding eigenvalues are

$$e_{2i}(t) = \frac{\cos(\theta_{2i}t)}{\sqrt{2 + \frac{\sin(2\theta_{2i}a)}{2\theta_{2i}}}}, e_{2i-1}(t) = \frac{\sin(\theta_{2i-1}t)}{\sqrt{2 - \frac{\sin(2\theta_{2i-1}a)}{2\theta_{2i-1}}}}, \text{for all } i \geq 1,\ t \in [-a, a].$$

The θ_k's are solutions to the following transcendental equation

$$1 - l\theta \tan(a\theta) = 0 = l\theta + \tan(a\theta).$$

See p. 23, Section 2.3 of [155] or [245] for a derivation of such an expansion.

For more general forms of covariance functions $C(t, s)$, it may not be possible to find explicitly the eigenvectors and eigenvalues. The Karhunen-Loève expansion can be found numerically, and in practice only a finite number of terms in the expansion are required. Specifically, we usually perform a principal component analysis by truncating the sum at some N such that

$$\frac{\sum_{i=1}^{N} \lambda_i}{\sum_{i=1}^{\infty} \lambda_i} = \frac{\sum_{i=1}^{N} \lambda_i}{\int_a^b \mathbb{V}\mathrm{ar}[X(t)]\, dt} \geq \alpha.$$

Here α is typically taken as 0.9, 0.95, and 0.99. The eigenvalues and eigenfunctions are found by solving numerically the following eigenproblem (integral equation):

$$\int_a^b C(t, s) e_k(t)\, ds = \lambda_k e_k(s), \quad s \in [a, b] \text{ and } k = 1, 2, \ldots, N.$$

See, e.g., Section 2.3 of [155] or [419] for a Galerkin method for this problem. We can also apply the Nyström method or the quadrature method, where the integral is replaced with a representative weighted sum. Several numerical methods for representing a stochastic process with a given covariance kernel are presented in [308, Chapter 7], where numerical methods are not based on Karhunen-Loève expansion, but based on Fourier analysis and other methods.

The decay of eigenvalues in the Karhunen-Loève expansion depends on the smoothness of covariance functions.

Definition 2.1.9 ([419]) *A covariance function $C : \mathcal{D} \times \mathcal{D} \to \mathbb{R}$ is said to be piecewise analytic/smooth/$H^{p,q}$ on $\mathcal{D} \times \mathcal{D}$, $0 \leq p, q \leq \infty$ if there exists a partition $\mathfrak{D} = \{D_j\}_{j=1}^J$ into a finite sequence of simplexes D_j and a finite family $\mathfrak{G} = \{G_j\}_{j=1}^J$ of open sets in $\mathbb{R}^d$ such that*

$$\bar{D} = \cup_{j=1}^J \bar{D}_j, \; \bar{D}_j \subseteq G_j, \quad 1 \leq j \leq J$$

and such that $C|_{D_j \times D_j'}$ has an extension to $G_j \times G_{j'}$ which is analytic in $D_j \times D_{j'}$/is smooth in $D_j \times D_{j'}$/is in $H^p(D_j) \otimes H^q(D_{j'})$ for any pair (j, j').

The following conclusions on the eigenvalues in the Karhunen-Loève expansion are from [419].

Theorem 2.1.10 *Assume that $C \in L^2(\mathcal{D} \times \mathcal{D})$ be a symmetric covariance function which leads to a compact and nonnegative operator from $L^2(\mathcal{D})$ defined by $\mathfrak{C}u(x) = \int_{\mathcal{D}} C(x, y)u(y)\, dy$. If C is piecewise analytic on $\mathcal{D} \times \mathcal{D}$, then the eigenvalues λ_k in the Karhunen-Loève expansion satisfy that*

$$0 \leq \lambda_k \leq K_1 e^{-K_2 k^{1/d}}, \quad k \geq 1.$$

The constants K_1 and K_2 depend only on the covariance function C and the domain $\mathcal{D}$. If C is piecewise $H^p(\mathcal{D}) \otimes L^2(\mathcal{D})$ with $p \geq 1$, then the eigenvalues λ_k in the Karhunen-Loève expansion decay algebraically fast

$$0 \leq \lambda_k \leq K_3 k^{-p/d}, \quad k \geq 1.$$

For the Gaussian covariance function, $C(x, y) = \sigma^2 \exp(-(|x - y|^2)/\gamma^2/\operatorname{diam}(\mathcal{D}))$. Then the eigenvalues λ_k in the Karhunen-Loève expansion decay exponentially fast:

$$0 \leq \lambda_k \leq K_4 \gamma^{-k^{1/d} - 2} / \Gamma(0.5 k^{1/d}), \quad k \geq 1.$$

An different approach to show the decay of the eigenvalues is presented in [308, Chapter 7] using Fourier analysis for isotropic covariance kernels (the two-point covariance kernel depends only on distances of two points).

Theorem 2.1.11 ([419]) *Assume that the process $a(x, \omega)$ has a covariance function C, which is piecewise analytic/in $H^{p,q}$ on $\mathcal{D} \times \mathcal{D}$. Then the eigenfunctions are analytic/in H^p in each $\bar{D}_j \in \mathfrak{D}$.*

With further conditions on the domain D_j in $\mathfrak{D}$, it can be shown that the derivatives of eigenfunctions $e_k(x)$'s decay at the speed of $|\lambda_k|^{-s}$ when C is piecewise smooth where $s > 0$ is an arbitrary number.

2.2 Brownian motion and white noise

Definition 2.2.1 (One-dimensional Brownian motion) *A one-dimensional continuous time stochastic process $W(t)$ is called a standard Brownian motion if*

- *$W(t)$ is almost surely continuous in t,*
- *$W(t)$ has independent increments,*
- *$W(t) - W(s)$ obeys the normal distribution with mean zero and variance $t - s$.*
- *$W(0) = 0$.*

It can be readily shown that $W(t)$ is Gaussian process. We then call $\dot{W}(t) = \frac{d}{dt}W$, formally the first-order derivative of $W(t)$ in time, *white noise.*

By Example 2.1.6 and Exercise 2.7.7, then the Brownian motion $W(t)$, $t \in [0,1]$ can be represented by

$$W(t) = \sqrt{2}\sum_{i=1}^{\infty} \xi_i \frac{\sin\left(\left(i-\frac{1}{2}\right)\pi t\right)}{\left(i-\frac{1}{2}\right)\pi}, \quad t \in [0,1],$$

where ξ_i's are mutually independent standard Gaussian random variables. The Brownian motion and white noise can also be defined in terms of orthogonal expansions. Suppose that $\{m_k(t)\}_{k\geq 1}$ is a complete orthonormal system (CONS) in $L^2([0,T])$. The Brownian motion $W(t)$, $t \in [0,T]$ can be defined by (see, e.g., [315])

$$W(t) = \sum_{i=1}^{\infty} \xi_i \int_0^t m_i(s)\,ds, \ t \in [0,T], \tag{2.2.1}$$

where ξ_i's are mutually independent standard Gaussian random variables. It can be checked that the Gaussian process defined by (2.2.1) is indeed a standard Brownian motion by Definition 2.2.1. Correspondingly, the white noise is defined by

$$\dot{W}(t) = \sum_{i=1}^{\infty} \xi_i m_i(t), \ t \in [0,T]. \tag{2.2.2}$$

When $m_i(t) = \sqrt{2/T}\cos((i-1/2)\pi t/T)$, $i \geq 1$, then the representation (2.2.1) coincides with the *Karhunen-Loève expansion* of Brownian motion in Example 2.1.6 when $T = 1$.

Definition 2.2.2 (Multidimensional Brownian motion) *A continuous stochastic process $W_t = (W_1(t), \ldots, W_m(t))^\top$ is called an m-dimensional Brownian motion on $\mathbb{R}^m$ when $W_i(t)$ are mutually independent standard Brownian motions on $\mathbb{R}$.*

Definition 2.2.3 (Multidimensional Brownian motion, alternative definitions) *An $\mathbb{R}^d$-valued continuous Gaussian process $X(t)$ with mean function $\mu(t) = \mathbb{E}[X(t)]$ and the covariance function $C(t,s) = \mathbb{C}ov[(X(t), X(s))] = \mathbb{E}[(X(s)-\mu(s))(X(t)-\mu(t))^\top]$ is called a d-dimensional Brownian motion if for any $0 \leq t_0 < t_1 < \cdots < t_n$,*

- *$X(t_i)$ and $X(t_{i+1}) - X(t_i)$ are independent;*
- *the covariance function (a matrix) is a diagonal matrix with entries $\min(t_i, t_j)$, $0 \leq i, j \leq n$.*

When $\mu(t) = 0$ for all t and $C(t,s) = \min(t,s)$, the Gaussian process is called a standard Brownian motion.

2.2.1 Some properties of Brownian motion

Theorem 2.2.4 *The covariance $\mathbb{C}ov[(W(t), W(s))] = \mathbb{E}[W(t)W(s)] = \min(t,s)$.*

- *Time-homogeneity: For any $s > 0$, $\tilde{W}(t) = W(t+s) - W(s)$ is a Brownian motion, independent of $\sigma(W(u), u \leq s)$.*
- *Brownian scaling: For every $c > 0$, $cW(t/c^2)$ is a Brownian motion.*
- *Time inversion: Let $\tilde{W}(0) = 0$ and $\tilde{W}(t) = tW(1/t)$, $t > 0$. Then $\tilde{W}(t)$ is a Brownian motion.*

Corollary 2.2.5 (Strong law of large numbers for Brownian motion) *If $W(t)$ is a Brownian motion, then it holds almost surely that*

$$\lim_{t\to\infty} \frac{W(t)}{t} = 0.$$

Theorem 2.2.6 (Law of the iterated logarithm) *Let W_t be a standard Brownian motion. Then*

$$\mathbb{P}(\limsup_{t\to 0} \frac{W_t}{\sqrt{2t\,|\log\log(t)|}} = 1) = 1, \quad \mathbb{P}(\liminf_{t\to 0} \frac{W_t}{\sqrt{2t\,|\log\log(t)|}} = -1) = 1.$$

$$\mathbb{P}(\limsup_{t\to \infty} \frac{W_t}{\sqrt{2t\log\log(t)}} = 1) = 1, \quad \mathbb{P}(\liminf_{t\to \infty} \frac{W_t}{\sqrt{2t\log\log(t)}} = -1) = 1.$$

Example 2.2.7 (Ornstein-Uhlenbeck process) *Consider a centered one-dimensional Gaussian process with exponential covariance function $\exp(-\frac{|t-s|}{\sigma})$. The Gaussian process is usually called a Ornstein-Uhlenbeck process. Suppose that $W(t)$ is a standard Brownian motion. For $t \geq 0$, the Ornstein-Uhlenbeck process can be written as*

$$O(t) = e^{-\frac{t}{\sigma}} W(e^{\frac{2t}{\sigma}}).$$

Example 2.2.8 *The Brownian bridge $X(t)$ is a one-dimensional Gaussian process with time $t \in [0,1]$ and covariance* $\mathbb{C}ov[(X(t), X(s))] = \min(t,s) - ts = \begin{cases} s(1-t), 0 \le s \le t \le 1. \\ t(1-s), 0 \le t \le s \le 1. \end{cases}$

Suppose that $W(t)$ is a standard Brownian motion. Then $X(t)$ can be represented by

$$X(t) = W(t) - tW(1) = t\big(W(t) - W(1)\big) + (1-t)\big(W(t) - W(0)\big), \quad 0 \le t \le 1.$$

The process $X(t)$ bridges $W(t) - W(1)$ and $W(t) - W(0)$. It can be readily verified that $\mathbb{C}\mathrm{ov}[(X(t), X(s))] = \min(t,s) - ts$ and $X(t)$ is continuous and starting from 0. Moreover,

$$W(t) = (t+1)X(\frac{t}{t+1}), \quad X(t) = (1-t)W(\frac{t}{1-t}).$$

Regularity of Brownian motion

For deterministic functions, $f(x)$, $x \in \mathbb{R}$ is Hölder continuous of order α if and only if there exists a constant C such that

$$|f(x+h) - f(x)| \le Ch^{\alpha}, \quad \text{for all } h > 0 \text{ and all } x.$$

When $\alpha = 1$, we call it Lipschitz continuous. When C depends on x, then we call it locally Hölder continuous of order α

$$|f(x+h) - f(x)| \le C(x)h^{\alpha}, \quad \text{for all small enough } h > 0.$$

Definition 2.2.9 *Consider two stochastic processes, $X(t)$ and $Y(t)$, defined on the probability space $(\Omega, \mathcal{F}, \mathbb{P})$. We call $Y(t)$ a modification (or version) of $X(t)$ if for every $t \ge 0$, we have*

$$\mathbb{P}(X(t) = Y(t)) = 1.$$

Theorem 2.2.10 (Kolmogorov and Centsov continuity theorem, [255, Section 2.2.B]) *Given a stochastic process $X(t)$ with $t \in [a,b]$, if there exist constants $p > r$, $K > 0$ such that*

$$\mathbb{E}[|X(t) - X(s)|^p] \le K\,|t-s|^{1+r}, \text{ for } t, s \in [a,b],$$

then $X(t)$ has a modification $Y(t)$ which is almost everywhere (in ω) continuous: for all $t, s \in [a,b]$,

$$|Y(t,\omega) - Y(s,\omega)| \le C(\omega)\,|t-s|^{\alpha}, 0 < \alpha < \frac{r}{p}.$$

For $X(\boldsymbol{t})$, $\boldsymbol{t} \in T \subseteq \mathbb{R}^d$, if there exist constants $p > r$, K such that

$$\mathbb{E}[|X(\boldsymbol{t}) - X(\boldsymbol{s})|^p] \leq K \left|\boldsymbol{t} - \boldsymbol{s}\right|^{d+r}, \text{ for } \boldsymbol{t}, \boldsymbol{s} \in T,$$

then $X(\boldsymbol{t})$ has a modification $Y(\boldsymbol{t})$ which is almost everywhere in ω continuous: for all $\boldsymbol{t}$, $\boldsymbol{s} \in T$,

$$\mathbb{E}[(\sup_{\boldsymbol{s}\neq\boldsymbol{t}} \frac{|Y(\boldsymbol{t},\omega) - Y(\boldsymbol{s},\omega)|}{|\boldsymbol{t}-\boldsymbol{s}|}^{\alpha})^p] < \infty, 0 < \alpha < \frac{r}{p}.$$

Theorem 2.2.11 *For $\alpha < \frac{1}{2}$, the Brownian motion has a modification which is of locally Hölder continuous of order α.*

Proof. For integer $n \geq 1$, by Kolmogorov and Centsov continuity theorem, it only requires to show that

$$\mathbb{E}[|W(t) - W(s)|^{2n}] \leq C_n \left|t - s\right|^n.$$

Then recalling the conclusion from Exercise 2.7.1 leads to the conclusion.

Theorem 2.2.12 ([255, Section 2.9.D]) *The Brownian motion is nowhere differentiable: for almost all ω, the sample path (realization, trajectory) $W(t,\omega)$ is nowhere differentiable as function of t. Moreover, for almost all ω, the path (realization, trajectory) $W(t,\omega)$ is nowhere Hölder continuous with exponent $\alpha > \frac{1}{2}$.*

Definition 2.2.13 (p-variation) *The p-variation of a real-valued function f, defined on an interval $[a,b] \subset \mathbb{R}$, is the quantity*

$$|f|_{p,\mathrm{TV}} = \sup_{\Pi_n,\, |\pi_n|\to 0} \sum_{i=0}^{n-1} |f(x_{i+1}) - f(x_i)|^p,$$

where the supremum runs over the set of all partitions Π_n of the given interval.

Theorem 2.2.14 (Unbounded total variation of Brownian motion) *The paths (realizations, trajectories) of Brownian motion are of infinite total variation almost surely (a.s., with probability one).*

Proof. Without loss of generality, let's consider the interval $[0,1]$.

$$|W|_{1,\mathrm{TV}} = \sup_{\Pi_n} \sum_{i=0}^{n-1} |W(t_{i+1}) - W(t_i)| \geq \sum_{i=0}^{n-1} |W(\frac{i+1}{n}) - W(\frac{i}{n})| =: V_n.$$

Denote by $W(\frac{i+1}{n}) - W(\frac{i}{n}) = \frac{\xi_i}{\sqrt{n}}$. Then ξ_i's are i.i.d. $\mathcal{N}(0,1)$ random variables. Observe that $\mathbb{E}[V_n] = \sqrt{n}\mathbb{E}[|\xi_1|]$ and $\mathbb{V}\mathrm{ar}[V_n] = 1 - (\mathbb{E}[|\xi_1|])^2$.

Then it follows from the Chebyshev inequality (see Appendix D), we have

$$\begin{aligned}\mathbb{P}(V_n \geq \frac{1}{2}\mathbb{E}[|\xi_1|]\sqrt{n}) &= \mathbb{P}(V_n - \mathbb{E}[|\xi_1|]\sqrt{n} \geq -\frac{1}{2}\mathbb{E}[|\xi_1|]\sqrt{n}) \\ &\geq 1 - \mathbb{P}(\left|V_n - \mathbb{E}[|\xi_1|]\sqrt{n}\right| \geq \frac{1}{2}\mathbb{E}[|\xi_1|]\sqrt{n}) \\ &\geq 1 - \frac{\mathrm{Var}[V_n]}{\left(\frac{1}{2}\mathbb{E}[|\xi_1|]\sqrt{n}\right)^2} = 1 - 4\frac{1-(\mathbb{E}[|\xi_1|])^2}{n(\mathbb{E}[|\xi_1|])^2}.\end{aligned}$$

Thus we have

$$\mathbb{P}(|W|_{1,\mathrm{TV}} \geq \frac{\mathbb{E}[|\xi_1|]}{2}\sqrt{n}) \geq \mathbb{P}(V_n \geq \frac{\mathbb{E}[|\xi_1|]}{2}\sqrt{n}) = 1 - 4\frac{1-(\mathbb{E}[|\xi_1|])^2}{n(\mathbb{E}[|\xi_1|])^2}.$$

Letting $n \to \infty$, we obtain

$$\mathbb{P}(|W|_{1,\mathrm{TV}} = \infty) = 1.$$

2.2.2 Approximation of Brownian motion

According to the representation in Chapter 2.2, we have at least three approximate representations for Brownian motion by a finite number of random variables.

By Definition 2.2.1, the Brownian motion at time t_{n+1} can be approximated by

$$\sum_{i=0}^{n} \Delta W_i = \sum_{i=0}^{n} \sqrt{\Delta \mathsf{t}_i}\xi_i, \text{ where } \Delta W_i = W(\mathsf{t}_{i+1}) - W(\mathsf{t}_i), \text{ and } \Delta \mathsf{t}_i = \mathsf{t}_{i+1} - \mathsf{t}_i, \tag{2.2.3}$$

where ξ_i's are i.i.d. standard Gaussian random variables. A sample path (realization, trajectory) of Brownian motion is illustrated in Figure 2.1. Here is Matlab code for generating Figure 2.1.

Code 2.1. A sample path of Brownian motion.

```
% One realization of W(t) at time grids k*dt
clc, clear all
t =2.5;
n= 1000;
dt = t / n;
%  Increments of Brownian motion
Winc  = zeros ( n + 1,1);
% Declare the status of random number generator -- Mersenne
% Twister
rng(100, 'twister');
Winc(1:n) = sqrt ( dt ) * randn ( n, 1);
%  Brownian motion - cumulative sum of all previous
%  increments
```

```
W(2:n+1,1) = cumsum ( Winc(1:n) );
figure(10)
plot((1:n+1).'*dt,W,'b-','Linewidth',2);
xlabel('t')
ylabel('W(t)')
axis tight
```

One popular approximation of Brownian motion in continuous time is piecewise linear approximation (also known as polygonal approximation, see, e.g., [457, 481, 482] or [241, p. 396]), i.e.,

$$W^{(n)}(t) = W(\mathsf{t}_i) + (W(\mathsf{t}_{i+1}) - W(\mathsf{t}_i))\frac{t-\mathsf{t}_i}{\mathsf{t}_{i+1}-\mathsf{t}_i}, \quad t \in [\mathsf{t}_i, \mathsf{t}_{i+1}). \tag{2.2.4}$$

Another way to approximate Brownian motion is by a truncated *orthogonal expansion*:

$$W^{(\mathsf{n})}(t) = \sum_{i=1}^{\mathsf{n}} \xi_i \int_0^t m_i(s)\,ds, \quad \xi_j = \int_0^T m_j(t)\,dW, \quad t \in [0,T], \tag{2.2.5}$$

where $\{m_i(t)\}$ is a CONS in $L^2([0,T])$ and ξ_j are mutually independent standard Gaussian random variables.

In this book, we mostly use the cosine basis $\{m_l(s)\}_{l\geq 1}$ given by

Fig. 2.1. An illustration of a sample path of Brownian motion using cumulative summation of increments.

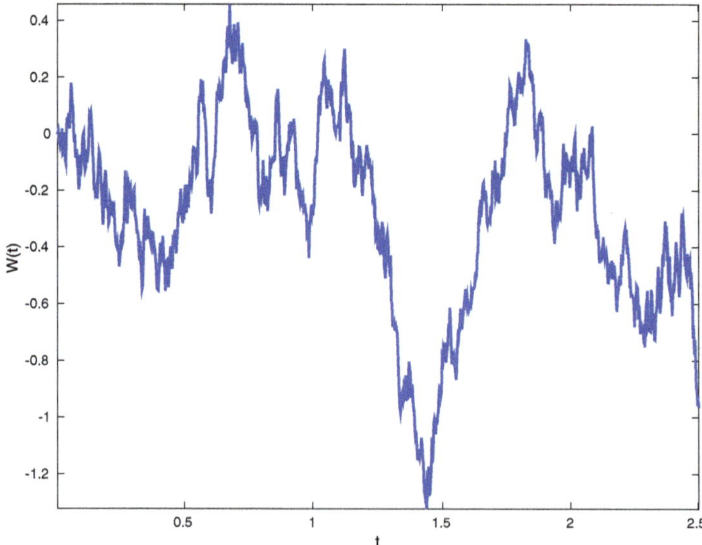

$$m_1(s) = \frac{1}{\sqrt{T}}, \quad m_l(s) = \sqrt{\frac{2}{T}} \cos(\frac{\pi(l-1)s}{T}), \quad l \geq 2, \quad 0 \leq s \leq T, \tag{2.2.6}$$

or a piecewise version of spectral expansion (2.2.5) by taking a partition of $[0,T]$, e.g., $0 = \mathsf{t}_0 < \mathsf{t}_1 < \cdots \mathsf{t}_{\mathsf{K}-1} < \mathsf{t}_\mathsf{K} = T$. We then have

$$W^{(\mathsf{n},\mathsf{K})}(t) = \sum_{k=1}^{\mathsf{K}} \sum_{l=1}^{\mathsf{n}} M_{k,l}(t)\xi_{k,l}, \quad \xi_{k,l} = \int_{I_k} m_{k,l}(s)\, dW(s) \tag{2.2.7}$$

where $M_{k,l}(t) = \int_{\mathsf{t}_{k-1}}^{\mathsf{t}_k \wedge t} m_{k,l}(s)\, ds$ ($\mathsf{t}_k \wedge t = \min(\mathsf{t}_k, t)$), $\{m_{k,l}\}_{l=1}^{\infty}$ is a CONS in $L^2(I_k)$ and $I_k = [\mathsf{t}_{k-1}, \mathsf{t}_k)$. The random variables $\xi_{k,l}$ are i.i.d. standard Gaussian random variables. Sometimes (2.2.7) is written as

$$W^{(\mathsf{n},\mathsf{K})}(t) = \sum_{k=1}^{\mathsf{K}} \sum_{l=1}^{\mathsf{n}} \int_0^t \mathbf{1}_{I_k}(s) m_{k,l}(s)\, ds \xi_{k,l}, \tag{2.2.8}$$

where $\mathbf{1}.$ is the indicator function.

Here different choices of CONS lead to different representations. The orthonormal piecewise constant basis over time interval $I_k = [\mathsf{t}_{k-1}, \mathsf{t}_k)$, with $h_k = (\mathsf{t}_k - \mathsf{t}_{k-1})/\sqrt{\mathsf{n}}$, is

$$m_{k,l}(t) = \frac{1}{\sqrt{h}} \chi_{[\mathsf{t}_{k-1}+(l-1)h_k, \mathsf{t}_{k-1}+lh_k)}, \quad l = 1, 2, \cdots, \mathsf{n}. \tag{2.2.9}$$

When $\mathsf{n} = 1$, this basis gives the classical piecewise linear interpolation (2.2.4). The orthonormal Fourier basis gives Wiener's representation (see, e.g., [259, 358, 391]):

$$m_{k,1} = \frac{1}{\sqrt{\mathsf{t}_k - \mathsf{t}_{k-1}}}, \quad m_{k,2l} = \sqrt{\frac{2}{\mathsf{t}_k - \mathsf{t}_{k-1}}} \sin\left(2l\pi \frac{t - \mathsf{t}_{k-1}}{\mathsf{t}_k - \mathsf{t}_{k-1}}\right),$$

$$m_{k,2l+1}(t) = \sqrt{\frac{2}{\mathsf{t}_k - \mathsf{t}_{k-1}}} \cos\left(2l\pi \frac{t - \mathsf{t}_{k-1}}{\mathsf{t}_k - \mathsf{t}_{k-1}}\right), \ t \in [\mathsf{t}_{k-1}, \mathsf{t}_k). \tag{2.2.10}$$

Note that taking only $m_{k,1}$ ($\mathsf{n} = 1$) in (2.2.10) again leads to the piecewise linear interpolation (2.2.4). Besides, we can also use the Haar wavelet basis, which gives the Levy-Ciesielsky representation [255].

Remark 2.2.15 *Once we have a formal representation (approximation) of Brownian motion, we then can readily obtain a formal representation (approximation) of white noise and thus we skip the formulas for white noise.*

Lemma 2.2.16 *Consider a uniform partition of* $[0,T]$, *i.e.,* $\mathsf{t}_k = k\Delta$, $\mathsf{K}\Delta = T$. *For* $t \in [0,T]$, *there exists a constant* $C > 0$ *such that*

$$\mathbb{E}[(W(t) - W^{(\mathsf{n},\mathsf{K})}(t))^2] \leq C\frac{\Delta}{n},$$

and for sufficient small $\epsilon > 0$

$$\left|W(t) - W^{(\mathsf{n},\mathsf{K})}(t)\right| \leq \mathcal{O}\left(\left(\frac{\Delta}{n}\right)^{1/2-\epsilon}\right).^{1} \tag{2.2.11}$$

For $t = \mathsf{t}_k$, *we have*

$$W(\mathsf{t}_k) - W^{(\mathsf{n},\mathsf{K})}(\mathsf{t}_k) = 0, \tag{2.2.12}$$

if the CONS $\{m_{k,l}\}_{l=1}^{\infty}$ *contains* $\frac{1}{\sqrt{\mathsf{t}_k - \mathsf{t}_{k-1}}}$ *as its elements, i.e.,* $\int_{\mathsf{t}_{k-1}}^{\mathsf{t}_k} m_{k,l}(s)$ $\frac{1}{\sqrt{\mathsf{t}_k - \mathsf{t}_{k-1}}} \, ds = \delta_{l,1}$.

Proof. By the spectral approximation of $W(t)$ (2.2.7) and the fact that $\xi_{k,l,i}$ are i.i.d., we have

$$\begin{aligned} \mathbb{E}[(W(t) - W^{(\mathsf{n},\mathsf{K})}(t))^2] &= \sum_{k=1}^{\mathsf{K}} \sum_{l=\mathsf{n}+1}^{\infty} \left(\int_{\mathsf{t}_{k-1}}^{t \wedge \mathsf{t}_k} m_{k,l}(s) \, ds\right)^2 \\ &= \sum_{k=1}^{\mathsf{K}} \sum_{l=\mathsf{n}+1}^{\infty} \left(\int_{\mathsf{t}_{k-1}}^{\mathsf{t}_k} \chi_{[0,t]}(s) m_{k,l}(s) \, ds\right)^2 \\ &\leq C \frac{\Delta}{n}, \end{aligned}$$

where $\mathsf{t}_k \wedge t = \min(\mathsf{t}_k, t)$ and we have applied the standard estimate for L^2-projection using piecewise orthonormal basis $m_{k,l}(s)$, see, e.g., [417] and have used the fact that the indicator function $\chi_{[0,t]}(s)$ belongs to the Sobolev space $H^{1/2}((0,T))$ for any $T > t$.

Once we have the L^2-estimate, we can apply the Borel-Cantelli lemma (see Appendix D) to obtain the almost sure (a.s.) convergence (2.2.11).

If $t = \mathsf{t}_k$, we have $\int_{\mathsf{t}_{k-1}}^{\mathsf{t}_k} m_{k,l}(s) \, ds = 0$ for any $l \geq 2$ if $m_{k,1} = \frac{1}{\sqrt{\mathsf{t}_k - \mathsf{t}_{k-1}}}$. and thus (2.2.12) holds.

Though any CONS in $L^2([0,T])$ can be used in the spectral approximation (2.2.5), we use a CONS containing a constant in the basis. Consequently, we have the following relation

$$\int_{t_n}^{t_{n+1}} dW^{(\mathsf{n})}(t) = \Delta W_n, \quad \Delta W_n = W(t_{n+1}) - W(t_n). \tag{2.2.13}$$

We will use these approximations in most of the chapters in the book for Wong-Zakai approximation.

---

[1] The big "$\mathcal{O}$" implies that the error is bounded by a positive constant times the term in the parenthesis.

2.3 Brownian motion and stochastic calculus

As Brownian motion $W(t)$ is not a process of bounded variation, the integral $\int_0^t f(t)\,dW(t)$ cannot be interpreted using Riemann-Stieltjes integration or Lebesgue-Stieltjes integration, even for very smooth stochastic process f. However, it can be understood as Ito integral or Stratonovich integral. For an adapted process $f(t)$ (with respect to the natural filtration of Brownian motion), the Ito integral is defined as, see, e.g., [388], for all partitions of the interval $[0, T]$,

$$\lim_{|\Pi_n|\to 0} \mathbb{E}[(\int_0^T f(t)\cdot dW - \sum_{i=1}^{n-1} f(t_{i-1})\Delta W_i)^2] = 0,$$

where $\Pi_n = \{0 = t_0 < t_1 < t_2 < \cdots < t_n = T\}$ is a partition of the interval $[0, T]$ and $|\Pi_n| = \max_{0\le i\le n-1} |t_{i+1} - t_i|$. The finite sum in this definition is defined at the *left-hand points* in each subinterval of the partition. For Stratonovich calculus, the finite sum is defined at the *midpoints* in each subinterval of the partition. i.e.,

$$\int_0^T f(t)\circ dW = \lim_{|\Pi_n|\to 0} \sum_{i=1}^{n-1} f(\frac{t_{i-1}+t_i}{2})\Delta W_i.$$

Again, the limit is understood in the mean-square sense, see, e.g., [388, 431].

Example 2.3.1 *It can be readily checked that*

$$\int_0^T W(t)\,dW(t) = \frac{W^2(T)-T}{2}, \quad \int_0^T W(t)\circ dW(t) = \frac{W^2(T)}{2}. \tag{2.3.1}$$

Let us show that the first formula holds. By simple calculation and the properties of increments of Brownian motion, for $0 = t_0 < t_1 < t_2 < \cdots < t_n = T$, we have

$$\begin{aligned}
&\mathbb{E}[(\frac{W^2(T)-T}{2} - \sum_{i=0}^{n-1} W(t_i)\Delta W_i)^2] \\
&= \mathbb{E}[(\frac{W^2(T)-T}{2} - \sum_{i=0}^{n-1} \frac{W(t_i)+W(t_{i+1})}{2}\Delta W_i + \frac{1}{2}\sum_{i=1}^{n}(\Delta W_i)^2)^2] \\
&= \mathbb{E}[(\frac{W^2(T)-T}{2} - \sum_{i=0}^{n-1} \frac{W^2(t_{i+1})-W^2(t_i)}{2} + \frac{1}{2}\sum_{i=1}^{n}(\Delta W_i)^2)^2] \\
&= \mathbb{E}[(\frac{-T}{2} + \frac{1}{2}\sum_{i=1}^{n}(\Delta W_i)^2)^2] \to 0, \quad n\to\infty.
\end{aligned}$$

The second formula can be derived similarly.

Moreover, we have the following conversion formula.

Theorem 2.3.2 (Conversion of a Stratonovich integral to an Ito integral) *A Stratonovich integral can be computed via the Ito integral:*

$$\int_0^T f(t, W(t) \circ dW(t) = \int_0^T f(t, W(t)) \, dW(t) + \frac{1}{2} \int_0^T \partial_x f(t, W(t)) \, dt.$$

Here $f(t, W(t))$ is a scalar function and $\partial_x f$ is the derivative with respect to the second argument of f. When $f \in \mathbb{R}^{m \times n}$ is a matrix function, then

$$[\int_0^T f(t, W(t) \circ dW(t)]_i = [\int_0^T f(t, W(t)) \, dW(t)]_i + \frac{1}{2} \int_0^T \sum_{j=1}^n \partial_{x_j} f_{i,j}(t, W(t)) \, dt, \quad i = 1, 2, \cdots, m.$$

Here v_i means the i-th component of a vector v.

The proof can be done using the definition of two integrals and mean value theorem. We leave the proof to interested readers.

Definition 2.3.3 (Quadratic covariation) *The quadratic covariation of two processes X and Y is*

$$[X, Y]_t = \lim_{|\Pi_n| \to 0} \sum_{k=1}^n (X(t_k) - X(t_{k-1})) (Y(t_k) - Y(t_{k-1})).$$

Here $\Pi_n = \{0 = t_0 < t_1 < \cdots < t_{n-1} < t_n = t\}$ is an arbitrary partition of the interval $[0, t]$.

When $X = Y$, the quadratic covariation becomes the quadratic variation:

$$[X]_t = [X, X]_t = \lim_{|\Pi_n| \to 0} \sum_{k=1}^n X(t_k) - X(t_{k-1})^2.$$

The quadratic covariation can be computed by the polarization identity:

$$[X, Y]_t = \tfrac{1}{4}([X + Y]_t - [X - Y]_t).$$

With the definition of quadratic covariation, we have

$$\int_0^T f(t, W(t) \circ dW(t) = \int_0^T f(t, W(t)) \, dW(t) + \frac{1}{2} \int_0^T \partial_x f(t, W(t)) \, d[W, W]_t.$$

More generally, we have the following conversion rule

$$\int_0^t Y(s) \circ dX(s) = \int_0^t Y(s) \, dX(s) + \frac{1}{2}[X, Y]_t. \tag{2.3.2}$$

For Ito integral, we have the following properties. Define

$$\mathbb{L}^2_{\mathrm{ad}}(\Omega; L^2([a,b])) = \left\{ f_t(\omega) | f_t(\omega) \text{ is } \mathcal{F}_t\text{-measurable and } \mathbb{E}[\int_a^b f_s^2\, ds] < \infty \right\}.$$

Theorem 2.3.4 *For $f, g \in \mathbb{L}^2_{ad}(\Omega; L^2([0,T]))$, we have*

- *(linear property)* $\int_0^t (af(s) + bg(s))\, dW(s) = a \int_0^t f(s)\, dW(s) + b \int_0^t g(s)$ $dW(s)$, $\quad a, b \in \mathbb{R}$,
- *(Ito isometry)* $\mathbb{E}[(\int_0^t f(s)\, dW(s))^2] = \int_0^t \mathbb{E}[f^2(s)]\, ds$,
- *(Generalized Ito isometry)* $\mathbb{E}[\int_0^t f(s)\, dW(s) \int_0^t g(s)\, dW(s)] = \int_0^t \mathbb{E}[f(s)$ $g(s)]\, dt$,
- $M_t = \int_0^t f(s)\, dW(s)$ *is a continuous martingale. Moreover, the quadratic variation of* M_t *is* $|M|_{2,\mathrm{TV}} = \int_0^t f^2(s)\, ds$, $0 \le t \le T$ *and*

$$\mathbb{E}[\sup_{0 \le t \le T} (\int_0^t f(s)\, dW(s))^2] \le 4\mathbb{E}[\int_0^t \mathbb{E}[f^2(s)]\, ds].$$

Example 2.3.5 (Multiple Ito integral) *Assuming that $W(t)$ is a standard Brownian Motion, show that*

$$\frac{1}{n!} \int_0^t \int_0^{t_n} \cdots \int_0^{t_2} dW(t_1) \cdots dW(t_n) = t^{n/2} H_n(\frac{W(t)}{\sqrt{t}}).$$

Here H_n is the n-th Hermite polynomial:

$$H_n(x) = (-1)^n e^{x^2/2} \frac{d^n}{dx^n} e^{-x^2/2}. \qquad (2.3.3)$$

When $n = 0$, $H_n(x) = 1$ and we use the convention that when $n < 1$ the integral is defined as 1. When $n = 1$, $\int_0^t dW(t_1) = W(t) = t^{1/2} H_1(\frac{W(t)}{\sqrt{t}})$ as $H_1(x) = x$. Then it can be shown by induction that the integrand in the left-hand side is in $\mathbb{L}^2_{\mathrm{ad}}(\Omega; L^2([0,t]))$ and thus the multiple integral is indeed an Ito integral and is equal to the right-hand side.

When we want to define the integral $\int_0^t f(t)\, dW(t)$ via a spectral representation of Brownian motion instead of using increments of Brownian motion, we have to use the so-called Ito-Wick product (Wick product) and Stratonovich product.

The use of Ito-Wick product relies on two facts: the integrand f can be expressed as Hermite polynomial expansion of some random variables ("random basis") and the product is well defined for these basis. The basis and the Ito-Wick product will be shown and defined shortly. Specifically, let $\{\xi_k\}_{k=1}^{\infty}$ be a sequence of mutually independent standard Gaussian random variables from a spectral representation of Brownian motion and let $\mathcal{F} = \sigma(\xi_k)_{k\geq 1}$. The following Cameron-Martin theorem states that any element in $\mathbb{L}^2(\Omega, \mathcal{F}, \mathbb{P})$ can be represented by a linear combination of elements in the Cameron-Martin basis (2.3.4).

Theorem 2.3.6 (Cameron-Martin [57]) *Let $(\Omega, \mathcal{F}, \mathbb{P})$ be a complete probability space. The collection $\Xi = \{\xi_\alpha\}_{\alpha\in\mathcal{J}}$ is an orthonormal basis in $\mathbb{L}^2(\Omega, \mathcal{F}, \mathbb{P})$, where ξ_α's are defined as*

$$\xi_\alpha := \prod_\alpha \left(\frac{H_{\alpha_l}(\xi_l)}{\sqrt{\alpha_l!}} \right), \; \xi_l = \int_0^t m_l(s)\, dW(s), \quad \alpha \in \mathcal{J}, \tag{2.3.4}$$

where $\{m_l\}$ is a complete orthonormal basis in $L^2([0,t])$ and $\mathcal{J}$ is the collection of of multi-indices $\alpha = (\alpha_l)_{l\geq 1}$ of finite length, i.e.,

$$\mathcal{J} = \left\{ \alpha = (\alpha_l)_{l\geq 1}, \quad \alpha_l \in \{0,1,2,\ldots\}, \; |\alpha| := \sum_l \alpha_l < \infty \right\}.$$

Any $\eta \in \mathbb{L}^2(\Omega, \mathcal{F}, \mathbb{P})$ can be represented as the following Wiener chaos expansion

$$\eta = \sum_{\alpha\in\mathcal{J}} \eta_\alpha \xi_\alpha, \quad \eta_\alpha = \mathbb{E}[\eta\xi_\alpha], \quad \text{and} \quad \mathbb{E}[\eta^2] = \sum_{\alpha\in\mathcal{J}} \eta_\alpha^2. \tag{2.3.5}$$

The collection Ξ of random variables $\{\xi_\alpha, \alpha \in \mathcal{J}\}$ is called the Cameron-Martin basis. It can be readily shown that $\mathbb{E}[\xi_\alpha \xi_\beta] = 1$ if $\alpha = \beta$ and 0 otherwise. See some specific examples of the Cameron-Martin basis in Chapter 2.5.3.

Following [223, Section 2.5] and [316], under certain conditions on $f(t)$ (continuous semi-martingale with respect to the natural filtration of Brownian motion), we have

$$\int_0^t f(t)\, dW(t) = \int_0^t f(t) \diamond \dot{W}(t)\, dt, \tag{2.3.6}$$

where the definition of Ito-Wick product "$\diamond$" is based on the product for elements of the Cameron-Martin basis: $\xi_\alpha \diamond \xi_\beta = \sqrt{\frac{(\alpha+\beta)!}{\alpha!\beta!}}\xi_{\alpha+\beta}$.

With the approximation (2.2.5), Ogawa [386] defined the following so-called Ogawa integral,

$$\int_0^t f(t) * dW(t) = \sum_{i=1}^{\infty} \int_0^t f(s)m_i(s)\,ds \int_0^t m_i(s)\,dW(s), \quad t \in [0,T] \quad (2.3.7)$$

where $\{m_i(t)\}$ is the CONS on $L^2([0,T])$. (Note that Ogawa's original integral is only defined when $t = T$.) Nualart and Zakai [385] proved that the Ogawa integral is equivalent to the Stratonovich integral if the Ogawa integral exists, with the Stratonovich integral defined as

$$\int_0^t f(t) \circ dW(t) = \lim_{|\Pi_n| \to 0} \sum_{i=1}^{n} \frac{1}{t_{i+1} - t_i} \int_{t_i}^{t_{i+1}} f(s)\,ds(W(t_{i+1}) - W(t_i)). \quad (2.3.8)$$

Moreover, if the integrand $f(t)$ is a continuous semi-martingale (e.g., Ito process in Definition 2.4.2) on the natural filtration of $W(t)$, then the Ogawa integral coincides with the Stratonovich integral defined at the mid-points.

$$\int_0^t f(t) \circ dW(t) = \lim_{|\Pi_n| \to 0} \sum_{i=1}^{n} f\left(\frac{t_i + t_{i+1}}{2}\right)(W(t_{i+1}) - W(t_i)). \quad (2.3.9)$$

As application of stochastic integrals, the fractional Brownian motion $B_H(t)$, $t \geq 0$, can be introduced. It is a centered Gaussian process with the following covariance function

$$\mathbb{E}[B_H(t)B_H(s)] = \tfrac{1}{2}(|t|^{2H} + |s|^{2H} - |t-s|^{2H}), \quad 0 < H < 1.$$

The constant H is called the Hurst index or Hurst parameter. The fractional Brownian motion can be represented by

$$B_H(t) = B_H(0) + \frac{1}{\Gamma(H+1/2)} \left\{ \int_{-\infty}^{0} \left[(t-s)^{H-1/2} - (-s)^{H-1/2}\right] dW(s) + \int_0^t (t-s)^{H-1/2}\,dW(s) \right\}.$$

2.4 Stochastic chain rule: Ito formula

One motivation of Ito formula is to evaluate Ito integral with a complicated integrand.

Theorem 2.4.1 (Ito formula in the simplest form) *If f and its first two derivatives are continuous on $\mathbb{R}$, then it holds with probability one (almost surely, a.s.) that*

$$f(W(t)) = f(W(t_0)) + \int_{t_0}^{t} f'(W(s))\, dW(s) + \frac{1}{2}\int_{t_0}^{t} f''(W(s))\, ds.$$

From the theorem, we can compute the Ito integral $\int_{t_0}^{t} f'(W(s))\, dW(s)$ by

$$\int_{t_0}^{t} f'(W(s))\, dW(s) = f(W(t)) - f(W(t_0)) - \frac{1}{2}\int_{t_0}^{t} f''(W(s))\, ds.$$

Definition 2.4.2 (Ito process) *An Ito process is a stochastic process of the form*

$$X_t = X(t_0) + \int_{t_0}^{t} a(s)\, ds + \int_{t_0}^{t} \sigma(s)\, dW(s),$$

where $X(t_0)$ is $\mathcal{F}_{t_0}$-measurable, a_s and σ_s are adapted w.r.t. $\mathcal{F}_s$ and

$$\int_{t_0}^{t} |a(s)|\, ds < \infty, \quad \int_{t_0}^{t} \|\sigma(s)\|^2\, ds < \infty \text{ a.s..}$$

The filtration $\{\mathcal{F}_s,\, t_0 \le s \le t\}$ is defined such that

- for any s, B_s, $a(s)$ and $\sigma(s)$ are $\mathcal{F}_s$-measurable;
- for any $t_1 \le t_2$, $B_{t_2} - B_{t_1}$ is independent of $\mathcal{F}_{t_1}$.

Suppose that $X(t)$ exists a.s. such that

$$X(t) = X(t_0) + \int_{t_0}^{t} a(s, X(s))\, ds + \sum_{r=1}^{m} \int_{t_0}^{t} \sigma_r(s, X(s))\, dW_r(s).$$

Here $X(t)$, $X(t_0), a$, $\sigma_r \in \mathbb{R}^d$ and $\sigma \in \mathbb{R}^{d\times m}$. Also, $W_r(s)$'s are mutually independent Brownian motions, $a(s)$ and $\sigma(s)$ are adapted w.r.t. $\mathcal{F}_s$, and

$$\int_{t_0}^{t} |a(s, X(s))|\, ds < \infty, \quad \sum_{r=1}^{m} \int_{t_0}^{t} |\sigma_r(s, X(s))|^2\, ds < \infty \text{ a.s..}$$

The filtration $\{\mathcal{F}_s,\, t_0 \le s \le t\}$ is defined such that for any s, $W_r(s)$ is $\mathcal{F}_s$-measurable and for any $t_1 \le t_2$, $W_r(t_2) - W_r(t_1)$ is independent of $\mathcal{F}_{t_1}$. Ito formula for a $C^1([0,T]; C^2(\mathbb{R}^d))$ function $f(t,\cdot)$ is

$$f(t, X(t)) = f(t_0, X(t_0)) + \sum_{r=1}^{m} \int_{t_0}^{t} \Lambda_r f(s, X(s))\, dW_r(s) + \int_{t_0}^{t} \mathcal{L} f(s, X(s))\, ds,$$

where

$$\Lambda_r = \sigma_r^\top \nabla = \sum_{i=1}^{d} \sigma_{i,r} \frac{\partial}{\partial x_i}, \quad \nabla = \left(\frac{\partial}{\partial x_1}, \frac{\partial}{\partial x_2}, \cdots, \frac{\partial}{\partial x_d}\right),$$

$$\mathcal{L} = \frac{\partial}{\partial t} + a^\top \nabla + \frac{1}{2} \sum_{r=1}^{m} \sum_{i,j=1}^{d} \sigma_{i,r}\sigma_{j,r} \frac{\partial^2}{\partial x_i \partial x_j}.$$

Remark 2.4.3 *For the multi-dimensional Ito formula, we can use the following table to memorize the formula when $W_j(t)$ are mutually independent Brownian motions.*

$\times$	$dW_j(t)$	dt
$dW_i(t)$	$\mathbf{1}_{\{i=j\}}dt$	0
dt	0	0

Theorem 2.4.4 (Integration by parts formula) *Let $X(t)$ and Y_t be Ito processes defined in Definition 2.4.2. Then the following integration by parts formula holds*

$$X(t)Y(t) = X(t_0)Y(t_0) + \int_{t_0}^{t} X(s)\,dY(s) + \int_{t_0}^{t} Y(s)\,dX(s) + \int_{t_0}^{t} dX(s)\,dY(s).$$

Here $dX(s)\,dY(s)$ can be formally computed using the table in Remark 2.4.3.

This integration by parts formula is a corollary of multi-dimensional Ito formula for Ito processes.

Consider two Ito processes, X and Y: $dX = a^X(t)\,dt + \sigma^X(t)\,dW(t)$ and $dY = a^Y(t)\,dt + \sigma^Y(t)\,dW(t)$. Then we have from Remark 2.4.3 that

$$dXdY = (a^X(s)\,dt + \sigma^X(t)\,dW(t)(a^Y(s)\,dt + \sigma^Y(t)\,dW(t))) = \sigma^X(t)\sigma^Y(t)\,dt.$$

2.5 Integration methods in random space

Numerical SODEs and SPDEs are usually dependent on the Monte Carlo method and its variants to obtain the desired statistics of the solutions. The fundamental quantity of interest is of the following form

$$\frac{1}{(\sqrt{2\pi})^d}\int_{\mathbb{R}^d} f(x)e^{-\frac{|x|^2}{2}}\,dx.$$

This is a standard numerical integration problem. Let us consider $d = 1$ in a general setting. An integral can be treated as an expectation of random variable

$$\int_{\mathbb{R}} f(x)p(x)\,dx = \mathbb{E}[f(X)], \tag{2.5.1}$$

where X has a probability density $p(x) \geq 0$ and $\int_{\mathbb{R}} p(x)\,dx = 1$ (Here $p(x) = \frac{1}{\sqrt{2\pi}}e^{-x^2/2}$).

2.5.1 Monte Carlo method and its variants

A simple (standard, brute force) Monte Carlo estimator of this integration is

$$\mu = \mathbb{E}[f(X)] \approx \frac{1}{n}\sum_{i=1}^{n} f(X_i) =: \hat{\mu} \tag{2.5.2}$$

where X_i are copies of X (i.i.d. with X). The convergence of (2.5.2) is guaranteed by the law of large numbers. Observe that $\hat{\mu}$ is a random variable and the mean of this estimator is $\mathbb{E}[\hat{\mu}] = \mu$, the variance of the estimator is

$$\mathbb{V}\mathrm{ar}[\hat{\mu}] = \frac{\mathbb{V}\mathrm{ar}[f(X)]}{n}. \tag{2.5.3}$$

The error of the Monte Carlo estimator is measured by the following *confidence interval* (95%)

$$(\mu - 1.96\sqrt{\frac{\mathbb{V}\mathrm{ar}[f(X)]}{n}}, \mu + 1.96\sqrt{\frac{\mathbb{V}\mathrm{ar}[f(X)]}{n}}),$$
$$\sigma^2(f(X)) = \mathbb{E}[f^2(X)] - (\mathbb{E}[f(X)])^2.$$

In practice, we use the following since we don't know μ:

$$(\hat{\mu} - 1.96\sqrt{\frac{\mathbb{V}\mathrm{ar}[f(X)]}{n}}, \hat{\mu} + 1.96\sqrt{\frac{\mathbb{V}\mathrm{ar}[f(X)]}{n}}),$$
$$\sigma^2(f(X)) = \mathbb{E}[f^2(X)] - (\mathbb{E}[f(X)])^2.$$

The error estimate of the Monte Carlo estimator is based on the central limit theorem. Specifically, when n is large, the Monte Carlo estimator $\hat{\mu}$ is treated as a Gaussian random variable and

$$Z =: \frac{\hat{\mu} - \mu}{\sqrt{\mathbb{V}\mathrm{ar}[f(X)]/n}} = \frac{\hat{\mu} - \mu}{\sqrt{\mathbb{V}\mathrm{ar}[\hat{\mu}]}}$$

is a standard Gaussian random variable. Here the number 1.96[2] is the approximate z value such that $\mathbb{P}(Z < z) = 0.95$ and the number $1.96\frac{\mathbb{V}\mathrm{ar}[f(X)]}{\sqrt{n}}$ is called the statistical error.

The convergence rate of the Monte Carlo estimator can be also shown by the Chebyshev inequality, the fact that $\mathbb{E}[\hat{\mu}] = \mu$ and (2.5.3):

$$\mathbb{P}(|\mathbb{E}[f(X)] - \hat{\mu}| \geq \sqrt{\frac{\mathbb{V}\mathrm{ar}[\hat{\mu}]}{\varepsilon}}) = \mathbb{P}(|\mathbb{E}[f(X)] - \hat{\mu}| \geq n^{-1/2}\sqrt{\frac{\mathbb{V}\mathrm{ar}[f(X)]}{\varepsilon}}) \leq \varepsilon.$$

For any fixed ε, the error of the Monte Carlo estimator decreases in the rate $\mathcal{O}(n^{-1/2})$ when we increase the number of samples.

In practice, the random numbers X_i are replaced with pseudorandom numbers using pseudorandom number generators, see, e.g., [358, 376]. Also, the variance $\mathbb{V}\mathrm{ar}[f(X)]$ is replaced by its numerical value (called empirical variance):

$$\sqrt{\frac{1}{n-1}(\sum_{i=0}^{n} f^2(X_i) - \hat{\mu}^2)} \text{ or } \sqrt{\frac{1}{n-1}(\sum_{i=0}^{n} (f(X_i) - \hat{\mu})^2}. \tag{2.5.4}$$

In Matlab (Matlab 2011b and later version), the estimator of $\mathbb{E}[X^p]$ ($p \geq 1$, $X \sim \mathcal{N}(0,1)$) and the error can be implemented as follows.

[2]The value 2 is used sometimes in this book.

Code 2.2. Estimation of the second moment of a standard Gaussian random variable.

```
% Declare the status of random number generators -- Mersenne
% Twister
 rng(100, 'twister');
 n= 1000;  % the number of sample points
 X =  randn ( n, 1);
 p=2;
 mu_hat =  mean (X.^p);
 mu =1;   % X is a standard Gaussian random variable
 mc_int_err = mu_hat - mu;   %  integration error
 stat_err = 1.96*sqrt(var (X.^2)/n);  %   statistical error
```

When the variance $\mathbb{V}\mathrm{ar}[f(X)]$ is large, the Monte Carlo estimator $\hat{\mu}$ is not accurate: the empirical variance in (2.5.4) may be large and the statistical error in the confidence interval can be too large that the interval is not trusted; moreover, the empirical variance may not be reliable either.

To have an accurate Monte Carlo estimator, i.e., a small confidence interval, e.g., when we want $\sqrt{\frac{\mathbb{V}\mathrm{ar}[f(X)]}{n}} = 10^{-2}$ but $\mathbb{V}\mathrm{ar}[f(X)] = 10$, we then need $n = 10^4\mathbb{V}\mathrm{ar}[f(X)]) = 10^5$ Monte Carlo sample points. To reduce this number of Monte Carlo sample points, we need "the variance $\mathbb{V}\mathrm{ar}[f(X)]$ small." To "reduce" the variance, there are several methods available, such as importance sampling (change of measure), control variates, stratified sampling (decomposing the sampling domain into smaller sub-domains). For such variance reduction methods, one can refer to many books on this topic, such as [259, 264, 376].

For one-dimensional integration, the convergence rate of Monte Carlo method is simply $O(n^{-1/2})$ and too slow to compete with deterministic numerical integration methods. Actually, the advantage of Monte Carlo methods is their efficiency for high dimensional integration (large d). The statistical error of a Monte Carlo estimator does not depend on the dimensionality.

Multilevel Monte Carlo method

One of the recently developed variance reduction methods, the so-called multilevel Monte Carlo method, see, e.g., [156] has attracted a lot of attention for numerical SODEs and SPDEs. The idea of the multilevel Monte Carlo method is to write the desired statistics in a telescoping sum and then to sample the difference terms (between terms defined on two different mesh sizes) in the telescoping sum with a small number of sampling paths, where the corresponding "variance" is small. The multilevel Monte Carlo method for (2.5.1) is based on the following formulation.[3]

---

[3]As a method of control variate, it can be written in the following form

$$\int_{\mathbb{R}} f(x)p(x)\,dx = \int_{\mathbb{R}} [f(x) - \lambda\big(g(x) - \mathbb{E}[g]\big)]p(x)\,dx,$$

$$\int_{\mathbb{R}} f(x)p(x)\,dx = \int_{\mathbb{R}} [f(x) - f_0(x)]p(x)\,dx + \int_{\mathbb{R}} f_L(x)p(x)\,dx$$
$$+ \sum_{l=0}^{L-1} \int_{\mathbb{R}} [f_l(x) - f_{l+1}(x)]p(x)\,dx, \; L \geq 0.^4$$

The control variate f_0 is chosen such that $f_0 \approx f$ and it is much cheaper to simulate f_L (and $\mathbb{E}[f_L(x)]$), and $f_l - f_{l+1}$ has smaller variances than f's. Consider $L = 0$. The Monte Carlo estimate is then

$$\frac{1}{N_0} \sum_{k=1}^{N_0} f_0(X_k) + \frac{1}{N_1} \sum_{k=1}^{N_1} [f(X_{N_0+k}) - f_0(X_{N_0+k})].$$

Let C_0 be the cost for the Monte Carlo estimate of $\mathbb{E}[f_0]$ and C_1 the cost for the Monte Carlo estimate of $\mathbb{E}[f - f_0]$. The total cost of the estimate is $N_0C_0 + N_1C_1$ while the variance is $N_0^{-1}\mathrm{Var}[f_0] + N_1^{-1}\mathrm{Var}[f - f_0]$. For a fixed cost, we can minimize the variance by choosing

$$\frac{C_0}{C_1} = \frac{V_0/N_0^2}{V_1/N_1^2}, \text{ i.e.,} \frac{N_1}{N_0} = \frac{\sqrt{V_1/C_1}}{\sqrt{V_0/C_0}}.$$

For $L > 0$, the idea is similar and one can take N_l proportional to $\sqrt{V_l/C_l}$. Applying to simulation of statistics of solutions to SDEs, the difference terms can be defined on finer meshes (smaller time step sizes) which would admit smaller variances and thus require a smaller number of sampling paths. The computational cost is thus reduced.

2.5.2 Quasi-Monte Carlo methods

Quasi-Monte Carlo methods (QMC) were originally designed as deterministic integration methods in random space and allowed only moderately high dimensional integrations, see, e.g., [376, 423]. QMC has a similar estimator as the Monte Carlo estimator

$$\mathrm{QMC}_n(f) = \frac{1}{n} \sum_{k=1}^{n} f(\mathbf{x}_k).$$

However, one significant difference is that the sequence x_k is deterministic instead of a random or pseudo-random sequence. The sequence is designed to provide better uniformity (measured in *discrepancy*) than a random sequence and to satisfy the basic (worse-case) bound

---

where the control variate $g(x)$ has known expectation $\mathbb{E}[g]$ (w.r.t. $p(x)$) and g is well correlated with f and optimal value for λ can be estimated by a few samples.

[4]We use the convention that when $L = 0$, the summation is zero.

$$|\mathbb{E}[f] - \text{QMC}_n(f)| \leq C(\log n)^k n^{-1} \left| \frac{\partial^d f}{\partial x_1 \cdots \partial x_d} \right|_{1,\text{TV}},$$

where the constants $C > 0$ and $k \geq 0$ do not depend on N but may depend on the dimension d. Here it is required that the mixed derivative of f has a bounded total variation while MC requires only a bounded variance. Compared to Monte Carlo methods, the convergence rate is approximately proportional to n^{-1} instead of $n^{-1/2}$ and there is no statistical error since a deterministic sequence is used. However, the convergence rate is smaller if the dimension d is large (usually less than 40). To overcome the dependence on dimension, one can use randomized quasi-Monte Carlo methods.

Some commonly used quasi-Monte Carlo sequences are Halton and Sobol sequences, Niederreiter sequences, etc. In an example at Chapter 2.5.5 and in Chapter 9.4, we will use randomized quasi-Monte Carlo sequences, Halton sequence, and Sobol sequence, where the Matlab code for these sequences is provided.

We refer to [56] for an introduction to the Monte Carlo method and quasi-Monte Carlo methods and [115] for recent development in deterministic and randomized quasi-Monte Carlo methods.

Compared to the Monte Carlo type method, the following two methods have no statistical errors and allow efficient short-time integration of SPDEs.

2.5.3 Wiener chaos expansion method

The Cameron-Martin theorem (Theorem 2.3.6) provides a spectral representation of square-integrable stochastic processes defined on the complete probability space $(\Omega, \mathcal{F}, \mathbb{P})$. This representation is also known as Wiener chaos expansion, see, e.g., [155, 479, 485].

Let us review what we have in Chapter 2.3. The Cameron-Martin basis is listed in Table 2.2 where there is only one Wiener process in (2.3.4).

Table 2.2. Some elements of the Cameron-Martin basis ξ_α in (2.3.4).

$\lvert\alpha\rvert$	α	ξ_α
0	$\alpha = (0, 0, \ldots)$	1
1	$\alpha = (0, \ldots, 0, 1, 0, \ldots)$	$H_1(\xi_i) = \xi_i$
2	$\alpha = (0, \ldots, 0, 2, 0, \ldots)$	$H_2(\xi_i)/\sqrt{2} = \frac{1}{\sqrt{2}}\left(\xi_i^2 - 1\right)$
2	$\alpha = (0, \ldots, 0, 1, 0, \ldots, 0, 1, 0, \ldots)$	$H_1(\xi_i)H_1(\xi_j) = \xi_i \xi_j$

We need in practice to truncate the number of random variables, i.e., let the elements of α be zero with large indexes. To be precise, we introduce the following notation: the order of multi-index α:

$$d(\alpha) = \max\left\{l \geq 1 : \alpha_{k,l} > 0 \text{ for some } k \geq 1\right\}.$$

Also, we need to limit the number of $|\alpha|$. We actually define truncated set of multi-indices

$$\mathcal{J}_{\mathsf{N},\mathsf{n}} = \{\alpha \in \mathcal{J} : \ |\alpha| \leq \mathsf{N}, \ \ d(\alpha) \leq \mathsf{n}\}.$$

In this set, there is a finite number of n dimensional random variables and the number is

$$\sum_{i=0}^{\mathsf{N}} \binom{\mathsf{n}+i-1}{i} = \binom{\mathsf{n}+\mathsf{N}}{\mathsf{N}} = \binom{\mathsf{n}+\mathsf{N}}{\mathsf{n}}.$$

In Table 2.3, we list the elements in a truncated Cameron-Martin basis. More

Table 2.3. Elements of a truncated Cameron-Martin basis ξ_α for a finite dimensional random space where $\alpha \in \mathcal{J}_{\mathsf{N},\mathsf{n}}$, $\mathsf{N} = 2$ and $\mathsf{n} = 2$.

$\|\alpha\|$	α	ξ_α
0	$\alpha = (0,0)$	1
1	$\alpha = (1,0)$	$H_1(\xi_1) = \xi_1$
1	$\alpha = (0,1)$	$H_1(\xi_2) = \xi_2$
2	$\alpha = (2,0)$	$H_2(\xi_1)/\sqrt{2} = \frac{1}{\sqrt{2}}(\xi_1^2 - 1)$
2	$\alpha = (0,2)$	$H_2(\xi_2)/\sqrt{2} = \frac{1}{\sqrt{2}}(\xi_2^2 - 1)$
2	$\alpha = (1,1)$	$H_1(\xi_1)H_1(\xi_2) = \xi_1\xi_2$

examples of the basis can be generated using the representation of the Hermite polynomial (2.3.3). Here are the first seven Hermite polynomials:

$$\begin{aligned} H_0(x) &= 1, \quad H_1(x) = x, \\ H_2(x) &= x^2 - 1, \\ H_3(x) &= x^3 - 3x, \\ H_4(x) &= x^4 - 6x^2 + 3, \\ H_5(x) &= x^5 - 10x^3 + 15x, \\ H_6(x) &= x^6 - 15x^4 + 45x^2 - 15. \end{aligned}$$

The Hermite polynomials can be represented (computed) by the three-term recurrence relation

$$H_{n+1}(x) = xH_n(x) - nH_{n-1}(x),\ n \geq 1, \quad H_0(x) = 1,\ H_1(x) = x. \tag{2.5.5}$$

Let's now consider the Wiener chaos expansion for computing the integration (2.5.1). The method is based on the Wiener chaos expansion of $f(X)$, where X is a standard Gaussian random variable. Suppose that

$$f(X) = \sum_{j=0}^{\infty} f_j \xi_j, \text{ where } \xi_j's \text{ are from the Cameron-Martin basis.}$$

Once we find f_j, we compute the integration (2.5.1) by

$$\mathbb{E}[f(X)] = \mathbb{E}[\sum_{j=0}^{\infty} f_j \xi_j] = f_0.$$

In practice, the coefficients of the Wiener chaos expansion, f_j's can be numerically computed using the governing stochastic equation. Note that ξ_j are orthonormal and thus $\mathbb{E}[\xi_j] = 0$ for $j \geq 1$ and $f_j = \mathbb{E}[f(X)\xi_j]$. Moreover, we can numerically compute $\mathbb{E}[g(f(X))]$ by

$$\mathbb{E}[g(f(X))] \approx \mathbb{E}[g(\sum_{j=0}^{\mathsf{N}} f_j \xi_j)].$$

We will illustrate the idea of Wiener chaos method as a numerical method in Chapter 2.5.5. We will see that the Wiener chaos method is essentially a spectral Galerkin method in the random space.

2.5.4 Stochastic collocation method

In the framework of deterministic integration methods for SPDEs in random space, another solution for nonlinear SPDEs or linear SPDEs with random coefficient is to employ collocation techniques in random space. Here by stochastic collocation methods, we mean the sampling strategies using high dimensional deterministic quadratures (with certain polynomial exactness) to evaluate desired expectations of solutions to SPDEs.

Let's now consider the stochastic collocation method (SCM) for computing the integration (2.5.1). Note that the deterministic integral can be computed by any quadrature rule. Here we consider a sequence of one-dimensional Gauss–Hermite quadrature rules Q_n with number of nodes $n \in \mathbb{N}$ for univariate functions $\psi(\mathsf{y})$, $\mathsf{y} \in \mathbb{R}$:

$$Q_n \psi(\mathsf{y}) = \sum_{\alpha=1}^{n} \psi(\mathsf{y}_{n,\alpha}) \mathsf{w}_{n,\alpha}, \tag{2.5.6}$$

where $\mathsf{y}_{n,1} < \mathsf{y}_{n,2} < \cdots < \mathsf{y}_{n,n}$ are the roots of the Hermite polynomial $H_n(\mathsf{y}) = (-1)^n e^{\mathsf{y}^2/2} \frac{d^n}{d\mathsf{y}^n} e^{-\mathsf{y}^2/2}$ and $\mathsf{w}_{n,\alpha} = n!/(n^2[H_{n-1}(\mathsf{y}_{n,\alpha})]^2)$ are the associated weights. It is known that $Q_n\psi$ is exactly equal to the integral $I_1\psi$ when ψ is a polynomial of degree less than or equal to $2n-1$, i.e., the polynomial degree of exactness of Gauss–Hermite quadrature rules Q_n is equal to $2n-1$. The integration (2.5.1) can be computed by

$$\mathbb{E}[f(X)] = \int_{\mathbb{R}} f(x)p(x)\, dx \approx \sum_{k=0}^{\mathsf{N}} f(\mathsf{y}_{\mathsf{N},k}) \mathsf{w}_k.$$

The Gauss-Hermite quadrature points $\mathsf{y}_{\mathsf{N},k}$ are the zeros (roots) of the $\mathsf{N}+1$-th order Hermite polynomial $H_{\mathsf{N}+1}$. The weights w_i's are the corresponding quadrature weights. The statistics $\mathbb{E}[g(f(X))]$ can be approximated by

$$\mathbb{E}[g(f(X))] \approx \sum_{k=0}^{\mathsf{N}} g(f(\mathsf{y}_{\mathsf{N},k}))\mathsf{w}_k.$$

The quadrature rule here is a starting point of stochastic collocation methods.

As we are usually looking for numerical approximation of some statistics like $\mathbb{E}[f(X)]$, we can simply look for "function values" on some deterministic quadrature points corresponding to certain quadrature rules. The stochastic collocation methods are then collocating a stochastic equation at these deterministic quadrature points using the classical collocation method in random space and seeking the "function values" at the quadrature points by solving the resulting equations they satisfy.

We will illustrate the idea of stochastic collocation method as a numerical method in Chapter 2.5.5. The stochastic collocation method is a spectral collocation method in the random space.

Smolyak's sparse grid

Sparse grid quadrature is a certain reduction of product quadrature rules, which decreases the number of quadrature nodes and allows effective integration in moderately high dimensions [425] (see also [154, 381, 477]).

Consider a d-dimensional integral of a function $\varphi(y)$, $y \in \mathbb{R}^d$, with respect to a Gaussian measure:

$$I_d\varphi := \frac{1}{(2\pi)^{d/2}} \int_{\mathbb{R}^d} \varphi(y) \exp\left(-\frac{1}{2}\sum_{i=1}^{d} y_i^2\right) dy_1 \cdots dy_d. \tag{2.5.7}$$

We can approximate the multidimensional integral $I_d\varphi$ by a tensor product quadrature rule

$$\begin{aligned} I_d\varphi \approx \bar{I}_d\varphi &:= Q_n \otimes Q_n \cdots \otimes Q_n \varphi(y_1, y_2, \cdots, y_d) \\ &= Q_n^{\otimes d} \varphi(y_1, y_2, \cdots, y_d) \\ &= \sum_{\alpha_1=1}^{n} \cdots \sum_{\alpha_d=1}^{n} \varphi(\mathsf{y}_{n,\alpha_1}, \ldots, \mathsf{y}_{n,\alpha_d}) \mathsf{w}_{n,\alpha_1} \cdots \mathsf{w}_{n,\alpha_d}, \end{aligned} \tag{2.5.8}$$

where for simplicity we use the same amount on nodes in all the directions. The quadrature $\bar{I}_d\varphi$ is exact for all polynomials from the space $\mathcal{P}_{k_1} \otimes \cdots \otimes \mathcal{P}_{k_d}$ with $\max_{1 \le i \le d} k_i = 2n - 1$, where $\mathcal{P}_k$ is the space of one-dimensional polynomials of degree less than or equal to k (we note in passing that this fact is easy to prove using probabilistic representations of $I_d\varphi$ and $\bar{I}_d\varphi$). Computational costs of quadrature rules are measured in terms of a number of function evaluations, which is equal to n^d in the case of the tensor product (2.5.8), i.e., the computational cost of (2.5.8) grows exponentially fast with dimension.

The sparse grid of Smolyak [425] reduces the computational complexity of the tensor product rule (2.5.8) by exploiting the difference quadrature formulas:

$$A(\mathsf{L},d)\varphi := \sum_{d\le|\mathbf{i}|\le \mathsf{L}+d-1} (Q_{i_1}-Q_{i_1-1})\otimes\cdots\otimes(Q_{i_d}-Q_{i_d-1})\varphi,$$

where $Q_0 = 0$ and $\mathbf{i} = (i_1, i_2, \ldots, i_d)$ is a multi-index with $i_k \geq 1$ and $|\mathbf{i}| = i_1 + i_2 + \cdots + i_d$. The number L is usually referred to as the level of the sparse grid. The sparse grid rule (2.5.9) can also be written in the following form [477]:

$$A(\mathsf{L},d)\varphi = \sum_{\mathsf{L}\le|\mathbf{i}|\le \mathsf{L}+d-1} (-1)^{\mathsf{L}+d-1-|\mathbf{i}|}\binom{d-1}{|\mathbf{i}|-\mathsf{L}} Q_{i_1}\otimes\cdots\otimes Q_{i_d}\varphi. \qquad (2.5.9)$$

The quadrature $A(\mathsf{L},d)\varphi$ is exact for polynomials from the space $\mathcal{P}_{k_1}\otimes\cdots\otimes\mathcal{P}_{k_d}$ with $|\mathbf{k}| = 2\mathsf{L}-1$, i.e., for polynomials of total degree up to $2\mathsf{L}-1$ [381, Corollary 1].

Denote the set of sparse grid points $\mathsf{x}_\kappa = (\mathsf{x}_\kappa^1, \cdots, \mathsf{x}_\kappa^d)$ by $\mathcal{H}_\mathsf{L}^{\mathsf{n}q}$, where x_κ^j $(1 \le j \le d)$ belongs to the set of points used by the quadrature rule Q_{i_j}. According to (2.5.9), we only need to know the function values at the sparse grid $\mathcal{H}_\mathsf{L}^{\mathsf{n}q}$:

$$A(\mathsf{L},d)\varphi = \sum_{\kappa=1}^{\eta(\mathsf{L},\mathsf{n}q)} \varphi(\mathsf{x}_\kappa)\mathsf{W}_\kappa, \quad \mathsf{x}_\kappa = (\mathsf{x}_\kappa^1,\cdots,\mathsf{x}_\kappa^d) \in \mathcal{H}_\mathsf{L}^{\mathsf{n}q}, \qquad (2.5.10)$$

where W_κ are determined by (2.5.9) and the choice of the quadrature rules Q_{i_j} and they are called the sparse grid quadrature weights. Due to (2.5.9), the total number of nodes used by this sparse grid rule is estimated by

$$\#\mathcal{H}_\mathsf{L}^{\mathsf{n}q} = \eta(\mathsf{L},d) \le \sum_{\mathsf{L}\le|\mathbf{i}|\le \mathsf{L}+d-1} i_1\times\cdots\times i_d.$$

Table 2.4 lists the number of sparse grid points up to level 5 when the level is not greater than d.

Table 2.4. The number of sparse grid points for the sparse grid quadrature (2.5.9) using the one-dimensional Gauss-Hermite quadrature rule (2.5.6), when the sparse grid level $\mathsf{L} \le d$.

	$\mathsf{L}=1$	$\mathsf{L}=2$	$\mathsf{L}=3$	$\mathsf{L}=4$	$\mathsf{L}=5$
$\eta(\mathsf{L},d)$	1	$2d+1$	$2d^2+2d+1$	$\frac{4}{3}d^3+2d^2+\frac{14}{3}d+1$	$\frac{2}{3}d^4+\frac{4}{3}d^3+\frac{22}{3}d^2+\frac{8}{3}d+1$

The quadrature $\bar{I}_d\varphi$ from (2.5.8) is exact for polynomials of total degree $2\mathsf{L}-1$ when $n=\mathsf{L}$. It is not difficult to see that if the required polynomial exactness (in terms of total degree of polynomials) is relatively small, then the sparse grid rule (2.5.9) substantially reduces the number of function evaluations compared with the tensor-product rule (2.5.8). For instance, suppose that the dimension $d=40$ and the required polynomial exactness is equal to 3. Then the cost of the tensor product rule (2.5.8) is $3^{40} \doteq 1.2158 \times 10^{19}$ while the cost of the sparse grid rule (2.5.9) based on the one-dimensional rule (2.5.6) is 3281.

2.5.5 Application to SODEs

Let's consider the stochastic differential equation (1.1.2) where

$$dX = W(t)X\,dt, \quad 0 < t \leq 1 \quad X(0) = X_0 = 1, \tag{2.5.11}$$

where $W(t)$ is a standard Brownian motion.

We employ the simplest time discretization – forward Euler scheme. For a uniform partition of $[0,1]$, $t_k = kh$, $1 \leq k \leq N$, and $Nh = 1$. The forward Euler scheme is

$$X_{k+1} = X_k + W(t_k)X_k h, \quad 0 \leq k \leq N-1. \tag{2.5.12}$$

The goal here is to numerically compute the mean and variance of the solution or simply $\mathbb{E}[X_N^p]$, $p=1,2$.

Here we notice that $W(t_k)$ need further discretization. We recall that in Chapter 2.2.2, there are two ways to approximate $W(t_k)$. The first one is to use increments of Brownian motions (2.2.3) and the forward Euler scheme becomes

$$X_{k+1} = X_k + \sqrt{h}\sum_{i=0}^{k} \xi_i X_k h, \quad 0 \leq k \leq N-1. \tag{2.5.13}$$

Here $\xi_0 = 0$. Then we have that the solution can be represented by $X_N(\xi_1, \xi_2, \ldots, \xi_{N-1})$, and the desired statistics are of the form (2.5.7). We can then use the methods described in the last section.

For Monte Carlo methods, we can use pseudo-random number generators. In Matlab, the Monte Carlo method for (2.5.13) can be implemented as follows:

Code 2.3. Monte Carlo method with the forward Euler scheme for Equation (2.5.11) using (2.5.13).

```
clc, clear all
rng(100, 'twister');    % for repeatable pseudo random
% sequences.
t_final = 1;   x_ini= 1;
```

```
  N= 1000;  h =t_final/N;
  num_sample_path = 1e4;
  % time marching, Euler scheme
  W_k = 0;
  X_k = x_ini* ones(num_sample_path,1);
  for  k= 1: N-1
       W_k = W_k + sqrt(h)* randn(num_sample_path,1);
       X_k = X_k + W_k.*X_k*h;
  end
  X_mean = mean(X_k);
  X_second_moment = mean(X_k.^2);
  X_mean_stat_error = 1.96 * sqrt(var(X_k)/ num_sample_path);
  X_second_moment_stat_error = 1.96 * sqrt (var (X_k.^2)/
                                  num_sample_path);
```

For quasi-Monte Carlo method, we use the scrambled Halton sequence [262],

Code 2.4. Quasi-Monte Carlo method with the forward Euler scheme for Equation (2.5.11) using (2.5.13).

```
clc, clear all
t_final = 1;   x_ini= 1;
N= 1000;  h =t_final/N;
num_sample_path = 1e4;
rng(100, 'twister');    % for repeatable  randomized
% quasi-random sequences.
qmc_sequence= haltonset(N-1 ,'Skip',1e3,'Leap',20);
% Halton sequence
qmc_sequence= scramble( qmc_sequence, 'RR2');   % scramble,
% randomizing
qmc_sequence = net( qmc_sequence, num_sample_path);
qmc_sequence=erfinv(2*qmc_sequence-1)*sqrt(2);  % an inverse
% transformation for Gaussian
% time marching, Euler scheme
W_k = 0;
X_k = x_ini* ones(num_sample_path,1);
for  k= 1: N-1
    W_k = W_k + sqrt(h)* qmc_sequence(:,k);
    X_k = X_k + W_k.*X_k*h;
end
X_mean = mean(X_k);
X_second_moment = mean(X_k.^2);
X_mean_stat_error = 1.96 * sqrt(var(X_k)/ num_sample_path);
X_second_moment_stat_error = 1.96 * sqrt( var (X_k.^2)/
                                  num_sample_path);
```

In this code, the scrambled Halton sequence is used, see [262]. We can also use the scrambled Sobol sequence [340] instead of a scrambled Halton sequence.

Code 2.5. A scrambled Sobol sequence of quasi-Monte Carlo method.

```
qmc_sequence = sobolset(N-1,'Skip',1e3,'Leap',20);
% Sobol sequence
qmc_sequence = scramble(qmc_sequence,'MatousekAffineOwen');
qmc_sequence = erfinv(2*qmc_sequence-1)*sqrt(2);
% inverse transformation
```

A stochastic collocation method requires the values of X_k at quadrature points (in random space) according to (2.5.8). To find these values, we apply the collocation method in random space: for $\kappa = 1, \ldots, \eta(\mathsf{L}, N-1)$,

$$\mathbb{E}[X_{k+1}(\xi_1, \ldots, \xi_k)\delta((\xi_1, \ldots, \xi_k) - (\mathsf{x}_\kappa^1, \cdots, \mathsf{x}_\kappa^k))]$$
$$= \mathbb{E}[(1 + \sqrt{h}\sum_{i=0}^{k} \xi_i h))X_k(\xi_1, \ldots, \xi_{k-1})\delta((\xi_1, \ldots, \xi_k) - (\mathsf{x}_\kappa^1, \cdots, \mathsf{x}_\kappa^k))].$$

By the property of the delta function, we have a system of deterministic and decoupled equations:

$$X_{k+1}(\mathsf{x}_\kappa^1, \cdots, \mathsf{x}_\kappa^k) = (1 + \sqrt{h}\sum_{i=0}^{k} \mathsf{x}_\kappa^i h)X_k(\mathsf{x}_\kappa^1, \cdots, \mathsf{x}_\kappa^{k-1}), \quad \kappa = 1, \ldots, \eta(\mathsf{L}, N-1).$$

Here we use the sparse grid code 'nwspgr.m' at `http://www.sparse-grids.de/`. We now list the code for sparse grid collocation methods.

Code 2.6. Sparse grid collocation with the forward Euler scheme for Equation (2.5.11) using (2.5.13).

```
clc, clear all
t_final = 1;    X_ini= 1;
N= 40;   h =t_final/N;
sparse_grid_dim = N-1;
sparse_grid_level = 2;
[sparse_grid_nodes, sparse_grid_weights]=nwspgr('GQN', ...
      sparse_grid_dim,  sparse_grid_level);
num_sample_path=size(sparse_grid_weights,1);
% time marching, Euler scheme
W_k = 0;
X_k = X_ini* ones(num_sample_path,1);
for  k= 1: N-1
      W_k = W_k + sqrt(h)* sparse_grid_nodes(:,k);
      X_k = X_k + W_k.*X_k*h;
```

```
end
X_mean = sum(X_k.*sparse_grid_weights);
X_second_moment = sum(X_k.^2.*sparse_grid_weights);
```

Consider now the Wiener chaos method for (2.5.13). Suppose that $X_{k+1} = \sum_{\alpha\in\mathcal{J}_{\mathsf{N},k}} x_{\alpha,k}\xi_\alpha$ for $1 \leq k \leq N-1$. We first apply a Galerkin method in random space – multiplying by the Cameron-Martin basis ξ_β, $\beta \in\in \mathcal{J}_{\mathsf{N},k}$ over both sides of (2.5.13) and taking expectation (integration over the random space). We then have

$$\mathbb{E}[\xi_\beta \sum_{\alpha\in\mathcal{J}_{\mathsf{N},k}} x_{\alpha,k}\xi_\alpha] = \mathbb{E}[\xi_\beta(1+\sqrt{h}\sum_{i=1}^{k}\xi_i h)) \sum_{\alpha\in\mathcal{J}_{\mathsf{N},k-1}} x_{\alpha,k-1}\xi_\alpha].$$

By the orthonormality of the Cameron-Martin basis, we have

$$x_{\beta,k} = x_{\beta,k-1}\mathbf{1}_{\beta_k=0} + h^{3/2}\sum_{i=1}^{k}\sum_{\alpha\in\mathcal{J}_{\mathsf{N},k-1}} x_{\alpha,k-1}\mathbb{E}[\xi_i\xi_\alpha\xi_\beta].$$

This turns the discrete stochastic equation from the forward Euler scheme into a system of deterministic equations of the Wiener chaos expansion coefficients $x_{\beta,k+1}$. This system of deterministic equation of the coefficients is called *propagator* of the discrete stochastic equation (2.5.13).

To solve for $\xi_{\beta,k+1}$, one needs to find the expectations of the triples $\mathbb{E}[\xi_i\xi_\alpha\xi_\beta]$. Recalling the recurrence relation (2.5.5) and orthogonality of Hermite polynomials, the triples are zero unless $\alpha = \beta \pm \varepsilon_i$, where ε_i is a multiindex with $|\varepsilon_i| = 1$ and its only nonzero element is the i-th one. Recalling the (2.5.5) and when $\beta_i = \alpha_i + 1$, we have

$$\begin{aligned}\mathbb{E}[\xi_i\xi_{\alpha_i}\xi_{\beta_i}] &= \mathbb{E}[\xi_i H_{\alpha_i}(\xi_i)H_{\beta_i}(\xi_i)]/\sqrt{\alpha_i!}/\sqrt{\beta_i!}\\ &= \mathbb{E}[H_{\alpha_i+1}(\xi_i)H_{\beta_i}(\xi_i)]/\sqrt{\alpha_i!}\sqrt{\beta_i!} = \sqrt{\alpha_i+1}.\end{aligned}$$

Then, the triples can be computed as

$$\mathbb{E}[\xi_i\xi_\alpha\xi_\beta] = \sqrt{\alpha_i+1}\mathbf{1}_{\alpha+\varepsilon_i=\beta} + \sqrt{\beta_i+1}\mathbf{1}_{\beta+\varepsilon_i=\alpha}. \quad (2.5.14)$$

We have a *propagator* ready for implementation: for $\beta \in \mathcal{J}_{\mathsf{N},k}$,

$$\begin{aligned}x_{\beta,k} = \big(x_{\beta,k-1} + h^{3/2}\sum_{i=1}^{k-1}\big(\sqrt{\beta_i}x_{\beta-\varepsilon_i,k-1} + \sqrt{\beta_i+1}x_{\beta+\varepsilon_i,k-1}\mathbf{1}_{|\beta|\leq \mathsf{N}-1}\big)\big)\\ \mathbf{1}_{\beta_k=0} + h^{3/2}x_{\alpha,k-1}\mathbf{1}_{\beta_k=1}\mathbf{1}_{\alpha=(\beta_1,\ldots,\beta_{k-1})}.\end{aligned}$$

Now let's consider the spectral approximation of Brownian motion (2.2.5). Applying the spectral approximation (2.2.5), the problem (2.5.11) can be discretized as follows:

$$d\tilde{X} = W^{(\mathsf{n})}(t)\tilde{X}\,dt, \quad \tilde{X}_0 = 1. \quad (2.5.15)$$

Different from (2.5.13), Equation (2.5.15) is still continuous in time. We need a further discretization in time. Let us again use the forward Euler scheme:

$$\tilde{X}_{k+1} = \tilde{X}_k + W^{(\mathsf{n})}(t_k)\tilde{X}_k h, \quad \tilde{X}_0 = 1. \tag{2.5.16}$$

With the approximation (2.2.5) and the cosine basis (2.2.6), we have

$$W^{(\mathsf{n})}(t_k) = \sum_{i=1}^{\mathsf{n}} \xi_i M_i(t_k), \quad M_1(t) = t,\ M_i(t) = \int_0^t m_i(s)\,ds$$
$$= \frac{\sqrt{2}}{(i-1)\pi}\sin(\pi(i-1)t), i \geq 2.$$

The Monte Carlo method is similar to what we have before. The only difference between Code 2.3 and the code here is in computing the approximated Brownian motion.

Code 2.7. Monte Carlo method with the forward Euler scheme for Equation (2.5.11) using (2.5.16) (WZ).

```
clc, clear all
rng(100, 'twister');     % for repeatable pseudo random
% sequences.
t_final  = 1;    x_ini= 1;
N= 1000;   h =t_final/N;
num_sample_path  = 1e4;
% time marching, Euler scheme
W_k  = 0;
X_k  = x_ini* ones(num_sample_path,1);
n=40;  % truncation of specral approximation
xi=randn(num_sample_path,n);
for  k= 1: N-1
     t_k=k*h;
     W_k  = t_k*xi(:,1);
     for i=2:n
        W_k  = W_k+ sqrt(2)*xi(:,i)*sin(pi*(i-1)*t_k)/(i-1)/pi;
     end
     X_k  = X_k  + W_k.*X_k*h;
end
X_mean  = mean(X_k);
X_second_moment  = mean(X_k.^2);
X_mean_stat_error  = 1.96 * sqrt( var(X_k)/num_sample_path);
X_second_moment_stat_error  = 1.96 * sqrt( var (X_k.^2)/
                                    num_sample_path);
```

Similarly, the sparse grid collocation method here differs from Code 2.6 in computing the approximated Brownian motion.

Code 2.8. Sparse grid collocation with the forward Euler scheme for Equation (2.5.11) using (2.5.16) (WZ).

```
clc, clear all
t_final = 1;    X_ini= 1;
N= 10000;  h =t_final/N;
n=40;  % truncation of specral approximation
sparse_grid_dim = n;     % n
sparse_grid_level = 2;  %  less than 5
[sparse_grid_nodes, sparse_grid_weights]=nwspgr('GQN', ...
    sparse_grid_dim,  sparse_grid_level);
num_sample_path=size(sparse_grid_weights,1);
% time marching, Euler scheme
W_k = 0;
X_k = X_ini* ones(num_sample_path,1);
for  k= 1: N-1
    t_k=k*h;
    W_k = t_k*sparse_grid_nodes(:,1);
    for i=2:n
        W_k = W_k+ sqrt(2)* sparse_grid_nodes(:,i)
              *sin(pi*(i-1)*t_k)/(i-1)/pi;
    end
    X_k = X_k + W_k.*X_k*h;
end
X_mean = sum(X_k.*sparse_grid_weights);
X_second_moment = sum(X_k.^2.*sparse_grid_weights);
```

Let us derive the Wiener chaos method for (2.5.16). Suppose that $\tilde{X}_k = \sum_{\alpha\in\mathcal{J}_{\mathsf{N},\mathsf{n}}} x_{\alpha,k}\xi_\alpha$ for $k \geq 1$. We first apply a Galerkin method in random space – multiplying the Cameron-Martin basis ξ_β, $\beta \in \mathcal{J}_{\mathsf{N},,\mathsf{n}}$ over both side of (2.5.13) and taking expectation (integration over the random space). We then have

$$\mathbb{E}[\xi_\beta \sum_{\alpha\in\mathcal{J}_{\mathsf{N},,\mathsf{n}}} x_{\alpha,k+1}\xi_\alpha] = \mathbb{E}[\xi_\beta(1+\sum_{i=1}^{\mathsf{n}}\xi_i M_i(t_k)) \sum_{\alpha\in\mathcal{J}_{\mathsf{N},\mathsf{n}}} x_{\alpha,k}\xi_\alpha].$$

By the orthonormality of the Cameron-Martin basis, we have the following *propagator* of the discrete stochastic equation (2.5.16):

$$x_{\beta,k+1} = x_{\beta,k} + h\sum_{i=1}^{\mathsf{n}} M_i(t_k) \sum_{\alpha\in\mathcal{J}_{\mathsf{N},\mathsf{n}}} x_{\alpha,k}\mathbb{E}[\xi_i\xi_\alpha\xi_\beta].$$

Recall the fact (2.5.14) and we have a *propagator* ready for implementation

$$x_{\beta,k+1} = x_{\beta,k} + h\sum_{i=1}^{\mathsf{n}} M_i(t_k)\big(\sqrt{\beta_i}x_{\beta-\varepsilon_i,k} + \sqrt{\beta_i+1}x_{\beta+\varepsilon_i,k}\mathbf{1}_{|\beta|\leq \mathsf{N}-1}\big),$$
$$\beta\in\mathcal{J}_{\mathsf{N},\mathsf{n}}.$$

In Table 2.5, we present first two moments (mean and second moment) obtained by the Monte Carlo method and quasi-Monte Carlo method and stochastic collocation methods using different approximation of Brownian motion for Equation (2.5.11). We recall from Chapter 1.1 that the first moment at $t = 1$ is $\mathbb{E}[X(1)] = X(0)\exp(1/2) \approx 1.6487$ and the second moment is $\mathbb{E}[X^2(1)] = X^2(0)\exp(2 \approx 7.3891$. From the table, we can see that quasi-Monte Carlo methods give the most accurate values for both the mean and the second moment while the stochastic collocation method is the least accurate. One possible reason for the failure of the stochastic collocation method is the high dimensionality in random space. It is believed that beyond 40 dimension the stochastic collocation method is empirically inefficient.

2.6 Bibliographic notes

Other approximations of Brownian motion. Besides the piecewise linear approximation of Brownian motion (2.2.4), there are several other approximations, e.g., the mollifier approximation (see, e.g., [241, p. 397] and [387]), or the random walk by Donsker's theorem (see, e.g., [255]). However, we omit the discussion of construction of Brownian motion here as only the first two approaches are used in practice and in this book.

Table 2.5. The first two moments using different methods in random space and Euler scheme in time for Equation (2.5.11) at $t = 1$. For Wong-Zakai (WZ) approximation, we use $\mathsf{n} = 40$. The time step size is 10^{-4}. The exact moments are given by (1.1.4): the mean is $\exp(1/3) \approx 1.3956$ and the second moment is $\exp(3/2) \approx 1.9477$.

Methods	Mean	Second moment	Approximation of Brownian motion
Monte Carlo (MC)	1.1856 ± 0.0148	1.9771 ± 0.0717	Increments
Quasi-Monte Carlo (QMC)	1.2237 ± 0.0152	2.0969 ± 0.0673	Increments
Stochastic collocation method (SCM)	1.1545	1.6325	Increments
Monte Carlo (MC)	1.1777 ± 0.0145	1.9353 ± 0.0609	WZ
Quasi-Monte Carlo (QMC)	1.5662 ± 0.0173	3.2312 ± 0.0903	WZ
Stochastic collocation method (SCM)	1.1695	1.7141	WZ

Mollification of Brownian motion is also known as the mollifier approximation and is used in the approximation of stochastic integrals and SODEs, see, e.g., [117, 177, 239, 323, 325, 387, 397, 510, 511]

$$\tilde{W}(t) = \int_{t_n}^{t} \int_{\mathbb{R}} K(\theta, s)\, dW(s)\, d\theta, \quad t \in [\mathrm{t}_n, \mathrm{t}_{n+1}), \tag{2.6.1}$$

where K is symmetric. This type of approximation was proposed for a method of lines for SODEs in [387], where no numerical results were presented. When this approximation is applied to SODEs, consistency (convergence without order) has been proved in [117, 177, 179, 326], etc. In [241], the approaches of piecewise linear approximation and mollification have been unified with proved convergence order, known as Ikeda-Nakao-Yamato-type approximations, see also [195].

Spectral approximation of Brownian motion. With a piecewise constant basis in (2.2.5), the use of multiple Ito integrals (Wiener chaos expansion) and multiple Stratonovich integrals was addressed in [53, 54]. When the spectral basis is chosen as Haar wavelets, the approximation is also known as Levy-Ciesielsky approximation, see, e.g., [255]). The expansion (2.2.1) is an extension of Fourier expansion of Brownian motion proposed by Wiener [391, Chapter IX]. It is also known as Levy-Ciesielski approximation [81, 255, 278], Ito-Niso approximation [243], or Fourier approximation [391]; see [240] for a historical review of this approximation.

The approximation with trigonometric orthonormal basis has been used in Wiener chaos methods (see, e.g., [55, 225, 315, 316, 318, 505]) and will be the approximation for our Wong-Zakai approximation throughout this book. See Chapter 4 for the Wong-Zakai approximation using (2.2.5) for SODEs with time delay.

For the multilevel Monte Carlo method for numerical SPDEs, see, e.g., [1, 25, 75, 82, 440, 441] for elliptic equations with random coefficients, [363–365] for stochastic hyperbolic equations, [24, 158] for stochastic parabolic equations, and [160] for a stochastic Brinkman problem. Ref. [1] proposed time discretization schemes with large stability regions to further reduce the cost of the multilevel Monte Carlo method. However, it has been shown that the performance of multilevel Monte Carlo methods is *not robust*, see, e.g., [264, Chapter 4] when the variances of the desired stochastic processes are large.

Quasi-Monte Carlo methods have also been investigated for numerical SPDEs. Some randomized quasi-Monte Carlo methods have been successfully applied to solve stochastic elliptic equations with random coefficients, see, e.g., [164, 165, 281, 282] where the solution is analytic in random space (parameter space). For a good review on quasi-Monte Carlo methods, see [115].

Wiener chaos expansion methods. As numerical methods, they have been summarized in [155, 439, 487]. This idea of representing a random variable (process) with orthogonal polynomial of the random variable (process) with respect to its corresponding probability density function is not limited to Gaussian random variables (processes) and has been extended to more general cases, see, e.g., [415, 488]. Xiu and Karniadakis [488] developed the Wiener-Askey polynomial chaos using a broad class of Askey's orthogonal polynomials [10]. Soize and Ghanem [427] discuss chaos expansions with respect to an arbitrary probability measure, see also Wan and Karniadakis [468]. These methods are known as generalized polynomial chaos, or gPC, see reviews in, e.g., [484, 485].

For linear problems driven by white noise in space or in time, the Wiener chaos expansion method has been investigated both theoretically (see, e.g., [315–318]) and numerically (see, e.g., [469, 505]). The advantage of Wiener chaos expansion method is that the resulting system of PDEs is linear, lower triangular, and deterministic. Also, the Wiener chaos expansion method can be of high accuracy. However, there are two main difficulties for the Wiener chaos expansion as a numerical method. The first is the *efficiency of long-time integration*. Usually, this method is only efficient for short-time integration, see, e.g., [53, 315]. This limitation can be somewhat removed when a recursive procedure is adopted for computing certain statistics, e.g., first two moments of the solution, see, e.g., [505, 507]. The second is *nonlinearity*. When SPDEs are nonlinear, Wiener chaos expansion methods result in fully coupled systems of deterministic PDEs while the interactions between different Wiener chaos expansion terms necessitate exhaustive computation. This effect has been shown numerically through the stochastic Burgers equation and the Navier-Stokes equations [225].

One remedy for nonlinear problems is to introduce the *Wick-Malliavin approximation* for nonlinear terms. Wick-Malliavin approximations can be seen as a perturbation of a Wick product formulation by adding high-order Malliavin derivatives of the nonlinear terms to the Wick product formulation, see [346, 462] and Chapter 11 for details. Basically, lower level Wick-Malliavin approximation (with lower-order Malliavin derivatives) allows weaker nonlinear interactions between the Wiener chaos expansion terms. Let us consider the Burgers equation with additive noise, for example. When only the Wick product is used (zeroth-order Malliavin derivatives only), the resulting system is lower triangular and contains only one nonlinear equation. When Malliavin derivatives of up to first-order are used, the resulting system of PDEs is only weakly coupled and contains only two nonlinear equations. This approach has been shown to be very efficient for short-time integration of equations with quadratic nonlinearity and small noise, see, e.g., [462].

The Wick product had been formulated in [223] for various SPDEs before the Wick-Malliavin approximation was introduced. The Wick product formulation has been explored with finite element methods in physical space, see, e.g., [254, 327–333, 443–445, 498] and also [469] for a brief review on SPDEs equipped with Wick product. In Chapter 11, we will discuss the Wick-Malliavin approximation for linear elliptic equations with multiplicative noise and some nonlinear equations with quadratic nonlinearity and additive noise.

A stochastic collocation method (SCM) was first introduced in [439] and later on by other authors, see, e.g., [11] and [486]. While WCE (Wiener Chaos Expansion) is a spectral Galerkin method in random space, see, e.g., [155, 488]), SCM is a stochastic version of deterministic collocation methodology. As collocation methods for deterministic problems, see, e.g., [163]), the stochastic collocation methods exhibit high accuracy comparable to the WCE performance, see, e.g., [123] for elliptic equations with random coefficients.

For stochastic differential equations with color noise, it has been demonstrated in a number of works (see, e.g., [11, 12, 35, 370, 378, 486, 500] and references therein) that stochastic collocation methods (Smolyak's sparse grid collocation, SGC) can be a competitive alternative to the Monte Carlo technique and its variants in the case of differential equations. The success of these methods relies on the solution smoothness in the random space and can usually be achieved when it is sufficient to consider only a limited number of random variables (i.e., in the case of a low dimensional random space). As Wiener chaos methods, stochastic collocation methods are limited in practice as they can be used for a small number of random variables. *Based on empirical evidence (see, e.g., [393]), the use of SGC is limited to problems with random space dimensionality of less than* 40.

More efficient algorithms might be built using *anisotropic SGC methods* [172, 379] or goal-oriented quadrature rules, which employ more quadrature points along the "most important" direction, e.g., [373, 389, 390]. Here we consider only isotropic SGC with predetermined quadrature rules. In fact, the effectiveness of adaptive sparse grids relies heavily on the order of importance in random dimension of numerical solutions to stochastic differential equations, which is not always easy to reach. Furthermore, all these sparse grids as integration methods in random space grow quickly with random dimensions and thus cannot be used for longer time integration (usually with large random dimensions).

For SODEs driven by white noise in time, the stochastic collocation method has been known as cubature on Wiener space (e.g., [202, 300, 322, 373, 377]), optimal quantization (e.g., [389, 390]) to solve SODEs in random space, sparse grid of Smolyak type (e.g., [153, 154, 172, 398]), or particle simulation (e.g., [132]). For stochastic collocation methods for equations with color noise, see, e.g., [11, 486].

The stochastic collocation methods result in decoupled systems of equations as Monte Carlo method and its variants do, which can be of great advantage in parallel computation. High accuracy and fast convergence can be also observed for stochastic evolution equations, e.g., [153, 154, 172, 398] where the sparse grid of Smolyak type was used.

However, the fundamental limitation of these collocation methods is the exponential growth of sampling points with an increasing number of random parameters, see, e.g., [172], and thus a failure for longer time integration, see error estimates for cubature on Wiener space (e.g., [28, 70, 116]) and conclusions for optimal quantization (e.g., [389, 390]).

2.7 Suggested practice

Exercise 2.7.1 *Show that for any centered Gaussian random variable,* $\xi \sim \mathcal{N}(0,\sigma^2)$, $\mathbb{E}[\xi^{2n}] = \sigma^{2n}(2n-1)!!$ *and* $\mathbb{E}[\xi^{2n-1}] = 0$, *for any* $n \geq 1$. *Here* $(2n-1)!! = 1 \times 3 \times \cdots \times (2n-1)$.

Exercise 2.7.2 *Consider a Gaussian vector* (X, Y) *with* X, Y *satisfying* $\mathbb{C}ov[(X, Y)] = 0$ *(uncorrelated). Then* X *and* Y *are independent.*

Exercise 2.7.3 *For a Gaussian random variable* $X \sim \mathcal{N}(0,\sigma^2)$ *and a Bernoulli* Z *with* $\mathbb{P}(Z = \pm 1) = \frac{1}{2}$, X *and* Z *are independent. Show that*

a) the product ZX *is a Gaussian random variable;*
b) and X *and* ZX *are uncorrelated;*
c) but X *and* ZX *are not independent.*

Check whether $(X, ZX)^\top$ *is a Gaussian random vector or not.*

Exercise 2.7.4 *Assume that* X *and* Y *are Gaussian random variables, then* $X+Y$ *independent of* $X-Y$ *implies* X *independent of* Y *if and only if* (X, Y) *is a Gaussian random vector.*

Exercise 2.7.5 *Show that the covariance function* C *is nonnegative definite, i.e., for all* $t_1, \ldots t_k \in T$ *and all* $z_1, z_2, \ldots, z_k \in \mathbb{R}$

$$\sum_{i=1}^{k}\sum_{j=1}^{k} C(t_i, t_j) z_i z_j \geq 0.$$

Hint: assume that $\mathbb{E}[X(t)] = 0$ and write the formula as $\mathbb{E}[\big(\sum_{i=1}^{k} z_i X(t_i)\big)^2]$.

Exercise 2.7.6 *Assume that* $W(t)$ *is a standard Brownian motion. Show that the covariance of* $W(t)$ *is* $\mathbb{C}ov[(W(t), W(s))] = \min(t, s)$.

Exercise 2.7.7 *If* $X(t)$ *is a one-dimensional Gaussian process with covariance* $\mathbb{C}ov[X(t), X(s)] = \min(t, s)$, *then it is a one-dimensional Brownian motion.*

Exercise 2.7.8 *Use the definition of Brownian motion to show that the process in* (2.2.1) *is indeed a Brownian motion on* $[0, T]$.

Exercise 2.7.9 *Compute the Hölder exponent* α *for the following functions:* $f(t) = \sqrt{t}$, $f(t) = t^\beta$ $(\beta > 0)$ *for* $t \in [0, 1]$ *and for* $t \in [\varepsilon, 1]$ $(0 < \varepsilon < 1)$.

Exercise 2.7.10 *Show that if there exists a constant* C *such that for any* x,

$$|f(x+h) - f(x)| \leq C h^\beta, \ \beta > 1,$$

then $f(x)$ *is a constant.*

Exercise 2.7.11 *For integer $n \geq 1$, verify that*

$$\mathbb{E}[|B_t - B_s|^{2n}] \leq C_n |t-s|^n .$$

Exercise 2.7.12 *If f is differentiable and its derivative is Riemann-integrable, its total variation over the interval $[a, b]$ is*

$$|f|_{1,\mathrm{TV},[\mathrm{a,b}]} = \int_a^b |f'(t)|\, dt.$$

Exercise 2.7.13 *Show that a Brownian motion $W(t)$ has a bounded quadratic variation and any p-variation of $W(t)$ is zero when $p \geq 3$.*

Exercise 2.7.14 *Assume that V_n is from Theorem 2.2.14. Show that $\mathbb{E}[V_n] = \sqrt{n}\mathbb{E}[|\xi_1|]$ and $\mathbb{V}ar[V_n] = 1 - (\mathbb{E}[|\xi_1|])^2$.*

Exercise 2.7.15 *Compute $\mathbb{E}[|\xi_1|]$ in Theorem 2.2.14.*

Exercise 2.7.16 *Show that $\xi_l = \int_0^T m_l(s)\, dW(s)$'s are i.i.d. standard Gaussian random variables, where $\{m_l\}$ is a complete orthonormal basis in $L^2([0,T])$.*

Exercise 2.7.17 *Show by definition the second formula in Example 2.3.1.*

Exercise 2.7.18 *Using Taylor's expansion and bounded quadratic variation of Brownian motion, prove Theorem 2.4.1 when f has bounded first two derivatives.*

3

Numerical methods for stochastic differential equations

In this chapter, we discuss some basic aspects of stochastic differential equations (SDEs) including stochastic ordinary (SODEs) and partial differential equations (SPDEs).

In Chapter 3.1, we first present some basic theory of SODEs including the existence and uniqueness of strong solutions and solution methods for SODEs such as integrating factor methods and the method of moment equations of solutions. We then discuss in Chapter 3.2 numerical methods for SODEs and strong and weak convergence of numerical solutions as well as linear stability theory of numerical SODEs. We summarize basic aspects of numerical SODEs in Chapter 3.2.5.

We present some basic estimates of regularity of solutions to SPDEs in Chapter 3.3. Then, we introduce the solutions in several senses: strong solution, variational solution, mild solution, and Wiener chaos solution. Existence and uniqueness of variational solutions are presented. In Chapter 3.4, we briefly review numerical methods for parabolic SPDEs including different techniques for numerical solutions aiming at strong and weak convergence. A comparison of numerical stability between PDEs and SPDEs is also presented in Chapter 3.4. We summarize basic aspects of numerical methods for SPDEs in Chapter 3.4.6. Numerical methods for other type of equations such as stochastic hyperbolic equations are presented in Chapter 3.5 of bibliographic notes. Some exercises are provided at the end of this chapter.

3.1 Basic aspects of SODEs

Let us consider the following simple stochastic ordinary equation:

$$dX(t) = -\lambda X(t)\, dt + dW(t), \quad \lambda > 0. \tag{3.1.1}$$

Z. Zhang, G.E. Karniadakis, *Numerical Methods for Stochastic Partial Differential Equations with White Noise*,
Applied Mathematical Sciences 196, DOI 10.1007/978-3-319-57511-7_3

It can be readily verified by Ito's formula (Theorem 2.4.1) that the following process

$$X(t) = e^{-\lambda t}x_0 + \int_0^t e^{-\lambda(t-s)}\, dW(s) \tag{3.1.2}$$

satisfies Equation (3.1.1). By the Kolmogorov continuity theorem (Theorem 2.2.10), the solution is Hölder continuous of order less than 1/2 in time since

$$\mathbb{E}[|X(t) - X(s)|^2] \leq (t-s)^2(\frac{2}{\lambda} + x_0^2) + |t-s|\,. \tag{3.1.3}$$

This simple model shows that the solution to a stochastic differential equation is Hölder continuous of order less than 1/2 and thus does not have derivatives in time. This low regularity of solutions leads to different concerns in SODEs and their numerical methods.

3.1.1 Existence and uniqueness of strong solutions

Let $(\Omega, \mathcal{F}, \mathbb{P})$ be a probability space and $(W(t), \mathcal{F}_t^W) = ((W_1(t), \ldots, W_m(t))^\top, \mathcal{F}_t^W)$ be an m-dimensional standard Wiener process, where $\mathcal{F}_t^W$, $0 \leq t \leq T$, is an increasing family of σ-subalgebras of $\mathcal{F}$ induced by $W(t)$. Consider the system of Ito SODEs

$$dX = a(t,X)dt + \sum_{r=1}^m \sigma_r(t,X)dW_r(t), \quad t \in (t_0, T],\ X(t_0) = x_0, \tag{3.1.4}$$

where X, a, σ_r are m-dimensional column-vectors and x_0 is independent of w. We assume that $a(t,x)$ and $\sigma(t,x)$ are sufficiently smooth and globally Lipschitz.

The SODEs (3.1.4) can be rewritten in Stratonovich sense under mild conditions. With the relation (2.3.2), the equation (3.1.4) can be written as

$$dX = [a(t,X) - c(t,X)]dt + \sum_{r=1}^m \sigma_r(t,X)dW_r(t), \quad t \in (t_0, T],\ X(t_0) = x_0, \tag{3.1.5}$$

where

$$c(t,X) = \frac{1}{2}\sum_{r=1}^m \frac{\partial \sigma_r(t,X)}{\partial x}\sigma_r(t,X),$$

and $\frac{\partial \sigma_r}{\partial x}$ is the Jacobi matrix of the column-vector σ_r:

$$\frac{\partial \sigma_r}{\partial x} = \begin{bmatrix} \frac{\partial \sigma_r}{\partial x_1} & \cdots & \frac{\partial \sigma_r}{\partial x_m} \end{bmatrix} = \begin{bmatrix} \frac{\partial \sigma_{1,r}}{\partial x_1} & \cdots & \frac{\partial \sigma_{1,r}}{\partial x_m} \\ \vdots & \ddots & \vdots \\ \frac{\partial \sigma_{m,r}}{\partial x_1} & \cdots & \frac{\partial \sigma_{m,r}}{\partial x_m} \end{bmatrix}.$$

We denote $f \in \mathbb{L}_{\rm ad}(\Omega; L^2([a,b]))$ if $f(t)$ is adapted to $\mathcal{F}_t$ and $f(t,\omega) \in L^2([a,b])$, i.e.,

$$f \in \mathbb{L}_{\rm ad}(\Omega; L^2([a,b])) = \Big\{ f(t,\omega) | f(t,\omega) \text{ is } \mathcal{F}_t\text{-measurable} \\ \text{and } \mathbb{P}(\int_a^b f_s^2\, ds < \infty) = 1 \Big\}.$$

Here $\{\mathcal{F}_t;\, a \le t \le b\}$ is a filtration such that

- for each t, $f(t)$ and $W(t)$ are $\mathcal{F}_t$-measurable, i.e., $f(t)$ and $W(t)$ are adapted to the filtration $\mathcal{F}_t$.
- for any $s \le t$, $W(t) - W(s)$ is independent of the σ-filed $\mathcal{F}_s$.

Definition 3.1.1 (A strong solution to a SODE) *We say that $X(t)$ is a (strong) solution to SDE* (3.1.4) *if*

- $a(t, X(t)) \in \mathbb{L}_{\rm ad}(\Omega, L^1([c,d]))$,
- $\sigma(t, X(t)) \in \mathbb{L}_{\rm ad}(\Omega, L^2([c,d]))$,
- *and $X(t)$ satisfies the following integral equation a.s.*

$$X(t) = x + \int_0^t a(s, X(s))\, ds + \int_0^t \sigma(s, X(s))\, dW(s). \tag{3.1.6}$$

In general, it is difficult to give a necessary and sufficient condition for existence and uniqueness of strong solutions. Usually we can give sufficient conditions.

Theorem 3.1.2 (Existence and uniqueness) *If X_0 is $\mathcal{F}_0$-measurable and $\mathbb{E}[X_0^2] < \infty$. The coefficients a, σ satisfy the following conditions.*

- *(Lipschitz condition) a and σ are Lipschitz continuous, i.e., there is a constant $K > 0$ such that*

$$|a(x) - a(y)| + \sum_{r=1}^{m} |\sigma_r(x) - \sigma_r(y)| \le K|x - y|.$$

- *(Linear growth) a and σ grow at most linearly i.e., there is a $C > 0$ such that*

$$|a(x)| + |\sigma(x)| \le C(1 + |x|),$$

then the SDE above has a unique strong solution and the solution has the following properties

- *$X(t)$ is adapted to the filtration generated by X_0 and $W(s)$ $(s \le t)$.*
- $\mathbb{E}[\int_0^t X^2(s)\, ds] < \infty$.

Here are some examples where the conditions in the theorem are satisfied.

- (Geometric Brownian motion) For μ, $\sigma \in \mathbb{R}$,

$$dX(t) = \mu X(t)\,dt + \sigma X(t)\,dW(t), \quad X_0 = x.$$

- (Sine process) For $\sigma \in \mathbb{R}$,

$$dX(t) = \sin(X(t))\,dt + \sigma\,dW(t), \quad X_0 = x.$$

- (modified Cox-Ingersoll-Ross process) For θ_1, $\theta_2 \in \mathbb{R}$,

$$dX(t) = -\theta_1 X(t)\,dt + \theta_2\sqrt{1 + X(t)^2}\,dW(t), \quad X_0 = x. \quad \theta_1 + \frac{\theta_2^2}{2} > 0.$$

Remark 3.1.3 *The condition in the theorem is also known as global Lipschitz condition. A straightforward generalization is one-sided Lipschitz condition (global monotone condition)*

$$(x-y)^\top(a(x)-a(y)) + p_0\sum_{r=1}^m |\sigma_r(x)-\sigma_r(y)|^2 \le K|x-y|^2, \quad p_0 > 0,$$

and the growth condition can also be generalized as

$$x^\top a(x) + p_1\sum_{r=1}^m |\sigma_r(x)|^2 \le C(1+|x|^2).$$

We will discuss in detail this condition in Chapter 5.

Theorem 3.1.4 (Regularity of the solution) *Under the conditions of Theorem 3.1.2, the solution is continuous and there exists a constant $C > 0$ depending only on t that*

$$\mathbb{E}[|X(t)-X(s)|^2] \le C\,|t-s|\,.$$

Then by the Kolmogorov continuity theorem (Theorem 2.2.10), we can conclude that the solution is only Hölder continuous with exponent less than 1/2, which is the same as Brownian motion.

3.1.2 Solution methods

This process (3.1.2) here is a special case of the Ornstein-Uhlenbeck process, which satisfies the equation

$$dX(t) = \kappa(\theta - X(t))\,dt + \sigma\,dW(t). \tag{3.1.7}$$

where κ, $\sigma > 0$, $\theta \in \mathbb{R}$. The solution to (3.1.7) can be obtained by *the method of change-of-variable*: $Y(t) = \theta - X(t)$. Then by Ito's formula we have

$$dY(t) = -\kappa Y(t)\,dt + \sigma\,d(-W(t)).$$

Similar to (3.1.2), the solution is

$$Y(t) = e^{-\kappa t}Y_0 + \sigma \int_0^t e^{-\kappa(t-s)}\,d(-W(s)). \tag{3.1.8}$$

Then by $Y(t) = \theta - X(t)$, we have

$$X(t) = X_0 e^{-\kappa t} + \theta(1 - e^{-\kappa t}) + \sigma \int_0^t e^{-\kappa(t-s)}\,dW(s).$$

In a more general case, we can use similar ideas to find explicit solutions to SODEs.

The integrating factor method

We apply the integrating factor method to solve nonlinear SDEs of the form

$$dX(t) = f(t, X(t))\,dt + \sigma(t)X(t)\,dW(t), \quad X_0 = x. \tag{3.1.9}$$

where f is a continuous deterministic function defined from $\mathbb{R}^+ \times \mathbb{R}$ to $\mathbb{R}$.

- Step 1. Solve the equation

  $$dG(t) = \sigma(t)G(t)\,dW(t).$$

  Then we have

  $$G(t) = \exp\Big(\int_0^t \sigma(s)\,dW(s) - \frac{1}{2}\int_0^t \sigma^2(s)\,ds\Big).$$

  The integrating factor function is defined by $F(t) = G^{-1}(t)$. It can be readily verified that $F(t)$ satisfies

  $$dF(t) = -\sigma(t)F(t)\,dW(t) + \sigma^2(t)F(t)\,dt.$$

- Step 2. Let $X(t) = G(t)C(t)$ and then $C(t) = F(t)X(t)$. Then by the product rule, (3.1.9) can be written as

  $$d(F(t)X(t)) = F(t)f(t, X(t))\,dt.$$

Then C_t satisfies the following "deterministic" ODE

$$dC(t) = F(t)f(t, G(t)C(t)). \tag{3.1.10}$$

- Step 3. Once we obtain $C(t)$, we get $X(t)$ from $X(t) = G(t)C(t)$.

Remark 3.1.5 *When* (3.1.10) *cannot be explicitly solved, we may use some numerical methods to obtain* $C(t)$.

Example 3.1.6 *Use the integrating factor method to solve the SDE*

$$dX(t) = (X(t))^{-1}\,dt + \alpha X(t)\,dW(t), \quad X_0 = x > 0,$$

where α *is a constant.*

Solution. Here $f(t,x) = x^{-1}$ and $F(t) = \exp(-\alpha W(t) + \frac{\alpha^2}{2}t)$. We only need to solve

$$dC(t) = F(t)[G^{-1}(t)C(t)]^{-1} = F^2(t)/C(t).$$

This gives $d(C(t))^2 = 2F^2(t)\,dt$ and thus

$$(C(t))^2 = 2\int_0^t \exp(-2\alpha W(s) + \alpha^2 s)\,ds + x^2.$$

Since the initial condition is $x > 0$, we take $Y(t) > 0$ such that

$$X(t)=G(t)Y(t)=\exp(\alpha W(t)-\frac{\alpha^2}{2}t)\sqrt{2\int_0^t \exp(-2\alpha W(s) + \alpha^2 s)\,ds + x^2} > 0.$$

Moment equations of solutions

For a more complicated SODE, we cannot obtain a solution that can be written explicitly in terms of $W(t)$. For example, the modified Cox-Ingersoll-Ross model (3.1.11) does not have an explicit solution:

$$dX(t) = \kappa(\theta - X(t))dt + \sigma\sqrt{X(t)}dW(t), \quad X_0 = x, \tag{3.1.11}$$

However, we can say a bit more about **the moments of the process** $X(t)$. Write (3.1.11) in its integral form:

$$X(t) = x + \kappa\int_0^t (\theta - X(s))ds + \sigma\int_0^t \sqrt{X(s)}\,dW(s) \tag{3.1.12}$$

and using Ito's formula gives

$$X^2(t)=x^2+(2\kappa\theta+\sigma^2)\int_0^t X(s)\,ds - 2\kappa\int_0^t X(s)^2\,ds + 2\sigma\int_0^t (X(s))^{3/2}\,dW(s). \tag{3.1.13}$$

From this equation and the properties of Ito's integral, we can obtain the moments of the solution. *The first moment* can be obtained by taking expectation over both sides of (3.1.12):

$$m_t := \mathbb{E}[X(t)] = x + \kappa \left(\theta t - \int_0^t \mathbb{E}[X(s)]\, ds \right),$$

because the expectation of the stochastic integral part is zero.[1] We can then solve the following ODE:

$$dm_t = \kappa(\theta - m_t)dt.$$

The solution is given by:

$$m_t = \theta + (x - \theta)e^{-\kappa t}.$$

For *the second moment*, we get from (3.1.13) that

$$\mathbb{E}[X^2(t)] = x^2 + (2\kappa\theta + \sigma^2) \int_0^t \mathbb{E}[X(s)]ds - 2\kappa \int_0^t \mathbb{E}[X^2(s)]ds.$$

This is again an ODE similar to the one for m_t to solve:

$$\mathbb{E}[X^2(t)] = x^2 + (2\kappa\theta + \sigma^2)\left(\theta t + (x - \theta)\frac{(1 - e^{-\kappa t})}{\kappa}\right) - 2\kappa \int_0^t \mathbb{E}[X^2(s)]ds.$$

Here we also assume that we have $\int_0^t \mathbb{E}[|X(s)|^3]\, ds < \infty$ so that $\int_0^t (X(s))^{3/2}\, dW(s)$ is an Ito integral with a square-integrable integrand and thus $\mathbb{E}[\int_0^t (X(s))^{3/2}\, dW(s)] = 0$.

Remark 3.1.7 *It can be shown using Feller's test [255, Theorem 5.29] that the solution to* (3.1.13) *exists and is unique when $2\kappa\theta > \sigma^2$ and $X_0 \geq 0$. Moreover, the solution is strictly positive when $X_0 > 0$. If $\mathbb{E}[|X_0|^3] < \infty$, then $\mathbb{E}[|X(t)|^p] < \infty$, $1 \leq p \leq 3$.*

Unfortunately, even the first few moments are difficult to obtain in general. For example, we cannot get a closure for the second-order moment of the following SDE

$$dX(t) = \kappa(\theta - X(t))\, dt + \big(X(t)\big)^\alpha\, dW(t), \quad \frac{1}{2} < \alpha < 1.$$

We cannot even obtain the first-order moment of the following SDE

$$dX(t) = \sin(X(t))\, dt + dW(t), \quad X_0 = x.$$

[1] Here we need to verify that $\int_0^t \sqrt{X(s)}\, dW(s)$ is indeed Ito's integral with a square-integrable integrand, by showing that $\int_0^t \mathbb{E}[|X(s)|]\, ds < \infty$. See Remark 3.1.7.

3.2 Numerical methods for SODEs

As explicit solutions to SODEs are usually hard to find, we seek numerical approximation of solutions.

3.2.1 Derivation of numerical methods based on numerical integration

A starting point for numerical SODEs is numerical integration. Consider the SODE (3.1.4) over $[t, t+h]$:

$$X(t+h) = X(t) + \int_t^{t+h} a(s, X(s))\, ds + \sum_{r=1}^{m} \int_t^{t+h} \sigma_r(s, X(s))\, dW_r.$$

The simplest scheme for (3.1.4) is the forward Euler scheme. In the forward Euler scheme, we replace (approximate)

$$\int_t^{t+h} a(s, X(s))\, ds \text{ with } \int_t^{t+h} a(t, X(t))\, ds = a(t, X(t))h$$

and

$$\int_t^{t+h} \sigma_r(s, X(s))\, dW_r \text{ with } \int_t^{t+h} \sigma_r(t, X(t))\, dW_r$$
$$= \sigma_r(t, X(t))(W_r(t+h) - W_r(t)).$$

Then we obtain the forward Euler scheme (also known as Euler-Maruyama scheme):

$$X_{k+1} = X_k + a(t_k, X_k)h + \sum_{r=1}^{m} \sigma_l(t_k, X_k)\Delta_k W_r, \qquad (3.2.1)$$

where $h = (T - t_0)/N$, $t_k = t_0 + kh$, $k = 0, \ldots, N$. $X_0 = x_0$ and $\Delta_k W_r = W_r(t_{k+1}) - W_r(t_k)$. The Euler scheme can be implemented by replacing the increments $\Delta_k W_r$ with Gaussian random variables:

$$X_{k+1} = X_k + a(t_k, X_k)h + \sum_{r=1}^{m} \sigma_l(t_k, X_k)\sqrt{h}\xi_{l,k+1}, \qquad (3.2.2)$$

where $\xi_{r,k+1}$ are i.i.d. $\mathcal{N}(0,1)$ random variables.

Replacing (approximating) the drift term with its value at $t+h$, we have

$$\int_t^{t+h} a(s, X(s))\, ds \approx \int_t^{t+h} a(t+h, X(t+h))\, ds = a(t+h, X(t+h))h.$$

The resulting scheme is called backward Euler scheme (also known as drift-implicit Euler scheme)

$$X_{k+1} = X_k + a(t_{k+1}, X_{k+1})h + \sum_{r=1}^{m} \sigma_l(t_k, X_k)\Delta_k W_r, \quad k = 0, 1, \ldots, N-1. \tag{3.2.3}$$

The following schemes can be considered as extensions of forward and backward Euler schemes

$$X_{k+1} = X_k + [(1-\lambda)a(t_k, X_k) + \lambda a(t_{k+1}, X_{k+1})]h + \sum_{r=1}^{m} \sigma_r(t_k, X_k)\sqrt{h}\xi_{r,k+1}, \tag{3.2.4}$$

where $\lambda \in [0, \lambda]$, or similarly

$$X_{k+1}=X_k+a((1-\lambda)t_k+\lambda t_{k+1},\ (1-\lambda)X_k+\lambda X_{k+1})h+\sum_{r=1}^{m} \sigma_r(t_k, X_k)\sqrt{h}\xi_{r,k+1}. \tag{3.2.5}$$

We can also derive numerical methods for (3.1.4) in order to get high-order convergence. For example, in (3.1.4), we can approximate the diffusion terms σ_r using their half-order Ito-Taylor's expansion which leads to the Milstein scheme [354]. Let us illustrate the derivation of the Milstein scheme for an autonomous SODE (a and σ do not explicitly depend on t) when $m = 1$ and $r = 1$, i.e., for scalar equation with single noise. With the following approximation,

$$\begin{aligned}
\int_t^{t+h} a(X(s))\,ds &\approx \int_t^{t+h} a(X_t)\,ds = a(X(t))h \\
\int_t^{t+h} \sigma(X(s))\,dW(s) &\approx \int_t^{t+h} [\sigma(X(t)) + \int_t^s \sigma'(X(t))\sigma(X_t)\,dW(\theta)]\,dW(s) \\
&= \sigma(X(t))(W(t+h) - W(t)) + \sigma'(X(t))\sigma(X(t)) \\
&\quad \int_t^{t+h}\int_t^s dW(\theta)\,dW(s),
\end{aligned}$$

we can obtain the Milstein scheme

$$\begin{aligned}
X_{k+1} = X_k &+ a(X_k)h + \sigma(X_k)(W(t_{k+1}) - W(t_k)) \\
&+ \frac{1}{2}\sigma(X_k)\sigma'(X_k)[(W(t_{k+1}) - W(t_k))^2 - h].
\end{aligned}$$

One can also derive a drift-implicit Milstein scheme as follows:

$$\begin{aligned}
X_{k+1} = X_k &+ a(X_{k+1})h + \sigma(X_k)(W(t_{k+1}) - W(t_k)) \\
&+ \frac{1}{2}\sigma(X_k)\sigma'(X_k)[(W(t_{k+1}) - W(t_k))^2 - h].
\end{aligned}$$

For (3.1.4), the Milstein scheme is as follows, see, e.g., [259, 358],

$$X_{k+1} = X_k + a(t_k, X_k)h + \sum_{r=1}^{m} \sigma_r(t_k, X_k)\xi_{rk}\sqrt{h}$$
$$+ \sum_{i,l=1}^{m} \Lambda_i \sigma_l(t, X_k) I_{i,l,t_k}, \tag{3.2.6}$$

where $I_{i,l,t_k} = \int_{t_k}^{t_{k+1}} \int_{t_k}^{s} dW_i \, dW_l$. To efficiently evaluate this double integral, see Chapter 4 and some bibliographic notes.The scheme (3.2.6) is of first-order mean-square convergence. For commutative noises, i.e.

$$\Lambda_i \sigma_l = \Lambda_l \sigma_i, \quad \Lambda_l = \sigma_l^\top \frac{\partial}{\partial x}, \tag{3.2.7}$$

we can use only increments of Brownian motions of the double Ito integral in (3.2.6) since

$$I_{i,l,t_k} + I_{l,i,t_k} = (\xi_{ik}\xi_{lk} - \delta_{il})h/2,$$

where δ_{il} is the Kronecker delta function. In this case, we have a simplified version of (3.2.6):

$$X_{k+1} = X_k + a(t_k, X_k)h + \sum_{r=1}^{m} \sigma_r(t_k, X_k)\xi_{lk}\sqrt{h}$$
$$+ \frac{1}{2} \sum_{i,l=1}^{m} \Lambda_i \sigma_l(t, X_k)(\xi_{ik}\xi_{lk} - \delta_{il})h. \tag{3.2.8}$$

There has been an extensive literature on numerical methods for SODES. We refer to [217, 416] for introduction to numerical methods for SODEs and to [259, 354] for a systematic construction of numerical methods for SODEs. In chapter 4 we will present three different methods for SODEs with delay, which can also be applied to standard SODEs if we set the delay equal to zero.

For numerical methods for SODEs and SPDEs, the key issues are whether a numerical method converges and in what sense and whether it is stable in some sense, as well as how fast it converges.

3.2.2 Strong convergence

Definition 3.2.1 (Strong convergence in L^p) *A method (scheme) is said to have a strong convergence order γ in L^p if there exists a constant $K > 0$ independent of h such that*

$$\mathbb{E}[|X_k - X(t_k)|^p] \leq K h^{p\gamma}$$

for any $k = 0, 1, \ldots, N$ and $Nh = T$ and sufficiently small h.

In many applications and in this book, a strong convergence refers to convergence in the mean-square sense, i.e., $p = 2$.

If the coefficients of (3.1.4) satisfy the conditions in Theorem 3.1.2, the forward Euler scheme (3.2.1) and the backward Euler scheme (3.2.3) are convergent with half-order ($\gamma = 1/2$) in the mean-square sense (strong convergence order half), i.e.,

$$\max_{1\leq k\leq N} \mathbb{E}[|X(t_k) - X_k|^2] \leq Kh,$$

where K is positive constant independent of h. When the noise is additive, i.e., the coefficients of noises are functions of time instead of functions of the solutions, these schemes are of first-order convergence.

Under the conditions in Theorem 3.1.2, the Milstein scheme (3.2.6) can be shown to have a strong convergence order one, i.e., $\gamma = 1$.

Note that all these schemes are one-step schemes. One can use the following Milstein's fundamental theorem to derive their mean-square convergence order. Introduce the one-step approximation $\bar{X}_{t,x}(t+h)$, $t_0 \leq t < t+h \leq T$, for the solution $X_{t,x}(t+h)$ of (3.1.4), which depends on the initial point (t,x), a time step h, and $\{W_1(\theta) - W_1(t), \ldots, W_m(\theta) - W_m(t),\ t \leq \theta \leq t+h\}$ and which is defined as follows:

$$\bar{X}_{t,x}(t+h) = x + A(t,x,h; W_i(\theta) - W_i(t),\ i = 1,\ldots,m,\ t \leq \theta \leq t+h). \quad (3.2.9)$$

Using the one-step approximation (3.2.9), we recurrently construct the approximation $(X_k, \mathcal{F}_{t_k})$, $k = 0,\ldots,N$, $t_{k+1} - t_k = h_{k+1}$, $T_N = T$:

$$\begin{aligned} X_0 = X(t_0),\ X_{k+1} &= \bar{X}_{t_k,\bar{X}_k}(t_{k+1}) \quad (3.2.10)\\ &= X_k + A(t_k, X_k, h_{k+1}; W_i(\theta) - W_i(t_k),\ i = 1,\ldots,m,\ t_k \leq \theta \leq t_{k+1}). \end{aligned}$$

For simplicity, we will consider a uniform time step size, i.e., $h_k = h$ for all k. The proof of the following theorem can be found in [353] and [354, 358, Chapter 1].

Theorem 3.2.2 (Fundamental convergence theorem of one-step numerical methods) *Suppose that*

(i) the coefficients of (3.1.4) *are Lipschitz continuous;*
(ii) the one-step approximation $\bar{X}_{t,x}(t+h)$ *from* (3.2.9) *has the following orders of accuracy: for some* $p \geq 1$ *there are* $\alpha \geq 1$, $h_0 > 0$, *and* $K > 0$ *such that for arbitrary* $t_0 \leq t \leq T-h$, $x \in \mathbb{R}^d$, *and all* $0 < h \leq h_0$:

$$|\mathbb{E}[X_{t,x}(t+h) - \bar{X}_{t,x}(t+h)]| \leq K(1+|x|^2)^{1/2} h^{q_1},$$

$$\left[\mathbb{E}|X_{t,x}(t+h) - \bar{X}_{t,x}(t+h)|^{2p}\right]^{1/(2p)} \leq K(1+|x|^{2p})^{1/(2p)} h^{q_2} \quad (3.2.11)$$

with

$$q_2 \geq \frac{1}{2},\ q_1 \geq q_2 + \frac{1}{2};$$

Then for any N *and* $k = 0, 1, \ldots, N$ *the following inequality holds:*

$$\left[\mathbb{E}|X_{t_0,X_0}(t_k) - \bar{X}_{t_0,X_0}(t_k)|^{2p}\right]^{1/(2p)} \leq K(1 + \mathbb{E}|X_0|^{2p})^{1/(2p)} h^{q_2 - 1/2}, \tag{3.2.12}$$

where $K > 0$ *do not depend on* h *and* k, *i.e., the order of accuracy of the method* (3.2.10) *is* $q = q_2 - 1/2$.

Many other schemes of strong convergence based on Ito-Taylor's expansion have been developed, such as Runge-Kutta methods, predictor-corrector methods, and splitting (split-step) methods, see, e.g., [259, 358]. Here we assume that the coefficients are Lipschitz continuous while in practice the coefficients may be non-Lipschitz continuous. We will discuss this issue in Chapter 5.

3.2.3 Weak convergence

The weak integration of SDEs is computing the expectation

$$\mathbb{E}[f(X(T))], \tag{3.2.13}$$

where $f(x)$ is a sufficiently smooth function with growth at infinity not faster than a polynomial:

$$|f(x)| \leq K(1 + |x|^{\varkappa}) \tag{3.2.14}$$

for some $K > 0$ and $\varkappa \geq 1$.

Definition 3.2.3 (Weak convergence) *A method (scheme) is said to have a weak convergence order* γ *if there exists a constant* $K > 0$ *independent of* h *such that*

$$|\mathbb{E}[f(X_k)] - \mathbb{E}[f(X(t_k))]| \leq K h^{\gamma}$$

for any $k = 0, 1, \ldots, N$ *and* $Nh = T$ *and sufficiently small* h.

Under the conditions of Theorem 3.1.2, the following error estimate holds for the forward Euler scheme (3.2.2) (see, e.g., [358, Chapter 2]):

$$|\mathbb{E}f(X_k) - \mathbb{E}f(X(t_k))| \leq Kh, \tag{3.2.15}$$

where $K > 0$ is a constant independent of h. The backward Euler scheme (3.2.3) and the Milstein scheme (3.2.6), are all of weak convergence order 1.

This first-order weak convergence of the forward Euler scheme can also be achieved by replacing $\xi_{l,k+1}$ with discrete random variables [358], e.g., the weak Euler scheme has the form

$$\tilde{X}_{k+1} = \tilde{X}_k + ha(t_k, \tilde{X}_k) + \sqrt{h} \sum_{r=1}^{m} \sigma_r(t_k, \tilde{X}_k)\zeta_{r,k+1}, \qquad k = 0, \ldots, N-1, \tag{3.2.16}$$

where $\tilde{X}_0 = x_0$ and $\zeta_{r,k+1}$ are i.i.d. random variables with the law

$$P(\zeta = \pm 1) = 1/2. \tag{3.2.17}$$

The following error estimate holds for (3.2.16)–(3.2.17) (see, e.g., [358, Chapter 2]):

$$|\mathbb{E}f(\tilde{X}_N) - \mathbb{E}f(X(T))| \leq Kh, \tag{3.2.18}$$

where $K > 0$ can be a different constant than that in (3.2.15).

Introducing the function $\varphi(y)$, $y \in \mathbb{R}^{mN}$, so that

$$\varphi(\xi_{1,1}, \ldots, \xi_{r,1}, \ldots, \xi_{1,N}, \ldots, \xi_{m,N}) = f(X_N), \tag{3.2.19}$$

we have

$$\begin{aligned} \mathbb{E}[f(X(T)] &\approx \mathbb{E}f(X_N) = \mathbb{E}\varphi(\xi_{1,1}, \ldots, \xi_{r,1}, \ldots, \xi_{1,N}, \ldots, \xi_{m,N}) \quad (3.2.20)\\ &= \frac{1}{(2\pi)^{mN/2}} \int_{\mathbb{R}^{rN}} \varphi(y_{1,1}, \ldots, y_{m,1}, \ldots, y_{1,N}, \ldots, y_{m,N}) \\ &\quad \exp\left(-\frac{1}{2}\sum_{i=1}^{mN} y_i^2\right) dy. \end{aligned}$$

Further, it is not difficult to see from (3.2.16) to (3.2.17) and (2.5.8) that

$$\begin{aligned} \mathbb{E}[f(X(T)] &\approx \mathbb{E}f(\tilde{X}_N) = \mathbb{E}\varphi(\zeta_{1,1}, \ldots, \zeta_{m,1}, \ldots, \zeta_{1,N}, \ldots, \zeta_{m,N}) \quad (3.2.21)\\ &= Q_2^{\otimes mN} \varphi(y_{1,1}, \ldots, y_{m,1}, \ldots, y_{1,N}, \ldots, y_{m,N}), \end{aligned}$$

where Q_2 is the Gauss-Hermite quadrature rule with nodes ± 1 and equal weights $1/2$, see, e.g., [2, Equation 25.4.46]. We note that $\mathbb{E}[f(\tilde{X}_N)]$ can be viewed as an approximation of $\mathbb{E}[f(X_N)]$ and that (cf. (3.2.15) and (3.2.18)) $\left|\mathbb{E}[f(X_N)] - \mathbb{E}[f(\tilde{X}_N)]\right| = \mathcal{O}(h)$.

Remark 3.2.4 *Let $\zeta_{l,k+1}$ in (3.2.16) be i.i.d. random variables with the law*

$$P(\zeta = \mathsf{y}_{n,j}) = \mathsf{w}_{n,j}, \quad j = 1, \ldots, n, \tag{3.2.22}$$

where $\mathsf{y}_{n,j}$ are nodes of the Gauss-Hermite quadrature Q_n and $\mathsf{w}_{n,j}$ are the corresponding quadrature weights (see (2.5.6)). Then

$$\mathbb{E}f(\tilde{X}_N) = \mathbb{E}\varphi(\zeta_{1,N}, \ldots, \zeta_{m,N}) = Q_n^{\otimes mN} \varphi(y_{1,1}, \ldots, y_{m,N}),$$

which can be a more accurate approximation of $\mathbb{E}[f(X_N)]$ than $\mathbb{E}[f(\tilde{X}_N)]$ from (3.2.21) but the weak-sense error for the SDEs approximation $\mathbb{E}f(\tilde{X}_N) - \mathbb{E}f(X(T))$ remains of order $\mathcal{O}(h)$.

We can also use the second-order weak scheme (3.2.23) for (3.1.4) (see, e.g., [358, Chapter 2]):

$$X_{k+1} = X_k + ha(t_k, X_k) + \sqrt{h}\sum_{i=1}^{m} \sigma_i(t_k, X_k)\xi_{i,k+1} + \frac{h^2}{2}\mathfrak{L}a(t_k, X_k) \quad (3.2.23)$$
$$+h\sum_{i=1}^{m}\sum_{j=1}^{r} \Lambda_i\sigma_j(t_k, X_k)\eta_{i,j,k+1} + \frac{h^{3/2}}{2}\sum_{i=1}^{m}(\Lambda_i a(t_k, X_k)$$
$$+\mathfrak{L}\sigma_i(t_k, X_k))\xi_{i,k+1},$$
$$k = 0, \ldots, N-1,$$

where $X_0 = x_0$; $\eta_{i,j} = \frac{1}{2}\xi_i\xi_j - \gamma_{i,j}\zeta_i\zeta_j/2$ with $\gamma_{i,j} = -1$ if $i < j$ and $\gamma_{i,j} = 1$ otherwise;

$$\Lambda_l = \sum_{i=1}^{m} \sigma_l^i \frac{\partial}{\partial x_i}, \quad \mathfrak{L} = \frac{\partial}{\partial t} + \sum_{i=1}^{m} a^i \frac{\partial}{\partial x_i} + \frac{1}{2}\sum_{r=1}^{m}\sum_{i,j=1}^{m} \sigma_l^i\sigma_l^i \frac{\partial^2}{\partial x_i \partial x_j};$$

and $\xi_{i,k+1}$ and $\zeta_{i,k+1}$ are mutually independent random variables with Gaussian distribution or with the laws $P(\xi = 0) = 2/3$, $P(\xi = \pm\sqrt{3}) = 1/6$ and $P(\zeta = \pm 1) = 1/2$. The following error estimate holds for (3.2.23) (see, e.g., [358, Chapter 2]):

$$|\mathbb{E}f(X(T)) - Ef(X_N)| \leq Kh^2.$$

We again refer to [259, 358] for more weakly convergent numerical schemes.

3.2.4 Linear stability

To understand the stability of time integrators for SODEs, we consider the following linear model:

$$dX = \lambda X\, dt + \sigma\, dW(t), \quad X_0 = x, \quad \lambda < 0. \quad (3.2.24)$$

Consider one-step methods of the following form:

$$X_{n+1} = A(z)X_n + B(z)\sqrt{h}\sigma\xi_n, \quad z = \lambda h. \quad (3.2.25)$$

Here h is the time step size and $A(z)$ and $B(z)$ are analytic functions, $\delta W_n = \sqrt{h}\xi_n$ are i.i.d. Gaussian random variables.

For (3.2.24), we have $X(t)$ is a Gaussian random variable and

$$\lim_{t\to\infty} \mathbb{E}[X(t)] = 0, \quad \lim_{t\to\infty} \mathbb{E}[X^2(t)] = \frac{\sigma^2}{2\lambda}.$$

It can be readily shown that X_n is also a Gaussian random variable with $\mathbb{E}[X_n] = A^n(z)x$ and

$$\lim_{n\to\infty} \mathbb{E}[|X_n|^2] = \frac{\sigma^2}{2\lambda}R(z), \quad R(z) = -\frac{2zB^2(z)}{1-A^2(z)}.$$

Here are some examples:

- Euler scheme: $A(z) = 1 + z$, $B(z) = 1$, and $R(z) = \frac{2}{2+z}$.
- Backward Euler scheme: $A(z) = B(z) = \frac{1}{1-z}$ and $R(z) = \frac{2}{2-z}$.
- Trapezoidal rule, $A(z) = \frac{1+z/2}{1-z/2}$ and $B(z) = \frac{1}{1-z/2}$ and $R(z) = 1$.

When $R(z) = 1$ and $|A(z)| < 1$, then we can obtain the exact distribution of $X(\infty)$. In this case, we call the one-step scheme is A-stable.

For long-time integration, *L-stability* is also helpful when a stiff problem (e.g., λ is large) is investigated. The L-stability requires A-stability and that $A(-\infty) = 0$ such that $\lim_{z\to-\infty} \mathbb{E}[X_n] = \lim_{z\to-\infty} A^n(z)x = 0$ for any fixed n. When λh is large (e.g., λ is too large to have such a practical h that λh is small), L-stable schemes can still obtain the decay of the solution $\mathbb{E}[X(t)] = 0$ even with moderately small time step sizes while A-stable schemes usually require very small h. For example, the trapezoidal rule is A-stable but not L-stable since $A(-\infty) = 1$. In practice, this means that for an extremely large λ, the trapezoidal rule damps the mean since $\mathbb{E}[X_{n+1}] = \lim_{z\to-\infty} A(z)\mathbb{E}[X_n] = \mathbb{E}[X_n]$ while $\mathbb{E}[X(t_{n+1})] = \lim_{z\to-\infty} \exp(-z)\mathbb{E}[X_{t_n}] = 0$, where $t_{n+1}-t_n = h$.

However, when $A(z)$ and $B(z)$ are rational functions of z, it is impossible to have both A-stability and L-stability since when $R(z) = 1$, it holds that $A(-\infty) = 1$. The claim can be proved by the argument of contradiction.

It is still possible to have a scheme such that it is L-stable and A-stable. Define

$$\tilde{X}_n = C(z)X_n + D(z)\sigma\sqrt{h}\xi_n, \quad z = \lambda h,$$

where X_n is from (3.2.25). For example, for the backward Euler scheme, $A(z) = B(z) = \frac{1}{1-z}$ and

$$\lim_{n\to\infty} \mathbb{E}[\left|\tilde{X}_n\right|^2] = \frac{\sigma^2}{2\lambda}(C^2(z)R(z) - 2zD^2(z)).$$

The limit is exactly the same as the variance of $X(\infty)$ when $C(z) = 1$ and $D(z) = (1 - z/2)^{-1}$. Such a scheme with $\tilde{X}$ approximating X is called a *post-processing* scheme or a *predictor-corrector* scheme.

The linear model (3.2.24) is designed for additive noise. For multiplicative noise, we can consider the following geometric Brownian motion.

$$dX = \lambda X\, dt + \sigma X\, dW(t), \quad X(0) = 1. \tag{3.2.26}$$

Here we assume that $\lambda, \sigma \in \mathbb{R}$. The solution to (3.2.26) is

$$X = \exp((\lambda - \frac{1}{2}\sigma^2)t + \sigma W(t)).$$

The solution is mean-square stable if $\lambda + \frac{\sigma^2}{2} < 0$, i.e., $\lim_{t\to\infty} \mathbb{E}[X^2(t)] = 0$. It is asymptotic stable ($\lim_{t\to\infty} |X(t)| = 0$) if $\lambda - \frac{\sigma^2}{2} < 0$. The mean-square stability implies asymptotic stability. Here we consider only mean-square stability.

Applying the forward Euler scheme (3.2.2) to the linear model (3.2.26), we have

$$X_{k+1} = (1 + \lambda h + \sqrt{h}\xi_k)X_k.$$

The second moment is $\mathbb{E}[X_{k+1}^2] = \mathbb{E}[X_k^2]\mathbb{E}[(\lambda h + \sqrt{h}\xi_k)^2] = \mathbb{E}[X_k^2]\big((1+\lambda h)^2 + h\sigma^2\big)$. For $\lim_{k\to\infty} \mathbb{E}[X_{k+1}] = 0$, we need $(1 + \lambda h)^2 + h\sigma^2 < 1$. Similarly, for the backward Euler scheme (3.2.3), we need $1 + \sigma^2 h < (1 - \lambda h)^2$.

We call the region of $(\lambda h, \sigma^2 h)$ where a scheme is mean-square stable the mean-square stability region of the scheme. To allow relative large h for stiff problems, e.g., when λ is large, we need a large stability region. Usually, explicit schemes have smaller stability regions than implicit (including semi-implicit and drift-implicit) schemes do.

Both schemes (3.2.2) and (3.2.6) are explicit and hence they have small stability regions. To improve the stability region, we can use some semi-implicit (drift-implicit) schemes, e.g., (3.2.3) and drift-implicit Milstein scheme. Fully implicit schemes are also used in practice because of their symplecticity-preserving property and effectiveness in long-term integration. The following fully implicit scheme is from [358, 448]:

$$\begin{aligned} X_{k+1} = X_k &+ a(t_{k+\lambda}, (1-\lambda)X_k + \lambda X_{k+1})h \\ &-\lambda \sum_{r=1}^{m}\sum_{j=1}^{d} \frac{\partial \sigma_r}{\partial x^j}(t_{k+\lambda}, (1-\lambda)X_k + \lambda X_{k+1})\sigma_r^j(t_{k+\lambda}, (1-\lambda)X_k \\ &\quad +\lambda X_{k+1})h \\ &+\sum_{r=1}^{m} \sigma_r(t_{k+\lambda}, (1-\lambda)X_k + \lambda X_{k+1})\,(\zeta_{rh})_k\,\sqrt{h}, \end{aligned} \tag{3.2.27}$$

where $0 < \lambda \leq 1$, $t_{k+\lambda} = t_k + \lambda h$ and $(\zeta_{rh})_k$ are i.i.d. random variables so that

$$\zeta_h = \begin{cases} \xi, \ |\xi| \leq A_h, \\ A_h, \ \xi > A_h, \\ -A_h, \ \xi < -A_h, \end{cases} \tag{3.2.28}$$

with $\xi \sim \mathcal{N}(0, 1)$ and $A_h = \sqrt{2l|\ln h|}$ with $l \geq 1$. We recall [358, Lemma 1.3.4] that

$$\left|\mathbb{E}[(\xi^2 - \zeta_h^2)]\right| \leq (1 + 2\sqrt{2l|\ln h|})h^l. \tag{3.2.29}$$

All these semi-implicit and fully implicit schemes are of half-order convergence in the mean-square sense, see, e.g., [358, Chapter 1]. When the noise is additive, i.e., the coefficients of noises are functions of time instead of functions of the solutions, these schemes are of first-order convergence.

3.2.5 Summary of numerical SODEs

Numerical methods for (3.1.4) with Lipschitz continuous coefficients have been investigated extensively, see, e.g., [238, 259, 354, 358]. As the solution to (3.1.4) is usually Hölder continuous with exponent $1/2-\epsilon$ $(0<\epsilon<1/2)$, the convergence order in the mean-square sense is usually half. In fact, by Ito's formula, we can readily have

$$\mathbb{E}[|X_{t_0,X_0}(t+h)-X_{t_0,X_0}(t)|^2]\leq \exp(Ct)h.$$

Then we can conclude from the Kolmogorov continuity theorem (Theorem 2.2.10 or see e.g., [255, Chapter 2.2.A]) that the solution is Hölder continuous with exponent $1/2-\epsilon$. First-order schemes of mean-square convergence can be also derived using the Ito-Taylor expansion and they require significant computational effort, see, e.g., the Milstein scheme (3.2.6). When the coefficients of noises satisfy the *commutative conditions*, the computational effort in simulating the double integrals in (3.2.6) can be significantly reduced, see (3.2.8). When a weak convergence is desired, i.e., expectations of functionals of solutions or simply moments of solutions are desired, one can further approximate the Gaussian random variables in the schemes of strong convergence with simple symmetric random walks. For long-time integration of SDEs, some structure-preserving schemes should be used, e.g., the mid-point method (3.2.27)–(3.2.28) as one of symplectic methods (see [358, Chapter 4] for more symplectic schemes for SDEs).

However, numerical methods for (3.1.4) with non-Lipschitz continuous coefficients are far from being mature. When the coefficients are of polynomial growth, several numerical schemes have been proposed, see, e.g., [218, 233, 235, 448] for schemes of strong convergence and [359] for schemes of weak convergence. Compared to numerical methods of SDEs with Lipschitz continuous coefficients, the key ingredient is the *moment stability* of the numerical schemes for SDEs with non-globally Lipschitz coefficients, see [448] and Chapter 5. An extension was made for numerical schemes where numerical solutions have exponential moments, see, e.g., [232, 237].

3.3 Basic aspects of SPDEs

Let us discuss the regularity of a simple SPDE – one-dimensional heat equation with additive space-time noise. As will be shown, the regularity is low and depends on the smoothness of the driving space-time noise.

Example 3.3.1 (Heat equation with random forcing) *Consider the following one-dimensional heat equation driven by some space-time forcing:*

$$\partial_t u(t,x)=D\partial_x^2 u(t,x)+F(t,x),\quad (t,x)\in(0,\infty)\times(0,l), \tag{3.3.1}$$

with vanishing Dirichlet boundary conditions and deterministic initial condition $u_0(x)$. Here D is a physical constant depending on the conductivity of the thin wire and $F(t,x) = \sum_{k=1}^{\infty} \sqrt{q_k} m_k(x) \dot{W}_k(t)$. Here $\sum_{k=1}^{\infty} q_k < \infty$, $\{m_k(x)\}_k$ is a complete orthonormal basis in $L^2([0,l])$ and W_k's are i.i.d. standard Brownian motion.

To find a solution, we apply the *method of eigenfunction expansion.* Let $\{e_k(x)\}_k$ be eigenfunctions of the operator ∂_x^2 with vanishing Dirichlet boundary conditions) and the corresponding eigenvalues are λ_k:

$$-\partial_x^2 e_k = \lambda_k e_k, \quad e_k|_{x=0,l} = 0, \quad k = 1, 2, \ldots. \tag{3.3.2}$$

with $0 \leq \lambda_1 \leq \lambda_2 \leq \cdots \leq \lambda_k \leq \cdots$ and $\lim_{k\to\infty} \lambda_k = +\infty$. Actually, they can be computed explicitly

$$\lambda_k = k^2 (\frac{\pi}{l})^2, \quad e_k(x) = \sqrt{\frac{2}{l}} \sin(k\frac{\pi}{l}x). \tag{3.3.3}$$

For simplicity, we let $D = 1$ and $m_k = e_k$ for any k. We look for a formal solution of the following form

$$u(t,x) = \sum_{k=1}^{\infty} u_k(t) e_k(x).$$

Plugging this formula into (3.3.1) and multiplying by e_i before integrating over both sides of the equation, we have

$$du_i(t) = -\lambda_i u_i(t)\, dt + \sqrt{q_i}\, dW_i(t), \quad u_i(0) = \int_0^l u_0(x) e_i(x)\, dx.$$

This is the Ornstein-Uhlenbeck process 3.1.7 and the solution is

$$u_i(t) = u_{0,i} e^{-\lambda_i t} + \sqrt{q_i} \int_0^t e^{-\lambda_i (t-s)}\, dW_i(s), \quad u_{0,i} = \int_0^l u_0(x) e_i(x)\, dx.$$

Thus the solution is

$$u(t,x) = \sum_{k=1}^{\infty} [\int_0^l u_0(x) e_k(x)\, dx e^{-\lambda_k t} + \sqrt{q_k} \int_0^t e^{-\lambda_k (t-s)}\, dW_k(s)] e_k(x). \tag{3.3.4}$$

Now we show the regularity of the solution. When $\sum_{k=1}^{\infty} q_k < \infty$, we have $\mathbb{E}[\|u(t,x) - u(s,x)\|^2] \leq C(t-s)$ ($t-s$ small) and then the solution is Hölder continuous in time, by Kolmogorov's continuity theorem. In fact,

$$\begin{aligned}
\mathbb{E}[\|u(t,x)-u(s,x)\|^2] &= \sum_{k=1}^{\infty}\mathbb{E}[|u_k(t)-u_k(s)|^2]\\
&= \sum_{k=1}^{\infty}\left|u_{0,k}(e^{-\lambda_k t}-e^{\lambda_k s})\right|^2 + q_k\mathbb{E}[\left|\int_0^t e^{-\lambda_k(t-\theta)}\,dW_k(\theta)\right.\\
&\quad \left.-\int_0^s e^{-\lambda_k(s-\theta)}\,dW_k(\theta)\right|^2]\\
&\le \sum_{k=1}^{\infty}[\lambda_k\left|u_{0,k}\right|^2(t-s)^2 + q_k\frac{1-e^{-2\lambda_k(t-s)}}{\lambda_k}(t-s)]\\
&\le C(t-s)^2\sum_{k=1}^{\infty}\left|u_{0,k}\right|^2\lambda_k + (t-s)\sum_{k=1}^{\infty}q_k\frac{1-e^{-2\lambda_k(t-s)}}{\lambda_k}\\
&\le C(t-s),
\end{aligned}$$

where we require that $u_0\in H^1([0,l])$, i.e., $\sum_{k=1}^{\infty}u_{0,k}^2k^2<\infty$. In the second last line, we also have used the following conclusion

$$\mathbb{E}[\left|\int_0^t e^{-\lambda(t-\theta)}\,dW(\theta)-\int_0^s e^{-\lambda(s-\theta)}\,dW(\theta)\right|^2]\le\frac{1-e^{-2\lambda(t-s)}}{\lambda},\quad \text{for any } t,\ \lambda>0. \tag{3.3.5}$$

The conclusion is left as an exercise (Exercise 3.6.10).

We can show that the solution is Hölder continuous with exponent less than 1. By the fact $|e_k(x)-e_k(y)|\le\sqrt{\frac{2}{l}}k\frac{\pi}{l}\,|x-y|$, it can be readily checked that

$$\begin{aligned}
\mathbb{E}[\|u(t,x)-u(t,y)\|^2] &= \sum_{k=1}^{\infty}\mathbb{E}[|u_k(t)|^2]\,|e_k(x)-e_k(y)|^2\\
&\le \frac{2\pi^2}{l^3}\,|x-y|^2\sum_{k=1}^{\infty}\mathbb{E}[|u_k(t)|^2]k^2\\
&= \frac{2\pi^2}{l^3}\,|x-y|^2\sum_{k=1}^{\infty}(\left|u_{0,k}^2\right|e^{-2\lambda_k t}+q_k\frac{1-e^{-2\lambda_k t}}{2\lambda_k})k^2.
\end{aligned}$$

Recalling (3.3.3) and $u_0\in H^1([0,l])$, we have

$$\mathbb{E}[\|u(t,x)-u(t,y)\|^2]\le(C+\sum_{k=1}^{\infty}q_k)\,|x-y|^2. \tag{3.3.6}$$

The regularity in x follows from Kolmogorov's continuity theorem.

By the Burkholder-Davis-Gundy inequality (see Appendix D), we have

$$\mathbb{E}[\sup_{0\le t\le T}|u^p(t,x)|]\le C_p\left|\sum_{k=1}^{\infty}q_ke_k^2(x)\frac{1-e^{-2\lambda_k t}}{2\lambda_k}\right|^{p/2},\quad p\ge 1. \tag{3.3.7}$$

As long as $\sum_{k=1}^{\infty} \frac{q_k}{\lambda_k}$ converges, $\mathbb{E}[\sup_{0 \le t \le T} |u^p(t,x)|] < \infty$. However, the second-order derivative of the solution in x should be understood as a distribution instead of a function. For simplicity, let's suppose that $u_0(x) = 0$. The solution becomes

$$u(t,x) = \sum_{k=1}^{\infty} \sqrt{q_k} \int_0^t e^{-\lambda_k (t-s)} \, dW_k(s) e_k(x). \tag{3.3.8}$$

The second derivative of $u(t,x)$ in x is

$$\partial_x^2 u(t,x) = \sum_{k=1}^{\infty} \lambda_k \sqrt{q_k} \int_0^t e^{-\lambda_k (t-s)} \, dW_k(s) e_k(x). \tag{3.3.9}$$

As a Gaussian process, this process exists a.s. and requires a bounded second-order moment, i.e.,

$$\mathbb{E}[(\partial_x^2 u(t,x))^2] = \sum_{k=1}^{\infty} q_k \lambda_k \frac{1 - e^{-2\lambda_k t}}{2} e_k^2(x) \ge \frac{1 - e^{-2\lambda_1 t}}{2} \sum_{k=1}^{\infty} q_k \lambda_k e_k^2(x).$$

Thus if q_k is proportional to $1/k^p$, $p < 3$, $\sum_{k=1}^{\infty} q_k \lambda_k$ diverges. The condition on $\sum_{k=1}^{\infty} q_k < \infty$ will not give us second-order derivatives in a classical sense.

In conclusion, the solution to (3.3.1) is not smooth and in general does not have second-order derivatives unless the space-time noise is very smooth in space. For example, when $q_k = 0$ for $k \ge N > 1$, we have a finite dimensional noise, we can expect second-order derivatives in space.

Example 3.3.2 (Multiplicative noise) *Consider the following equation*

$$du = a u_{xx} dt + \sigma u_x \, dW(t), \quad x \in (0, 2\pi)$$

with periodic boundary condition. Then by Fourier transform, when $2a - \sigma^2 > 0$, the solution has second-order moments.

3.3.1 Functional spaces

Consider a domain $\mathcal{D} \subseteq \mathbb{R}^d$. When k is a positive integer, we denote

$$\mathcal{C}^k(\mathcal{D}) = \{u : \mathcal{D} \to \mathbb{R} | D^\alpha u \text{ are continuous}, \quad |\alpha| \le k\}.$$

When $k = 1$, we simply denote $\mathcal{C}(\mathcal{D})$ instead of $\mathcal{C}^1(\mathcal{D})$. The space $\mathcal{C}^\infty(\mathcal{D}) = \cap_{k=1}^{\infty} \mathcal{C}^k(\mathcal{D})$. The space $\mathcal{C}_0^k(\mathcal{D})$ denotes functions in $\mathcal{C}^k(\mathcal{D})$ with compact support. Recall that the compact support of a function f is the closure of the subset of X where f is nonzero: $\overline{\{x \in X \mid f(x) \neq 0\}}$. The Hölder space $\mathcal{C}_b^{r+1}(\mathcal{D})$ is equipped with the following norm

$$\|f\|_{\mathcal{C}_b^r} = \max_{0 \le |\beta| \le \lfloor r \rfloor} \left\| D^\beta f \right\|_{L^\infty} + \sup_{\substack{x,y \in \mathcal{D} \\ |\beta| = \lfloor r \rfloor, r > \lfloor r \rfloor}} \frac{\left| D^\beta f(x) - D^\beta f(y) \right|}{|x-y|^{r - \lfloor r \rfloor}},$$

and $\lfloor r \rfloor$ is the integer part of the positive number r.

For $1 \leq p \leq \infty$, denote by $L^p(\mathcal{D})$ the space of Lebesgue measurable functions with finite L^p-norm, i.e., when $f \in L^p(\mathcal{D})$, then $\|f\|_{L^p(\mathcal{D})} = \left(\int_{\mathcal{D}} |f|^p \, dx\right)^{1/p} < \infty$ if $1 \leq p < \infty$. When $p = \infty$, $\|f\|_\infty = \operatorname{ess\,sup}_{\mathcal{D}} |f|$. In the Sobolev space $W^{k,p}(\mathcal{D})$, $k = 0, 1, 2, \ldots$, $1 \leq p \leq \infty$, the Sobolev norm is defined as

$$\|u\|_{W^{k,p}(\Omega)} := \begin{cases} \left(\sum_{|\alpha| \leq k} \|D^\alpha u\|^p_{L^p(\Omega)}\right)^{\frac{1}{p}}, & 1 \leq p < +\infty; \\ \max_{|\alpha| \leq k} \|D^\alpha u\|_{L^\infty(\Omega)}, & p = +\infty. \end{cases}$$

If $p = 2$, $W^{k,p}(\mathcal{D}) = H^k(\mathcal{D})$ is a Sobolev-Hilbert space. When k is not an integer, we need the following Slobodeckij semi-norm $|\cdot|_{\theta,p,\mathcal{D}}$, defined by

$$\int_{\mathcal{D}} \int_{\mathcal{D}} \frac{|f(x) - f(y)|^p}{|x - y|^{p\theta + d}} \, dx \, dy)^{1/p}. \tag{3.3.10}$$

The Sobolev-Slobodeckij space $W^{k,p}(\mathcal{D})$ is defined as

$$W^{k,p}(\mathcal{D}) = \left\{ f \in W^{\lfloor k \rfloor, p}(\mathcal{D}) \mid \sup_{|\alpha| = \lfloor s \rfloor} [D^\alpha f]_{\theta,p,\mathcal{D}} < \infty \right\}$$

associated with the norm

$$\|f\|_{W^{s,p}(\mathcal{D})} = \|f\|_{W^{\lfloor s \rfloor, p}(\mathcal{D})} + \sup_{|\alpha| = \lfloor s \rfloor} [D^\alpha f]_{\theta,p,\mathcal{D}}.$$

The space $W_0^{k,p}(\mathcal{D})$ is defined as the closure of $\mathcal{C}_0^\infty(\mathcal{D})$ with respect to the norm $\|\cdot\|_{W^{s,p}(\mathcal{D})}$. When $p = 2$ and k is a nonnegative integer, we write $H_0^k(\mathcal{D}) = W_0^{k,p}(\mathcal{D})$.

For $k \geq 0$, $W^{-k,p}(\mathcal{D})$ is defined as the dual space of $W_0^{k,q}(\mathcal{D})$, where p' is the conjugate of p ($1/p + 1/q = 1$, $p, q \geq 1$). In particular, $H^{-k}(\mathcal{D})$ is the dual space of $H_0^k(\mathcal{D})$ and for $f \in H^{-1}(\mathcal{D})$,

$$\|f\|_{H^{-1}(\mathcal{D})} = \sup_{v \in H_0^1(\mathcal{D})} \frac{\langle f, v \rangle}{\|v\|_{H^1}(\mathcal{D})}.$$

We will drop the domain $\mathcal{D}$ in norms if no confusion arises. For example, $\|f\|_{H^{-1}(\mathcal{D})}$ will be written as $\|f\|_{H^{-1}}$.

3.3.2 Solutions in different senses

Definition 3.3.3 (Strong solution) *A predictable L^2-valued process $\{u(t)\}_{t \in [0,T]}$ is called a strong solution to* (3.3.23)–(3.3.24) *if*

$$u(t) = u_0 + \int_0^t \mathcal{L}u(t) + f \, ds + \sum_{k \geq 1} \int_0^t (\mathcal{M}_k u(t) + g_k) dW_k(t).$$

As shown in the beginning of this section, the solution in this sense is restrictive for infinite dimensional noises, especially for space-time white noise, even in one dimension. For finite dimensional noises, the sense of solution can be still used.

We now introduce solutions to stochastic partial differential equations in three senses: *variational solution*, *mild solution*, and *Wiener chaos solution.*

Definition 3.3.4 (Variational solution) *A variational solution $u(t,x)$ is a predictable L^2-valued process such that for any $v \in C_0^\infty(\mathcal{D})$ (C^∞ functions with compact support) for each $t \in (0,T]$*

$$(u,v) = (u_0,v) + \int_0^t \int_{\mathcal{D}} \mathcal{L}u, v\,dx + (f,v)\,dt + \int_0^t \sum_{k=1}^\infty (\mathcal{M}_k u + g_k\,dW_k, v), \tag{3.3.11}$$

where $\mathcal{L} - \frac{1}{2}\sum_{k=1}^\infty \mathcal{M}_k\mathcal{M}_k > 0$ and for $\mathcal{L} - \frac{1}{2}\sum_{k=1}^\infty \mathcal{M}_k\mathcal{M}_k = 0$ (fully degenerate)

$$(u,v) = (u_0,v) + \int_0^t \int_{\mathcal{D}} u\mathcal{L}^* v, dx + (f,v)\,dt + \int_0^t \sum_{k=1}^\infty [\mathcal{M}_k u + g_k]\,dW_k, v), \tag{3.3.12}$$

where $\mathcal{L}^*$ is the adjoint operator of $\mathcal{L}$ with respect to the inner product in H.

A variational solution to the stochastic partial differential equation (3.3.1) can be defined as a random process u such that for all $v \in C_0^\infty([0,l])$,

$$\int_0^l u(t,x)v(x)\,dx = \int_0^l u_0(x)v(x)\,dx - \int_0^t \int_0^l u'(x)v'(x)\,dx\,ds + \int_0^t \int_0^L F(t,x)v(x)\,dx\,ds$$

or in an even weaker sense,

$$\int_0^l u(t,x)v(x)\,dx = \int_0^l u_0(x)v(x)\,dx + \int_0^t \int_0^L u(x)v''(x)\,dx\,ds + \int_0^t \int_0^l F(t,x)v(x)\,dx\,ds.$$

Definition 3.3.5 (Mild solution) *When $\mathcal{L}$ is deterministic and time-independent, the mild sense of the solution for Equation* (3.3.23) *is*

$$u = e^{\mathcal{L}t}u_0 + \int_0^t e^{\mathcal{L}(t-s)} f\,ds + \int_0^t e^{\mathcal{L}(t-s)} \sum_{k=1}^\infty [\mathcal{M}_k u + g_k]\,dW_k(s). \tag{3.3.13}$$

Let $G(t, x, y)$ be the Green's function for the linear equation (3.3.1) with vanishing Dirichlet boundary conditions:

$$G(t; x, y) = \sum_{n=-\infty}^{\infty} \Phi(t; x - y - 2nl) + \Phi(t; x + y - 2nl),$$

where $\Gamma(t; z) = \frac{1}{\sqrt{4\pi t}} e^{-z^2/(4t)}$. The solution to (3.3.1) *of mild form* is

$$u(t, x) = \int_0^l G(t; x, y) u_0(y)\, dy + \int_0^t \int_0^l G(t - s; x, y) F(t, y)\, dy\, ds.$$

As $F(t, x)$ is defined by a q-cylindrical Wiener process, denoted as $\dot{W}^Q(t, x)$, then

$$u(t, x) = \int_0^l G(t; x, y) u_0(y)\, dy + \int_0^t \int_0^l G(t - s; x, y)\, dW^Q(s, y).$$

For stochastic elliptic equations in Chapter 10, we use a solution in the mild sense, which we will define similarly.

Now we present a *Wiener chaos solution* to the linear SPDE (3.3.23)–(3.3.24) (see, e.g., [316, 318, 345]). Denote by $\mathcal{J}$ the set of multi-indices $\alpha = (\alpha_{k,l})_{k,l\geq 1}$ of finite length $|\alpha| = \sum_{i,k=1}^{\infty} \alpha_{k,l}$, i.e.,

$$\mathcal{J} = \{\alpha = (\alpha_{k,l},\ k, l \geq 1),\ \ \alpha_{k,l} \in \{0, 1, 2, \ldots\},\ |\alpha| < \infty\}.$$

Here k denotes the number of Wiener processes and l the number of Gaussian random variables approximating each Wiener process as will be shown shortly. We represent the solution of (3.3.23)–(3.3.24) as

$$u(t, x) = \sum_{\alpha \in \mathcal{J}} \frac{1}{\sqrt{\alpha!}} \varphi_\alpha(t, x) \xi_\alpha, \tag{3.3.14}$$

where $\{\xi_\alpha\}$ is a complete orthonormal system (CONS) in $L^2(\Omega, \mathcal{F}_t, P)$ (Cameron-Martin basis), $\alpha! = \prod_{k,l}(\alpha_{k,l}!)$. To obtain the coefficients $\varphi_\alpha(t, x)$, we rewrite the SPDE (3.3.23) in the following form using the Ito-Wick product

$$du(t,x) = [\mathcal{L}u(t,x) + f(x)]\, dt + \sum_{k\geq 1} [\mathcal{M}_k u(t,x) + g_k(x)] \diamond \dot{W}_k\, dt,\ \ (t,x) \in (0,T] \times \mathcal{D},$$

$$u(0, x) = u_0(x),\ \ x \in \mathcal{D}, \tag{3.3.15}$$

where $\dot{W}_k$ is formally the first-order derivative of W_k in time, i.e., $\dot{W})_k \frac{d}{dt} W_k$. Then we substitute the representation (3.3.14) into (3.3.23) and take expectation on both sides of (3.3.23) after multiplying ξ_α on both sides (3.3.23). With the properties of the Ito-Wick product $\xi_\alpha \diamond \xi_\beta = \sqrt{\frac{(\alpha+\beta)!}{\alpha!\beta!}} \xi_{\alpha+\beta}$ and

$\mathbb{E}[\xi_\alpha \xi_\beta] = \delta_{\alpha=\beta}$, we then obtain that φ_α satisfies the following system of equations (the *propagator*):

$$\begin{aligned}
\frac{\partial \varphi_\alpha(s,x)}{\partial s} &= \mathcal{L}\varphi_\alpha(s,x) + f(x)\mathbf{1}_{\{|\alpha|=0\}} \\
&\quad + \sum_{k,l} \alpha_{k,l} m_l(s) \left[\mathcal{M}_k \varphi_{\alpha^-(k,l)}(s,x) + g_k(x)\mathbf{1}_{\{|\alpha|=1\}}\right], \\
& 0 < s \le t, \ x \in \mathcal{D}, \\
\varphi_\alpha(0,x) &= u_0(x)\mathbf{1}_{\{|\alpha|=0\}}, \ x \in \mathcal{D},
\end{aligned} \tag{3.3.16}$$

where $\alpha^-(k,l)$ is the multi-index with components

$$\left(\alpha^-(k,l)\right)_{i,j} = \begin{cases} \max(0, \alpha_{i,j} - 1), & \text{if } i = k \text{ and } j = l, \\ \alpha_{i,j}, & \text{otherwise.} \end{cases} \tag{3.3.17}$$

Here we also use the spectral approximation of Brownian motion (2.2.1).

Remark 3.3.6 *Since the Cameron-Martin basis is complete, see Theorem 2.3.6, a truncation of the WCE (3.3.14) can present a consistent (convergent) numerical methods for SPDEs.*

3.3.3 Solutions to SPDEs in explicit form

The first one is the *stochastic advection-diffusion equation* with periodic boundary condition, written in the Stratonovich form as

$$\begin{aligned}
du(t,x) &= \epsilon u_{xx}(t,x)dt + \sigma u_x(t,x) \circ dW(t), \quad t > 0, \ x \in (0, 2\pi), \\
u(0,x) &= \sin(x),
\end{aligned} \tag{3.3.18}$$

or in the Itô form as

$$du(t,x) = a u_{xx}(t,x)\, dt + \sigma u_x(t,x)\, dW(t), \quad u(0,x) = \sin(x).$$

Here $W(t)$ is a standard one-dimensional Wiener process, $\sigma > 0$, $\epsilon \ge 0$ are constants, and $a = \epsilon + \sigma^2/2$. The solution of (3.3.18) is

$$u(t,x) = e^{-\epsilon t} \sin(x + \sigma W(t)), \tag{3.3.19}$$

and its first and second moments are

$$\mathbb{E}[u(t,x)] = e^{-at}\sin(x), \quad \mathbb{E}[u^2(t,x)] = e^{-2\epsilon t}\left(\frac{1}{2} - \frac{1}{2}e^{-2\sigma^2 t}\cos(2x)\right).$$

We note that for $\epsilon = 0$ the equation (3.3.18) becomes degenerate.

The second model problem is the following Ito *reaction-diffusion equation* with periodic boundary condition:

$$\begin{aligned}
du(t,x) &= a u_{xx}(t,x)\, dt + \sigma u(t,x)\, dW(t), \quad t > 0, \ x \in (0, 2\pi), \\
u(0,x) &= \sin(x),
\end{aligned} \tag{3.3.20}$$

where $\sigma > 0$ and $a \geq 0$ are constants. Its solution is

$$u(t,x) = \exp\left(-(a+\frac{\sigma^2}{2})t + \sigma W(t)\right)\sin(x), \tag{3.3.21}$$

and its first and second moments are

$$\mathbb{E}[u(t,x)] = e^{-at}\sin(x), \quad \mathbb{E}[u^2(t,x)] = \exp\left(-(2a-\sigma^2)t\right)\sin^2(x).$$

3.3.4 Linear stochastic advection-diffusion-reaction equations

Consider the following SPDE written in Itô's form:

$$du(t,x) = [\mathcal{L}u(t,x) + f(x)]\,dt + \sum_{k\geq 1}[\mathcal{M}_k u(t,x) \tag{3.3.22}$$
$$+g_k(x)]\,dW_k(t), \ (t,x) \in (0,T] \times \mathcal{D},$$
$$u(0,x) = u_0(x), \quad x \in \mathcal{D}, \tag{3.3.23}$$

where

$$\mathcal{L}u(t,x) = \sum_{i,j=1}^{d} a_{ij}(x)\,D_i D_j u(t,x) + \sum_{i=1}^{d} b_i(x) D_i u(t,x) + c(x)\,u(t,x),$$
$$\mathcal{M}_k u(t,x) = \sum_{i=1}^{d} b_i^k(x) D_i u(t,x) + h^k(x)\,u(t,x), \tag{3.3.24}$$

and $D_i := \partial_{x_i}$ and $\mathcal{D}$ be an open domain in $\mathbb{R}^d$. We assume that $\mathcal{D}$ is either bounded with a regular boundary or that $\mathcal{D} = \mathbb{R}^d$. In the former case we will consider periodic boundary conditions and in the latter the Cauchy problem. Let $(W(t), \mathcal{F}_t) = (\{W_k(t), k \geq 1\}, \mathcal{F}_t)$ be a system of one-dimensional independent standard Wiener processes defined on a complete probability space $(\Omega, \mathcal{F}, \mathcal{P})$, where $\mathcal{F}_t$, $0 \leq t \leq T$, is a filtration satisfying the usual hypotheses.

Remark 3.3.7 *The problem* (3.3.23) *with* (3.3.24) *can be regarded as a problem driven by a cylindrical Wiener process. Consider a cylindrical Wiener process* $W(t,x) = \sum_{k=1}^{\infty} \lambda_k W_k(t) e_k(x)$, *where* $\sum_{k=1}^{\infty} \lambda_k^2 < \infty$, $\{W_k(t)\}$ *are independent Wiener processes, and* $\{e_k(x)\}_{k=1}^{\infty}$ *is a complete orthonormal system (CONS) in* $L^2(\mathcal{D})$, *see, e.g., [94, 408]. Thus, we can view* (3.3.23)–(3.3.24) *as SPDEs driven by this cylindrical Wiener process when* $\mathcal{M}_k u = e_k(x)\mathcal{M}u$ *and* $\mathcal{M}$ *is first-order or zeroth order differential operator.*

3.3.5 Existence and uniqueness

We assume the *coercivity condition* that there exist a constant $\delta_{\mathcal{L}} > 0$ and a real number $C_{\mathcal{L}}$ such that for any $v \in H^1(\mathcal{D})$,

$$\langle \mathcal{L}v, v\rangle + \frac{1}{2}\sum_{k\geq 1}\|\mathcal{M}_k v\|^2 + \delta_{\mathcal{L}}\|v\|_{H^1}^2 \leq C_{\mathcal{L}}\|v\|^2, \tag{3.3.25}$$

where $\langle \cdot, \cdot \rangle$ is the duality between the Sobolev spaces $H^{-1}(\mathcal{D})$ and $H^1(\mathcal{D})$ associated with the inner-product over $L^2(\mathcal{D})$ and $\|\cdot\|$ is the $L^2(\mathcal{D})$-norm. A necessary condition for (3.3.25) is that the coefficients satisfy

$$\sum_{i,j=1}^{d} \left(2a_{i,j}(x) - \sum_{k \geq 1} \sigma_{i,k}(x)\sigma_{k,j}(x) \right) y_i y_j \geq 2\delta_{\mathcal{L}} |y|^2, \quad x, y \in \mathcal{D}.$$

With these assumptions, we have a unique square-integrable (variational) solution of (3.3.23)–(3.3.24) if we also have the following conditions:

- the coefficients of operators $\mathcal{L}$ and $\mathcal{M}$ in (3.3.24) are uniformly bounded and predictable for every $x \in \mathcal{D}$. The coefficients $a_{i,j}(x)$'s are Lipschitz continuous;
- For $\phi \in H^1(\mathcal{D})$, $\sum_{k \geq 1} \mathbb{E}[\|\mathcal{M}_k \phi(t)\|_{L^2}^2] < \infty$;
- the initial condition $u_0(x) \in \mathbb{L}^2(\Omega; L^2)$ is $\mathcal{F}_0$-measurable;
- $f(t, \omega)$ and $g_k(t, \omega)$ are adapted and $\int_0^T \|f(t)\|_{H^{-1}}^2 \, dt < \infty$, $\sum_{k \geq 1} \int_0^T \|g_k(t)\|_{L^2}^2 \, dt < \infty$.

Then for each $\phi \in H^1$ or a dense subset of H^1 and all $t \in [0, T]$, the adapted process $u(t)$ is a variational solution to (3.3.23)–(3.3.24). With the coercivity condition and $\|\mathcal{L}\phi\|_{H^{-1}} \leq C_0 \|\phi\|_{H^1}$, there exists a unique solution $u \in \mathbb{L}^2(\Omega, C((0, T)); L^2(\mathcal{D})$ and satisfies

$$\mathbb{E}[\sup_{0<t<T} \|u(t)\|_{L^2}^2] + \frac{\delta_L}{2} \mathbb{E}[\int_0^T \|u(t)\|_V^2 \, dt] \leq C\mathbb{E}[\|u_0\|_{L^2}^2]$$
$$+ C\mathbb{E}[\int_0^T \|f(t)\|_{H^{-1}}^2 \, dt] + C \sum_{k \geq 1} \int_0^T \|g_k(t)\|_{L^2}^2 \, dt.$$

Here C depends on C_0, $C_{\mathcal{L}}$, $\delta_{\mathcal{L}}$, and T. see, e.g., [318, 345] for proofs.

3.3.6 Conversion between Ito and Stratonovich formulation

In Stratonovich form, (3.3.23) and (3.3.24) are written as

$$du(t,x) = [\tilde{\mathcal{L}} u(t,x) + f(x)] \, dt + \sum_{k=1}^{q} [\mathcal{M}_k u(t,x) + g_k(x)] \circ \dot{W}_k \, dt, \; (t,x) \in (0,T] \times \mathcal{D},$$
$$u(0,x) = u_0(x), \quad x \in \mathcal{D}, \tag{3.3.26}$$

where $\tilde{\mathcal{L}} u = \mathcal{L} u - \frac{1}{2} \sum_{1 \leq k \leq q} \mathcal{M}_k [\mathcal{M}_k u + g_k]$.

Example 3.3.8 *Consider the following one-dimensional equation for* $(t, x) \in (0, T] \times (0, 2\pi)$*:*

$$du = [(\epsilon + \frac{1}{2}\sigma^2)\partial_x^2 u + \beta \sin(x)\partial_x u] \, dt + \sigma \partial_x u \, dW(t), \tag{3.3.27}$$

where $W(t)$ *is a standard scalar Brownian motion (Wiener process),* $\epsilon > 0$*,* β*,* σ *are constants.*

In the Stratonovich form, Equation (3.3.27) can be written as

$$du = [\epsilon \partial_x^2 u + \beta \sin(x) \partial_x u] \, dt + \sigma \partial_x u \circ dW(t). \tag{3.3.28}$$

The problems (3.3.23) and (3.3.24) are said to have *commutative noises* if

$$\mathcal{M}_k \mathcal{M}_j = \mathcal{M}_j \mathcal{M}_k, \quad 1 \leq k, j \leq q, \tag{3.3.29}$$

and to have noncommutative noises otherwise. When $q = 1$, (3.3.29) is satisfied and thus this is a special case of commutative noises. When $\mathcal{M}_k$ are zeroth-order operators, ($\sigma_{i,k} = 0$), (3.3.29) is satisfied and the problem also has commutative noises. The definition is consistent with that of commutative and noncommutative noises for stochastic ordinary differential equations, see, e.g., [259, 358].

Example 3.3.9 *Consider the following one-dimensional equation for* $(t, x) \in (0, T] \times (0, 2\pi)$*:*

$$\begin{aligned} du = & [(\epsilon + \frac{1}{2}\sigma_1^2 \cos^2(x)) \partial_x^2 u + (\beta \sin(x) - \frac{1}{4}\sigma_1^2 \sin(2x)) \partial_x u] \, dt \\ & + \sigma_1 \cos(x) \partial_x u \, dW_1(t) + \sigma_2 u \, dW_2(t), \end{aligned} \tag{3.3.30}$$

where $(W_1(t), W_2(t))$ *is a standard two-dimensional Wiener process,* $\epsilon > 0$*,* β*,* σ_1*,* σ_2 *are constants.*

In the Stratonovich form, Equation (3.3.30) is written as

$$du = [\epsilon \partial_x^2 u + \beta \sin(x) \partial_x u] \, dt + \sigma_1 \cos(x) \partial_x u \circ dW_1(t) + \sigma_2 u \circ dW_2(t). \tag{3.3.31}$$

The problem has commutative noises (3.3.29):

$$\sigma_1 \cos(x) \partial_x \sigma_2 \text{Id} u = \sigma_2 \text{Id} \sigma_1 \cos(x) \partial_x = \sigma_1 \sigma_2 \cos(x) \partial_x.$$

Here Id is the identity operator.

Example 3.3.10 *Consider the following one-dimensional equation for* $(t, x) \in (0, T] \times (0, 2\pi)$*:*

$$\begin{aligned} du = & [(\epsilon + \frac{1}{2}\sigma_1^2) \partial_x^2 u + \beta \sin(x) \partial_x u + \frac{1}{2}\sigma_2^2 \cos^2(x) u] \, dt \\ & + \sigma_1 \partial_x u \, dW_1(t) + \sigma_2 \cos(x) u \, dW_2(t), \end{aligned} \tag{3.3.32}$$

where $(W_1(t), W_2(t))$ *is a standard Wiener process,* $\epsilon > 0$*,* β*,* σ_1*,* σ_2 *are constants.*

In the Stratonovich form, Equation (3.3.32) is written as

$$du = [\epsilon \partial_x^2 u + \beta \sin(x) \partial_x u] \, dt + \sigma_1 \partial_x u \circ dW_1(t) + \sigma_2 \cos(x) u \circ dW_2(t). \tag{3.3.33}$$

The problem has noncommutative noises as the coefficients do not satisfy (3.3.29).

3.4 Numerical methods for SPDEs

In this section, we briefly review numerical methods for SPDEs and broadly classify the numerical methods in literature into three categories:

- *Direct semi-discretization methods.* In this category, we usually discretize the underlying SPDEs in time and/or in space, applying classical techniques from time-discretization methods of stochastic ordinary differential equations (SODEs) and/or from spatial discretization methods of partial differential equations (PDEs).
- *Wong-Zakai approximation.* In this category, we first discretize the space-time noise before any discretization in time and space and thus we need further spatial-temporal discretizations.
- *Preprocessing methods.* In this category, we first transform the underlying SPDE into some equivalent form before we discretize the SPDEs.

We start by considering the following SPDE over the physical domain $\mathcal{D} \subseteq \mathbb{R}^d$,

$$dX = [\mathcal{A}X + f(X)]\,dt + g(X)\,dW^Q, \tag{3.4.1}$$

where the Q-Wiener process W^Q is defined in (1.2.3). The physical space is one-dimensional, i.e., $d = 1$, unless otherwise stated. When $\mathcal{D}$ is bounded, we consider periodic boundary conditions (with further requirements on the domain) or Dirichlet boundary conditions.

The leading operator $\mathcal{A}$ can be second-order or fourth-order differential operators, which are positive definite. The nonlinear functions f, g are usually Lipschitz continuous. The problem (3.4.1) is endowed either with only initial conditions in the whole space ($\mathcal{D} = \mathbb{R}^d$) or with initial and boundary conditions in a bounded domain ($\mathcal{D} \subsetneq \mathbb{R}^d$).

Let us introduce the stability and convergence of numerical schemes. We denote $\delta t_k = (t_{k+1} - t_k)$ ($k = 1, 2, \ldots, K$, $\sum_{k=1}^{K} \delta t_k = T$) are the time step sizes. Sometimes we simply use the time step size δt when all δt_k's are equal. We also denote by $N > 0$ the number of orthogonal modes in spectral methods or discretization steps in space ($Nh = |\mathcal{D}|$, $|\mathcal{D}|$ is the length of the interval $\mathcal{D} \subset \mathbb{R}^d$ when $d = 1$) for finite difference methods or finite element methods. We denote a numerical solution to (3.4.1) by $X_{N,K}$.

Let H be a separable Hilbert space (Hilbert space with a countable basis) with corresponding norm $\|\cdot\|_H$. We usually take $H = L^2(\mathcal{D})$,

Definition 3.4.1 (Convergence) *Assume that $X_{N,K}$ is a numerical solution to* (3.4.1) *and $X(x,T)$ is a solution to* (3.4.1) *at time T.*

- ***Mean-square convergence (Strong convergence).*** *If there exists a constant C independent of h and δ such that*

$$\mathbb{E}[\|X_{N,K} - X(\cdot,T)\|^2_{L^2(\mathcal{D})}] \leq C\big(h^{2p_1} + (\delta t)^{2p_2}\big), \quad p_1, p_2 > 0, \tag{3.4.2}$$

then the numerical solution is convergent in the mean-square sense to the solution to (3.4.1)*. The mean-square convergence order in time is* p_1 *and the convergence order in physical space is* p_2*.*

- ***Almost sure convergence (Pathwise convergence).*** *If there is a finite random variable* $C(\omega) > 0$ *independent of* h *and* δ *such that*

$$\|X_{N,K} - X(\cdot, T)\|_{L^2(\mathcal{D})} \leq C(\omega)\big(h^{p_1} + (\delta t)^{p_2}\big), \tag{3.4.3}$$

  then the numerical solution is convergent almost surely to the solution to (3.4.1)*.*

- ***Weak convergence.*** *If there exists a constant* C *independent of* h *and* δ *such that*

$$\|\mathbb{E}[\phi(X_{N,K})] - \mathbb{E}[\phi(X(\cdot, T))]\|_{L^2(\mathcal{D})} \leq C\big(h^{p_1} + (\delta t)^{p_2}\big), \tag{3.4.4}$$

  then the numerical solution is weakly convergent to the solution to (3.4.1)*.*

We say the convergence order (in mean-square sense, almost sure sense or weak sense) in time is p_1 *and the convergence order (in mean-square sense, almost sure sense or weak sense) in physical space is* p_2*.*

Remark 3.4.2 *Here we do not specify what the sense of solutions is. The definition is universal for strong solutions, variational solutions, and mild solutions for SPDEs.*

We do not consider the effect of truncation of infinite dimensional process W^Q*. In general, the convergence of the truncated finite dimensional process to* W^Q *depends on the decay rate of* q_i (1.2.3) *as well as on the smoothing effect of the inverse of the leading operator* $\mathcal{A}$*.*

The following zero-stability is concerned with whether the numerical solution can be controlled by initial values.

Definition 3.4.3 (Zero-Stability) *Assume that* $X_{N,K}$ *is a numerical solution to* (3.4.1) *and* $X(x, T)$ *is a solution to* (3.4.1) *at time* T*.*

- ***Mean-square zero-stability.*** *If there exists a constant* C *independent of* N *and* K *such that*

$$\mathbb{E}[\|X_{N,K}\|^2] \leq C \max_{0 \leq k \leq m} \mathbb{E}[\|X_{N,k}\|^p], \text{ for some nonnegative integers } m, p, \tag{3.4.5}$$

  then the numerical solution is stable in the mean-square sense.

- ***Almost sure zero-stability.*** *If there is a finite random variable* $C(\omega) > 0$ *independent of* N *and* K *such that*

$$\|X_{N,K}\| \leq C(\omega) \max_{0 \leq k \leq m} \|X_{N,k}\|, \text{ for some nonnegative integer } m, \tag{3.4.6}$$

  then the numerical solution is stable almost surely.

Remark 3.4.4 *In most cases, we use $m = 1$, which is appropriate for one-step numerical methods. The case $m \geq 2$ is for m-step numerical methods, where the first m-steps cannot be obtained from the m-step numerical methods but can be obtained from some other numerical methods (usually one-step methods) with smaller time step sizes.*

We can also define linear stability, which is concerned with asymptotic behavior of numerical solutions with K goes to infinity while the time step size δt is fixed. The linear stability for evolutionary SPDEs is a straightforward extension of linear stability for SODEs, which can also be defined in the mean-square, almost sure or weak sense. The concern of linear stability often leads to the Courant-Friedrichs-Lewy condition (often abbreviated as CFL condition): the mesh size in time has to be proportional to a certain power (depending on the order of leading operator $\mathcal{A}$) of the mesh size in space. This is similar to linear stability of PDEs. An example of the linear stability for a linear stochastic advection-diffusion equation is presented in Chapter 3.4.5.

3.4.1 Direct semi-discretization methods for parabolic SPDEs

The *time-discretization* methods for (3.4.1) can be seen as a straightforward application of numerical methods for SODEs, where increments of Brownian motions are used. After performing a truncation in physical space, we will obtain a system of finite dimensional SODEs, and subsequently we can apply standard numerical methods for SODEs, e.g., those from [259, 354, 358]. It is very convenient to simply extend the methods of numerical PDEs and SODEs to solve SPDEs. One can select the optimal numerical methods for the underlying SPDEs with carefully analyzing the characteristics of related PDEs and SODEs.

However, it is not possible to derive high-order schemes with direct time discretization methods as the solutions to SPDEs have very low regularity. For example, for the heat equation with additive space-time white noise in one dimension (3.3.1), see Exercise 3.6.11, the sample paths of the solution is Hölder continuous with exponent $1/4 - \epsilon$ ($\epsilon > 0$ is arbitrarily small) in time and is Hölder continuous with exponent $1/2 - \epsilon$ in space.

Second-order equations

For finite dimensional noise, we can directly apply those time-discretization methods for SODEs to SPDEs as solutions are usually smooth in space. Ref. [167] considered Euler and other explicit schemes for a scalar Wiener process and Ref. [261] further considered linear-implicit schemes in time under the same problem setting. Specifically, both papers considered (3.4.1) with $W^Q = W(t)$ being one-dimensional Brownian motion. After discretizing in physical space, we obtain a system of SODEs:

$$dX_N = [\mathcal{A}_N X_N + f(X_N)]\, dt + g(X_N)\, dW, \tag{3.4.7}$$

where X_N can be derived from finite difference schemes, finite element schemes, spectral Galerkin/collocation schemes, finite volume schemes, and many other schemes for discretization in space. The explicit Euler scheme for (3.4.7) [167] is

$$X_N^{k+1} = X_N^k + [\mathcal{A}_N^k X_N^k + f(X_N^k)]\delta t_k + g(X_N^k)\Delta W_k, \tag{3.4.8}$$

where δt_k's are the time step sizes and X_N^k is an approximation of X_N at $t_k = \sum_{i=1}^k \delta t_k$. In [167], the first-order Milstein scheme was also applied. However, for explicit schemes, the CFL condition requires small time step sizes even when the mesh size in physical space is relatively large. When $\mathcal{A} = \Delta$, we need $\delta t N^2$ less than some constant which depends on the size of the domain. To avoid such a severe restriction, Ref. [261] applied the drift-implicit (linear-implicit) Euler scheme

$$X_N^{k+1} = X_N^k + [\mathcal{A}_N^{k+1} X_N^{k+1} + f(X_N^k)]\delta t_k + g(X_N^k)\Delta W_k, \tag{3.4.9}$$

which does not require a severe CFL condition when g and f are Lipschitz continuous. However, the same conclusion is not true when the coefficient of the noise involves first-order derivatives of the solution X. Recall that we do not have such an issue for deterministic PDEs [178]. See, e.g., Chapter 3.4.5 for such an example.

In a similar setting, Ref. [139] proposed the Milstein scheme for the Kolmogorov Petrovskii-Piskunov (KPP) equation with multiplicative noise using a finite difference scheme in space. See [158, 366, 404, 422, 426] for more numerical results.

For infinite dimensional noise but with fast decaying q_i in (1.2.3), Hausenblas (2003) [210] considered the mean-square convergence of linear-implicit and explicit Euler scheme, and Crank-Nicolson scheme in time for (3.4.1) with certain smooth f and g and proved half-order convergence for these schemes. Here, the Crank-Nicolson scheme means linear-implicit Crank-Nicolson, where the nonlinear terms and coefficients of noise are treated explicitly as in the linear-implicit Euler scheme (3.4.9):

$$X_N^{k+1} = X_N^k + [\mathcal{A}_N^{k+1/2} X_N^{k+1/2} + f(X_N^k)]\delta t + g(X_N^k)\Delta W_k^N(x), \tag{3.4.10}$$

where $\Delta W_k^N(x)$ is a discretization of $dW^Q(t,x)$ and

$$\mathcal{A}_N^{k+1/2} X_N^{k+1/2} = \frac{\mathcal{A}_N^{k+1} X_N^{k+1} + \mathcal{A}_N^k X_N^k}{2}.$$

Here $\Delta W_k^N(x)$ can be $\frac{W^Q(t_{k+1},x_j) - W^Q(t_k,x_j)}{\delta t}$ in a finite difference scheme or $\frac{P_h W^Q(t_{k+1},x) - P_h W^Q(t_k,x)}{\delta t}$ in a finite element scheme where P_h is a L^2-projection into a finite dimensional solution space.

The author remarked that for the Crank-Nicolson scheme the convergence order can be improved to one for linear equations with additive noise, as in the case of SODEs. Also, Hausenblas (2003) [211] proved the first-order weak convergence of these numerical schemes for (3.4.1) with additive noise. Millet & Morien [350] considered (3.4.1) with space-time noise, where q_i and e_i are eigenvalues and eigenfunctions of a specific isotropic kernel. Hausenblas (2002) [209] considered a slightly different equation

$$dX = [\mathcal{A}X + f(t, X)]\, dt + \sum_j g_i(t, X)\, dW_j(t), \tag{3.4.11}$$

where $\sum_j \|g_i(t, \cdot)\|^2_{H^2(\mathcal{D})} < \infty$ and some boundedness of f and g is imposed. Half-order convergence in time is proved for the linear-implicit and explicit Euler schemes and the Crank-Nicolson scheme.

However, if **space-time white noise** is considered ($q_i = 1$ in (1.2.3)), the regularity in time is shown to be less than 1/4, see Exercise 3.6.11 for a linear equation. Thus, the optimal order of convergence in time is $1/4 - \epsilon$ if only increments of Brownian motion (with equispaced time steps) are used, see, e.g., [6, 95] for the case of linear equations.

Gyöngy and Nualart introduced an implicit numerical scheme in time for the SPDE (3.4.1) with additive noise and proved convergence in probability without order in [196] and for (3.4.1) with mean-square order $1/8 - \epsilon$ in time [197]. Gyöngy [186, 188] also applied finite differences in space to the SPDE (3.4.1) and then used several temporal implicit and explicit schemes, including the linear-implicit Euler scheme. The author showed that these schemes converge with order $1/2 - \epsilon$ in space and with order $1/4 - \epsilon$ in time for multiplicative noise with Lipschitz nonlinear terms similar to the linear equations in [6, 95]. Refs. [371, 372] proposed an implicit Euler scheme on nonuniform time grid for (3.4.1) with $f = 0$ to reduce the computational cost; the upper bound estimate was presented in [371] while the lower bound was presented in [372], of the mean-square errors in terms of computational cost.

As we mentioned before, *the solution to* (3.4.1) *is of low regularity and thus it is not possible to derive high-order schemes with direct time discretization methods.* See, e.g., [464] for discussion on first-order schemes (Milstein type schemes) for (3.4.1) and also [249] for a review of numerical approximation of (3.4.1) along this line.

For *spatial semi-discretization* methods for solving SPDEs (including but not limited to (3.4.1)), see finite difference methods, see, e.g., [6, 313, 420, 495]; finite element methods, see, e.g., [6, 22, 152, 464, 469, 491, 494]; finite volume methods for hyperbolic problems, see, e.g., [274, 367]; spectral methods, see, e.g., [65, 78, 242, 302]). See also [149, 193, 206, 207] for acceleration schemes in space using Richardson's extrapolation method. As spatial discretizations are classical topics in numerical methods for PDEs, we refer the readers to standard textbooks, such as [178] for finite difference schemes, [216] for spectral methods, and [296] for finite volume methods. In most cases, especially for linear problems (or essential linear problems, e.g., linear leading operator with Lipschitz nonlinearity), one can simply apply the classical spatial discretization techniques. One caveat is for solutions of extremely low regularity, see, e.g., numerical methods for Burgers equations with additive space-time white noise in Chapter 3.4.4, where different spatial discretization methods indeed make significant difference.

Fourth-order equations

Now we consider fourth-order equations, i.e., $\mathcal{A}$ is a fourth-order differential operator, which have been investigated in [265–267, 271, 291], etc. As the kernels associated with fourth-order operators can have more smoothing effects than those associated with second-order differential operator, we can expect better convergence in space and also in time.

Ref. [265] considered fully discrete finite element approximations for a fourth-order linear stochastic parabolic equation with additive space-time white noise in one space dimension where strong convergence with order $3/8$ in time and $3/2-\epsilon$ in space was proved. Ref. [271] proved the convergence of finite element approximation of the nonlinear stochastic Cahn-Hilliard-Cook equation by additive space-time color noise

$$dX = \Delta^2 X + \Delta f(X) + dW^Q. \tag{3.4.12}$$

Ref. [71] presented some numerical results of a semi-implicit backward differentiation formula in time for nonlinear Cahn-Hilliard equation while no convergence analysis is given. For the linearized Cahn-Hilliard-Cook equation ($f = 0$) with additive space-time color noise, Ref. [291] applied a standard finite element method and an implicit Euler scheme in time and obtained quasi-optimal convergence order in space. Kossioris and Zouris considered an implicit Euler scheme in time and finite elements in space for the linear Cahn-Hilliard equation with additive space-time white noise in [267] and the same equation but with even rougher noise which is the fist-order spatial derivative of the space-time white noise in [266]. In [267], they proved that the strong convergence order is $(4-d)/8$ in time and $(4-d)/2-\epsilon$ in space for $d = 2, 3$.

3.4.2 Wong-Zakai approximation for parabolic SPDEs

In this approach, we first truncate the Brownian motion with a smooth process of bounded variation yielding a PDE with finite dimensional noise. Thus, after truncating the Brownian motion, we have to discretize a deterministic PDE both in time and in space to obtain fully discrete schemes.

The most popular approximation Brownian motion in this approach is piecewise linear approximation of Brownian motion (2.2.4), see, e.g., [481]. Piecewise linear approximation for SPDEs has been well studied in theory, see, e.g., [142, 180, 227, 442, 454, 455, 457] (for mean-square convergence), [46, 181, 200, 201] (for pathwise convergence), [18, 30, 72, 79, 182–185, 199, 351, 456] (for support theorem, the relation between the support of distribution of the solution and that of its Wong-Zakai approximation). For mean-square convergence of Wong-Zakai approximation for (3.4.13) with $\mathcal{M}_k$ having no differential operator, Ref. [227] proved a half-order convergence, see also [53, 54]. For pathwise convergence, Ref. [200] proved a $1/4-\epsilon$-order convergence and Ref. [201] proved a $1/2-\epsilon$-order convergence when $\mathcal{M}_k$ is a first-order differential operator.

All the aforementioned papers were on the convergence of the Wong-Zakai approximation itself, i.e., without any further discretization of the resulting PDEs. *Numerical solutions of SPDEs based on the Wong-Zakai approximation have not yet been well explored for SPDEs.* Even for SODEs, Ref. [307] seems to be the first attempt to obtain numerical solutions from Wong-Zakai approximation, where the authors considered a stiff ODE solver instead of presenting new discretization schemes.

In this book, we will derive fully discrete schemes based on Wong-Zakai approximations and show the relationships between the derived schemes and the classical schemes (e.g., those in [259, 354, 358]); see Chapter 4 for details.

3.4.3 Preprocessing methods for parabolic SPDEs

In this type of methods, the underlying equation is first transformed into an equivalent form, which may bring some benefits in computation, and then is dealt with time discretization techniques. For example, splitting techniques split the underlying equation into stochastic part and deterministic part and save computational cost if either part can be efficiently solved either numerically or even analytically. In the splitting methods, we also have the freedom to use different schemes for different parts.

We will only review two methods in this class: splitting techniques and exponential integrator methods. In addition to these two methods, there are other preprocessing methods such as methods of averaging-over-characteristics, e.g., [361, 396, 428]; particle methods, e.g., [87–91, 285]; algebraic method, e.g., [405]; filtering on space-time noise [310]; etc.

Splitting methods

Splitting methods are also known as fractional step methods, see, e.g., [162], and sometimes as predictor-corrector methods, see, e.g., [131]. They have been widely used for their computational convenience, see, e.g., [33, 85, 86, 189, 191, 192, 244, 293, 304, 305]. Typically, the splitting is formulated by the following Lie-Trotter splitting, which splits the underlying problem, say (3.4.13), into two parts: 'stochastic part' (3.4.14a) and 'deterministic part' (3.4.14b). Consider the following Cauchy problem (see, e.g., [131, 190, 191])

$$du(t,x) = \mathcal{L}u(t,x)\,dt + \sum_{k=1}^{d_1} \mathcal{M}_k u(t,x) \circ dW_k, \quad (t,x) \in (0,T] \times \mathcal{D}, \quad (3.4.13)$$

where $\mathcal{L}$ is linear second-order differential operator, $\mathcal{M}_k$ is linear differential operator up to first order, and $\mathcal{D}$ is the whole space $\mathbb{R}^d$. The typical Lie-Trotter splitting scheme for (3.4.13) reads, over the time interval $(t_n, t_{n+1}]$, in integral form

$$\tilde{u}_n(t,x) = u_n(t_n,x) + \int_{t_n}^{t} \sum_{k=1}^{d_1} \mathcal{M}_k \tilde{u}_n(s,x) \circ dW_k(s), \quad t \in (t_n, t_{n+1}], \tag{3.4.14a}$$

$$u_n(t,x) = \tilde{u}_n(t_{n+1}) + \int_{t_n}^{t} \mathcal{L} u_n(s,x)\, ds, \quad t \in (t_n, t_{n+1}]. \tag{3.4.14b}$$

When $\mathcal{M}_k$ is a zeroth order differential operator, Ref. [433] presented results for pathwise convergence with half-order in time under the L^2-norm in space when $d_1 = 1$. Under similar settings as in [433], Ref. [244] proved that a normalization of numerical density in the Zakai equation in a splitting scheme is equivalent to solving the Kushner equation (nonlinear SPDE for the normalized density, see, e.g., [286]) by a similar splitting scheme (first order in the mean-square sense). When $\mathcal{M}_k$ is a first-order differential operator, Ref. [131] proved half-order mean-square convergence in time under the L^2-norm in space. Gyöngy and Krylov managed to provide the first-order mean-square convergence in time under higher-order Sobolev-Hilbert norms [191], and under even stronger norm in space [190].

Other than finite dimensional noise, Refs. [32, 33] considered semilinear parabolic equations (3.4.1) with multiplicative space-time color noises. With the Lie-Trotter splitting, they established strong convergence of the splitting scheme and proved half-order mean-square convergence in time. Cox and van Neerven [85] obtained mean-square and pathwise convergence order of Lie-Trotter splitting methods for Cauchy problems of linear stochastic parabolic equations with additive space-time noise. Other than the problems (3.4.1) and (3.4.13), the Lie-Trotter splitting techniques have been applied to different problems, such as stochastic hyperbolic equations (e.g., [7, 27, 412]), rough partial differential equations (e.g., [138]), stochastic Schrödinger equation (e.g., [47, 168, 304, 305, 338]), etc.

Integrating factor (exponential integrator) techniques

In this approach, we first write the underlying SPDE in mild form (integration-factor) and then combine different time-discretization methods to derive fully discrete schemes. It was first proposed in [309, 369], under the name of *exponential Euler scheme* and was further developed to derive higher-order scheme, see, e.g., [29, 246–250, 252].

In this approach, it is possible to derive high-order schemes in the strong sense since we may incorporate the dynamics of the underlying problems as shown for ODEs with smooth random inputs in [257]. By formulating Equation (3.4.1) with additive noise in mild form, we have

$$X(t) = e^{\mathcal{A}t}X_0 + \int_0^t e^{\mathcal{A}(t-s)} f(X(s))\,ds + \int_0^t e^{\mathcal{A}(t-s)}\,dW^Q(s), \qquad (3.4.15)$$

then we can derive an exponential Euler scheme [309, 369]:

$$X_{k+1} = e^{\mathcal{A}h}[X_k + hf(X_k) + W^Q(t_{k+1}) - W^Q(t_k)], \qquad (3.4.16)$$

or as in [250, 369]

$$X_{k+1} = e^{\mathcal{A}h}X_k + \mathcal{A}^{-1}(e^{\mathcal{A}h} - I)f(X_k) + \int_{t_k}^{t_{k+1}} e^{\mathcal{A}(t_{k+1}-s)}\,dW^Q(s), \; (3.4.17)$$

where $t_k = kh$, $k = 0, \cdots, N$, $Nh = T$.

In certain cases, the total computational cost for the exponential Euler scheme can be reduced when $\eta_k = \displaystyle\int_{t_k}^{t_{k+1}} e^{\mathcal{A}(t_{k+1}-s)}\,dW^Q(s)$ is simulated as a whole instead of using increments of Brownian motion. For example, when $\mathcal{A}e_i = -\lambda_i e_i$, we observe that η_k solves the following equation

$$Y = \sum_{i=1}^{\infty} \int_{t_k}^{t_{k+1}} \mathcal{A}Y\,ds + \sum_{i=1}^{\infty} \int_{t_k}^{t_{k+1}} \sqrt{q_i}e_i\,dW_i(s), \qquad (3.4.18)$$

and thus η_k can be represented by

$$\begin{aligned} \eta_k &= \sum_{i=1}^{\infty} \sqrt{\gamma_i}e_i(x)\xi_{k,i}, \quad \xi_{k,i} = \frac{1}{\sqrt{\gamma_i}} \int_{t_k}^{t_{k+1}} e^{\lambda_i(t_{k+1}-s)}\,dW_i(s), \\ \gamma_i &= \frac{q_i}{2\lambda_i}(1 - \exp(2\lambda_i h)). \end{aligned} \qquad (3.4.19)$$

In this way, we incorporate the interaction between the dynamics and the noise, and thus we can have first-order mean-square convergence [249, 250]. See [248, 258, 311, 362] for further discussion on additive noise. For multiplicative noise, a first-order scheme (Milstein scheme) has been derived under this approach [252], where commutative conditions on diffusion coefficients for equations with infinite dimension noises were identified and a one-and-a-half order scheme in the mean-square sense has been derived in [29]. See also [4, 21, 23, 275, 312, 473] for further discussion on exponential integration schemes for SPDEs with multiplicative noises.

3.4.4 What could go wrong? Examples of stochastic Burgers and Navier-Stokes equations

As a special class of parabolic SPDEs, stochastic Burgers and Navier-Stokes equations require more attention for their strong interactions between the strong nonlinearity and the noises. Similar to linear heat equation with additive noise, the convergence for time-discretization of one-dimensional Burgers equations is no more than $1/4$, see [400] for multiplicative space-time noise with convergence in probability, and [38] for additive space-time noise with pathwise convergence. The convergence in space is less than $1/4$, see [5] for additive space-time white noise with pathwise convergence, and [37] for additive space-time color noise with pathwise convergence.

Because of the strong nonlinearity, the discretization in space and in time may cause some effects, such as "a spatial version of the Ito-Stratonovich correction" [203, 205]. Hairer et al considered finite difference schemes for the Burgers equation with additive space-time noise in [205]:

$$\partial_t u = \nu \partial_x^2 u + (\nabla G(u))\partial_x u + \sigma \dot{W}^Q, \quad x \in [0, 2\pi]. \tag{3.4.20}$$

If we only consider the discretization of the first-order differential operator, e.g.,

$$\partial_t u^\varepsilon = \nu \partial_x^2 u^\varepsilon + (\nabla G(u^\varepsilon))\partial_x u^\varepsilon + \sigma \dot{W}^Q, \quad \partial_x u^\varepsilon =: \frac{u(x+a\varepsilon) - u(x-b\varepsilon)}{(a+b)\varepsilon}, \; a, b > 0, \tag{3.4.21}$$

then it can be proved that this equation converges to (see [203])

$$\partial_t v = \nu \partial_x^2 v + (\nabla G(v))\partial_x v - \frac{\sigma^2}{4\nu}\frac{a-b}{a+b}\Delta G(v) + \sigma \dot{W}^Q, \quad x \in [0, 2\pi], \tag{3.4.22}$$

if $\dot{W}^Q$ is space-time white noise; and no correction term appears if $\dot{W}^Q$ is more regular than space-time white noise, e.g., white in time but correlated in space. Effects of some other standard discretizations in space, e.g., Galerkin methods, and fully discretizations were also discussed in [203].

Now we consider the stochastic incompressible Navier-Stokes (1.2.4). When the noise is color noise in space: $\mathbb{E}[W^Q(x,t)W^Q(y,s)] = q(x,y)\min(s,t)$ and $q(x,x)$ is square-integrable over the physical domain, Ref. [44] showed the existence and strong convergence of the solutions for the fully discrete schemes in two-dimensional case. Ref. [69] considered three semi-implicit Euler schemes in time and standard finite elements methods for the two-dimensional (1.2.4) with periodic boundary conditions. They presented the solution convergence in probability with order $1/4$ in time similar to the one-dimensional stochastic Burgers equation with additive noise. They also showed that for the corresponding Stokes problem, the fully discrete scheme converges in the strong sense with order half in time and order one in physical space.

For (1.2.4) in the bounded domain with Dirichlet boundary condition, Ref. [502] considered the backward Euler scheme and proved half-order strong convergence when the multiplicative noise is space-time color noise. Ref. [493] considered an implicit-explicit scheme and proved a convergence order depending on the regularity index of initial condition. Ref. [120] considered finite elements methods and a semi-implicit Euler for stochastic Navier-Stokes equation (1.2.4) and Ref. [121] considered similar fully discrete schemes for stochastic Navier-Stokes introduced in [348]. Ref. [492] provided *a posteriori* error estimates for stochastic Navier-Stokes equation. See [41] (recursive approximation), [128] (implicit scheme), [169] (Wong-Zakai approximation), [414, 496] (Galerkin approximation), [122] (Wiener chaos expansion) for more discussion on numerical methods and, e.g., [108] for existence and uniqueness of (1.2.4). See also [175] for strong convergence of Fourier Galerkin methods for the hyperviscous Burgers equation and some numerical results for stochastic Burgers equation equipped with the Wick product [445].

Many other evolution equations have also been explored, such as *stochastic KdV equations* (see, e.g., [102, 105, 106, 110, 111, 215], *Ginzburg-Landau equation* (see, e.g., [302]), *stochastic Schrödinger equations* (see, e.g., [26, 99–101, 103, 104, 369]), *stochastic age-dependent population* (see, e.g., [214]), etc. For steady stochastic partial differential equations, especially for stochastic elliptic equation, see, e.g., [6, 12, 34, 64, 118, 140, 194, 469]. See further discussion in Chapter 10.

3.4.5 Stability and convergence of existing numerical methods

There are various aspects to be considered for numerical methods for SPDEs, e.g., the sense of existence of solutions, the sense of convergence, the sense of stability, etc. Here the existence of solutions and numerical solutions to SPDEs are usually interpreted as mild solutions or as variational solutions. We focus on weak convergence and pathwise convergence in this subsection. For strong convergence, we refer to [276] for an optimal convergence order of finite element methods and linear-implicit Euler scheme in time for (3.4.1); see also the aforementioned papers for strong convergence in different problem settings.

Weak convergence

Similar to the weak convergence of numerical methods for SODEs, the main tool for the weak convergence is the *Kolmogorov equation* associated with the functional and the underlying SPDE [92, 94]. For linear equations, the Kolmogorov equation for SPDEs is sufficient to obtain optimal weak convergence, see, e.g., [112, 147, 421]. Ref. [421] considered weak convergence of the θ-method in time and spectral method in physical space for the heat equation with additive space-time noise, and showed that the weak convergence order is twice that of strong convergence for a finite dimensional functional. Ref. [147] obtained a similar conclusion for more general functionals, the restriction on which was further removed in [112]. More recently, there have been more works following this approach [269, 270, 299, 402] for linear equations. For the linear Cahn-Hilliard equation with additive noise, Ref. [269] obtained the weak error for the semidiscrete schemes by linear finite elements with order $h^{2\beta}|\log(h)|)$, where h^β is the strong convergence order and β is determined by q_i's and the smoothness of the initial condition. Ref. [270] provided weak convergence order for the same problem but with further time discretization and proved that the weak convergence order is twice the strong convergence order.

For nonlinear equations, *Malliavin calculus* for SPDEs has also been used for optimal weak convergence, see, e.g., [109, 211, 475]. Ref. [211] applied Malliavin calculus to a parabolic SPDE to obtain the weak convergence the of linear-implicit Euler and Crank-Nicolson schemes in time for additive noise, where the first-order weak convergence (with certain condition on the functional) is obtained. Ref. [213] showed that the order of weak convergence of the leap-frog scheme, both in space and time, is twice that of strong convergence for wave equation with additive noise as shown for heat equations, see, e.g., [109, 112, 147]. Ref. [109] established weak convergence order for the semilinear heat equation with multiplicative space-time noise and showed that the weak convergence order is twice the strong convergence order in time. Ref. [475] obtained weak convergence order of the linear-implicit Euler scheme in time for (3.4.1) with additive noise and obtained similar conclusions. For exponential Euler schemes for SODEs, it was proved that the weak convergence order is one (see, e.g., [369]), which is the same as the mean-square convergence order.

For weak convergence of numerical methods for elliptic equations, we can use multivariate calculus to compute the derivatives with respect to (random) parameters and Taylor's expansion, see, e.g., [73, 74] and also Chapter 10.

Pathwise convergence

There are two approaches to obtain pathwise convergence. The first is via mean-square convergence. By the Borel-Cantelli lemma (see, e.g., [186]), it can be shown that pathwise convergence order is the same as mean-square convergence order (up to an arbitrarily small constant $\epsilon > 0$). For example, Ref. [95] first deduced a pathwise convergence on schemes from the mean-square convergence order established in [188]. Refs. [21, 23, 86, 289, 290] first obtained the mean-square convergence order and then documented the pathwise convergence. The second approach is without knowing the mean-square convergence. In [433], the authors required pathwise boundedness (uniformly boundedness in time step sizes) to have a pathwise convergence with order $1/2 - \epsilon$. In [248], it was shown that it is crucial to establish the pathwise regularity of the solution to obtain pathwise convergence order.

Finally, we note that there are some other senses of convergence, see, e.g., [19] for convergence in probability using several approximations of white noise.

Stability

Here we will not review the stability of numerical methods for SPDEs as they are usually accompanied by a convergence study. We refer to [464] for the stability of the fully discrete schemes for (3.4.1). We also refer to the following two papers for some general framework on stability and convergence. Ref. [288] proposed a version of Lax equivalence theorem for (3.4.1) with additive and multiplicative noise while W^Q is replaced with a càdlàg (right continuous with a left limit) square-integrable martingale. Ref. [275] suggested a general framework for Galerkin methods for (3.4.1) and applied them to Milstein schemes.

It is known that the mean-square stability region of a numerical scheme in time for SPDEs with multiplicative noise is smaller than that of the scheme for PDEs, e.g., Crank-Nicolson scheme for (3.4.1) with multiplicative noise [464], or alternating direction explicit scheme for heat equation with multiplicative noise [426].

To illustrate different stability requirements of numerical PDEs and SPDEs, we summarize a recent work on the mean-square stability of Milstein scheme for one-dimensional advection-diffusion equation with multiplicative scalar noise [158, 404]. Ref. [404] analyzed the linear stability (proposed in [49]) of the first-order σ-θ-scheme and Ref. [219] for SODEs. For a specific equation of the form (3.4.13) with periodic boundary conditions:

$$dv = -\mu \partial_x v dt + \frac{1}{2}\partial_x^2 v dt - \sqrt{\rho}\partial_x v dW_t, \quad 0 \leq \rho < 1, x \in (0,1) \tag{3.4.23}$$

the σ-θ scheme reads, with time step size δt and space step size δx,

$$\begin{aligned} V_{n+1} = V_n &- \frac{\theta}{2}(\frac{\delta t}{\delta x}\mu D_1 - \frac{\delta t}{\delta x^2}D_2)V_{n+1} - \frac{1-\theta}{2}(\frac{\delta t}{\delta x}\mu D_1 \\ &- \frac{\delta t}{\delta x^2}D_2)V_n \quad (\text{'deterministic part'}) \\ &- \frac{\delta t}{\delta x^2}\rho[\sigma D_2 V_{n+1} + (1-\sigma)D_2 V_n] \\ &(\text{'correction term due to stochastic part'}) \\ &- \frac{\sqrt{\rho}}{2}\frac{\sqrt{\delta t}}{\delta x}D_1 V_n \xi_n + \frac{\rho}{2}\frac{\delta t}{\delta x^2}D_2 V_n \xi_n^2, \quad (\text{'stochastic part'}) \end{aligned} \tag{3.4.24}$$

where ξ_n are i.i.d. independent standard Gaussian random variables, $\theta \in [0,1]$ and D_1 and D_2 are the first and second central difference operators:

$$D_1 V_n = \frac{V_{n+1} - V_{n-1}}{2\delta t}, \quad D_2 V_n = \frac{V_{n+1} - 2V_n + V_{n-1}}{\delta t^2}.$$

It was shown that when $\sigma = -1, \theta > 1/2$ the scheme is unconditionally stable as we have, by Fourier stability analysis,

$$\frac{\delta t}{\delta x^2}[1 - 2(\theta - \rho\sigma - \rho^2)] < 1. \tag{3.4.25}$$

When $\sigma = 0$, $\theta = 0$, the scheme becomes the Milstein discretization in time in conjunction with finite difference schemes in physical space introduced in [158], requiring that $\mu^2\delta t \leq 1 - \rho$ in addition to (3.4.25). In Table 3.1, we summarize the CFL conditions for Equation (3.4.23) with various ρ and different discretization parameters θ and σ.

3.4.6 Summary of numerical SPDEs

For SPDEs driven by space-time noise, their solutions are usually of low regularity, especially when the noise is space-time white noise. Hence, it is difficult to obtain efficient high-order schemes.

Generally speaking, numerical methods for SPDEs are classified into three classes: direct discretization where numerical methods for PDEs and SODEs are directly applied, Wong-Zakai approximation where the noise is discretized before any space-time discretization, and preprocessing methods where the SPDEs are first reformulated equivalently or approximately before discretization of SPDEs.

Table 3.1. Stability region of the scheme (3.4.24) for Equation (3.4.23).

ρ	θ	σ	CFL condition	Scheme
0	0	–	$\frac{\delta t}{\delta x^2} < 1$	Explicit
0	$(0, \frac{1}{2})$	–	$\frac{\delta t}{\delta x^2}(1-2\theta) < 1$	Implicit
0	$[\frac{1}{2}, 1]$	–	–	implicit
$(0,1)$	0	0	$\frac{\delta t}{\delta x^2}(1+2\rho^2) < 1$	Explicit
$(0,1)$	$(0, \min(\frac{1}{2}+\rho^2, 1))$	0	$\frac{\delta t}{\delta x^2}[1-2(\theta-\rho^2)] < 1$	Implicit
$(0, \frac{\sqrt{2}}{2}]$	$(\min(\frac{1}{2}+\rho^2, 1), 1)$	0	–	Implicit
$(0,1)$	$[\frac{1}{2}-\rho+\rho^2, 1]$	-1	–	implicit
$[1, \infty)$	–	–	–	Not mean-square stable

The convergence and stability theory of numerical SODEs can be extended to numerical SPDEs. Difference senses of convergence and stability can be implemented such as mean-square convergence/stability, almost sure convergence/stability and weak convergence/stability. We do not discuss in detail here and in this book the convergence in probability. The stability for numerical SPDEs usually require no less than the corresponding PDEs (when noises vanish). For multiplicative noises, numerical SPDEs usually require more on the CFL condition than corresponding PDEs, see Chapter 3.4.5.

Because of low regularity of solutions to SPDEs, it is helpful to make full use of specific properties of the underlying SPDEs and preprocessing techniques to derive higher-order schemes while keeping the computational cost low. For example, we can use the exponential Euler scheme (3.4.17) with (3.4.19) when the underlying SPDEs are driven by additive noise and their leading differential operators are independent of randomness and time. When SPDEs (with multiplicative noises) have commutative noises (see e.g. (3.3.29) for a definition), we can use the Milstein scheme (first-order strong convergence, see, e.g., [252, 276, 366]) while only sampling increments of Brownian motions.

Another issue for numerical methods of SPDEs is to reduce their computational cost in high-dimensional random space as there are usually infinite dimensional stochastic processes whose truncations converge very slowly. This is the case even when high-order schemes like (3.4.19) can be used. Efficient infinite-dimensional integration methods should be employed to obtain the desired statistics with reasonable computational cost. See Chapter 2.5 for a brief review of numerical integration methods in random space.

SPDEs are usually solved with Monte Carlo methods and many similar deterministic equations have to be solved. In some special cases, however, SPDEs can be solved in a very efficient way. For example, for some SPDEs with periodic boundary conditions, we can transform the equation to a deterministic one, which can then be solved once with deterministic solvers, see Appendix B.

3.5 Summary and bibliographic notes

We have presented some basic aspects of SODEs and SPDEs and numerical methods of stochastic differential equations driven by Gaussian white noises. Solutions to SODEs and SPDEs are usually obtained numerically. Convergence and stability theory of numerical methods for SODEs and SPDEs are presented. The commutative conditions of the coefficients of noises may significantly reduce the computational cost.

We summarize the main points in this chapter.

- For SODEs, strong solutions and solution methods for analytical solutions are introduced in Chapter 3.1. Several senses of solutions to SPDEs are presented in Chapter 3.3.
- Conversion between Ito and Stratonovich formulation of SODEs and SPDEs is presented.
- For both SODEs and SPDEs, numerical methods are often used to obtain approximate solutions.
- Strong and weak convergence for numerical SODEs and SPDEs are introduced. Schemes of strong convergence for numerical SODEs are derived and presented in Chapter 3.2. A brief review of numerical schemes for SPDEs focusing on strong and weak convergence is presented in Chapter 3.4.
- Linear stability theory is presented for numerical SODEs in Chapter 3.2.4. Linear stability of a specific numerical scheme for an advection-diffusion equation with Gaussian white noise is discussed in Chapter 3.4.5.

With basics of numerical SODEs and SPDEs, we are ready to discuss more about numerical SODEs and SPDEs in the following chapters.

Bibliographic notes. Mean-square convergence has its own area of applicability, e.g., for simulating scenarios, visualization of stochastic dynamics, filtering, etc., see further discussion on this in [233, 259, 358] and references therein. Furthermore, the mean-square approximation is of theoretical interest and it also provides a guidance in constructing schemes of convergence in weak sense (see, e.g., [259, 354, 358]).

Partial differential equations (PDEs) driven by white noise have different interpretation of stochastic products and lead to different numerical approximations, unlike the PDEs driven by color noise. Specifically, stochastic products for white noise are usually interpreted with two different products: the Ito product and the Stratonovich product, see, e.g., [8]. Under certain conditions, these two products can be used to formulate the same problem. However, different products lead to different performance of numerical solvers for SPDEs driven by white noise, see Chapter 8.

Compared to parabolic equations, *stochastic wave equations* of second order can have better smoothing in time: the solutions are Hölder continuous with exponent $1/2 - \epsilon$ in time, and thus the optimal order of convergence in time is half if only increments of Brownian motion are used, see [465] for the one-dimensional wave equation with multiplicative noise. Ref. [9] considered the linear wave equation with additive single white noise in time using integration factor techniques, where the convergence of two-step finite difference schemes in time is of first-order. Ref. [476] applied exponential integration with (3.4.19) for the semilinear wave equation with additive space-time noise and obtained first-order mean-square convergence in time and half-order in space. Ref. [403] considered finite difference schemes in space for the stochastic semilinear wave equation with multiplicative space-time white noise and obtained optimal mean-square convergence with order less than $1/3$ in space given smooth initial conditions. Finite element methods were investigated in [272] and their convergence order was connected with the regularity of the solution. Ref. [65] considered semi-discretization using spectral Galerkin methods in physical space. Other than strong approximation of stochastic wave equations, Ref. [213] obtained second-order weak convergence both in space and in time for a leap-frog scheme in both space and time solving the one-dimensional semilinear wave equation driven by additive space-time white noise. Ref. [402] considered weak convergence of full discrete finite element methods for the linear stochastic elastic equation driven by additive space-time noise and showed that the weak order is twice the strong order both in time and in space.

Among *stochastic hyperbolic problems*, stochastic conservation laws have also attracted increasing interest, see, e.g., [76, 113, 126, 224, 413] for some theoretical results and, e.g., [27, 274, 367, 412] for some numerical studies. In Chapter 9 we present a practical example of a stochastic conservation law for a stochastic piston.

3.6 Suggested practice

Exercise 3.6.1 *Show that the solution to* (3.1.1) $X(t)$ *is a Gaussian process.*

Exercise 3.6.2 *Prove* (3.1.3). *You may try to prove a stronger conclusion:*

$$\mathbb{E}[|X(t) - X(s)|^2] \leq C\,|t - s|\,.$$

Here C *does not depend on* t *and* s.

Exercise 3.6.3 (Conversion of an Ito SDE to a Stratonovich SDE) *With the relation* (2.3.2), *show that the equation* (3.1.4) *can be written as* (3.1.5) *when* σ_r*'s are continuous in* t *and continuously differentiable in* x.

Exercise 3.6.4 *Suppose that* $X(t)$ *is adapted and* $\mathbb{E}[\int_0^t X^2(s)\,ds] < \infty$, *verify that* $X(t)$ *in Theorem 3.1.2 is a solution using Definition 3.1.1.*

Exercise 3.6.5 *Verify that the following SDE has coefficients satisfying the conditions here with* $1 \leq p_0 \leq \lambda/\sigma^2 + 1/2$:

$$dX = \kappa X(\theta - X) + \sigma\,|X|^{3/2}\,dW(t), \quad X_0 > 0.$$

Find a range of p_1 *that the growth condition in Remark 3.1.3 holds.*

Exercise 3.6.6 *Use the integrating factor method to solve the SDE*

$$dX(t) = (X(t))^\gamma\,dt + \alpha X(t)\,dW(t), \quad X_0 = x > 0,$$

where α *is a constant and* $\gamma \leq 1$.

Exercise 3.6.7 *Find* $\mathbb{E}[X(t)X(s)]$ *when* $X(t)$ *is a solution to* (3.1.11).

Hint. Use the relation $\mathbb{E}[X(t)X(s)] - \mathbb{E}[X_s^2] = \mathbb{E}[(X(t) - X(s))X(s)] = \mathbb{E}[(\mathbb{E}[X(t) - X(s)|\mathcal{F}_s])X(s)]$.

Exercise 3.6.8 *Explain why we need to use*

$$\int_t^{t+h} \sigma_r(s, X(s))\,dW_r \approx \int_t^{t+h} \sigma_r(t, X(t))\,dW_r$$

instead of using $\sigma_r(s, X(s))$ *at other time instants in between* t *and* $t+h$ *in both forward and backward Euler schemes.*

Exercise 3.6.9 *Show that*

$$\int_t^{t+h}\int_t^s dW(\theta)\,dW(s) = \frac{1}{2}[(W(t+h) - W(t))^2 - h].$$

Exercise 3.6.10 *Show that* (3.3.5) *holds.*

Exercise 3.6.11 *Suppose that* $u_0(x) = 0$ *and* $q_k = 1$, *for all* $k \geq 1$. *Show that the solution* $u(t,x)$ *of Equation* (3.3.1) *is Hölder continuous of order less than* $1/4$ *in* t, *and is Hölder continuous of order less than* $1/2$ *in* x.

Hint. By Kolmogorov's continuity theorem and use the following facts that $\sum_{k=1}^{\infty} k^{2\beta-2} < \infty$ when $\beta < 1/2$ and for $0 < \beta \leq 1$,

$$|e_k(x)-e_k(y)| = |e_k(x)-e_k(y)|^{1-\beta} |e_k(x)-e_k(y)|^{\beta} \leq 2^{1-\beta}\sqrt{\frac{2}{l}}k^{\beta}(\frac{\pi}{l})^{\beta} |x-y|^{\beta},$$

$$1 - e^{k^2(t_2-t_1)} \leq \left(1 - e^{-\lambda_k(t_2-t_1)}\right)^{\beta} \leq \lambda_k^{\beta}(t_2 - t_1)^{\beta}.$$

Part I

Numerical Stochastic Ordinary Differential Equations

We can understand stochastic partial differential equations (SPDEs) as stochastic ordinary differential equations (SODEs) in infinite dimensions, see, e.g., [94, 145]. It is then important to investigate numerical methods for SODEs before studying SPDEs. In this first part of the book, which includes two chapters, we will present the Wong-Zakai approximation for SODEs in Chapter 4, which is somewhat less investigated in the literature. As we discussed in the previous chapter, the Wong-Zakai approximation refers to the truncation of Brownian motion with a bounded variation process and thus requires further discretization in time. We show that errors of schemes based on the Wong-Zakai approximation are also determined by discretization in time, in addition to the choice of bounded variation processes.

In Chapter 4, SODEs have global Lipschitz coefficients. We also present some numerical methods for SODEs with non-global Lipschitz coefficients in Chapter 5. With the coefficients of polynomial growth, we show the key elements to convergence (rates) of the numerical schemes. The key point of such schemes is to control the fast growth of numerical solutions though the solutions blow up with a very small probability without control.

The methodology presented for SODEs in this part can be extended to SPDEs with care. In Part II, we will apply Wong-Zakai approximation to SPDEs. The basic findings in Chapter 4 still apply, but careful choices of discretization in time and also in space are needed. Solving SPDEs with coefficients of polynomial growth is a more delicate matter but the numerical schemes in Chapter 5 can be essentially applied.

4

Numerical schemes for SDEs with time delay using the Wong-Zakai approximation

Will a spectral approximation of Brownian motion lead to higher-order numerical methods for stochastic differential equations as the Karhunen-Loeve truncation of a smooth stochastic process does?

In classical numerical methods for stochastic differential equations, the Brownian motion is typically approximated by its piecewise linear interpolation simultaneously when a time discretization is performed. Another approximation, i.e., the spectral approximation of Brownian motion, has been proposed in theory for years but has not yet been investigated extensively in the context of numerical methods.

In Chapter 4.2, we show how to derive three numerical schemes for stochastic delay differential equations (SDDEs) using the Wong-Zakai (WZ) approximation. By approximating the Brownian motion with its truncated spectral expansion and then using different discretizations in time, we will obtain three schemes: a *predictor-corrector* scheme, a *midpoint* scheme, and a *Milstein-like* scheme. We prove that the predictor-corrector scheme converges with order half in the mean-square sense while the Milstein-like scheme converges with order one. In Chapter 4.3, we discuss the linear stability of these numerical schemes for SDDEs. Numerical results in Chapter 4.4 confirm the theoretical prediction and demonstrate that the midpoint scheme is of half-order convergence. Numerical results also show that the predictor-corrector and midpoint schemes can be of first-order convergence under commutative noises when there is no delay in the diffusion coefficients.

All these conclusions are summarized in Chapter 4.5, where we also present a review on the Wong-Zakai approximation for SODEs, SDDEs, and SPDEs with Gaussian or non-Gaussian noises. The simulation of double Ito integrals in Milstein-type schemes is also discussed for SDDEs as well as for SODEs. Some exercises are provided for readers to practice implementation of numerical methods for SDDEs at the end of this chapter.

Z. Zhang, G.E. Karniadakis, *Numerical Methods for Stochastic Partial Differential Equations with White Noise*,
Applied Mathematical Sciences 196, DOI 10.1007/978-3-319-57511-7_4

4.1 Wong-Zakai approximation for SODEs

Let us first illustrate the Wong-Zakai approximation by considering the piecewise linear approximation (2.2.4) of the one-dimensional Brownian motion $W(t)$ for the following Ito SODEs, see, e.g., [481, 482]

$$dX = b(t,X)dt + \sigma(t,X)dW(t), \quad X(0) = X_0, \tag{4.1.1}$$

and obtain the following ODE with smooth random inputs

$$dX^{(n)} = b(t,X^{(n)})dt + \sigma(t,X^{(n)})dW^{(n)}(t), \quad X(0) = X_0. \tag{4.1.2}$$

It is proved in [481, 482] that (4.1.2) converges in the mean-square sense to

$$dX = \left(b(t,X) + \frac{1}{2}\sigma(t,X)\sigma_x(t,X)\right)dt + \sigma(t,X)dW(t), \quad X(0) = X_0, \tag{4.1.3}$$

under mild assumptions, which can be written in Stratonovich form [430]

$$dX = b(t,X)dt + \sigma(t,X)\circ dW(t), \quad X(0) = X_0, \tag{4.1.4}$$

where '$\circ$' indicates the Stratonovich product. The term $\frac{1}{2}\sigma(t,X)\sigma_x(t,X)$ in (4.1.3) is called the standard Wong-Zakai correction term.

It is essential to identify the Wong-Zakai correction term (or the equation that the resulting equation from Wong-Zakai approximation converges to) in various cases. For SODEs with scalar noise, e.g., (4.1.1), when the Brownian motion is approximated by a process of *bounded variation* (rather than by piecewise linear approximation), Ref. [434] proved that the convergence to (4.1.3) holds in the pathwise sense (almost surely) if the drift b is locally Lipschitz continuous and is of linear growth and the diffusion σ is continuous with bounded first-order derivatives. However, this conclusion does not hold if σ does not have bounded first-order derivative in x [434] or the approximation of Brownian motion is not differentiable [342].

For SODEs with multiple noises, Sussmann [435] derived a generic Wong-Zakai correction term for multiple noises. Refs. [283, 284] provided a practical criterion to verify whether a general approximation of Brownian motions (even general semi-martingales) may lead to a standard Wong-Zakai correction term (e.g., $1/2\sigma_x\sigma$ for (4.1.4)) or other Wong-Zakai correction terms. To have a standard Wong-Zakai correction term, the essential condition for the approximation of Brownian motion is

$$\lim_{n\to\infty} \mathbb{E}[\int_0^T W^{(n)}\, dW^{(n)} - \int_0^T W(t)\circ dW(t)] = 0. \tag{4.1.5}$$

The convergence of Wong-Zakai approximation for SODEs has been established in different senses, e.g., pathwise convergence (e.g., [434, 435]), support theorem (the relation between the support of distribution of the solution and that of its Wong-Zakai approximation, e.g., [17, 351, 432, 457]), mean-square convergence (e.g., [15, 195, 241, 457]), and convergence in probability (e.g., [16]).

Observe that Equation (4.1.2) is still continuous in time though the Brownian motion is approximated by a finite dimensional smooth process. Some proper time-discretization schemes should be applied for the equation (4.1.2). We will get back to this important issue shortly after we introduce SDDEs and we will treat SODEs as a special case of SDDEs with vanishing delay.

4.1.1 Wong-Zakai approximation for SDDEs

Consider the following SDDE with constant delay in Stratonovich form:

$$\begin{aligned} \mathrm{d}X(t) &= f(X(t), X(t-\tau))\mathrm{d}t + \sum_{l=1}^{r} g_l(X(t), X(t-\tau)) \circ \mathrm{d}W_l(t), \quad t \in (0, T], \\ X(t) &= \phi(t), \quad t \in [-\tau, 0], \end{aligned} \tag{4.1.6}$$

where $\tau > 0$ is a constant, $(W(t), \mathcal{F}_t) = (\{W_l(t),\ 1 \le l \le r\}, \mathcal{F}_t)$ is a system of one-dimensional independent standard Wiener process, the functions $f : \mathbb{R}^d \times \mathbb{R}^d \to \mathbb{R}^d$, $g_l : \mathbb{R}^d \times \mathbb{R}^d \to \mathbb{R}^d$, $\phi(t) : [-\tau, 0] \to \mathbb{R}^d$ are continuous with $\mathbb{E}[\|\phi\|_{L^\infty}^2] < \infty$. We also assume that $\phi(t)$ is $\mathcal{F}_0$-measurable.

For the mean-square stability of Equation (4.1.6), we assume that f, g_l , $\partial_x g_l g_l$ and $\partial_{x_\tau} g_l g_l$, ($\partial_x$ and ∂_{x_τ} denote the derivatives with respect to the first and second variables, respectively), $l = 1, 2, \cdots, r$ in Equation (4.1.6) satisfy the following Lipschitz conditions:

$$|v(x_1, y_1) - v(x_2, y_2)|^2 \le L_v(|x_1 - x_2|^2 + |y_1 - y_2|^2), \tag{4.1.7}$$

and the linear growth conditions

$$|v(x_1, y_1)|^2 \le K(1 + |x_1|^2 + |y_1|^2) \tag{4.1.8}$$

for every x_1, y_1, x_2, $y_2 \in \mathbb{R}^d$, where L_v, K are positive constants, which depend only on v. Under these conditions, Equation (4.1.6) has a unique sample-continuous and $\mathcal{F}_t$-adapted strong solution $X(t) : [-\tau, +\infty) \to \mathbb{R}^d$, see, e.g., [334, 368].

Now we present the WZ approximation of Equation (4.1.6) using the spectral approximation (2.2.8) with piecewise constant basis (2.2.9) and Fourier basis (2.2.10). With these orthogonal approximations, we have the following WZ approximation for

$$\begin{aligned} \mathrm{d}\tilde{X}(t) &= f(\tilde{X}(t), \tilde{X}(t-\tau))\mathrm{d}t + \sum_{l=1}^{r} g_l(\tilde{X}(t), \tilde{X}(t-\tau))\mathrm{d}\widetilde{W}_l(t),\ t \in [0, T], \\ \tilde{X}(t) &= \phi(t), t \in (-\tau, 0], \end{aligned} \tag{4.1.9}$$

where $\widetilde{W}_l(t)$ can be any approximation of $W_l(t)$ described above.

For the piecewise linear interpolation (2.2.4), we have the following consistency condition of the WZ approximation (4.1.9) to Equation (4.1.6).

Theorem 4.1.1 (Consistency, [453]) *Suppose f and g_l in Equation (4.1.6) are Lipschitz continuous and satisfy conditions (4.1.7) and have second-order continuous and bounded partial derivatives. Suppose also the initial segment $\phi(t), t \in [-\tau, 0]$ to be on the probability space $(\Omega, \mathcal{F}, P)$ and $\mathcal{F}_0$-measurable and right continuous, and $\mathbb{E}[\|\phi\|^2_{L^\infty}] < \infty$. For $\tilde{X}(t)$ in (4.1.9) with piecewise linear approximation of Brownian motion (2.2.4), we have for any $t \in (0, T]$,*

$$\lim_{n\to\infty} \sup_{0\le s\le t} \mathbb{E}[|X(s) - \tilde{X}(s)|^2] = 0. \tag{4.1.10}$$

The consistency of the WZ approximation with spectral approximation (2.2.8) can be established by the argument of integration by parts as in [200, 241], under similar conditions on the drift and diffusion coefficients.

4.2 Derivation of numerical schemes

We further discretize Equation (4.1.9) in time and derive several numerical schemes for (4.1.6). To this end, we take a uniform time step size h, which satisfies $\tau = mh$ and m is a positive integer; $N_T = T/h$ (T is the final time); $t_n = nh$, $n = 0, 1, \cdots, N_T$. For simplicity, we take the same partition for the WZ approximation exactly as the time discretization, i.e.,

$$\mathbf{t}_n = t_n,\ n = 0, 1, \cdots, N_T \ \text{ and } \ \Delta =: \mathbf{t}_n - \mathbf{t}_{n-1} = t_n - t_{n-1} = h.$$

For Equation (4.1.9), we have the following integral form over $[t_n, t_{n+1}]$:

$$\begin{aligned}\int_{t_n}^{t_{n+1}} \mathrm{d}\tilde{X}(t) &= \int_{t_n}^{t_{n+1}} f(\tilde{X}(t), \tilde{X}(t-\tau))\mathrm{d}t \\ &\quad + \sum_{l=1}^{r} \int_{t_n}^{t_{n+1}} g_l(\tilde{X}(t), \tilde{X}(t-\tau))\mathrm{d}\widetilde{W}_l(t) \\ &= \int_{t_n}^{t_{n+1}} f(\tilde{X}(t), \tilde{X}(t-\tau))\mathrm{d}t \\ &\quad + \sum_{l=1}^{r} \int_{t_n}^{t_{n+1}} g_l(\tilde{X}(t), \tilde{X}(t-\tau)) \sum_{j=1}^{N_h} m_j^{(n)}(t)\xi_{l,j}^{(n)}\mathrm{d}t.\end{aligned} \tag{4.2.1}$$

Here we emphasize that *the time-discretization for the diffusion term has to be at least half-order.* Otherwise, the resulting scheme is not consistent, e.g., Euler-type schemes, in general, converge to the corresponding SDDEs in the Ito sense instead of those in the Stratonovich sense. In fact, if $g_l(\tilde{X}(t), \tilde{X}(t-\tau))$ $(l = 1, \cdots, r)$ is approximated by $g_l(\tilde{X}(t_n), \tilde{X}(t_n - \tau))$ in

Equation (4.2.1), then we have, for both Fourier basis (2.2.10) and piecewise constant basis (2.2.9),

$$\int_{t_n}^{t_{n+1}} \mathrm{d}\tilde{X}(t) = \int_{t_n}^{t_{n+1}} f(\tilde{X}(t), \tilde{X}(t-\tau))\mathrm{d}t + \sum_{l=1}^{r} g_l(\tilde{X}(t_n), \tilde{X}(t_n - \tau)\Delta W_{l,n},$$

where $\Delta W_{l,n} = W_l(t_{n+1}) - W_l(t_n)$. This will lead to an Euler-type scheme which converges to the following SDDE in the Ito sense, see, e.g., [14, 306], instead of (4.1.6):

$$\mathrm{d}X(t) = f(X(t), X(t-\tau))\mathrm{d}t + \sum_{l=1}^{r} g_l(X(t), X(t-\tau))\mathrm{d}W_l(t).$$

In the following, three numerical schemes for solving Equation (4.1.6) are derived using Taylor expansion and different discretizations in time in (4.2.1).

4.2.1 A predictor-corrector scheme

Taking $N_h = 1$, we have that both bases, (2.2.9) and (2.2.10), have only one term $m_1^{(n)} = 1/\sqrt{h}$ over each subinterval. Using the trapezoidal rule to approximate the integrals on the right-hand side of (4.2.1), we get

$$\begin{aligned} X_{n+1} = X_n &+ \frac{h}{2}\big[f(X_n, X_{n-m}) + f(X_{n+1}, X_{n-m+1})\big] \\ &+ \frac{1}{2}\sum_{l=1}^{r}\big[g_l(X_n, X_{n-m}) + g_l(X_{n+1}, X_{n-m+1})\big]\Delta W_{l,n}, \end{aligned} \tag{4.2.2}$$

where X_n is an approximation of $\tilde{X}(t_n)$ (thus an approximation of $X(t_n)$). The initial conditions are $X_n = \phi(nh)$, when $n = -m, -m+1, \cdots, 0$. Note that the scheme (4.2.2) is fully implicit and is not solvable as $\Delta W_{l,n}$ can take any values in the real line. To resolve this issue, we further apply the left rectangle rule on the right side of (4.2.1) to obtain a predictor for X_{n+1} in (4.2.2) so that the resulting scheme is explicit. Consequently, we arrive at a predictor-corrector scheme for SDDE (4.1.6):

$$\begin{aligned} \overline{X}_{n+1} &= X_n + hf(X_n, X_{n-m}) + \sum_{l=1}^{r} g_l(X_n, X_{n-m})\Delta W_{l,n}, \\ X_{n+1} &= X_n + \frac{h}{2}\big[f(X_n, X_{n-m}) + f(\overline{X}_{n+1}, X_{n-m+1})\big] \\ &\quad + \frac{1}{2}\sum_{l=1}^{r}\big[g_l(X_n, X_{n-m}) + g_l(\overline{X}_{n+1}, X_{n-m+1})\big]\Delta W_{l,n}, \\ &\qquad n = 0, 1, \cdots, N_T - 1. \end{aligned} \tag{4.2.3}$$

Taking $N_h = 1$ is sufficient for half-order schemes, such as the predictor-corrector scheme (4.2.3) and the following midpoint scheme. Both schemes employ $\int_{t_n}^{t_{n+1}} \sum_{j=1}^{N_h} m_j^{(n)}(t)\xi_{l,j}^{(n)} dt$, which is equal to $\Delta W_{l,n}$ for any $N_h \geq 1$, according to (2.2.8) and our choices of orthonormal bases (2.2.9) and (2.2.10).

Theorem 4.2.1 *Assume that f, g_l, $\partial_x g_l g_q$ and $\partial_{x_\tau} g_l g_q$ $(l, q = 1, 2, \cdots, r)$ satisfy the Lipschitz condition* (4.1.7) *and also g_l have bounded second-order partial derivatives with respect to all variables. If $\mathbb{E}[\|\phi\|_{L^\infty}^p] < \infty$, $p \geq 4$, then we have for the predictor-corrector scheme* (4.2.3),

$$\max_{1 \leq n \leq N_T} \mathbb{E}|X(t_n) - X_n|^2 = O(h). \tag{4.2.4}$$

When $\tau = 0$ both in drift and diffusion coefficients, the scheme (4.2.3) degenerates into one family of the predictor-corrector schemes in [42], which can have larger stability region than the explicit Euler scheme and some other one-step schemes, especially for stochastic differential equations with multiplicative noises, see Chapter 4.3. Moreover, we will numerically show that if the time delay only exists in the drift term in SDDE with commutative noise (for one-dimensional case, i.e., $d = 1$, the commutative condition is $g_l \partial_x g_q - g_q \partial_x g_l = 0$, $1 \leq l, q \leq r$), the predictor-corrector scheme can be convergent with order one in the mean-square sense.

Proof of Theorem 4.2.1. We present the proof for $d = 1$ in (4.1.6), which can be extended to multi-dimensional case $d > 1$ without difficulty. We recall that for the Milstein scheme (4.2.26), see [228], $\max_{1 \leq n \leq N_T} \mathbb{E}|X(t_n) - X_n^M|^2 = O(h^2)$. Then by the triangle inequality, it suffices to prove

$$\max_{1 \leq n \leq N_T} \mathbb{E}|X_n^M - X_n|^2 = O(h). \tag{4.2.5}$$

We denote that $f_n = f(X_n, X_{n-m})$ and $g_{l,n} = g_l(X_n, X_{n-m})$ and also

$$\begin{aligned} \rho_{f_n} &= f(\overline{X}_{n+1}, X_{n-m+1}) - f_n, \\ \rho_{g_{l,n}} &= g_l(\overline{X}_{n+1}, X_{n-m+1}) - [g_{l,n} + \partial_x g_{l,n} \sum_{q=1}^{r} g_{q,n} \Delta W_{q,n} \\ &\quad + \partial_{x_\tau} g_{l,n} \sum_{q=1}^{r} g_{q,n-m} \Delta W_{q,n-m}]. \end{aligned} \tag{4.2.6}$$

With (4.2.6), we can rewrite (4.2.3) as follows

$$\begin{aligned} X_{n+1} &= X_n + h f_n + \sum_{l=1}^{r} g_{l,n} \Delta W_{l,n} + \frac{1}{2} \sum_{l=1}^{r} \sum_{q=1}^{r} \partial_x g_{l,n} \Delta W_{q,n} \Delta W_{l,n} \\ &\quad + \frac{1}{2} \sum_{l=1}^{r} \sum_{q=1}^{r} \partial_{x_\tau} g_{l,n}\, g_{q,n-m} \Delta W_{q,n-m} \Delta W_{l,n} + \rho_n, \end{aligned} \tag{4.2.7}$$

where $\rho_n = h \rho_{f_n} + \frac{1}{2} \sum_{l=1}^{r} \rho_{g_{l,n}} \Delta W_{l,n}$.

It can be readily checked that if f, g_l satisfy the Lipschitz condition (4.1.7), and g_l has bounded second-order derivatives ($l = 1, \cdots, r$), then by the predictor-corrector scheme (4.2.3) and Taylor expansion of $g_l(\overline{X}_{n+1}, X_{n-m+1})$, we have $h^2\mathbb{E}[\rho_{f_n}^2] \leq Ch^3$, $\mathbb{E}[(\rho_{g_{l,n}}\Delta W_{l,n})^2] \leq Ch^3$, and thus by the triangle inequality,

$$\mathbb{E}[\rho_n^2] \leq Ch^3, \tag{4.2.8}$$

where the constant C depends on r and Lipschitz constants, independent of h.

Subtracting (4.2.7) from (4.2.26) and taking the expectation after squaring over both sides, we have

$$\begin{aligned}\mathbb{E}[(X_{n+1}^M - X_{n+1})^2] = \mathbb{E}[(X_n^M - X_n)^2] + 2\mathbb{E}[(X_n^M - X_n)(\sum_{i=0}^{4} R_i - \rho_n)] \\ -2\sum_{i=0}^{4}\mathbb{E}[\rho_n R_i] + \sum_{i,j=0}^{4}\mathbb{E}[R_i R_j] + \mathbb{E}[\rho_n^2],\end{aligned} \tag{4.2.9}$$

where we denote $f_n^M = f(X_n^M, X_{n-m}^M)$ and $g_{l,n}^M = g_l(X_n^M, X_{n-m}^M)$ and

$$\begin{aligned}
R_0 &= h(f_n^M - f_n) + \sum_{l=1}^{r}(g_{l,n}^M - g_{l,n})\Delta W_{l,n},\\
R_1 &= \sum_{l=1}^{r}\sum_{q=1}^{r}\left[\partial_x g_{l,n}^M g_{q,n}^M - \partial_x g_{l,n} g_{q,n}\right]\frac{\Delta W_{q,n}\Delta W_{l,n}}{2},\\
R_2 &= \sum_{l=1}^{r}\sum_{q=1}^{r}\left[\partial_{x_\tau} g_{l,n}^M g_{q,n-m}^M - \partial_{x_\tau} g_{l,n} g_{q,n-m}\right]\frac{\Delta W_{q,n-m}\Delta W_{l,n}}{2},\\
R_3 &= \sum_{l=1}^{r}\sum_{q=1}^{r}\partial_x g_{l,n}^M g_{q,n}^M (I_{q,l,t_n,t_{n+1},0} - \frac{\Delta W_{q,n}\Delta W_{l,n}}{2}),\\
R_4 &= \sum_{l=1}^{r}\sum_{q=1}^{r}\partial_{x_\tau} g_{l,n}^M g_{q,n-m}^M (I_{q,l,t_n,t_{n+1},\tau} - \frac{\Delta W_{q,n-m}\Delta W_{l,n}}{2}).
\end{aligned}$$

By the Lipschitz condition for f and g_l, and adaptedness of X_n, X_n^M, we have

$$\mathbb{E}[R_0^2] \leq C(h^2 + h)(\mathbb{E}[(X_n^M - X_n)^2] + \mathbb{E}[(X_{n-m}^M - X_{n-m})^2]). \tag{4.2.10}$$

To bound $\mathbb{E}[R_i^2]$ ($i = 1, 2, 3, 4$), we require that X_n and X_n^M have bounded (up to) fourth-order moments, which can be readily checked from the predictor-corrector scheme (4.2.3) and the Milstein scheme (4.2.26) under our assumptions. By the Lipschitz condition of g_l and $\partial_{x_\tau} g_l g_q$, we have

$$\mathbb{E}[R_2^2] \leq C\max_{1\leq l,q\leq r}\mathbb{E}[(\left|X_n^M - X_n\right| + \left|X_{n-m}^M - X_{n-m}\right|)^2(\Delta W_{q,n-m}\Delta W_{l,n})^2],$$

whence by the Cauchy inequality and the boundedness of $\mathbb{E}[X_n^4]$ and $\mathbb{E}[(X_n^M)^4]$, we have $\mathbb{E}[R_2^2] \le Ch^2$. Similarly, we have $\mathbb{E}[R_1^2] \le Ch^2$. By Lemma 4.2.4, and the linear growth condition (4.1.8) for $\partial_{x_\tau} g_l g_q$, we obtain

$$\mathbb{E}[R_4^2] \le C \max_{1\le l<q\le r} \mathbb{E}[(1 + \left|X_n^M\right|^2 + \left|X_{n-m}^M\right|^2)(I_{q,l,t_n,t_{n+1},\tau} \\ - \tfrac{\Delta W_{q,n-m}\Delta W_{l,n}}{2})^2] \le Ch^2,$$

since X_n^M, X_{n-m}^M have bounded fourth-order moments and by the Burkholder-Davis-Gundy inequality (see Appendix D), it holds that for $l \neq q$

$$\begin{aligned}
&\mathbb{E}[(I_{q,l,t_n,t_{n+1},\tau} - \frac{\Delta W_{q,n-m}\Delta W_{l,n}}{2})^4] \\
&= \mathbb{E}[(\int_{t_n}^{t_{n+1}} (W_q(t-\tau) - \frac{W_q(t_{n+1}-\tau) + W_q(t_n-\tau)}{2}) \circ dW_l)^4] \\
&\le C(\mathbb{E}[\int_{t_n}^{t_{n+1}} (W_q(t-\tau) - \frac{W_q(t_{n+1}-\tau) + W_q(t_n-\tau)}{2})^2\, ds])^2 \le Ch^4.
\end{aligned}$$

Similarly, we have $\mathbb{E}[R_3^2] \le Ch^2$. Thus, we have proved that

$$\mathbb{E}[R_i^2] \le Ch^2, \quad i = 1,2,3,4. \tag{4.2.11}$$

By the basic inequality $2ab \le a^2 + b^2$, we have

$$2\left|\mathbb{E}[(X_n^M - X_n)\rho_n]\right| \le h\mathbb{E}[(X_n^M - X_n)^2] + h^{-1}\mathbb{E}[\rho_n^2]. \tag{4.2.12}$$

By the fact that X_n and X_n^M are $\mathcal{F}_{t_n}$-measurable and Lipschitz condition for f,

$$\begin{aligned}
2\mathbb{E}[(X_n^M - X_n)R_0] &= 2h\mathbb{E}[(X_n^M - X_n)(f_n - f_{n-m})] \\
&\le Ch(\mathbb{E}[(X_n^M - X_n)^2] + \mathbb{E}[(X_{n-m}^M - X_{n-m})^2]).
\end{aligned} \tag{4.2.13}$$

Further, by the Lipschitz condition (4.1.7) for $\partial_x g_l g_l$, we have

$$\begin{aligned}
2\mathbb{E}[(X_n^M - X_n)R_1] &= \sum_{l=1}^{r} \mathbb{E}[(X_n^M - X_n)\partial_x g_{l,n}^M g_{l,n}^M - \partial_x g_{l,n} g_{l,n})]\mathbb{E}[(\Delta W_{l,n})^2] \\
&\le Ch(\mathbb{E}[(X_n^M - X_n)^2] + \mathbb{E}[(X_{n-m}^M - X_{n-m})^2]).
\end{aligned} \tag{4.2.14}$$

By the adaptedness of X_n, X_n^M and $\mathbb{E}[\Delta W_{l,n}] = \mathbb{E}[(I_{q,l,t_n,t_{n+1},0} - \frac{\Delta W_{q,n}\Delta W_{l,n}}{2})] = 0$, we have

$$\mathbb{E}[(X_n^M - X_n)R_i] = 0, \quad i = 2,3. \tag{4.2.15}$$

Again by the adaptedness of X_n and X_n^M, we can have

$$\mathbb{E}[(X_n^M - X_n)R_4] = 0. \tag{4.2.16}$$

In fact, by Lemma 4.2.4, we can represent $I_{q,l,t_n,t_{n+1},\tau}$ as

$$I_{q,l,t_n,t_{n+1},\tau} = \frac{h}{2}\xi_{q,1}^{(n-m)}\xi_{l,1}^{(n)} + \frac{h}{2\pi}\sum_{p=1}^{\infty}\frac{1}{p}[\xi_{q,2p+1}^{(n)}\xi_{l,2p}^{(n-m)} - \xi_{q,2p}^{(n-m)}\xi_{l,2p+1}^{(n)} - \sqrt{2}\xi_{q,1}^{(n-m)}\xi_{l,2p}^{(n)}]. \tag{4.2.17}$$

Then by the facts $\mathbb{E}[|(X_n^M - X_n)R_4|] \leq (\mathbb{E}[(X_n^M - X_n)^2])^{1/2}(\mathbb{E}[R_4^2])^{1/2} \leq Ch$ and $\mathbb{E}[(X_n^M - X_n)\xi_{l,k}^{(n)}] = 0$ for any $k \geq 1$, we obtain (4.2.16) from Lebesgue's dominated convergence theorem.

With (4.2.15)–(4.2.16) and Cauchy inequality, from (4.2.9) we have, for $n \geq m$,

$$\mathbb{E}[(X_{n+1}^M - X_{n+1})^2]$$
$$\leq \mathbb{E}[(X_n^M - X_n)^2] + 2\mathbb{E}[(X_n^M - X_n)(R_0 + R_1 - \rho_n)] + C\sum_{i=0}^{4}\mathbb{E}[R_i^2] + C\mathbb{E}[\rho_n^2]$$

and further by (4.2.8), (4.2.10)–(4.2.12), and (4.2.13)–(4.2.14), we obtain, for $n \geq m$,

$$\begin{aligned}\mathbb{E}[(X_{n+1}^M - X_{n+1})^2] &\leq (1 + Ch)\mathbb{E}[(X_n^M - X_n)^2] \\ &+ Ch\mathbb{E}[(X_{n-m}^M - X_{n-m})^2]) + (C + h^{-1})\mathbb{E}[\rho_n^2] \\ &+ C\sum_{i=0}^{4}\mathbb{E}[R_i^2] \leq (1 + Ch)\mathbb{E}[(X_n^M - X_n)^2] \\ &+ Ch\mathbb{E}[(X_{n-m}^M - X_{n-m})^2] + Ch^2, \end{aligned} \tag{4.2.18}$$

where C is independent of h. Similarly, we can also obtain that (4.2.18) holds for $n = 1, \cdots, m-1$. Taking the maximum over both sides of (4.2.18) and noting that $X_i^M - X_i = 0$ for $-m \leq i \leq 0$, we have

$$\max_{1\leq i\leq n+1} \mathbb{E}[(X_i^M - X_i)^2] \leq (1 + Ch)\max_{1\leq i\leq n} \mathbb{E}[(X_i^M - X_i)^2] + Ch^2.$$

Then (4.2.5) follows from the discrete Gronwall inequality (See Appendix D.s □

4.2.2 The midpoint scheme

Taking $N_h = 1$, applying the midpoint rule on the right side of (4.2.1) and by $X(t + \frac{h}{2}) \approx \frac{1}{2}(X(t+h) + X(t))$, we obtain the following midpoint scheme

$$\begin{aligned} X_{n+1} = X_n &+ hf\left(\frac{X_n + X_{n+1}}{2}, \frac{X_{n-m} + X_{n-m+1}}{2}\right) \\ &+ \sum_{l=1}^{r} g_l\left(\frac{X_n + X_{n+1}}{2}, \frac{X_{n-m} + X_{n-m+1}}{2}\right)\overline{\Delta W}_{l,n}, \\ n = 0, 1, &\cdots, N_T - 1, \end{aligned} \tag{4.2.19}$$

where we have truncated ΔW_n with $\overline{\Delta W}_n$ so that the solution to (4.2.19) has finite second-order moment and is solvable (see, e.g., [358, Section 1.3]). Here

$\overline{\Delta W}_n = \zeta^{(n)}\sqrt{h}$ instead of $\xi^{(n)}\sqrt{h}$, where $\zeta^{(n)}$ is a truncation of the standard Gaussian random variable $\xi^{(n)}$ (see, e.g., [358, pp. 39] and (3.2.28)):

$$\zeta^{(n)} = \xi^{(n)}\chi_{|\xi^{(n)}|\leq A_h} + \text{sgn}(\xi^{(n)})A_h\chi_{|\xi^{(n)}|>A_h}, \quad A_h = \sqrt{4|\log(h)|}.$$

When $\tau = 0$, this fully implicit midpoint scheme allows long-time integration for stochastic Hamiltonian systems [356]. As in the case of no delay, the midpoint scheme complies with the Stratonovich calculus without differentiating the diffusion coefficient.

Theorem 4.2.2 *Assume that f, g_l, $\partial_x g_l g_q$, and $\partial_{x_\tau} g_l g_q$ $(l, q = 1, 2, \cdots, r)$ satisfy the Lipschitz condition (4.1.7) and also g_l have bounded second-order partial derivatives with respect to all variables. If $\mathbb{E}[\|\phi\|^p_{L^\infty}] < \infty$, $p \geq 4$, then we have for the following midpoint scheme (4.2.19),*

$$\max_{1\leq n\leq N_T} \mathbb{E}|X(t_n) - X_n|^2 = O(h).$$

We refer the reader for the proof of the convergence rate of the midpoint scheme to [356]. The proof is almost the same if there is no delay in the diffusion coefficients. It has first-order convergence for Stratonovich stochastic differential equations with commutative noise when no delay arises in the diffusion coefficients. However, it has half-order convergence once the delay appears in the diffusion coefficients, and in this case, no commutative noises are defined.[1] The convergence rates in different cases will be shown numerically in Chapter 4.4.

Remark 4.2.3 *The relation (2.2.13) is crucial in the derivation of the schemes (4.2.3) and (4.2.19). If a CONS contains no constants, e.g., $\left\{\sqrt{\frac{2}{t_{n+1}-t_n}}\sin(\frac{k\pi(s-t_n)}{t_{n+1}-t_n})\right\}_{k=1}^{\infty}$, then from (4.2.1), $\Delta W_{l,n}$ in the scheme (4.2.3) should be replaced by*

$$\int_{t_n}^{t_{n+1}} d\widetilde{W}_l(t) = \sum_{j=1}^{N_h} \sqrt{\frac{2}{t_{n+1}-t_n}} \int_{t_n}^{t_{n+1}} \sin(\frac{j\pi(s-t_n)}{t_{n+1}-t_n})\, ds\xi^{(n)}_{l,j}, \tag{4.2.20}$$

which will be simulated with i.i.d. Gaussian random variables with zero mean and variance $\sum_{j=1}^{N_h} \frac{2h}{j^2\pi^2}(1-(-1)^j)^2$. According to the proof of Theorem 4.2.1, we require $N_h \sim O(h^{-1})$ so that

$$\mathbb{E}[\left|\int_{t_n}^{t_{n+1}} d\widetilde{W}_l(t) - \int_{t_n}^{t_{n+1}} dW_l(t)\right|^2] \sim O(h^2) \tag{4.2.21}$$

to make the corresponding scheme of half-order convergence. Numerical results show that the scheme (4.2.3) with ΔW_n replaced by (4.2.20) and $N_h \sim O(h^{-1})$ leads to similar accuracy and same convergence order with the predictor-corrector scheme (4.2.3) (numerical results are not presented).

[1] We can think that the noises are noncommutative, even when there is only a single noise.

4.2.3 A Milstein-like scheme

A first-order scheme, the Milstein scheme for SDDEs (4.1.6) can be derived based on the Ito-Taylor expansion [260] or anticipative calculus, see, e.g., [228]. Here we derive a similar scheme (called the Milstein-like scheme) based on WZ approximation (4.1.9) and Taylor expansion of the diffusion terms. When $s \in [t_n, t_{n+1}]$, we approximate $f(\tilde{X}(s), \tilde{X}(s-\tau))$ by $f(\tilde{X}(t_n), \tilde{X}(t_n-\tau))$ and by the Taylor expansion we have

$$\begin{aligned} g_l(\tilde{X}(s), \tilde{X}(s-\tau)) &\approx g_l(\tilde{X}(t_n), \tilde{X}(t_n-\tau)) + \partial_x g_l(\tilde{X}(t_n), \tilde{X}(t_n-\tau)) \\ &\quad [\tilde{X}(s) - \tilde{X}(t_n)] + \partial_{x_\tau} g_l(\tilde{X}(t_n), \tilde{X}(t_n-\tau)) \\ &\quad [\tilde{X}(s-\tau) - \tilde{X}(t_n-\tau)] \end{aligned} \tag{4.2.22}$$

Substituting the above approximations into (4.2.1) and omitting the terms whose order is higher than one in (4.2.1), we then obtain the following scheme:

$$\begin{aligned} X_{n+1} &= X_n + hf(X_n, X_{n-m}) + \sum_{l=1}^{r} g_l(X_n, X_{n-m})\tilde{I}_0 \\ &\quad + \sum_{l=1}^{r}\sum_{q=1}^{r} \partial_x g_l(X_n, X_{n-m}) g_q(X_n, X_{n-m}) \tilde{I}_{q,l,t_n,t_{n+1},0} \\ &\quad + \sum_{l=1}^{r}\sum_{q=1}^{r} \partial_{x_\tau} g_l(X_n, X_{n-m}) g_q(X_{n-m}, X_{n-2m}) \\ &\quad \chi_{t_n \geq \tau} \tilde{I}_{q,l,t_n,t_{n+1},\tau},\ n = 0, 1, \cdots, N_T - 1, \end{aligned} \tag{4.2.23}$$

where

$$\tilde{I}_0 = \int_{t_n}^{t_{n+1}} \mathrm{d}\widetilde{W}_l(t), \quad \tilde{I}_{q,l,t_n,t_{n+1},0} = \int_{t_n}^{t_{n+1}} \int_{t_n}^{t} \mathrm{d}\widetilde{W}_q(s)\mathrm{d}\widetilde{W}_l(t),\ t_n \geq 0;$$

$$\tilde{I}_{q,l,t_n,t_{n+1},\tau} = \int_{t_n}^{t_{n+1}} \int_{t_n-\tau}^{t-\tau} \mathrm{d}\widetilde{W}_q(s)\mathrm{d}\widetilde{W}_l(t),\ t_n \geq \tau.$$

Using the Fourier basis (2.2.10), the three stochastic integrals in (4.2.23) are computed by

$$\tilde{I}_0^F = \int_{t_n}^{t_{n+1}} m_i^{(1)}(t)\xi_{l,1}^{(n)}\mathrm{d}t = \Delta W_{l,n}, \tag{4.2.24}$$

$$\begin{aligned} \tilde{I}^F_{q,l,t_n,t_{n+1},0} &= \frac{h}{2}\xi_{q,1}^{(n)}\xi_{l,1}^{(n)} - \frac{\sqrt{2}h}{2\pi}\xi_{q,1}^{(n)}\sum_{p=1}^{s}\frac{1}{p}\xi_{l,2p}^{(n)} \\ &\quad + \frac{h}{2\pi}\sum_{p=1}^{s_1}\frac{1}{p}[\xi_{q,2p+1}^{(n)}\xi_{l,2p}^{(n)} - \xi_{q,2p}^{(n)}\xi_{l,2p+1}^{(n)}], \end{aligned}$$

$$\begin{aligned} \tilde{I}^F_{q,l,t_n,t_{n+1},\tau} &= \frac{h}{2}\xi_{q,1}^{(n-m)}\xi_{l,1}^{(n)} - \frac{\sqrt{2}h}{2\pi}\xi_{q,1}^{(n-m)}\sum_{p=1}^{s}\frac{1}{p}\xi_{l,2p}^{(n)} + \frac{h}{2\pi}\sum_{p=1}^{s_1}\frac{1}{p}[\xi_{q,2p+1}^{(n-m)}\xi_{l,2p}^{(n)} \\ &\quad - \xi_{q,2p}^{(n-m)}\xi_{l,2p+1}^{(n)}], \end{aligned}$$

where $s = [\frac{N_h}{2}]$ and $s_1 = [\frac{N_h - 1}{2}]$. When a piecewise constant basis (2.2.9) is used, these integrals are

$$\tilde{I}_0^L = \sum_{j=0}^{N_h-1} \Delta W_{l,n,j} = \Delta W_{l,n},$$

$$\tilde{I}^L_{q,l,t_n,t_{n+1},0} = \sum_{j=0}^{N_h-1} \Delta W_{l,n,j}[\frac{\Delta W_{q,n,j}}{2} + \sum_{i=0}^{j-1} \Delta W_{q,n,i}], \tag{4.2.25}$$

$$\tilde{I}^L_{q,l,t_n,t_{n+1},\tau} = \sum_{j=0}^{N_h-1} \Delta W_{l,n,j}[\frac{\Delta W_{q,n-m,j}}{2} + \sum_{i=0}^{j-1} \Delta W_{q,n-m,i}],$$

where $\Delta W_{k,n,j} = W_k(t_n + \frac{(j+1)h}{N_h}) - W_k(t_n + \frac{jh}{N_h})$, $k = 1, \cdots, r$, $j = 0, \cdots, N_h - 1$, and $\Delta W_{k,n,-1} = 0$. In Example 4.4.2, we will show that the piecewise linear interpolation is less efficient than the Fourier approximation for achieving the same order of accuracy.

The scheme (4.2.23) can be seen as further discretization of the Milstein scheme for Stratonovich SDDEs proposed in [228]:

$$\begin{aligned} X^M_{n+1} &= X^M_n + hf(X^M_n, X^M_{n-m}) + \sum_{l=1}^{r} g_l(X^M_n, X^M_{n-m})\Delta W_{l,n} \\ &+ \sum_{l=1}^{r}\sum_{q=1}^{r} \partial_x g_l(X^M_n, X^M_{n-m}) g_q(X^M_n, X^M_{n-m}) I_{q,l,t_n,t_{n+1},0} \\ &+ \sum_{l=1}^{r}\sum_{q=1}^{r} \partial_{x_\tau} g_l(X^M_n, X^M_{n-m}) g_q(X^M_{n-m}, X^M_{n-2m}) \chi_{t_n \geq \tau} I_{q,l,t_n,t_{n+1},\tau}, \\ &\qquad n = 0, 1, \cdots, N_T - 1. \end{aligned} \tag{4.2.26}$$

as the double integrals approximated by either the Fourier expansion or the piecewise linear interpolation: $\tilde{I}_0$, $\tilde{I}_{q,l,t_n,t_{n+1},0}$, and $\tilde{I}_{q,l,t_n,t_{n+1},\tau}$ are, respectively, approximation of the following integrals[2]:

$$I_0 = \int_{t_n}^{t_{n+1}} \circ dW_l(t), \qquad I_{q,l,t_n,t_{n+1},0} = \int_{t_n}^{t_{n+1}} \int_{t_n}^{t} \circ dW_q(s) \circ dW_l(t), \; t_n \geq 0$$

$$I_{q,l,t_n,t_{n+1},\tau} = \int_{t_n}^{t_{n+1}} \int_{t_n-\tau}^{t-\tau} \circ dW_q(s) \circ dW_l(t), \; t_n \geq \tau. \tag{4.2.27}$$

Actually, we have the following relations.

---

[2]The approximation of double integrals in the present context is similar to those using numerical integration techniques, which has been long explored, see, e.g., [259, 358].

Lemma 4.2.4 *For the Fourier basis (2.2.10) with $\Delta = t_{n+1} - t_n$, it holds that*

$$\tilde{I}_0^F = I_0, \tag{4.28}$$

$$\mathbb{E}[(\tilde{I}^F_{q,l,t_n,t_{n+1},0} - I_{q,l,t_n,t_{n+1},0})^2] = \varsigma(N_h)\frac{2\Delta^2}{(N_h\pi)^2} + \sum_{i=M}^{\infty}\frac{\Delta^2}{(i\pi)^2} \le c\frac{\Delta^2}{\pi^2 M}, \tag{4.29}$$

$$\mathbb{E}[(\tilde{I}^F_{q,l,t_n,t_{n+1},\tau} - I_{q,l,t_n,t_{n+1},\tau})^2] = \varsigma(N_h)\frac{2\Delta^2}{(N_h\pi)^2} + \sum_{i=M}^{\infty}\frac{\Delta^2}{(i\pi)^2} \le c\frac{\Delta^2}{\pi^2 M}, \tag{4.30}$$

where $\varsigma(N_h) = 0$ if N_h is odd and 1 otherwise, and M is the integer part of $(N_h+1)/2$.

Proof. From (4.2.24), the first formula (4.28) can be readily obtained. Now we consider (4.29). For $l = q$, it holds that

$$\tilde{I}_{l,l,t_n,t_{n+1},0} = I_{l,l,t_n,t_{n+1},0} = (\Delta W_{l,n})^2/2,$$

if (2.2.8) with piecewise constant basis (2.2.9) or Fourier basis (2.2.10) is used. For any orthogonal expansion (2.2.5), we have $\mathbb{E}[\int_{t_n}^{t_{n+1}}(\widetilde{W}_q(s) - W_q(s))\,dW_l \int_{t_n}^{t_{n+1}} \widetilde{W}_q(s)\,d(\widetilde{W}_l - W_l)] = 0$ and thus by $W_q(t_n) = \widetilde{W}_q(t_n)$, Ito's isometry and integration by parts, we have, when $l \neq q$,

$$\begin{aligned}
&\mathbb{E}[(\tilde{I}_{q,l,t_n,t_{n+1},0} - I_{q,l,t_n,t_{n+1},0})^2] \\
&= \mathbb{E}[(\int_{t_n}^{t_{n+1}} [\widetilde{W}_q(s) - W_q(s)] \circ dW_l + \int_{t_n}^{t_{n+1}} \widetilde{W}_q(s)\,d[\widetilde{W}_l - W_l])^2] \\
&= \mathbb{E}[(\int_{t_n}^{t_{n+1}} [\widetilde{W}_q(s) - W_q(s)]\,dW_l)^2] + \mathbb{E}[(\int_{t_n}^{t_{n+1}} \widetilde{W}_q(s)\,d[\widetilde{W}_l - W_l])^2] \\
&= \int_{t_n}^{t_{n+1}} \mathbb{E}[[\widetilde{W}_q(s) - W_q(s)]^2]\,ds + \mathbb{E}[(-\int_{t_n}^{t_{n+1}} [\widetilde{W}_l - W_l]\,d\widetilde{W}_q(s))^2].
\end{aligned}$$

Then by the mutual independence of all Gaussian random variables $\xi_{q,i}^{(n)}$, $i = 1, 2, \cdots$, $q = 1, 2, \cdots, r$, we obtain $\mathbb{E}[[\widetilde{W}_q(s) - W_q(s)]^2] = \sum_{i=N_h+1}^{\infty} M_i^2(s)$, where $M_i(s) = \int_{t_n}^{s} m_i(\theta)\,d\theta$ and for $l \neq q$,

$$\begin{aligned}
\mathbb{E}[(\int_{t_n}^{t_{n+1}} [\widetilde{W}_l(s) - W_l(s)]\,d\widetilde{W}_q)^2] &= \mathbb{E}[(\sum_{i=N_h+1}^{\infty}\sum_{j=1}^{N_h}\int_{t_n}^{t_{n+1}} M_i(s)m_j(s)\,ds\xi_{l,i}^{(n)}\xi_{q,j}^{(n)})^2] \\
&= \sum_{i=N_h+1}^{\infty}\sum_{j=1}^{N_h}(\int_{t_n}^{t_{n+1}} M_i(s)m_j(s)\,ds)^2.
\end{aligned}$$

Then we have

$$\begin{aligned}&\mathbb{E}[(\tilde{I}_{q,l,t_n,t_{n+1},0}-I_{q,l,t_n,t_{n+1},0})^2]\\&=\sum_{i=N_h+1}^{\infty}\int_{t_n}^{t_{n+1}}M_i^2(s)\,ds+\sum_{i=N_h+1}^{\infty}\sum_{j=1}^{N_h}(\int_{t_n}^{t_{n+1}}M_i(s)m_j(s)\,ds)^2\end{aligned} \tag{4.2.31}$$

In (4.2.31), we consider the Fourier basis (2.2.10). Since $M_i(s)$ $(i \geq 2)$ are also sine or cosine functions, we have

$$\sum_{i=N_h+1}^{\infty}\sum_{j=1}^{N_h}(\int_{t_n}^{t_{n+1}}M_i(s)m_j(s)\,ds)^2=(\int_{t_n}^{t_{n+1}}M_{N_h+1}(s)m_{N_h}(s)\,ds)^2 \tag{4.2.32}$$

when N_h is even and $\sum_{i=N_h+1}^{\infty}\sum_{j=1}^{N_h}(\int_{t_n}^{t_{n+1}}M_i(s)m_j(s)\,ds)^2=0$ when N_h is odd. Moreover, for $i \geq 2$, it holds from simple calculations that

$$\int_{t_n}^{t_{n+1}}M_i^2(s)\,ds=\frac{3\Delta^2}{(2\lfloor i/2\rfloor\pi)^2},\ i\text{ is even and }\frac{\Delta^2}{(2\lfloor i/2\rfloor\pi)^2}\text{ otherwise.} \tag{4.2.33}$$

Then by (4.2.31), (4.2.32), we have

$$\begin{aligned}&\mathbb{E}[(\tilde{I}^F_{q,l,t_n,t_{n+1},0}-I_{q,l,t_n,t_{n+1},0})^2]\\&=\sum_{i=N_h+1}^{\infty}\int_{t_n}^{t_{n+1}}M_i^2(s)\,ds+\sum_{i=N_h+1}^{\infty}\sum_{j=1}^{N_h}(\int_{t_n}^{t_{n+1}}M_i(s)m_j(s)\,ds)^2\\&=\varsigma(N_h)\frac{\Delta^2}{(N_h\pi)^2}+\sum_{i=N_h+1}^{\infty}\frac{3^{\varsigma(i)}\Delta^2}{(2[i/2]\pi)^2}=\varsigma(N_h)\frac{2\Delta^2}{(N_h\pi)^2}+\sum_{i=M}^{\infty}\frac{\Delta^2}{(i\pi)^2}.\end{aligned}$$

Hence, we arrive at (4.29) by the fact $\sum_{i=M}^{\infty}\frac{1}{i^2}\leq\frac{1}{M}$. Similarly, we can obtain (4.30).

With Lemma 4.2.4, we can show that the Milstein-like scheme (4.2.23) can be of first-order convergence in the mean-square sense.

Theorem 4.2.5 *Assume that f, g_l, $\partial_x g_l g_q$, and $\partial_{x_\tau} g_l g_q$ $(l,q=1,2,\cdots,r)$ satisfy the Lipschitz condition (4.1.7) and also g_l have bounded second-order partial derivatives with respect to all variables. If $\mathbb{E}[\|\phi\|^p_{L^\infty}]<\infty$, $p\geq 4$, then we have for the Milstein-like scheme (4.2.23),*

$$\max_{1\leq n\leq N_T}\mathbb{E}|X(t_n)-X_n|^2=O(h^2), \tag{4.2.34}$$

when the double integrals $\tilde{I}_{q,l,t_n,t_{n+1},0}$, $\tilde{I}_{q,l,t_n,t_{n+1},\tau}$ are computed by (4.2.24) and N_h is at the order of $1/h$.

Proof. We present the proof for $d = 1$ in (4.1.6), which can be extended to multi-dimensional case $d > 1$ without difficulty. Subtracting (4.2.23) from (4.2.26) and taking expectation after squaring over both sides, we have

$$\mathbb{E}[(X_{n+1}^M - X_{n+1})^2] = \mathbb{E}[(X_n^M - X_n)^2] + 2\sum_{i=0}^{4} \mathbb{E}[(X_n^M - X_n)R_i] + \sum_{i,j=0}^{4} \mathbb{E}[R_i R_j],$$

where we denote $f_n^M = f(X_n^M, X_{n-m}^M)$ and $g_{l,n}^M = g_l(X_n^M, X_{n-m}^M)$ and

$$R_0 = h(f_n^M - f_n) + \sum_{l=1}^{r} (g_{l,n}^M - g_{l,n})\Delta W_{l,n},$$

$$R_1 = \sum_{l=1}^{r}\sum_{q=1}^{r} \left[\partial_x g_{l,n}^M g_{q,n}^M - \partial_x g_{l,n} g_{q,n}\right] \tilde{I}_{q,l,t_n,t_{n+1},0}^F,$$

$$R_2 = \sum_{l=1}^{r}\sum_{q=1}^{r} \left[\partial_{x_\tau} g_{l,n}^M g_{q,n-m}^M - \partial_{x_\tau} g_{l,n} g_{q,n-m}\right] \tilde{I}_{q,l,t_n,t_{n+1},\tau}^F,$$

$$R_3 = \sum_{l=1}^{r}\sum_{q=1}^{r} \partial_x g_{l,n}^M g_{q,n}^M (I_{q,l,t_n,t_{n+1},0} - \tilde{I}_{q,l,t_n,t_{n+1},0}^F),$$

$$R_4 = \sum_{l=1}^{r}\sum_{q=1}^{r} \partial_{x_\tau} g_{l,n}^M g_{q,n-m}^M (I_{q,l,t_n,t_{n+1},\tau} - \tilde{I}_{q,l,t_n,t_{n+1},\tau}^F).$$

Similar to the proof of Theorem 4.2.1, we have

$$\mathbb{E}[R_0^2] \le C(h^2 + h)(\mathbb{E}[(X_n^M - X_n)^2] + \mathbb{E}[(X_{n-m}^M - X_{n-m})^2]), \tag{4.2.35}$$

$$\mathbb{E}[R_1^2] \le C \max_{1\le l,q\le r} \mathbb{E}[(\left|X_n^M - X_n\right|^2 + \left|X_{n-m}^M - X_{n-m}\right|^2)]\mathbb{E}[(\tilde{I}_{q,l,t_n,t_{n+1},0}^F)^2],$$

$$\mathbb{E}[R_2^2] \le C \max_{1\le l,q\le r} \mathbb{E}[(\left|X_n^M - X_n\right|^2 + \left|X_{n-m}^M - X_{n-m}\right|^2)(\tilde{I}_{q,l,t_n,t_{n+1},\tau}^F)^2],$$

$$\mathbb{E}[R_3^2] \le C \max_{1\le l<q\le r} \mathbb{E}[(1 + \left|X_n^M\right|^2 + \left|X_{n-m}^M\right|^2)]\mathbb{E}[(I_{q,l,t_n,t_{n+1},0} - \tilde{I}_{q,l,t_n,t_{n+1},0}^F)^2],$$

$$\mathbb{E}[R_4^2] \le C \max_{1\le l<q\le r} \mathbb{E}[(1 + \left|X_n^M\right|^2 + \left|X_{n-m}^M\right|^2)(I_{q,l,t_n,t_{n+1},\tau} - \tilde{I}_{q,l,t_n,t_{n+1},\tau}^F)^2].$$

First, we establish the following estimations:

$$\mathbb{E}[R_i^2] \le Ch^3, \quad i = 3, 4. \tag{4.2.36}$$

The case for $i = 3$ follows directly from Lemma 4.2.4 and boundedness of moments of X_n and X_n^M. By Lemma 4.2.4 and (4.2.24), we have

$$\begin{aligned}
&\mathbb{E}[(I_{q,l,t_n,t_{n+1},\tau} - \tilde{I}_{q,l,t_n,t_{n+1},\tau}^F)^4] \\
&= \mathbb{E}[(-\frac{\sqrt{2}h}{2\pi}\xi_{q,1}^{(n-m)} \sum_{p=s+1}^{\infty} \frac{1}{p}\xi_{l,2p}^{(n)} + \frac{h}{2\pi} \sum_{p=s_1+1}^{\infty} \frac{1}{p}[\xi_{q,2p+1}^{(n-m)}\xi_{l,2p}^{(n)} - \xi_{q,2p}^{(n-m)}\xi_{l,2p+1}^{(n)}])^4] \\
&\le Ch^4[(\sum_{p=s+1}^{\infty} \frac{1}{p^2})^2 + (\sum_{p=s_1+1}^{\infty} \frac{1}{p^2})^2] \le C\frac{h^4}{N_h^2},
\end{aligned}$$

where $s = [\frac{N_h}{2}]$ and $s_1 = [\frac{N_h-1}{2}]$. As N_h is at the order of $1/h$, we have

$$\mathbb{E}[(I_{q,l,t_n,t_{n+1},\tau} - \tilde{I}^F_{q,l,t_n,t_{n+1},\tau})^4] \leq Ch^6. \tag{4.2.37}$$

Then by the fact that X_n and X_n^M have bounded fourth-order moment, Cauchy inequality, and (4.2.37), we reach (4.2.36) when $i = 4$.

Second, we estimate $\mathbb{E}[R_i^2]$, $i = 1, 2$. By (4.2.24), the Lipschitz condition (4.1.7) and N_h is at the order of $1/h$, we have

$$\mathbb{E}[R_1^2] \leq Ch(\mathbb{E}[(X_n^M - X_n)^2] + \mathbb{E}[(X_{n-m}^M - X_{n-m})^2]). \tag{4.2.38}$$

Now we require to estimate $\mathbb{E}[R_2^2]$. By Lipschitz condition (4.1.7), the adaptedness of X_{n-m}^M and X_{n-m} and Cauchy inequality (twice), we have

$$\begin{aligned}
\mathbb{E}[R_2^2] &\leq C \max_{1\leq l,q\leq r} \left\{ \mathbb{E}[\left|X_n^M - X_n\right|^2 (\tilde{I}^F_{q,l,t_n,t_{n+1},\tau})^2] + \mathbb{E}[\left|X_{n-m}^M - X_{n-m}\right|^2 (\tilde{I}^F_{q,l,t_n,t_{n+1},\tau})^2] \right\}, \\
&\leq C \max_{1\leq l,q\leq r} (\mathbb{E}[\left|X_n^M - X_n\right|^4])^{1/4} (\mathbb{E}[(\tilde{I}^F_{q,l,t_n,t_{n+1},\tau})^8])^{1/4} (\mathbb{E}[\left|X_n^M - X_n\right|^2])^{1/2} \\
&\quad + Ch^2\mathbb{E}[(X_{n-m}^M - X_{n-m})^2].
\end{aligned}$$

It can be readily checked from (4.2.24) that $\mathbb{E}[(\tilde{I}^F_{q,l,t_n,t_{n+1},\tau})^8] \leq Ch^8$. Hence, from the boundedness of moments, we have

$$\mathbb{E}[R_2^2] \leq Ch^2(\mathbb{E}[(X_n^M - X_n)^2])^{1/2} + Ch^2\mathbb{E}[(X_{n-m}^M - X_{n-m})^2]. \tag{4.2.39}$$

Now estimate $\mathbb{E}[(X_n^M - X_n)R_i]$, $i = 0, 1, 2, 3, 4$. By the adaptedness of X_n and Lipschitz condition of f, we have

$$\mathbb{E}[(X_n^M - X_n)R_0] \leq Ch\mathbb{E}[(\left|X_n^M - X_n\right|^2 + \left|X_{n-m}^M - X_{n-m}\right|^2)]. \tag{4.2.40}$$

By the adaptedness of X_n and $\mathbb{E}[\tilde{I}_{q,l,t_n,t_{n+1},0}] = \delta_{q,l}h/2$ ($\delta_{q,l}$ is the Kronecker delta) and Lipschitz condition of $\partial_x g_l g_q$, we have

$$\mathbb{E}[(X_n^M - X_n)R_1] \leq Ch\mathbb{E}[(\left|X_n^M - X_n\right|^2 + \left|X_{n-m}^M - X_{n-m}\right|^2)]. \tag{4.2.41}$$

By the adaptedness of X_n and $\mathbb{E}[\tilde{I}_{q,l,t_n,t_{n+1},0} - I_{q,l,t_n,t_{n+1},0}] = 0$, we have

$$\mathbb{E}[(X_n^M - X_n)R_3] = 0. \tag{4.2.42}$$

Similar to the proof of (4.2.16), we have

$$\mathbb{E}[(X_n^M - X_n)R_4] = 0. \tag{4.2.43}$$

Then by (4.2.35), (4.2.36)–(4.2.39), (4.2.40)–(4.2.43), and Cauchy inequality, we have

$$\begin{aligned}
\mathbb{E}[(X_{n+1}^M - X_{n+1})^2] &\leq (1 + Ch)\mathbb{E}[(X_n^M - X_n)^2] + Ch\mathbb{E}[(X_{n-m}^M - X_{n-m})^2] \\
&\quad + Ch^2(\mathbb{E}[(X_n^M - X_n)^2])^{1/2} + Ch^3,
\end{aligned} \tag{4.2.44}$$

where $n \geq m$. Similarly, we have that (4.2.44) holds also for $1 \leq n \leq m - 1$. From here and by the nonlinear Gronwall inequality, we reach the conclusion (4.2.34).

Remark 4.2.6 *When a piecewise linear approximation of double integrals* (4.2.25) *is used in the Milstein-like scheme* (4.2.23)*, the first-order strong convergence can be proved similarly when N_h is of the order of $1/h$.*

The spectral truncations we use are from the piecewise linear interpolation and a Fourier expansion. Comparison between these two truncations will be presented for a specific numerical example in Chapter 4.4, where it is shown that the Fourier approach is faster than the piecewise constant approach, similar to the case when $\tau = 0$ (SODEs), see [358, Section 1.4].

4.3 Linear stability of some schemes

For fixed h, let's analyze the behavior of numerical schemes described above when $t_j \to \infty$. This is done by analyzing the simple but a revealing linear test problem

$$\begin{aligned} \mathrm{d}X(t) &= \lambda X(t)\mathrm{d}t + \sigma X(t-\tau)\, dW(t), \quad t \in (0, T], \\ X(t) &= \phi(t), \quad t \in [-\tau, 0], \end{aligned} \tag{4.3.1}$$

where $\lambda < 0$, σ, $\tau \geq 0$. It can be shown that when $2\lambda + \sigma^2 < 0$, the solution to (4.3.1) is asymptotically stable in the mean-square sense:

$$\lim_{t\to\infty} \mathbb{E}[X^2(t)] = 0.$$

Specifically, when $\tau = 0$, the solution is

$$\exp((\lambda - \frac{1}{2}\sigma^2)t + \sigma W(t)).$$

If $2\lambda + \sigma^2 < 0$, then

$$\lim_{t\to\infty} \mathbb{E}[X^2(t)] = \lim_{t\to\infty} \mathbb{E}[\exp((2\lambda - \sigma^2)t + 2\sigma W(t))] = \lim_{t\to\infty} \mathbb{E}[\exp((2\lambda + \sigma^2)t)] = 0.$$

Stability region of the forward Euler scheme

Applying the forward Euler scheme to the linear test equation (4.3.1), we have

$$X_{n+1} = (1+z)X_n + \sqrt{y}X_{n-m}\xi_n, \quad z = \tilde{\lambda}h, \text{ and } y = \sigma^2 h.$$

Then $\mathbb{E}[X_{n+1}^2] = (1+z)^2\mathbb{E}[X_n^2] + y\mathbb{E}[X_n X_{n-m}]$, and thus

$$\mathbb{E}[X_{n+1}^2] \leq ((1+z)^2 + y/2)\mathbb{E}[X_n^2] + y\mathbb{E}[X_{n-m}^2]/2 \leq ((1+z)^2 + y)\max(\mathbb{E}[X_n^2], \mathbb{E}[(X_{n-m})^2]).$$

By the discrete Gronwall inequality, we need $(1+z)^2 + y < 1$ so that $\mathbb{E}[X_{n+1}^2]$ is decreasing and is asymptotically mean-square stable. Thus, the stability region is

$$\left\{(z, y) \in (-2, 0) \times (0, 1) | (1+z)^2 + y < 1\right\}.$$

This mean-square stability region is illustrated in Figure 4.1. In the following, it is shown that the stability regions of the midpoint and the predictor-corrector schemes are larger than the forward Euler scheme's when $\tau > 0$.

Fig. 4.1. The mean-square stability region of the forward Euler scheme (4.2.3) when $\tau \geq 0$.

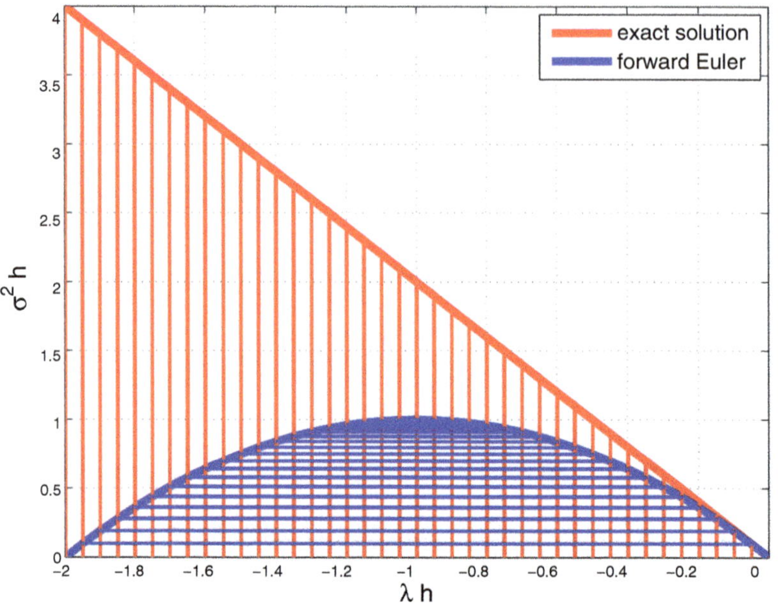

Stability analysis of the predictor-corrector scheme

In Stratonovich form, (4.3.1) is written as

$$\begin{aligned}&\mathrm{d}X(t) = \tilde{\lambda} X(t)\mathrm{d}t + \sigma X(t-\tau) \circ \mathrm{d}W_l(t), \quad t \in (0,\, T],\\ &X(t) = \phi(t), \quad t \in [-\tau, 0].\end{aligned} \tag{4.3.2}$$

where $\tilde{\lambda} = \lambda - \frac{\sigma^2}{2}\chi_{\{\tau=0\}}$. Applying (4.2.3) to (4.3.2), we have

$$\begin{aligned}\overline{X}_{n+1} &= X_n + \tilde{\lambda} h X_n + \sigma\sqrt{h} X_{n-m}\xi_n, \quad n \geq 0,\, m = \tau/h > 0\\ X_{n+1} &= X_n + \frac{\tilde{\lambda} h}{2}(X_n + \overline{X}_{n+1}) + \frac{1}{2}\sigma\sqrt{h}(X_{n-m} + X_{n-m+1})\xi_n.\end{aligned} \tag{4.3.3}$$

The scheme can be then simplified as

$$X_{n+1} = (1 + \tilde{\lambda} h + \frac{\tilde{\lambda}^2 h^2}{2})X_n + \frac{\sigma\sqrt{h}}{2}(1 + \tilde{\lambda} h)\xi_n X_{n-m} + \frac{\sigma\sqrt{h}}{2}\xi_n X_{n-m+1}.$$

By the Cauchy-Schwarz inequality, the second moment of X_{n+1} is bounded by

$$\mathbb{E}[X_{n+1}^2] \leq R^2(\tilde{\lambda} h)\mathbb{E}[X_n^2] + Q_1^2(\tilde{\lambda} h, \sigma\sqrt{h})\mathbb{E}[X_{n-m}^2] + Q_2^2(\tilde{\lambda} h, \sigma\sqrt{h})\mathbb{E}[X_{n-m+1}^2],$$

where $R(\tilde{\lambda}h) = 1 + \tilde{\lambda}h + \frac{\tilde{\lambda}^2h^2}{2}$, and

$$Q_1^2(\tilde{\lambda}h, \sigma\sqrt{h}) = \frac{\sigma^2 h}{4}(1+\tilde{\lambda}h)^2 + \frac{\sigma^2 h}{4}\left|1+\tilde{\lambda}h\right|, \quad Q_2^2(\tilde{\lambda}h, \sigma\sqrt{h}) = \frac{\sigma^2 h}{4} + \frac{\sigma^2 h}{4}\left|1+\tilde{\lambda}h\right|.$$

When

$$R^2(\tilde{\lambda}h) + Q_1^2(\tilde{\lambda}h, \sigma\sqrt{h}) + Q_2^2(\tilde{\lambda}h, \sigma\sqrt{h}) < 1,$$

i.e.,

$$(1 + \tilde{\lambda}h + \frac{\tilde{\lambda}^2h^2}{2})^2 + \frac{\sigma^2 h}{4}(\left|1+\tilde{\lambda}h\right| + 1)^2 < 1, \tag{4.3.4}$$

the second moment $\mathbb{E}[X_{n+1}^2]$ is asymptotically stable. According to (4.3.4), the stability region of the predictor-corrector scheme (4.2.3) when $\tau > 0$ is (here $z = \tilde{\lambda}h$ and $y = \sigma^2 h$)

$$\{(z,y) \in (-2,0) \times [0,3) | F(z,y) < 0 \text{ when } -1 < z < 0;$$

$$\text{and } G(z,y) > 0 \text{ when } -2 < z \leq -1 \text{ and } y < 3\},$$

where $F(z,y) = (2z+y)(1+z/2)^2 + z(2+z)\frac{z^2}{4}$ and $G(z,y) = \frac{z^3}{4} + z^2 + 2z + 2 + \frac{y}{4}z$. The stability region is illustrated in Figure 4.2. The stability region of the linear equation is plotted only for $\lambda h \geq -2.5$ while its stability region is $2\lambda + \sigma^2 < 0$.

Now let us derive the stability region of the predictor-corrector scheme (4.2.3) based on the condition (4.3.4). When $\sigma = 0$, we need $\left|R(\tilde{\lambda}h)\right| < 1$ which gives $-2 < \tilde{\lambda}h < 0$ (i.e., $\left|1+\tilde{\lambda}h\right| < 1$). Denote by $z = \tilde{\lambda}h$ and $y = \sigma^2 h$. If $1 + \tilde{\lambda}h > 0$, then (4.3.4) becomes

$$(1 + z + z^2/2)^2 + \frac{y}{4}(z+2)^2 < 1.$$

Let $F(z,y) = (1+z+z^2/2)^2 + \frac{y}{4}(z+2)^2 - 1$. It can be verified that if $1+z > 0$ and $|1+z| < 1$,

$$F(z,y) = (2z+y)(1+z/2)^2 + z(2+z)\frac{z^2}{4} < 0.$$

If $1 + z = 1 + \tilde{\lambda}h \leq 0$, then (4.3.4) becomes

$$(1 + z + z^2/2)^2 + \frac{y}{4}z^2 < 1.$$

Let $g(z,y) = (1+z+z^2/2)^2 + \frac{y}{4}z^2 - 1$. We want to find the range of z such that

$$g(z,y) < 0, \text{ when } -2 < z \leq -1.$$

Recall that

$$g(z,y) = z(\frac{z^3}{4} + z^2 + 2z + 2 + \frac{y}{4}z).$$

Fig. 4.2. The mean-square stability region of the predictor-corrector scheme (4.2.3) when $\tau > 0$.

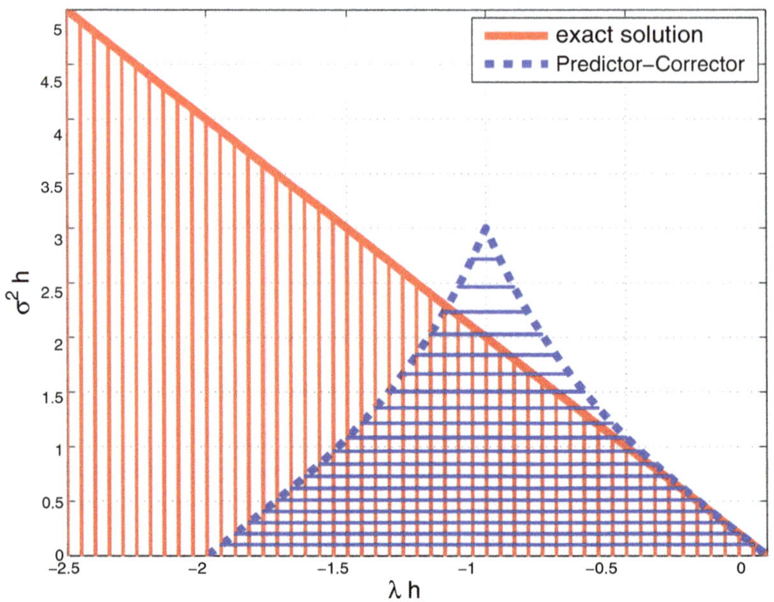

We then want $G(z,y) = g(z,y)/z = \frac{z^3}{4} + z^2 + 2z + 2 + \frac{y}{4}z > 0$. Note that

$$G(-1,y) = \frac{3-y}{4},\ G(-2,y) = -\frac{y}{2}, \quad \partial_z G(z,y) = \frac{3}{4}z^2 + 2z + 2 + \frac{y}{4} > 0.$$

Thus, only when $3 - y > 0$, there exists $-2 < z_0 < -1$ such that $0 = G(z_0, y) < G(-1, y)$.

Stability analysis of the midpoint scheme

The midpoint scheme applied to (4.3.2) reads

$$X_{n+1} = X_n + \tilde{\lambda}h(X_n + X_{n-1})/2 + \sigma(X_{n-m} + X_{n-m+1})/2\zeta^{(n)}\sqrt{h}, \quad n \geq 0,\ m = \tau/h \geq 0. \tag{4.3.5}$$

where $\zeta^{(n)}$ is a truncation of the standard Gaussian random variable $\xi^{(n)}$ as in (4.2.19). When $\tau > 0$, it holds that $X_{n+1} = R(\lambda h)X_n + Q(\lambda h, \sigma\sqrt{h})(X_{n-m} + X_{n-m+1})\zeta^{(n)}$ where

$$R(\lambda h) = \frac{1 + \lambda h/2}{1 - \lambda h/2}, \quad Q(\lambda h, \sigma\sqrt{h}) = \frac{1}{2}\frac{\sigma\sqrt{h}}{1 - \lambda h/2}.$$

Then we have

$$\begin{aligned}\mathbb{E}[X_{n+1}^2] &= R^2(\lambda h)\mathbb{E}[X_n^2] + Q^2(\lambda h, \sigma\sqrt{h})\mathbb{E}[(X_{n-m} + X_{n-m+1})^2]\mathbb{E}[(\varsigma^{(n)})^2] \\ &< R^2(\lambda h)\mathbb{E}[X_n^2] + Q^2(\lambda h, \sigma\sqrt{h})\mathbb{E}[(X_{n-m} + X_{n-m+1})^2] \qquad (4.3.6) \\ &< \big(R^2(\lambda h) + 2Q^2(\lambda h, \sigma\sqrt{h})\big)\max(\mathbb{E}[X_n^2], \mathbb{E}[(X_{n-m})^2], \mathbb{E}[X_{n-m+1})^2]).\end{aligned}$$

When $R^2(\lambda h) + 2Q^2(\lambda h, \sigma\sqrt{h}) < 1$, i.e.

$$(\frac{1+\lambda h/2}{1-\lambda h/2})^2 + \frac{1}{2}(\frac{\sigma\sqrt{h}}{1-\lambda h/2})^2 < 1,$$

then $\mathbb{E}[X_{n+1}^2]$ is strictly decreasing and is asymptotically stable. We conclude that the midpoint scheme is asymptotically stable for any $h > 0$ as long as $\lambda + \sigma^2/2 < 0$. In other word, the midpoint scheme has the same asymptotic mean-square stable region as the equation (4.3.1).

4.4 Numerical results

In this section, we test convergence rates of the aforementioned schemes and compare their numerical performance under commutative or noncommutative noises as well as the effect of delay. In the first example, we test the predictor-corrector scheme (4.2.3) and midpoint scheme (4.2.19) and show that both methods are of half-order mean-square convergence. Further, both schemes converge with order one in the mean-square sense for a SDDE with single white noise and no time delay in diffusion coefficients. In the second example, we investigate the Milstein-like scheme (4.2.23) and show that it is first-order convergent for SDDEs with multiple white noises.

Throughout this section, mean-square errors of numerical solutions are defined as

$$\rho_{h,T} = \big(\frac{1}{n_p}\sum_{i=1}^{n_p}|X_h(T,\omega_i) - X_{2h}(T,\omega_i)|^2\big)^{1/2},$$

where ω_i denotes the i-th single sample path and n_p is the total number of sample paths.

The numerical tests were performed using Matlab R2012a on a Dell Optiplex 780 computer with CPU (E8500 3.16 GHz). We used the Mersenne Twister random generator with seed 1 and took a large number of paths so that the statistical error can be ignored. Newton's method with tolerance $h^2/100$ was used to solve the nonlinear algebraic equations at each step of the implicit schemes.

We test the convergence rate for the predictor-corrector scheme (4.2.3) and the midpoint scheme (4.2.19) for SDDEs with different types of noises: noncommutative noise, single noise. We will show that the time delay in a diffusion coefficient keeps both methods only convergent at half-order, while for the SDDE with single noise, the two schemes can be of first-order accuracy in the mean-square sense if the time delay does not appear explicitly in the diffusion coefficients.

Example 4.4.1 *Consider Equation* (4.1.6) *in one dimension and assume the initial function* $\phi(t) = t + 0.2$, *with different diffusion terms:*

- *commutative (single) white noises without delay in the diffusion coefficients:*

$$dX = [-X(t) + \sin(X(t-\tau))]\,dt + \sin(X(t)) \circ dW(t); \tag{4.4.1}$$

- *noncommutative noises without delay in the diffusion coefficients:*

$$dX = [-X(t) + \sin(X(t-\tau))]\,dt + \sin(X(t)) \circ dW_1(t) + 0.5X(t) \circ dW_2(t), \tag{4.4.2}$$

  where the noises are noncommutative as $\partial_x(sin(x))0.5x - \partial_x(0.5x)$ $sin(x) \neq 0$;
- *single white noises with delay in the diffusion coefficients:*

$$dX = [-X(t) + \sin(X(t-\tau))]\,dt + \sin(X(t-\tau)) \circ dW(t). \tag{4.4.3}$$

For Equation (4.4.1), there is a single noise with no delay in the diffusion coefficient. The noise is a special case of commutative noises. The convergence order of these two schemes is one in the mean-square sense, see, e.g., [356]. The convergence order is numerically verified in Figure 4.3. The effect of the delay τ is small and does not affect the convergence order.

Fig. 4.3. Mean-square errors of the predictor-corrector (left) and midpoint schemes (right) for Example 4.4.1 at T=5 with different τ using $n_p = 10000$ sample paths: single white noise with a non-delayed diffusion term. These figures are adapted from [61].

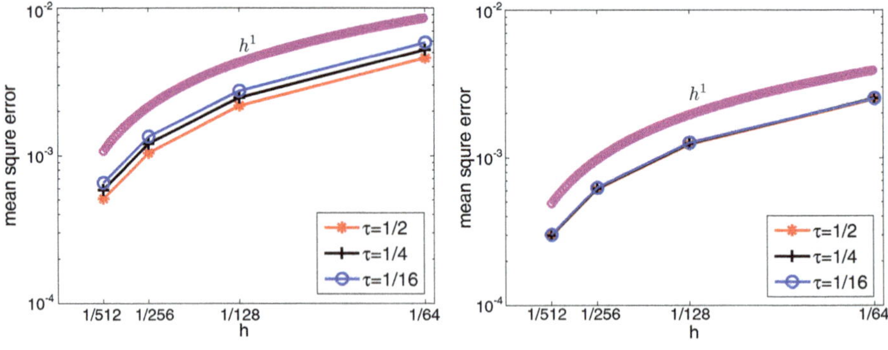

Fig. 4.4. Mean-square convergence test of the predictor-corrector (left column) and midpoint schemes (right column) on Example 4.4.1 at T=5 with different τ using n_p = 10000 sample paths: multi white noises with non-delayed diffusion terms. These figures are adapted from [61].

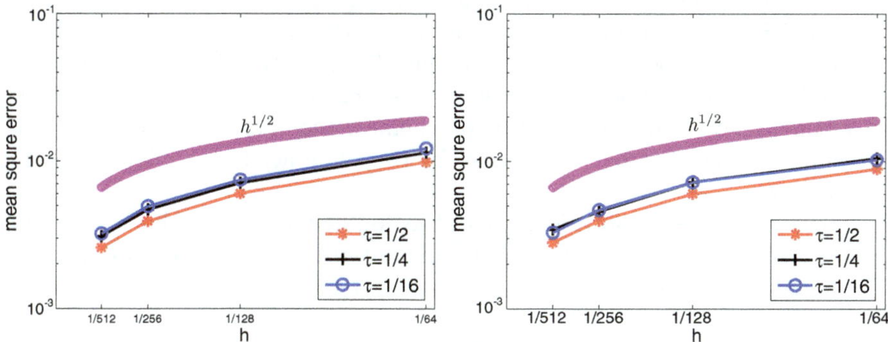

For Equation (4.4.2) with noncommutative noises, the convergence order is half for both schemes, see Figure 4.4.

For Equation (4.4.3), there is a single noise while the diffusion coefficient contains a delay term. In this case, we observe half-order strong convergence in Figure 4.5. This implies that the delay has affected the convergence order since the order is one if no delay appears in the diffusion coefficient, e.g., in (4.4.1).

Fig. 4.5. Mean-square convergence test of the predictor-corrector (left column) and midpoint schemes (right column) on Example 4.4.1 at T=5 with different τ using n_p = 10000 sample paths: single white noise with a delayed diffusion term. These figures are adapted from [61].

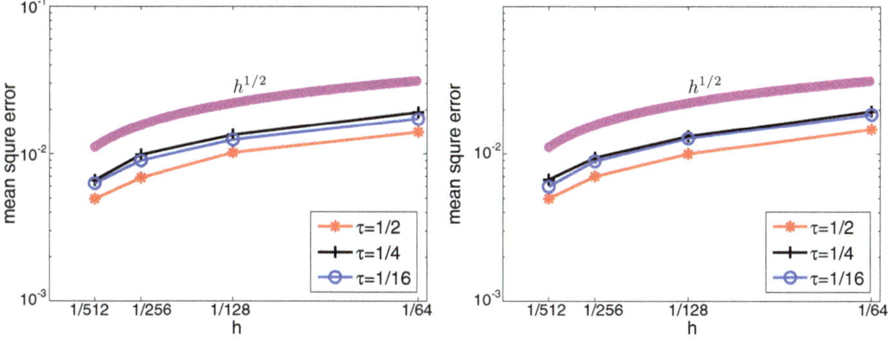

From this example, we conclude that *for the predictor-corrector and midpoint schemes, when the time delay only appears in the drift term, the convergence order is one for the equation with commutative noises and half for*

the one with noncommutative noises. However, *when the diffusion coefficients contain time delays, these two schemes are only half-order even for equations with a single white noise.*

In the following examples, we test the Milstein-like scheme (4.2.23) using different bases, i.e., piecewise constant basis (2.2.9) and Fourier approximation (2.2.10), and compare its numerical performance with the predictor-corrector and midpoint schemes. For the Milstein-like scheme, we show that for multiple noises, the computational cost for achieving the same accuracy is much higher than the other two schemes, see Example 4.4.2; while for single noise, the computational cost for the same accuracy is lower, see Example 4.4.3.

To reduce the computational cost, the double integrals in the Milstein-like scheme are computed by the Fourier expansion approximation (4.2.24) and the following relation

$$\tilde{I}_{q,l,t_n,t_{n+1},0} = \Delta W_{l,n} \Delta W_{q,n} - \tilde{I}_{l,q,t_n,t_{n+1},0}, \quad \tilde{I}_{l,l,t_n,t_{n+1},0} = \frac{(\Delta W_{l,n})^2}{2}. \tag{4.4.4}$$

We also use the following relations

$$\tilde{I}_{q,l,t_n,t_n+ph,0} = \sum_{j=0}^{p-1} \left[\tilde{I}_{q,l,t_n+jh,t_n+(j+1)h,0} + \Delta W_{l,n+j} \chi_{j\geq 1} \sum_{i=0}^{j-1} \Delta W_{q,n+i} \right],$$

$$\tilde{I}_{q,l,t_n,t_n+ph,\tau} = \sum_{j=0}^{p-1} \left[\tilde{I}_{q,l,t_n+jh,t_n+(j+1)h,\tau} + \Delta W_{l,n+j} \chi_{j\geq 1} \sum_{i=0}^{j-1} \Delta W_{q,n-m+i} \right].$$

Example 4.4.2 *We consider the Milstein-like scheme* (4.2.23) *for*

$$\begin{aligned} \mathrm{d}X(t) &= [-9X(t) + \sin(X(t-\tau))]\mathrm{d}t + [\sin(X(t)) + X(t-\tau)] \circ dW_1(t) \\ &\quad + [X(t) + \cos(0.5X(t-\tau))] \circ dW_2(t), \; t \in (0, T] \\ X(t) &= t + \tau + 0.1, \; t \in [-\tau, 0]. \end{aligned} \tag{4.4.5}$$

In Table 4.1, we show that for Equation (4.4.5), the Milstein-like scheme (4.2.23) converges with order one in the mean-square sense. Compared to the predictor-corrector scheme or the midpoint scheme, when the time step sizes are the same, the computational cost for the Milstein-like scheme (4.2.23) is several times higher. In the Milstein-like scheme, the extra computational cost comes from evaluating the double integrals $\tilde{I}^F_{q,l,t_n,t_{n+1},0}$ and $\tilde{I}^F_{q,l,t_n,t_{n+1},\tau}$ at each time step, which requires $7/(2h)(3r^2 - r)/2$ operations when we take the relation (4.4.4) into account.

We also test the Milstein-like scheme (4.2.23) using the piecewise constant basis (2.2.9). The computational cost is even higher than that of using the Fourier basis for the same time step size. Actually, the amount of operations for evaluating double integrals using (4.2.25) is $(1/(2h^2) + 5/(2h) - 1)(3r^2 - r)/2$, which is $O(1/h^2)$, much more than that of using the Fourier basis,

Table 4.1. Convergence rate of the Milstein-like scheme (left) for Equation (4.4.5) at $T = 1$ and comparison with the convergence rate of the predictor-corrector scheme (middle) and the midpoint scheme (right) using $n_p = 4000$ sample paths. The upper rows are with $\tau = 1/16$ and the lower are with $\tau = 1/4$.

h	$\rho_{h,T}$	Order	Time (s.)	$\rho_{h,T}$	Order	Time (s.)	$\rho_{h,T}$	Order	Time (s.)
2^{-5}	9.832e-02	–	1.0	7.164e-02	–	0.10	5.000e-02	–	0.29
2^{-6}	4.090e-02	1.27	1.7	3.734e-02	0.94	0.12	3.304e-02	0.60	0.41
2^{-7}	1.921e-02	1.09	3.3	2.308e-02	0.69	0.25	2.263e-02	0.55	0.79
2^{-8}	9.703e-03	0.99	6.4	1.616e-02	0.51	0.40	1.590e-02	0.51	1.54
2^{-5}	9.307e-02	–	0.93	6.956e-02	–	0.10	5.050e-02	–	0.22
2^{-6}	3.824e-02	1.28	1.6	3.582e-02	0.96	0.17	3.155e-02	0.68	0.39
2^{-7}	1.804e-02	1.08	2.8	2.205e-02	0.70	0.26	2.133e-02	0.56	0.78
2^{-8}	9.069e-03	0.99	5.5	1.434e-02	0.62	0.45	1.425e-02	0.58	1.59

$O(1/h)$. Our numerical tests (not presented here) confirmed the fast increase of the amount of operations.

However, the amount of operations of the Milstein-like scheme can be significantly reduced when there is just a single diffusion term.

Example 4.4.3 *We consider the Milstein-like scheme* (4.2.23) *for*

$$\begin{aligned} dX(t) &= [-2X(t) + 2X(t-\tau)]dt + [\sin(X(t)) + X(t-\tau)] \circ dW(t),\ t \in (0,T] \\ X(t) &= t + \tau,\ t \in [-\tau, 0]. \end{aligned} \tag{4.4.6}$$

In Table 4.2, we observe that the Milstein-like scheme for Equation (4.4.6) is still of first-order convergence but the predictor-corrector scheme and the midpoint scheme are only of half-order convergence. For the same accuracy, the computational cost for the Milstein-like scheme using the Fourier basis is less than that for the other two schemes. In fact, for single noise, we only need to compute one double integral $\tilde{I}_{1,1,t_n,t_{n+1},\tau}$. Moreover, when the coefficients of the diffusion term are small, a small number of Fourier modes is required for large time step sizes, i.e., N_h can be $O(1)$ instead of $O(1/h)$. The computational cost can thus be reduced somewhat, see, e.g., [358, Chapter 3] for such a discussion for equations with small noises without delay.

In summary, *the predictor-corrector scheme and midpoint scheme are convergent with half-order in the mean-square sense and can be of first-order if the underlying SDDEs with single noise (commutative noise) and the time delay is only in the drift coefficients*, see Example 4.4.1. In Example 4.4.2 the numerical tests show that the Milstein-like scheme is of first-order in the mean-square sense for SDDEs with noncommutative noise wherever the time delay appears, i.e., in the drift and/or diffusion coefficients. Compared to the other two schemes, the Milstein-like scheme is more accurate but is more expensive as it requires evaluations of double integrals, with cost inversely proportional to the time step size and proportional to the square of number of

Table 4.2. Convergence rate of the Milstein-like scheme (left) for Equation (4.4.6) (single white noise) at $T = 1$ and comparison with the convergence rate of the predictor-corrector scheme (middle) and midpoint scheme (right) using $n_p = 4000$ sample paths. The delay τ is taken as $1/4$.

h	$\rho_{h,T}$	Order	Time (s.)	h	$\rho_{h,T}$	Order	Time (s.)	$\rho_{h,T}$	Order	Time (s.)
2^{-5}	3.164e-02	–	0.28	2^{-8}	1.252e-02	–	0.37	1.263e-02	–	1.09
2^{-6}	1.688e-02	0.91	0.46	2^{-9}	9.219e-03	0.44	0.56	9.246e-03	0.45	2.05
2^{-7}	8.499e-03	0.99	0.79	2^{-10}	6.462e-03	0.51	1.03	6.471e-03	0.51	3.97
2^{-8}	4.570e-03	0.90	1.40	2^{-11}	4.617e-03	0.49	1.91	4.627e-03	0.48	7.58

noises. However, *for SDDEs with single noise in Example 4.4.3, the Milstein-like scheme (with the Fourier basis) can be superior to the predictor-corrector scheme and the midpoint scheme both in terms of accuracy and computational cost.*

4.5 Summary and bibliographic notes

We have presented three distinct schemes using different time-discretization techniques after approximating the Brownian motion by a spectral expansion (a type of Wong-Zakai approximation): midpoint scheme, predictor-corrector scheme, and a Milstein-like scheme. The mechanism is similar to the Wong-Zakai approximation for stochastic differential equation without delay. Further, if there is no delay in diffusion coefficients, the performance of all schemes is similar to that of schemes for stochastic ordinary differential equations without delay:

- The predictor-corrector scheme is of half order convergence in the mean-square sense, see Theorem 4.2.1 and numerical examples in Chapter 4.4.
- The midpoint scheme is of half order convergence in the mean-square sense, see numerical examples in Chapter 4.4.
- The Milstein-like scheme is of first order convergence in the mean-square sense, see Theorem 4.2.5 and Examples 4.4.2 and 4.4.3.

Also, when the diffusion coefficient satisfies the commutative conditions, both the predictor-corrector scheme and the midpoint scheme can be of first order convergence in the mean-square sense.

However, the numerical performance is a bit different if there is even a single diffusion coefficient with delay:

- The predictor-corrector scheme and midpoint scheme are still of half order convergence in the mean-square sense. This can be understood as the break of commutative conditions on diffusion coefficients for stochastic ordinary differential equations.
- The Milstein-like scheme is much more expensive as the double integrals involve the delayed history, which implies less use of this first-order scheme in practice, especially if there are high-dimensional Wiener processes involved. However, it is efficient when there is only one Brownian motion (Wiener processes).

Linear stability theory for basic numerical schemes is presented in Chapter 4.3.

A spectral approximation with different time discretizations does not lead to higher-order numerical methods unless we use higher-order time discretization. This is in accordance with classical theories on numerical methods for stochastic different equations, see, e.g., [259, 354, 358]. In the following chapters, we show that this conclusion is also true for parabolic equations with temporal white noise. However, in Chapter 10, we show that a spectral approximation of spatial white noise can lead to higher-order numerical methods of time-independent equations when the solutions are smooth, e.g., for elliptic and biharmonic equations with additive noise.

Bibliographic notes. Numerical solution of stochastic delay differential equations (SDDEs) has attracted increasing interest recently, as memory effects in the presence of noise are modeled with SDDEs in engineering and finance, e.g., [173, 221, 392, 449, 460]. Most of the numerical methods for SDDEs have focused on the convergence and stability of different time-discretization schemes since the early work [451, 452]. Currently, several time-discretization schemes have been well studied: the Euler-type schemes (the forward Euler scheme [14, 279] and the drift-implicit Euler scheme [230, 306, 483]), the Milstein schemes [49, 222, 228, 260], the split-step schemes [176, 472, 501], and also some multi-step schemes [50, 51, 59, 60]. These schemes are usually based on the Ito-Taylor expansion, see, e.g., [260] or anticipative calculus, see, e.g., [228].

The **Wong-Zakai (WZ) approximation** is different from these aforementioned approaches. The difference is that in WZ we first approximate the Brownian motion with an absolute continuous process, see, e.g., [324, 453, 458] for a piecewise linear approximation of Brownian motion in SDDEs. Subsequently, we apply proper time-discretization schemes for the resulting equation while the schemes mentioned in the last paragraph are ready for simulation without any further time discretization. The WZ approximation thus can be viewed as an intermediate step for deriving numerical schemes and can provide more flexibility of discretization of Brownian motion before performing any time discretization. Moreover, with the WZ approximation we can apply Taylor expansion rather than Ito-Taylor expansion and anticipative calculus.

The Wong-Zakai approximation for SODEs, see, e.g., [481, 482], is a semi-discretization method where Brownian motion is approximated by finite dimensional absolute continuous stochastic processes before any discretization in time or in physical space. There are different types of WZ approximation, see, e.g., [241, 386, 434, 510]. The Wong-Zakai approximation has been extended in various aspects since the seminal papers in [481, 482]:

- from single white noise to multiple white noise, see, e.g., [195, 435].
- from SODEs to SPDEs, hyperbolic equations (e.g., [352, 406, 407, 508]), parabolic equations (e.g., [3, 114]) including Burgers equation (e.g., [382, 383]) and Navier-Stokes equations (e.g., [79, 456, 457]), and equations on a manifold (e.g., [45, 198]), etc.
- from piecewise linear approximation to general approximation: mollifier type (e.g., [241, 323, 325]), Ikeda-Nakao-Yamato-type approximations (e.g., [143, 239, 323]) or their extensions (e.g., [195]), (Lévy-Ciesielski) spectral type (e.g., [54, 315, 508]), general color noises (e.g., [3, 434, 435]), etc.

- from SODEs driven by Gaussian white noise to those with general processes: general semi-martingales (e.g., [127, 142, 212, 263, 283, 284, 401]), fractional Brownian motion (e.g., [20, 450]), rough path (e.g., [137]), etc.

To the best of our knowledge, this is the first time that *fully discretized* numerical schemes via Wong-Zakai approximation have been presented. One of the reasons that Wong-Zakai approximation has not been popular as a numerical method is that it is difficult to establish error estimates. One difficulty is that in the Wong-Zakai approximation (not yet fully discretized in time), the solutions to the resulting equations are not adapted to the natural filtration of Wiener process. To have an adapted solution, Gyongy and his collaborators, e.g., [200, 201], modified the standard piecewise linear approximation (2.2.4) as

$$W^{(n)}(t) = W(\mathsf{t}_{i-1}) + (W(\mathsf{t}_i) - W(\mathsf{t}_{i-1}))\frac{t-\mathsf{t}_i}{\mathsf{t}_{i+1}-\mathsf{t}_i}, \quad t \in [\mathsf{t}_i, \mathsf{t}_{i+1}). \quad (4.5.1)$$

With such as modification, the authors of [200, 201] have proved a $1/4-\epsilon$-order convergence [200] and $1/2-\epsilon$-order convergence in pathwise sense for Wong-Zakai approximation of linear stochastic parabolic equations with multiplicative noise (white noise as coefficients of first-order differential operators).

Simulation of double integrals (4.2.27). In [228], $I_{q,l,t_n,t_{n+1},0}$ and $I_{q,l,t_n,t_{n+1},\tau}$ are approximated similarly. The Brownian motion W_q therein is approximated by the sum of $(t-t_n)/(t_{n+1}-t_n)W_q(t_n)$ and a truncated Fourier expansion of the Brownian bridge $W_q(t)-(t-t_n)/(t_{n+1}-t_n)W_q(t_n)$ for $t_n \le t \le t_{n+1}$, see also [159] and [259, Section 5.8]. It can be readily checked that this approximation is equivalent to the Fourier approximation (4.2.24). In numerical simulations (results are not presented), these two approximations lead to a little difference in computational cost and accuracy but the convergence order is the same.

As we note in the beginning of Chapter 2, the choice of complete orthonormal bases is arbitrary. However, the use of general spectral approximation may lead to different accuracy, see, e.g., [287] for a detailed comparison of some spectral approximation of multiple Stratonovich integrals.

In addition to the Fourier approximation, several methods of approximating $I_{q,l,t_n,t_{n+1},0}$ have been proposed: applying the trapezoid rule, see, e.g., [358, Section 1.4] and modified Fourier approximation, see, e.g., [480]. We note that the use of the trapezoid rule leads to a similar formula as in (4.2.25), which is shown to be less efficient than the Fourier approximation, see Example 4.4.2. In [480], $I_{q,l,t_n,t_{n+1},0}$ is approximated with the sum of a Fourier approximation and a tail process $A_{q,l,t_n,t_{n+1}}$, where the tail $A_{q,l,t_n,t_{n+1}}$ is modeled with the product of $r(r-1)/2$-dimensional i.i.d. Gaussian random variables and a functional of increments of Brownian motion $\Delta W_{l,n}$. It is shown in [159] that the modified Fourier approximation in [480] requires $O(r^4\sqrt{h})$ i.i.d. Gaussian random variables to maintain the first-order convergence while the Fourier approximation requires $O(r^2h^{-1})$ i.i.d.

Gaussian random variables. However, it is difficult to extend this approach to approximate $I_{q,l,t_n,t_{n+1},\tau}$ even when r is small because a tail $A_{q,l,t_n,t_{n+1},0}$ will be correlated with $A_{q,l,t_n,t_{n+1},\tau}$, which is difficult to identify and brings no computational benefits.

In practice, the cost of simulating double integrals is prohibitively expensive. However, there are cases where we can reduce or even avoid the simulation of double integrals. For example, when the diffusion coefficients are small and of the order ϵ and the coefficients at the double integrals are of order ϵ^2, we may take $\epsilon^2/N_h \sim O(h)$ to achieve an accuracy of $\mathcal{O}(h)$ in the mean-square sense, according to the proof of Theorem 4.2.5. Thus, only a small N_h (the number of terms in the truncated spectral approximation of Brownian motion) is required if $\epsilon \sim O(\sqrt{h})$. Also, when the diffusion coefficients contain no delay and satisfy the so-called commutative noises , i.e., $\frac{\partial g_l(x)}{\partial x} g_q(x) = \frac{\partial g_q(x)}{\partial x} g_l(x)$, the Milstein-like scheme can be rewritten as

$$
\begin{aligned}
X_{n+1}^M &= X_n^M + hf(X_n^M, X_{n-m}^M) + \sum_{l=1}^{r} g_l(X_n^M)\Delta W_{l,n} \\
&+ \frac{1}{2}\sum_{l=1}^{r}\sum_{q=1}^{r} \partial_x g_l(X_n^M) g_q(X_n^M)\Delta W_{l,n}\Delta W_{q,n}, \quad n = 0, 1, \cdots, N_T - 1.
\end{aligned}
$$

In this case, only Wiener increments are used and the Milstein scheme is of low cost.

4.6 Suggested practice

Exercise 4.6.1 *Derive and plot the mean-square stability region of the Milstein-like scheme* (4.2.23) *for Equation* (4.3.1).

Exercise 4.6.2 *Use the definitions of the Ito's and Stratonovich integrals in* (2.3) *to convert the equation* (4.1.6) *to an Ito's equation. In particular, Equation* (4.3.2) *can be written as* (4.3.1).

Hint. Apply the relation (2.3.2). Compare with the case $\tau = 0$ in Exercise 3.6.3.

Exercise 4.6.3 *Apply the backward (drift-implicit) Euler scheme to the linear test equation* (4.3.1) *and derive the mean-square stability region. Can you also apply the fully implicit Euler scheme and obtain the mean-square stability region?*

Exercise 4.6.4 (Single noise, commutative noises) *Write Matlab code for the Milstein-like scheme* (4.2.23), *the predictor-corrector scheme* (4.2.3), *and the midpoint scheme* (4.2.19) *for*

$$\begin{aligned} \mathrm{d}X(t) &= [-2X(t) + 2X(t-\tau)]\mathrm{d}t + [X(t) + X(t-\tau)] \circ dW(t),\ t \in (0, T] \\ X(t) &= t + \tau,\ t \in [-\tau, 0]. \end{aligned} \tag{4.6.1}$$

Check the pathwise (almost sure) convergence rate using

$$\rho_{h,T}^{a.s.} = |X_h(T, \omega) - X_{2h}(T, \omega)|,$$

where ω *represents a user-defined simulation path. The convergence rate should be close to* $1/2$, $1/2$, *and* 1, *which can be computed as follows (cf. the rate in Appendix E)*

$$\frac{\log(\rho_{h,T}^{a.s.}/\rho_{h/2,T}^{a.s.})}{\log(2)}.$$

Use the Fourier approximation (4.2.24) *for the Milstein-like scheme.*

Exercise 4.6.5 (Non-commutative noise) *Write Matlab code for the Milstein-like scheme* (4.2.23), *the predictor-corrector scheme* (4.2.3), *and the midpoint scheme* (4.2.19) *for Equation* (4.4.2) *and check the pathwise convergence rate as in Exercise 4.6.4.*

Exercise 4.6.6 *Write Matlab code for the Milstein-like scheme* (4.2.23), *the predictor-corrector scheme* (4.2.3), *and the midpoint scheme* (4.2.19) *for the following equation*

$$dX = [-X(t) + \sin(X(t-\tau))]\,dt + \sin(X(t-\tau)) \circ dW_1(t) + 0.5X(t) \circ dW_2(t), \tag{4.6.2}$$

and check the pathwise convergence rate as in Exercise 4.6.4. Remark on your observations on the difference of numerical results when $\tau = 0$ *and* $\tau \neq 0$ *for the Milstein-like scheme.*

5

Balanced numerical schemes for SDEs with non-Lipschitz coefficients

In this chapter, we discuss numerical methods for SDEs with coefficients of polynomial growth. The nonlinear growth of the coefficients induces instabilities, especially when the nonlinear growth is polynomial or even exponential. For stochastic differential equations (SDEs) with coefficients of polynomial growth at infinity and satisfying a one-sided Lipschitz condition, we prove a fundamental mean-square convergence theorem on the strong convergence order of a stable numerical scheme in Chapter 5.2. We apply the theorem to a number of existing numerical schemes. We present in Chapter 5.3 a special balanced scheme, which is explicit and of half-order mean-square convergence. Some numerical results are presented in Chapter 5.4. We summarize the chapter in Chapter 5.5 and present some bibliographic notes on numerical schemes for nonlinear SODEs. Three exercises are presented for interested readers.

5.1 A motivating example

Usually, in numerical analysis for SDEs [259, 354, 358], it is assumed that the SDEs coefficients are globally Lipschitz, which is a significant limitation as many models in applications of physical interest have coefficients growing faster at infinity than a linear function. If the global Lipschitz condition is violated, almost all standard numerical methods (explicit schemes) will fail to converge, see, e.g., [218, 226, 359, 437]. To see this, let us consider the following example.

$$dX = -X^3\,dt + dW(t), \quad X(0) = X_0. \tag{5.1.1}$$

Here the drift coefficient $-X^3$ is not global Lipschitz as it grows cubically. In the following, we show that the standard Euler scheme fails to converge.

Z. Zhang, G.E. Karniadakis, *Numerical Methods for Stochastic Partial Differential Equations with White Noise*,
Applied Mathematical Sciences 196, DOI 10.1007/978-3-319-57511-7_5

Suppose that we want to compute the solution at T, given a uniform partition of the time interval $[0,T]$ and a time step size h. We assume that $X(0) = X_0 = 1/h$.[1] Then by the Euler scheme (3.2.1), we have

$$X_1 = X_0 - X_0^3 h + \xi_1\sqrt{h} = \frac{1}{h} - \frac{1}{h^2} + \xi_1\sqrt{h}, \quad \xi_1 \sim \mathcal{N}(0,1).$$

With a large probability, X_1 is of the order $\frac{1}{h^2}$, as ξ_1 has a large probability taking small values, e.g., $\mathbb{P}(|\xi_1| \leq 3) = 0.98$. By one more step from the Euler scheme, we have

$$X_2 = X_1 - X_1^3 h + \xi_2\sqrt{h} = -\frac{1}{h^2} + \frac{1}{h^5} + \xi_2\sqrt{h}, \quad \xi_2 \sim \mathcal{N}(0,1),$$

which can be considered of the order $-\frac{1}{h^5}$ with a large probability. One more step gives us

$$X_3 = X_2 - X_2^3 h + \xi_3\sqrt{h} \approx -\frac{1}{h^{14}}, \quad \xi_3 \sim \mathcal{N}(0,1).$$

Even when h is relatively large, we may observe overflows in simulations. So the Euler scheme here is not convergent. It is proved in [234] that for (5.1.1) with any deterministic initial conditions X_0, the Euler-Maruyama scheme with a uniform time step size diverges in the strong sense over any compact interval $[0,T]$. For this example, there are several approaches to obtain convergent numerical schemes for (5.1.1):

- Euler schemes with variable step sizes. To avoid the possible quick growth of the solution, we can set the time step size small enough when needed. For example, in the above example, we can take the time step size h_x as $1/x^2$, when $|x| > 1/\sqrt{h}$ and x is the solution at the previous step.
- implicit schemes, e.g., the scheme (3.2.27).
- balanced implicit schemes, where the Euler scheme is modified with a penalty term, see Chapter 5.3.
- tamed schemes (balanced explicit schemes). The Euler scheme is modified so that the drift and diffusion coefficients are bounded by some polynomial growth of $1/h$, where h is the time step size.

Strong schemes for SDEs with non-globally Lipschitz coefficients have been considered in a number of recent works, see, e.g., [218, 219, 226, 233, 235, 236, 335, 336, 375, 410, 474] and the references therein); see an extended literature review on this topic in [233].

When one is interested in simulating averages $\mathbb{E}\varphi(X(T))$ of the solution to SDEs, weak-sense convergence cannot be guaranteed if the coefficients are not global Lipschitz, see, e.g., [359, 360] for simulation of averages at finite

[1] As Brownian motion can take values in $\mathbb{R}$, the (numerical) solution may reach any value at a certain time step. We can assume that we compute the solutions from such a step and denote it as the zeroth step.

time and also of ergodic limits when ensemble averaging is used. The concept of rejecting exploding trajectories, proposed and justified in [359], allows us to use any numerical method for solving SDEs with non-globally Lipschitz coefficients for estimating averages. Following this concept, we do not take into account the approximate trajectories $X(t)$, which leave a sufficiently large ball $S_R := \{x : |x| < R\}$ during the time T. See other approaches for resolving this problem in the context of computing averages, including the case of simulating ergodic limits via time averaging, e.g., in [40, 341, 437].

For SDEs under non-globally Lipschitz assumptions on the coefficients, the convergence of many standard numerical methods can fail, and this motivates the recent interest in both theoretical support of existing numerical methods and developing new methods.

In this chapter, we deal with mean-square (strong) approximation of SDEs with non-globally Lipschitz coefficients. We present a variant of the fundamental mean-square convergence theorem in the case of SDEs with non-globally Lipschitz coefficients proposed in [448], which is analogous to Milstein's fundamental theorem for the global Lipschitz case [353] (see also [354, 358]). More precisely, we assume that the SDEs coefficients can grow polynomially at infinity and satisfy a one-sided Lipschitz condition. The theorem is stated in Chapter 5.2. Its corollary on almost sure convergence is also given. In Chapter 5.2 we present a discussion on applicability of the fundamental theorem, including its application to the drift-implicit Euler scheme in [336] and thus establish its order of convergence.

Here we present a particular balanced method proposed in [448] and prove its convergence with order half in the non-globally Lipschitz setting in Chapter 5.3. Some numerical experiments supporting our results are presented in Chapter 5.4. Similar balanced methods have also been proposed elsewhere, see, e.g., [411, 436].

5.2 Fundamental theorem

Let $(\Omega, \mathcal{F}, \mathbb{P})$ be a complete probability space and $\mathcal{F}_t^W$ be an increasing family of σ-subalgebras of $\mathcal{F}$ induced by $W(t)$ for $0 \le t \le T$, where $(W(t), \mathcal{F}_t^W) = ((W_1(t), \ldots, W_m(t))^\top, \mathcal{F}_t^W)$ is an m-dimensional standard Wiener process. We consider the system of Ito stochastic differential equations (SDEs):

$$dX = a(t, X)dt + \sum_{r=1}^{m} \sigma_r(t, X)dW_r(t), \quad t \in (t_0, T], \; X(t_0) = X_0, \tag{5.2.1}$$

where X, a, σ_r are d-dimensional column-vectors and X_0 is independent of W. We assume that any solution $X_{t_0,X_0}(t)$ of (5.2.1) is regular on $[t_0, T]$, i.e., it is defined for all $t_0 \le t \le T$ [208].

Let $X_{t_0,X_0}(t) = X(t)$, $t_0 \le t \le T$, be a solution of the system (5.2.1). We will assume the following.

Assumption 5.2.1 *(i) The initial condition is such that*

$$\mathbb{E}|X_0|^{2p} \leq K < \infty, \quad \textit{for all} \ \ p \geq 1. \tag{5.2.2}$$

(ii) For a sufficiently large $p_0 \geq 1$ there is a constant $c_1 \geq 0$ such that for $t \in [t_0, T]$,

$$(x-y, a(t,x)-a(t,y)) + \frac{2p_0-1}{2}\sum_{r=1}^{m}|\sigma_r(t,x)-\sigma_r(t,y)|^2 \leq c_1|x-y|^2, \ x,y \in \mathbb{R}^d. \tag{5.2.3}$$

(iii) There exist $c_2 \geq 0$ and $\varkappa \geq 1$ such that for $t \in [t_0, T]$,

$$|a(t,x) - a(t,y)|^2 \leq c_2(1+|x|^{2\varkappa-2}+|y|^{2\varkappa-2})|x-y|^2, \quad x,y \in \mathbb{R}^d. \tag{5.2.4}$$

The condition (5.2.3) implies that

$$(x, a(t,x)) + \frac{2p_0 - 1 - \varepsilon}{2}\sum_{r=1}^{m}|\sigma_r(t,x)|^2 \leq c_0 + c_1'|x|^2, \quad t \in [t_0, T], \ x \in \mathbb{R}^d, \tag{5.2.5}$$

where $c_0 = |a(t,0)|^2/2 + \frac{(2p_0-1-\varepsilon)(2p_0-1)}{2\varepsilon}\sum_{r=1}^{m}|\sigma_r(t,0)|^2$ and $c_1' = c_1 + 1/2$. The inequality (5.2.5) together with (5.2.2) is sufficient to ensure finiteness of moments [208]: there is $K > 0$

$$\mathbb{E}|X_{t_0,X_0}(t)|^{2p} < K(1+\mathbb{E}|X_0|^{2p}), \ 1 \leq p \leq p_0 - 1, \ \ t \in [t_0, T]. \tag{5.2.6}$$

Also, (5.2.4) implies that

$$|a(t,x)|^2 \leq c_3 + c_2'|x|^{2\varkappa}, \quad t \in [t_0, T], \ x \in \mathbb{R}^d, \tag{5.2.7}$$

where $c_3 = 2|a(t,0))|^2 + 2c_2(\varkappa-1)/\varkappa$ and $c_2' = 2c_2(1+\varkappa)/\varkappa$.

Example 5.2.2 *Here is an example for Assumption 5.2.1 (ii):*

$$dX = -\mu X|X|^{r_1-1}dt + \lambda X^{r_2}dW,$$

where μ, $\lambda > 0$, $r_1 \geq 1$, and $r_2 \geq 1$. If $r_1 + 1 > 2r_2$ or $r_1 = r_2 = 1$, then (5.2.3) is valid for any $p_0 \geq 1$. If $r_1 + 1 = 2r_2$ and $r_1 > 1$ then (5.2.3) is valid for $1 \leq p_0 \leq \mu/\lambda^2 + 1/2$.

We introduce the one-step approximation $\bar{X}_{t,x}(t+h)$, $t_0 \leq t < t+h \leq T$, for the solution $X_{t,x}(t+h)$ of (5.2.1), which depends on the initial point (t,x), a time step h, and $\{W_1(\theta)-W_1(t), \ldots, W_m(\theta)-W_m(t), \ t \leq \theta \leq t+h\}$ and which is defined as follows:

$$\bar{X}_{t,x}(t+h) = x + A(t,x,h; W_i(\theta)-W_i(t), \ i=1,\ldots,m, \ t \leq \theta \leq t+h). \tag{5.2.8}$$

Using the one-step approximation (5.2.8), we recurrently construct the approximation $(X_k, \mathcal{F}_{t_k})$, $k = 0, \ldots, N$, $t_{k+1} - t_k = h_{k+1}$, $T_N = T$:

$$X_0 = X(t_0),\ X_{k+1} = \bar{X}_{t_k,\bar{X}_k}(t_{k+1}) \tag{5.2.9}$$
$$= X_k + A(t_k, X_k, h_{k+1}; W_i(\theta) - W_i(t_k),$$
$$i = 1, \ldots, m,\ t_k \le \theta \le t_{k+1}).$$

The following theorem is a generalization of Milstein's fundamental theorem [353] (see also [354, 358, Chapter 1]) from the global to non-globally Lipschitz case. It also has similarities with a strong convergence theorem in [218] proved for the case of non-globally Lipschitz drift, global Lipschitz diffusion, and Euler-type schemes.

For simplicity, we will consider a uniform time step size, i.e., $h_k = h$ for all k.

Theorem 5.2.3 ([448]) *Suppose (i) Assumption 5.2.1 holds;*

(ii) The one-step approximation $\bar{X}_{t,x}(t+h)$ from (5.2.8) has the following orders of accuracy: for some $p \ge 1$ there are $\alpha \ge 1$, $h_0 > 0$, and $K > 0$ such that for arbitrary $t_0 \le t \le T - h$, $x \in \mathbb{R}^d$, and all $0 < h \le h_0$:

$$|\mathbb{E}[X_{t,x}(t+h) - \bar{X}_{t,x}(t+h)]| \le K(1 + |x|^{2\alpha})^{1/2} h^{q_1}, \tag{5.2.10}$$

$$\left[\mathbb{E}|X_{t,x}(t+h) - \bar{X}_{t,x}(t+h)|^{2p}\right]^{1/(2p)} \le K(1 + |x|^{2\alpha p})^{1/(2p)} h^{q_2} \tag{5.2.11}$$

with

$$q_2 \ge \frac{1}{2},\ q_1 \ge q_2 + \frac{1}{2}; \tag{5.2.12}$$

(iii) The approximation X_k from (5.2.9) has finite moments, i.e., for some $p \ge 1$ there are $\beta \ge 1$, $h_0 > 0$, and $K > 0$ such that for all $0 < h \le h_0$ and all $k = 0, \ldots, N$:

$$\mathbb{E}|X_k|^{2p} < K(1 + \mathbb{E}|X_0|^{2p\beta}). \tag{5.2.13}$$

Then for any N and $k = 0, 1, \ldots, N$ the following inequality holds:

$$\left[\mathbb{E}|X_{t_0,X_0}(t_k) - \bar{X}_{t_0,X_0}(t_k)|^{2p}\right]^{1/(2p)} \le K(1 + \mathbb{E}|X_0|^{2\gamma p})^{1/(2p)} h^{q_2 - 1/2}, \tag{5.2.14}$$

where $K > 0$ and $\gamma \ge 1$ do not depend on h and k, i.e., the order of accuracy of the method (5.2.9) is $q = q_2 - 1/2$.

Corollary 5.2.4 ([448]) *In the setting of Theorem 5.2.3 for $p \ge 1/(2q)$ in (5.2.14), there is $0 < \varepsilon < q$ and an a.s. finite random variable $C(\omega) > 0$ such that*

$$|X_{t_0,X_0}(t_k) - X_k| \le C(\omega) h^{q-\varepsilon},$$

i.e., the method (5.2.9) for (5.2.1) converges with order $q - \varepsilon$ a.s.

The corollary is proved using the Borel-Cantelli lemma in Appendix D (see, e.g., [187, 361]).

Remark 5.2.5 *The assumptions and the statement of Theorem 5.2.3 include the famous fundamental theorem of Milstein [353] (see also Theorem 3.2.2) proved under the global conditions on the SDEs coefficients when the assumption* (5.2.13) *is naturally satisfied.*

The constant K in (5.2.14) depends on p, t_0, T as well as on the SDEs coefficients. The constant γ in (5.2.14) depends on α, β, and $\varkappa$.

5.2.1 On application of Theorem 5.2.3

Theorem 5.2.3 says that if moments of X_k are bounded and the scheme was proved to be convergent with order q in the global Lipschitz case then the scheme has the same convergence order q in the considered non-globally Lipschitz case.

However, checking the condition (5.2.13) on moments of a method X_k is often rather difficult. Usually, each scheme of non-globally Lipschitz SDEs requires a special consideration while for schemes for SDEs with global Lipschitz coefficients, boundedness of moments of X_k is just direct implication of the boundedness of moments of the SDEs solution and the one-step properties of the method, see [358, Lemma 1.1.5]). For a number of strong schemes, boundedness of moments in non-globally Lipschitz cases were proved, see, e.g., [218, 226, 233, 235, 437]. In Chapter 5.3 we show boundedness of moments for a balanced method. See also [448] for fully implicit methods (3.2.27).

Consider the drift-implicit Euler scheme [358, p. 30]:

$$X_{k+1} = X_k + a(t_{k+1}, X_{k+1})h + \sum_{r=1}^{m} \sigma_r(t_k, X_k)\xi_{rk}\sqrt{h}, \tag{5.2.15}$$

where $\xi_{rk} = (W_r(t_{k+1}) - W_r(t_k))/\sqrt{h}$ are Gaussian $\mathcal{N}(0,1)$ i.i.d. random variables. Assume that the coefficients $a(t,x)$ and $\sigma_r(t,x)$ have continuous first-order partial derivatives in t and the coefficient $a(t,x)$ also has continuous first-order partial derivatives in x^i and that all these derivatives and the coefficients themselves satisfy inequalities of the form (5.2.4). It is not difficult to show that the one-step approximation corresponding to (5.2.15) satisfies (5.2.10) and (5.2.11) with $q_1 = 2$ and $q_2 = 1$, respectively. Its boundedness of moments, in particular, under the condition (5.2.5) for time steps $h \leq 1/(2c_1)$, is proved in [233]. Then, due to Theorem 5.2.3, (5.2.15) converges with mean-square order $q = 1/2$ (note that for $q = 1/2$, it is sufficient to have $q_1 = 3/2$, which can be obtained under lesser smoothness of a).

In the case of additive noise (i.e., $\sigma_r(t,x) = \sigma_r(t)$, $r = 1, \ldots, m$), $q_1 = 2$ and $q_2 = 3/2$ and (5.2.15) converges with mean-square order 1 due to Theorem 5.2.3. We note that convergence of (5.2.15) with order half in the global Lipschitz case is well known [259, 354, 358]; in the case of non-globally Lipschitz drift and global Lipschitz diffusion was proved in [218, 226] (see also related results in [187, 437]); and under Assumption 5.2.1 strong convergence of (5.2.15) without order was proved in [233, 335] and its strong order half is established in [336].

Due to the bound (5.2.6) on the moments of the solution $X(t)$, it would be natural to require that β in (5.2.13) should be equal to 1. Indeed, (5.2.13) with $\beta = 1$ holds for the drift-implicit method (5.2.15) [233] and for fully implicit methods (see [448, Section 4] or (3.2.27).

5.2.2 Proof of the fundamental theorem

In this section we shall use the letter K to denote various constants, which are independent of h and k. The proof exploits the idea of the proof of this theorem in the global Lipschitz case [353].

We need the following lemma to prove the fundamental theorem. Lemma 5.2.6 is analogous to Lemma 1.1.3 in [358].

Lemma 5.2.6 *Suppose Assumption 5.2.1 holds. For the representation*

$$X_{t,x}(t+\theta) - X_{t,y}(t+\theta) = x - y + Z_{t,x,y}(t+\theta), \tag{5.2.16}$$

we have for $1 \le p \le (p_0 - 1)/\varkappa$ *:*

$$\mathbb{E}|X_{t,x}(t+h) - X_{t,y}(t+h)|^{2p} \le |x-y|^{2p}(1+Kh)\,, \tag{5.2.17}$$

$$\mathbb{E}\,|Z_{t,x,y}(t+h)|^{2p} \le K(1+|x|^{2\varkappa-2}+|y|^{2\varkappa-2})^{p/2}|x-y|^{2p}h^{p}\,. \tag{5.2.18}$$

Proof. Introduce the process $S_{t,x,y}(s) = S(s) := X_{t,x}(s) - X_{t,y}(s)$ and note that $Z(s) = S(s) - (x-y)$. We first prove (5.2.17). Using the Ito formula and the condition (5.2.3) (recall that (5.2.3) implies (5.2.6)), we obtain for $\theta \ge 0$:

$$\begin{aligned}
\mathbb{E}|S(t+\theta)|^{2p} &= |x-y|^{2p} + 2p\int_t^{t+\theta} \mathbb{E}|S|^{2p-2}\Bigg[S^{\mathsf{T}}(a(t,X_{t,x}(s)) - a(t,X_{t,y}(s))) \\
&\quad + \frac{1}{2}\sum_{r=1}^{m} |\sigma_r(t,X_{t,x}(s)) - \sigma_r(t,X_{t,y}(s))|^2\Bigg]\, ds \\
&\quad + 2p(p-1)\int_t^{t+\theta} \mathbb{E}|S|^{2p-4}\left|S^{\mathsf{T}}(s)\sum_{r=1}^{m}[\sigma_r(t,X_{t,x}(s))]\right. \\
&\quad - \left.[\sigma_r(t,X_{t,y}(s))]\right|^2 ds \\
&\le |x-y|^{2p} + 2p\int_t^{t+\theta} \mathbb{E}|S|^{2p-2}\Bigg[S^{\mathsf{T}}(a(t,X_{t,x}(s)) - a(t,X_{t,y}(s))) \\
&\quad + \frac{2p-1}{2}\int_t^{t+\theta}\sum_{r=1}^{m} |\sigma_r(t,X_{t,x}(s)) - \sigma_r(t,X_{t,y}(s))|^2\Bigg]\, ds \\
&\le |x-y|^{2p} + 2pc_1\int_t^{t+\theta} \mathbb{E}|S(s)|^{2p}\, ds
\end{aligned}$$

from which (5.2.17) follows after applying the Gronwall inequality.

Now we prove (5.2.18). Using the Ito formula and the condition (5.2.3), we obtain for $\theta \geq 0$:

$$
\begin{aligned}
\mathbb{E}\,|Z(t+\theta)|^{2p} &= 2p \int_t^{t+\theta} \mathbb{E}|Z|^{2p-2} \Big[Z^{\mathsf{T}}(a(t, X_{t,x}(s)) - a(t, X_{t,y}(s))) \\
&\quad + \frac{1}{2} \sum_{r=1}^{m} |\sigma_r(t, X_{t,x}(s)) - \sigma_r(t, X_{t,y}(s))|^2 \Big] \, ds \\
&\quad + 2p(p-1) \int_t^{t+\theta} \mathbb{E}|Z|^{2p-4} \left| Z^{\mathsf{T}} \sum_{r=1}^{m} [\sigma_r(t, X_{t,x}(s))] \right| \\
&\quad - |[\sigma_r(t, X_{t,y}(s))]|^2 \, ds \\
&\leq 2p \int_t^{t+\theta} \mathbb{E}|Z|^{2p-2}(s) \Big[S^{\mathsf{T}}(a(t, X_{t,x}(s)) - a(t, X_{t,y}(s))) \\
&\quad + \frac{2p-1}{2} \int_t^{t+\theta} \sum_{r=1}^{m} |\sigma_r(t, X_{t,x}(s)) - \sigma_r(t, X_{t,y}(s))|^2 \Big] \, ds \\
&\quad - 2p \int_t^{t+\theta} \mathbb{E}|Z|^{2p-2}(x - y, a(t, X_{t,x}(s)) - a(t, X_{t,y}(s)))ds \\
&\leq 2pc_1 \int_t^{t+\theta} \mathbb{E}|Z|^{2p-2}|S|^2 \, ds - 2p \int_t^{t+\theta} \mathbb{E}|Z|^{2p-2}(x - y, a \\
&\quad (t, X_{t,x}(s)) - a(t, X_{t,y}(s)))ds.
\end{aligned}
$$

Using the Young inequality, we get for the first term in the right-hand side of (5.2.19):

$$
\begin{aligned}
2pc_1 \int_t^{t+\theta} \mathbb{E}|Z|^{2p-2}|S|^2 \, ds &\leq 4pc_1 \int_t^{t+\theta} \mathbb{E}|Z|^{2p-2}(|Z|^2 + |x - y|^2) \, ds \quad (5.2.19) \\
&\leq K \int_t^{t+\theta} \mathbb{E}|Z|^{2p} ds + K|x - y|^2 \int_t^{t+\theta} \mathbb{E}|Z|^{2p-2} ds.
\end{aligned}
$$

Consider the second term in the right-hand side of (5.2.19). Using the Hölder inequality (twice), (5.2.4), (5.2.17) and (5.2.6), we obtain

$$
\begin{aligned}
&-2p \int_t^{t+\theta} \mathbb{E}|Z|^{2p-2}(x - y, a(t, X_{t,x}(s)) - a(t, X_{t,y}(s)))ds \qquad (5.2.20) \\
&\leq 2p \int_t^{t+\theta} \mathbb{E}|Z|^{2p-2}|a(t, X_{t,x}(s)) - a(t, X_{t,y}(s))||x - y|ds \\
&\leq K|x - y| \int_t^{t+\theta} \left[\mathbb{E}|Z|^{2p}\right]^{1-1/p} \left[\mathbb{E}|a(t, X_{t,x}(s)) - a(t, X_{t,y}(s))|^p\right]^{1/p} ds \\
&\leq K|x - y| \int_t^{t+\theta} \left[\mathbb{E}|Z|^{2p}\right]^{1-1/p} \\
&\quad \times (\mathbb{E}[(1 + |X_{t,x}(s)|^{2\varkappa-2} + |X_{t,y}(s)|^{2\varkappa-2})^{p/2}|X_{t,x}(s) - X_{t,y}(s)|^p])^{1/p} \, ds
\end{aligned}
$$

$$\leq K|x-y| \int_t^{t+\theta} \left[\mathbb{E}|Z|^{2p}\right]^{1-1/p} \left(\mathbb{E}[(1+|X_{t,x}(s)|^{2\varkappa-2}+|X_{t,y}(s)|^{2\varkappa-2})^p]\right)^{1/2p}$$
$$\left(\mathbb{E}[|X_{t,x}(s)-X_{t,y}(s)|^{2p}]\right)^{1/2p} ds$$
$$\leq K\,|x-y|^2\,(1+|x|^{2\varkappa-2}+|y|^{2\varkappa-2})^{1/2} \int_t^{t+\theta} \left[\mathbb{E}|Z|^{2p}\right]^{1-1/p} ds.$$

Substituting (5.2.19) and (5.2.20) in (5.2.19) and applying the Hölder inequality to $\mathbb{E}|Z|^{2p-2}$, we get

$$\mathbb{E}\,|Z(t+\theta)|^{2p} \leq K \int_t^{t+\theta} \mathbb{E}|Z|^{2p} ds$$
$$+ K\,|x-y|^2\,(1+|x|^{2\varkappa-2}+|y|^{2\varkappa-2})^{1/2} \int_t^{t+\theta} \left[\mathbb{E}|Z|^{2p}\right]^{1-1/p} ds$$

whence we obtain (5.2.18) for integer $p \geq 1$ using the Gronwall inequality, and then by the Jensen inequality for non-integer $p > 1$ as well.

Now we are ready to prove the fundamental theorem. Consider the error of the method $\bar{X}_{t_0,X_0}(t_{k+1})$ at the $(k+1)$-step:

$$\begin{aligned}\rho_{k+1} &:= X_{t_0,X_0}(t_{k+1}) - \bar{X}_{t_0,X_0}(t_{k+1}) = X_{t_k,X(t_k)}(t_{k+1}) - \bar{X}_{t_k,X_k}(t_{k+1}) \\ &= \left(X_{t_k,X(t_k)}(t_{k+1}) - X_{t_k,X_k}(t_{k+1})\right) + \left(X_{t_k,X_k}(t_{k+1}) - \bar{X}_{t_k,X_k}(t_{k+1})\right).\end{aligned} \tag{5.2.21}$$

The first difference in the right-hand side of (5.2.21) is the error of the solution arising due to the error in the initial data at time t_k, accumulated at the k-th step, which we can rewrite as

$$\begin{aligned}S_{t_k,X(t_k),X_k}(t_{k+1}) = S_{k+1} &:= X_{t_k,X(t_k)}(t_{k+1}) - X_{t_k,X_k}(t_{k+1}) \\ &= \rho_k + Z_{t_k,X(t_k),X_k}(t_{k+1}) \\ &= \rho_k + Z_{k+1},\end{aligned}$$

where Z is as in (5.2.16). The second difference in (5.2.21) is the one-step error at the $(k+1)$-step and we denote it as r_{k+1} :

$$r_{k+1} = X_{t_k,X_k}(t_{k+1}) - \bar{X}_{t_k,X_k}(t_{k+1}).$$

Let $p \geq 1$ be an integer. We have

$$\begin{aligned}\mathbb{E}|\rho_{k+1}|^{2p} &= \mathbb{E}\,|S_{k+1}+r_{k+1}|^{2p} = \mathbb{E}[(S_{k+1},S_{k+1}) + 2(S_{k+1},r_{k+1}) + (r_{k+1},r_{k+1})]^p \\ &\leq \mathbb{E}\,|S_{k+1}|^{2p} + 2p\mathbb{E}\,|S_{k+1}|^{2p-2}\,(\rho_k + Z_{k+1}, r_{k+1}) \\ &+ K\sum_{l=2}^{2p} \mathbb{E}\,|S_{k+1}|^{2p-l}\,|r_{k+1}|^l.\end{aligned}$$

Due to (5.2.17) of Lemma 5.2.6, the first term on the right-hand side of (5.2.22) is estimated as

$$\mathbb{E}\,|S_{k+1}|^{2p} \leq \mathbb{E}|\rho_k|^{2p}(1+Kh). \tag{5.2.22}$$

Consider the second term on the right-hand side of (5.2.22):

$$\begin{aligned}\mathbb{E}\left|S_{k+1}\right|^{2p-2}(\rho_k+Z_{k+1},r_{k+1}) &= \mathbb{E}\left|\rho_k\right|^{2p-2}(\rho_k,r_{k+1}) \\ &+\mathbb{E}\left(\left|S_{k+1}\right|^{2p-2}-\left|\rho_k\right|^{2p-2}\right)(\rho_k,r_{k+1}) \\ &+\mathbb{E}\left|S_{k+1}\right|^{2p-2}(Z_{k+1},r_{k+1}). \qquad (5.2.23)\end{aligned}$$

Due to $\mathcal{F}_{t_k}$-measurability of ρ_k and due to the conditional variant of (5.2.10), we get for the first term on the right-hand side of (5.2.23):

$$\mathbb{E}\left|\rho_k\right|^{2p-2}(\rho_k,r_{k+1}) \leq K\mathbb{E}\left|\rho_k\right|^{2p-1}(1+|X_k|^{2\alpha})^{1/2}h^{q_1}. \qquad (5.2.24)$$

Consider the second term on the right-hand side of (5.2.23) and first of all note that it is equal to zero for $p=1$. We have for integer $p \geq 2$:

$$\mathbb{E}\left(|S_{k+1}|^{2p-2}-|\rho_k|^{2p-2}\right)(\rho_k,r_{k+1}) \leq K\mathbb{E}\,|Z_{k+1}|\,|\rho_k||r_{k+1}|\sum_{l=0}^{2p-3}|S_{k+1}|^{2p-3-l}|\rho_k|^l.$$

Further, using $\mathcal{F}_{t_k}$-measurability of ρ_k and the conditional variants of (5.2.11), (5.2.17), and (5.2.18) and the Cauchy-Schwarz inequality (twice), we get for $p \geq 2$:

$$\begin{aligned}&\mathbb{E}\left(\left|S_{k+1}\right|^{2p-2}-\left|\rho_k\right|^{2p-2}\right)(\rho_k,r_{k+1}) \qquad (5.2.25)\\ &\leq K\mathbb{E}\left|\rho_k\right|^{2p-1}(1+|X(t_k)|^{2\varkappa-2}+|X_k|^{2\varkappa-2})^{1/4}h^{q_2+1/2}(1+|X_k|^{2\alpha})^{1/2}.\end{aligned}$$

Due to $\mathcal{F}_{t_k}$-measurability of ρ_k, the conditional variants of (5.2.11) and (5.2.18) and the Cauchy-Schwarz inequality (twice), we obtain for the third term on the right-hand side of (5.2.23):

$$\begin{aligned}&\mathbb{E}\left|S_{k+1}\right|^{2p-2}(Z_{k+1},r_{k+1}) \qquad (5.2.26)\\ &\leq \mathbb{E}[\mathbb{E}\left(S_{k+1}|^{4p-4}|\mathcal{F}_{t_k}\right)^{1/2}\mathbb{E}\left(|Z_{k+1}|^4|\mathcal{F}_{t_k}\right)^{1/4}\mathbb{E}\left(|r_{k+1}|^4|\mathcal{F}_{t_k}\right)^{1/4}] \\ &\leq K\mathbb{E}\left|\rho_k\right|^{2p-1}(1+|X(t_k)|^{2\varkappa-2}+|X_k|^{2\varkappa-2})^{1/4}h^{q_2+1/2}(1+|X_k|^{4\alpha})^{1/4}.\end{aligned}$$

Due to $\mathcal{F}_{t_k}$-measurability of ρ_k and due to the conditional variants of (5.2.11) and (5.2.17) and the Cauchy-Schwarz inequality, we estimate the third term on the right-hand side of (5.2.22):

$$\begin{aligned}K\sum_{l=2}^{2p}\mathbb{E}\left|S_{k+1}\right|^{2p-l}\left|r_{k+1}\right|^l &\leq K\sum_{l=2}^{2p}\mathbb{E}[\mathbb{E}(\left|S_{k+1}\right|^{4p-2l}|\mathcal{F}_{t_k})^{1/2} \\ &\qquad \mathbb{E}(|r_{k+1}|^{2l}|\mathcal{F}_{t_k})^{1/2}] \qquad (5.2.27)\\ &\leq K\sum_{l=2}^{2p}\mathbb{E}[|\rho_k|^{2p-l}h^{lq_2}(1+|X_k|^{2l\alpha})^{1/2}].\end{aligned}$$

Substituting (5.2.22)-(5.2.27) in (5.2.22) and recalling that $q_1 \geq q_2+1/2$, we obtain

$$
\begin{aligned}
\mathbb{E}|\rho_{k+1}|^{2p} \leq & \mathbb{E}|\rho_k|^{2p}(1+Kh) + K\mathbb{E}\,|\rho_k|^{2p-1}\,(1+|X_k|^{2\alpha})^{1/2}h^{q_2+1/2} \\
& + K\mathbb{E}\,|\rho_k|^{2p-1}(1+|X(t_k)|^{2\varkappa-2}+|X_k|^{2\varkappa-2})^{1/4}h^{q_2+1/2}(1+|X_k|^{2\alpha})^{1/2} \\
& + K\mathbb{E}\,|\rho_k|^{2p-1}(1+|X(t_k)|^{2\varkappa-2}+|X_k|^{2\varkappa-2})^{1/4}h^{q_2+1/2}(1+|X_k|^{4\alpha})^{1/4} \\
& + K\sum_{l=2}^{2p}\mathbb{E}[|\rho_k|^{2p-l}h^{lq_2}(1+|X_k|^{2l\alpha})^{1/2}] \\
\leq & \mathbb{E}|\rho_k|^{2p}(1+Kh) + K\mathbb{E}\,|\rho_k|^{2p-1}\,(1+|X(t_k)|^{2\varkappa-2} \\
& + |X_k|^{2\varkappa-2})^{1/4}h^{q_2+1/2}(1+|X_k|^{2\alpha})^{1/2} \\
& + K\sum_{l=2}^{2p}\mathbb{E}[|\rho_k|^{2p-l}h^{lq_2}(1+|X_k|^{2l\alpha})^{1/2}].
\end{aligned}
$$

Then using the Young inequality and the conditions (5.2.6) and (5.2.13), we obtain

$$
\mathbb{E}|\rho_{k+1}|^{2p} \leq \mathbb{E}|\rho_k|^{2p} + Kh\mathbb{E}|\rho_k|^{2p} + K(1+\mathbb{E}|X_0|^{\beta p(\varkappa-1)+2p\alpha\beta})h^{2p(q_2-1/2)+1}
$$

whence (5.2.14) with integer $p \geq 1$ follows from the Gronwall inequality. Then by the Jensen inequality (5.2.14) holds for non-integer p as well. □

5.3 A balanced Euler scheme

In this section we present a particular balanced scheme from the class of balanced methods introduced in [355] (see also [358, Chapter 1.3]) and prove its mean-square convergence with order half using Theorem 5.2.3. In Chapter 5.4, we test the balanced scheme, which is similar to the one in [233], on a model problem and demonstrate that it is more efficient than the tamed scheme in [233] (5.4.2), see Chapter 5.4.

The concept of balanced methods in [355] is introduced as Euler-type numerical methods with balance between approximating stochastic terms in stiff SDEs. Specifically, the balanced Euler schemes can be written as

$$
\begin{aligned}
X_{k+1} = X_k + a(t_k, X_k)h + \sum_{r=1}^{m}\sigma_r(t_k, X_k)(W_r(t_{k+1}) - W_r(t_k)) \\
+ P(t_k, t_{k+1}, X_k, X_{k+1}, h, W_r(t_{k+1}) - W_r(t_k)),
\end{aligned}
$$

where the term P is not zero unless $h = 0$. It can be considered as a *penalty method* or a *Lagrange multiplier method* for stiff SDEs. From the above formula, we know that the explicit Euler scheme is not a balanced method as it has the term P being 0.

Consider the following scheme for (5.2.1), which is a balanced Euler scheme:

$$
X_{k+1} = X_k + \frac{a(t_k, X_k)h + \sum_{r=1}^{m}\sigma_r(t_k, X_k)\xi_{rk}\sqrt{h}}{1 + h|a(t_k, X_k)| + \sqrt{h}\sum_{r=1}^{m}|\sigma_r(t_k, X_k)\xi_{rk}|}, \tag{5.3.1}
$$

where ξ_{rk} are Gaussian $\mathcal{N}(0,1)$ i.i.d. random variables. This scheme is a balanced method as it can be written as

$$X_{k+1} = X_k + a(t_k, X_k)h + \sum_{r=1}^{m} \sigma_r(t_k, X_k)\xi_{rk}\sqrt{h}$$
$$- \frac{a(t_k, X_k)h + \sum_{r=1}^{m} \sigma_r(t_k, X_k)\xi_{rk}\sqrt{h}}{1 + h|a(t_k, X_k)| + \sqrt{h}\sum_{r=1}^{m} |\sigma_r(t_k, X_k)\xi_{rk}|} h|(a(t_k, X_k)|$$
$$+\sqrt{h}\sum_{r=1}^{m} |\sigma_r(t_k, X_k)\xi_{rk}|).$$

Here the "extra term" (in the second line of the equation) is not vanishing unless $h = 0$.

We will prove two lemmas, which show that the scheme (5.3.1) satisfies the conditions of Theorem 5.2.3. The first lemma is on boundedness of moments and uses a stopping time technique (see also, e.g., [233, 359]).

Lemma 5.3.1 *Suppose Assumption 5.2.1 holds with sufficiently large* p_0. *For all natural* N *and all* $k = 0, \ldots, N$ *the following inequality holds for moments of the scheme* (5.3.1)*:*

$$\mathbb{E}|X_k|^{2p} \leq K(1 + \mathbb{E}|X_0|^{2p\beta}), \quad 1 \leq p \leq \frac{p_0 - 1}{3\varkappa - 1} - \frac{1}{2}, \tag{5.3.2}$$

with some constants $\beta \geq 1$ *and* $K > 0$ *independent of* h *and* k.

Remark 5.3.2 *It is common that* β *is larger than 1 in Theorem 5.2.3 for tamed-type methods (see [235] and the bibliographic notes at the end of this chapter) or the balanced Euler method* (5.3.1).

Proof. In the proof we shall use the letter K to denote various constants, which are independent of h and k.

The following elementary consequence of the inequalities (5.2.5) and (5.2.7) will be used in the proof: there exits a constant $K > 0$ such that

$$\sum_{r=1}^{m} |\sigma_r(t,x)|^2 \leq K(1 + |x|^{\varkappa+1}). \tag{5.3.3}$$

We observe from (5.3.1) that

$$|X_{k+1}| \leq |X_k| + 1 \leq |X_0| + (k+1). \tag{5.3.4}$$

Let $R > 0$ be a sufficiently large number. Introduce the events

$$\tilde{\Omega}_{R,k} := \{\omega : |X_l| \leq R, \ l = 0, \ldots, k\}, \tag{5.3.5}$$

and their compliments $\tilde{\Lambda}_{R,k}$. We first prove the lemma for integer $p \geq 1$. We have

$$\begin{aligned}
&\mathbb{E}\chi_{\tilde{\Omega}_{R,k+1}}(\omega)|X_{k+1}|^{2p} \leq \mathbb{E}\chi_{\tilde{\Omega}_{R,k}}(\omega)|X_{k+1}|^{2p} \qquad (5.3.6)\\
=&\mathbb{E}\chi_{\tilde{\Omega}_{R,k}}(\omega)|(X_{k+1}-X_k)+X_k|^{2p} \leq \mathbb{E}\chi_{\tilde{\Omega}_{R,k}}(\omega)|X_k|^{2p} + \mathbb{E}\chi_{\tilde{\Omega}_{R,k}}(\omega)\,|X_k|^{2p-2}\\
&\times\left[2p(X_k, X_{k+1}-X_k) + p(2p-1)|X_{k+1}-X_k|^2\right]\\
&+K\textstyle\sum_{l=3}^{2p}\mathbb{E}\chi_{\tilde{\Omega}_{R,k}}(\omega)\,|X_k|^{2p-l}\,|X_{k+1}-X_k|^l.
\end{aligned}$$

Consider the second term in the right-hand side of (5.3.6):

$$\begin{aligned}
&\mathbb{E}\chi_{\tilde{\Omega}_{R,k}}(\omega)\,|X_k|^{2p-2}\left[2p(X_k, X_{k+1}-X_k) + p(2p-1)|X_{k+1}-X_k|^2\right]\\
&= 2p\mathbb{E}\Bigg(\chi_{\tilde{\Omega}_{R,k}}(\omega)\,|X_k|^{2p-2}\,\mathbb{E}\Bigg[\Bigg(X_k, \frac{a(t_k,X_k)h+\sum_{r=1}^m \sigma_r(t_k,X_k)\xi_{rk}\sqrt{h}}{1+h|a(t_k,X_k)|+\sqrt{h}\sum_{r=1}^m|\sigma_r(t_k,X_k)\xi_{rk}|}\Bigg)\\
&\quad+\frac{2p-1}{2}\left|\frac{a(t_k,X_k)h+\sum_{r=1}^m \sigma_r(t_k,X_k)\xi_{rk}\sqrt{h}}{1+h|a(t_k,X_k)|+\sqrt{h}\sum_{r=1}^m|\sigma_r(t_k,X_k)\xi_{rk}|}\right|^2 \Bigg|\, \mathcal{F}_{t_k}\Bigg]\Bigg). \qquad (5.3.7)
\end{aligned}$$

Since ξ_{rk} are independent of $\mathcal{F}_{t_k}$ and the Gaussian density function is symmetric, we obtain

$$\chi_{\tilde{\Omega}_{R,k}}\mathbb{E}\left[\frac{\sum_{r=1}^m \sigma_r(t_k,X_k)\xi_{rk}\sqrt{h}}{1+h|a(t_k,X_k)|+\sqrt{h}\sum_{r=1}^m|\sigma_r(t_k,X_k)\xi_{rk}|}\,\Bigg|\,\mathcal{F}_{t_k}\right] = 0. \qquad (5.3.8)$$

Similarly, we analogously get for $l \neq r$:

$$\chi_{\tilde{\Omega}_{R,k}}\mathbb{E}\left[\frac{\sigma_r(t_k,X_k)\xi_{rk}\sqrt{h}\sigma_l(t_k,X_k)\xi_{lk}\sqrt{h}}{(1+h|a(t_k,X_k)|+\sqrt{h}\sum_{r=1}^m|\sigma_r(t_k,X_k)\xi_{rk}|)^2}\,\Bigg|\,\mathcal{F}_{t_k}\right] = 0. \quad (5.3.9)$$

Then the conditional expectation in (5.3.7) becomes

$$\begin{aligned}
A :=& \chi_{\tilde{\Omega}_{R,k}}\mathbb{E}\Bigg[\Bigg(X_k, \frac{a(t_k,X_k)h+\sum_{r=1}^m \sigma_r(t_k,X_k)\xi_{rk}\sqrt{h}}{1+h|a(t_k,X_k)|+\sqrt{h}\sum_{r=1}^m|\sigma_r(t_k,X_k)\xi_{rk}|}\Bigg) \qquad (5.3.10)\\
&+\frac{2p-1}{2}\left|\frac{a(t_k,X_k)h+\sum_{r=1}^m \sigma_r(t_k,X_k)\xi_{rk}\sqrt{h}}{1+h|a(t_k,X_k)|+\sqrt{h}\sum_{r=1}^m|\sigma_r(t_k,X_k)\xi_{rk}|}\right|^2\Bigg|\,\mathcal{F}_{t_k}\Bigg]\\
=& \chi_{\tilde{\Omega}_{R,k}}\mathbb{E}\Bigg[\frac{(X_k, a(t_k,X_k)h)}{1+h|a(t_k,X_k)|+\sqrt{h}\sum_{r=1}^m|\sigma_r(t_k,X_k)\xi_{rk}|}\\
&+\frac{2p-1}{2}\,\frac{a^2(t_k,X_k)h^2+h\sum_{r=1}^m(\sigma_r(t_k,X_k)\xi_{rk})^2}{\left(1+h|a(t_k,X_k)|+\sqrt{h}\sum_{r=1}^m|\sigma_r(t_k,X_k)\xi_{rk}|\right)^2}\,\Bigg|\,\mathcal{F}_{t_k}\Bigg]
\end{aligned}$$

$$\leq \chi_{\tilde{\Omega}_{R,k}} \mathbb{E}\left[\frac{(X_k, a(t_k, X_k)h)}{1 + h|a(t_k, X_k)| + \sqrt{h}\sum_{r=1}^m |\sigma_r(t_k, X_k)\xi_{rk}|} \right.$$
$$\left. + \frac{2p-1}{2} \frac{h\sum_{r=1}^m |\sigma_r(t_k, X_k)|^2 \xi_{rk}^2}{1 + h|a(t_k, X_k)| + \sqrt{h}\sum_{r=1}^m |\sigma_r(t_k, X_k)\xi_{rk}|} \right| \mathcal{F}_{t_k} \Bigg]$$
$$+ \frac{2p-1}{2} \chi_{\tilde{\Omega}_{R,k}} a^2(t_k, X_k)h^2$$
$$= \chi_{\tilde{\Omega}_{R,k}} \mathbb{E}\left[\frac{(X_k, a(t_k, X_k)h) + \frac{2p-1}{2} h\sum_{r=1}^m |\sigma_r(t_k, X_k)|^2}{1 + h|a(t_k, X_k)| + \sqrt{h}\sum_{r=1}^m |\sigma_r(t_k, X_k)\xi_{rk}|} \right.$$
$$\left. + \frac{2p-1}{2} \frac{h\sum_{r=1}^m |\sigma_r(t_k, X_k)|^2 (\xi_{rk}^2 - 1)}{1 + h|a(t_k, X_k)| + \sqrt{h}\sum_{r=1}^m |\sigma_r(t_k, X_k)\xi_{rk}|} \right| \mathcal{F}_{t_k} \Bigg]$$
$$+ \frac{2p-1}{2} \chi_{\tilde{\Omega}_{R,k}} a^2(t_k, X_k)h^2.$$

Using (5.2.5) and (5.2.7), we obtain

$$A \leq c_0 h + c_1'|X_k|^2 h \chi_{\tilde{\Omega}_{R,k}} \tag{5.3.11}$$
$$+ \frac{2p-1}{2} h \chi_{\tilde{\Omega}_{R,k}} \sum_{r=1}^m |\sigma_r(t_k, X_k)|^2$$
$$\mathbb{E}\left[\frac{(\xi_{rk}^2 - 1)}{1 + h|a(t_k, X_k)| + \sqrt{h}\sum_{r=1}^m |\sigma_r(t_k, X_k)\xi_{rk}|} \right| \mathcal{F}_{t_k} \Bigg]$$
$$+ K h^2 + K \chi_{\tilde{\Omega}_{R,k}} |X_k|^{2\varkappa} h^2.$$

Since $\mathbb{E}(\xi_{rk}^2 - 1) = 0$, moments of ξ_{rk} are bounded and ξ_{rk} are independent of $\mathcal{F}_{t_k}$, we obtain for the expectation in the second term in (5.3.11):

$$\chi_{\tilde{\Omega}_{R,k}} \mathbb{E}\left[\frac{(\xi_{rk}^2 - 1)}{1 + h|a(t_k, X_k)| + \sqrt{h}\sum_{r=1}^m |\sigma_r(t_k, X_k)\xi_{rk}|} \right| \mathcal{F}_{t_k} \Bigg] \tag{5.3.12}$$
$$= \chi_{\tilde{\Omega}_{R,k}} \mathbb{E}\left[\frac{(\xi_{rk}^2 - 1)}{1 + h|a(t_k, X_k)| + \sqrt{h}\sum_{r=1}^m |\sigma_r(t_k, X_k)\xi_{rk}|} - (\xi_{rk}^2 - 1) \right| \mathcal{F}_{t_k} \Bigg]$$
$$= -\chi_{\tilde{\Omega}_{R,k}} \mathbb{E}\left[(\xi_{rk}^2 - 1) \frac{h|a(t_k, X_k)| + \sqrt{h}\sum_{r=1}^m |\sigma_l(t_k, X_k)\xi_{lk}|}{1 + h|a(t_k, X_k)| + \sqrt{h}\sum_{l=1}^m |\sigma_r(t_k, X_k)\xi_{lk}|} \right| \mathcal{F}_{t_k} \Bigg]$$
$$\leq \chi_{\tilde{\Omega}_{R,k}} \mathbb{E}\left[|\xi_{rk}^2 - 1| \left(h|a(t_k, X_k)| + \sqrt{h}\sum_{r=1}^m |\sigma_l(t_k, X_k)||\xi_{lk}| \right) \right| \mathcal{F}_{t_k} \Bigg]$$
$$\leq \chi_{\tilde{\Omega}_{R,k}} K \left(h|a(t_k, X_k)| + \sqrt{h}\sum_{r=1}^m |\sigma_r(t_k, X_k)| \right).$$

Using (5.2.7) and (5.3.3), we get from (5.3.11)-(5.3.12):

$$
\begin{aligned}
A &\le c_0 h + c_1' \chi_{\tilde{\Omega}_{R,k}} |X_k|^2 h + K h \chi_{\tilde{\Omega}_{R,k}} \sum_{r=1}^{m} |\sigma_r(t_k, X_k)|^2 [h|a(t_k, X_k)| \\
&\quad + \sqrt{h} \sum_{r=1}^{m} |\sigma_r(t_k, X_k)|] \\
&\quad + K h^2 + K \chi_{\tilde{\Omega}_{R,k}} |X_k|^{2\varkappa} h^2 \qquad (5.3.13) \\
&\le \chi_{\tilde{\Omega}_{R,k}} K h (1 + |X_k|^2 + |X_k|^{2\varkappa+1} h + |X_k|^{3/2(1+\varkappa)} h^{1/2}).
\end{aligned}
$$

Now consider the last term in (5.3.6):

$$
\begin{aligned}
&\mathbb{E}\chi_{\tilde{\Omega}_{R,k}}(\omega) \, |X_k|^{2p-l} \, |X_{k+1} - X_k|^l \qquad (5.3.14) \\
&\le K \mathbb{E}\chi_{\tilde{\Omega}_{R,k}}(\omega) \, |X_k|^{2p-l} \left[h^l |a(t_k, X_k)|^l + h^{l/2} \sum_{r=1}^{m} |\sigma_r(t_k, X_k)|^l |\xi_{rk}|^l \right] \\
&\le K \mathbb{E}\chi_{\tilde{\Omega}_{R,k}}(\omega) \, |X_k|^{2p-l} \, h^{l/2} \left[1 + h^{l/2} |X_k|^{l\varkappa} + |X_k|^{l\frac{\varkappa+1}{2}} \right], \quad l \ge 3,
\end{aligned}
$$

where we used (5.2.7) and (5.3.3) again as well as the fact that $\chi_{\tilde{\Omega}_{R,k}}(\omega)$ and X_k are $\mathcal{F}_{t_k}$-measurable while ξ_{rk} are independent of $\mathcal{F}_{t_k}$.

Combining (5.3.6), (5.3.7), (5.3.10), (5.3.13), and (5.3.14), we obtain

$$
\begin{aligned}
&\mathbb{E}\chi_{\tilde{\Omega}_{R,k+1}}(\omega) |X_{k+1}|^{2p} \qquad (5.3.15) \\
&\le \mathbb{E}\chi_{\tilde{\Omega}_{R,k}}(\omega) |X_k|^{2p} + K h \mathbb{E}\chi_{\tilde{\Omega}_{R,k}}(\omega) \, |X_k|^{2p-2} \\
&\quad \left[1 + |X_k|^2 + |X_k|^{2\varkappa+1} h + |X_k|^{3/2(1+\varkappa)} h^{1/2} \right] \\
&\quad + K \sum_{l=3}^{2p} \mathbb{E}\chi_{\tilde{\Omega}_{R,k}}(\omega) \, |X_k|^{2p-l} \, h^{l/2} \left[1 + h^{l/2} |X_k|^{l\varkappa} + |X_k|^{l\frac{\varkappa+1}{2}} \right] \\
&\le \mathbb{E}\chi_{\tilde{\Omega}_{R,k}}(\omega) |X_k|^{2p} + K h \mathbb{E}\chi_{\tilde{\Omega}_{R,k}}(\omega) \, |X_k|^{2p} + K \sum_{l=2}^{2p} \mathbb{E}\chi_{\tilde{\Omega}_{R,k}}(\omega) \, |X_k|^{2p-l} \, h^{l/2} \\
&\quad + K h \mathbb{E}\chi_{\tilde{\Omega}_{R,k}}(\omega) \, |X_k|^{2p-2} \left[|X_k|^{2\varkappa+1} h + |X_k|^{3/2(1+\varkappa)} h^{1/2} \right] \\
&\quad + K \sum_{l=3}^{2p} \mathbb{E}\chi_{\tilde{\Omega}_{R,k}}(\omega) \, |X_k|^{2p-l} \, h^{l/2} \left[h^{l/2} |X_k|^{l\varkappa} + |X_k|^{l\frac{\varkappa+1}{2}} \right]
\end{aligned}
$$

Choosing

$$
R = R(h) = h^{-1/(3\varkappa-1)}, \quad \varkappa \ge 1, \qquad (5.3.16)
$$

we get, for $l = 3, \ldots, 2p$,

$$
\begin{aligned}
\chi_{\tilde{\Omega}_{R,k}}(\omega) \, |X_k|^{2p-2} \left[|X_k|^{2\varkappa+1} h + |X_k|^{3/2(1+\varkappa)} h^{1/2} \right] &\le \chi_{\tilde{\Omega}_{R(h),k}} 2 |X_k|^{2p}, \\
\chi_{\tilde{\Omega}_{R,k}}(\omega) \, |X_k|^{2p-l} \, h^{l/2} \left[h^{l/2} |X_k|^{l\varkappa} + |X_k|^{l\frac{\varkappa+1}{2}} \right] &\le \chi_{\tilde{\Omega}_{R(h),k}} 2 |X_k|^{2p}
\end{aligned}
$$

and hence we rewrite (5.3.15) as

$$\begin{aligned} &\mathbb{E}\chi_{\tilde{\Omega}_{R(h),k+1}}(\omega)|X_{k+1}|^{2p} \\ &\leq \mathbb{E}\chi_{\tilde{\Omega}_{R(h),k}}(\omega)|X_k|^{2p} \\ &\quad +Kh\mathbb{E}\chi_{\tilde{\Omega}_{R(h),k}}(\omega)\,|X_k|^{2p} + K\sum_{l=1}^{p}\mathbb{E}\chi_{\tilde{\Omega}_{R(h),k}}(\omega)\,|X_k|^{2(p-l)}\,h^l \\ &\leq \mathbb{E}\chi_{\tilde{\Omega}_{R(h),k}}(\omega)|X_k|^{2p} + Kh\mathbb{E}\chi_{\tilde{\Omega}_{R(h),k}}(\omega)\,|X_k|^{2p} + Kh, \end{aligned} \tag{5.3.17}$$

where in the last line we have used the Young inequality. From here, we get by the Gronwall inequality that

$$\mathbb{E}\chi_{\tilde{\Omega}_{R(h),k}}(\omega)|X_k|^{2p} \leq K(1+\mathbb{E}|X_0|^{2p}), \tag{5.3.18}$$

where $R(h)$ is from (5.3.16) and K does not depend on k and h but it depends on p.

It remains to estimate $\mathbb{E}\chi_{\tilde{\Lambda}_{R(h),k}}(\omega)|X_k|^{2p}$. We have

$$\begin{aligned} \chi_{\tilde{\Lambda}_{R,k}} = 1-\chi_{\tilde{\Omega}_{R,k}} = 1-\chi_{\tilde{\Omega}_{R,k-1}}\chi_{|X_k|\leq R} = \chi_{\tilde{\Lambda}_{R,k-1}} + \chi_{\tilde{\Omega}_{R,k-1}}\chi_{|X_k|>R} \\ = \cdots = \textstyle\sum_{l=0}^{k}\chi_{\tilde{\Omega}_{R,l-1}}\chi_{|X_l|>R}, \end{aligned}$$

where we put $\chi_{\tilde{\Omega}_{R,-1}} = 1$. Then, using (5.3.4), (5.3.18), (5.2.2), and Cauchy-Schwarz's and Markov's inequalities, we obtain

$$\begin{aligned} &\mathbb{E}\chi_{\tilde{\Lambda}_{R(h),k}}(\omega)|X_k|^{2p} = \mathbb{E}\textstyle\sum_{l=0}^{k}|X_k|^{2p}\chi_{\tilde{\Omega}_{R(h),l-1}}\chi_{|X_l|>R(h)} \\ &\leq \left(\mathbb{E}|X_0+k|^{4p}\right)^{1/2}\textstyle\sum_{l=0}^{k}\left(\mathbb{E}\left[\chi_{\tilde{\Omega}_{R(h),l-1}|X_l|>R(h)}\right]\right)^{1/2} \\ &= \left(\mathbb{E}|X_0+k|^{4p}\right)^{1/2}\textstyle\sum_{l=0}^{k}\left(P(\chi_{\tilde{\Omega}_{R(h),l-1}}|X_l|>R)\right)^{1/2} \\ &\leq \left(\mathbb{E}|X_0+k|^{4p}\right)^{1/2}\textstyle\sum_{l=0}^{k}\dfrac{\left(\mathbb{E}(\chi_{\tilde{\Omega}_{R(h),l-1}}|X_l|^{2(2p+1)(3\varkappa-1)})\right)^{1/2}}{R(h)^{(2p+1)(3\varkappa-1)}} \\ &\leq K\left(\mathbb{E}|X_0+k|^{4p}\right)^{1/2}\left(\mathbb{E}(1+|X_0|^{2(2p+1)(3\varkappa-1)})\right)^{1/2}kh^{2p+1} \\ &\qquad \leq K(1+\mathbb{E}|X_0|^{2(2p+1)(3\varkappa-1)})^{1/2}, \end{aligned}$$

which together with (5.3.18) implies (5.3.2) for integer $p \geq 1$. Then, by the Jensen inequality, (5.3.2) holds for non-integer p as well. □

The next lemma gives estimates for the one-step error of the balanced Euler scheme (5.3.1).

Lemma 5.3.3 *Assume that* (5.2.6) *holds. Assume that the coefficients* $a(t,x)$ *and* $\sigma_r(t,x)$ *have continuous first-order partial derivatives in* t *and that these derivatives and the coefficients satisfy inequalities of the form* (5.2.4). *Then the scheme* (5.3.1) *satisfies the inequalities* (5.2.10) *and* (5.2.11) *with* $q_1 = 3/2$ *and* $q_2 = 1$, *respectively.*

As in the global Lipschitz case [355, 358], the proof of Lemma 5.3.3 is routine one-step error analysis.

Proof. We recall an auxiliary result from [448] that for $\varphi(t,x)$ which have continuous first-order partial derivative in t and that the derivative and the function satisfy inequalities of the form (5.2.4). For $\alpha \geq 1$ and $s \geq t$, we have

$$\mathbb{E}\,|\varphi(s, X_{t,x}(s)) - \varphi(t,x)|^{\alpha} \leq K(1+|x|^{2\alpha\varkappa-\alpha})[(s-t)^{\alpha/2}+(s-t)^{\alpha}], \quad (5.3.19)$$

which, in particular, holds for the functions $a(t,x)$ and $\sigma_r(t,x)$ under the conditions of the lemma.

Now consider the one-step approximation of the SDE (5.2.1), which corresponds to the balanced Euler method (5.3.1):

$$X = x + \frac{a(t,x)h + \sum_{r=1}^{m}\sigma_r(t,x)\xi_r\sqrt{h}}{1 + h|a(t,x)| + \sqrt{h}\sum_{r=1}^{m}|\sigma_r(t,x)\xi_r|} \quad (5.3.20)$$

and the one-step approximation corresponding to the explicit Euler scheme:

$$\tilde{X} = x + a(t,x)h + \sum_{r=1}^{m}\sigma_r(t,x)\xi_r\sqrt{h}. \quad (5.3.21)$$

We start with analysis of the one-step error of the Euler scheme:

$$\tilde{\rho}(t,x) := X_{t,x}(t+h) - \tilde{X}.$$

Using (5.3.19), we obtain[2]

$$\begin{aligned} |\mathbb{E}\tilde{\rho}(t,x)| &= \left|\mathbb{E}\int_t^{t+h}(a(s,X_{t,x}(s)) - a(t,x))ds\right| \quad (5.3.22) \\ &\leq \mathbb{E}\int_t^{t+h}|a(s,X_{t,x}(s)) - a(t,x)|ds \\ &\leq Kh^{3/2}(1+|x|^{2\varkappa-1}). \end{aligned}$$

Further,

$$\begin{aligned} \mathbb{E}\tilde{\rho}^{2p}(t,x) \leq{}& K\mathbb{E}\left|\int_t^{t+h}(a(s,X_{t,x}(s)) - a(t,x))ds\right|^{2p} \quad (5.3.23) \\ &+K\sum_{r=1}^{q}\mathbb{E}\left|\int_t^{t+h}(\sigma_r(s,X_{t,x}(s)) - \sigma_r(t,x))\,dW_r(s)\right|^{2p}. \end{aligned}$$

[2] Assuming additional smoothness of $a(t,x)$, we can get an estimate for $\mathbb{E}\tilde{\rho}(t,x)$ of order h^2 but this will not improve the result of this lemma for the balanced Euler scheme (5.3.1).

Applying (5.3.19), we have

$$\mathbb{E}\left|\int_t^{t+h}(a(s,X_{t,x}(s))-a(t,x))ds\right|^{2p}\leq Kh^{3p}(1+|x|^{4p\varkappa-2p}) \quad (5.3.24)$$

and

$$\mathbb{E}\left|\int_t^{t+h}(\sigma_r(s,X_{t,x}(s))-\sigma_r(t,x))\,dW_r(s)\right|^{2p} \quad (5.3.25)$$

$$\leq Kh^{p-1}\int_t^{t+h}\mathbb{E}\left|\sigma_r(s,X_{t,x}(s))-\sigma_r(t,x)\right|^{2p}ds\leq Kh^{2p}(1+|x|^{4p\varkappa-2p}).$$

It follows from (5.3.23) to (5.3.25) that

$$\mathbb{E}\tilde{\rho}^{2p}(t,x)\leq Kh^{2p}(1+|x|^{4p\varkappa-2p}). \quad (5.3.26)$$

Now we compare the one-step approximations (5.3.20) of the balanced Euler scheme and (5.3.21) of the Euler scheme:

$$X=x+\frac{a(t,x)h+\sum_{r=1}^m\sigma_r(t,x)\xi_r\sqrt{h}}{1+h|a(t,x)|+\sqrt{h}\sum_{r=1}^m|\sigma_r(t,x)\xi_r|},\quad \tilde{X}=X=\rho(t,x), \quad (5.3.27)$$

$$\rho(t,x)=\left(a(t,x)h+\sum_{r=1}^m\sigma_r(t,x)\xi_r\sqrt{h}\right)\frac{h|a(t,x)|+\sqrt{h}\sum_{r=1}^m|\sigma_r(t,x)\xi_r|}{1+h|a(t,x)|+\sqrt{h}\sum_{r=1}^m|\sigma_r(t,x)\xi_r|}.$$

Using the equality (5.3.8) and the assumptions made on the coefficients (see (5.2.4)), we obtain

$$|\mathbb{E}\rho(t,x)|=\left|a(t,x)h\mathbb{E}\frac{h|a(t,x)|+\sqrt{h}\sum_{r=1}^m|\sigma_r(t,x)\xi_r|}{1+h|a(t,x)|+\sqrt{h}\sum_{r=1}^m|\sigma_r(t,x)\xi_r|}\right|\leq Kh^{3/2}(1+|x|^{3\varkappa/2}),$$

which together with (5.3.27) and (5.3.23) implies that (5.3.20) satisfies (5.2.10) with $q_1=3/2$. Further,

$$\mathbb{E}\rho^{2p}(t,x)\leq h^{2p}\mathbb{E}\left[\sqrt{h}|a(t,x)|+\sum_{r=1}^m|\sigma_r(t,x)\xi_r|\right]^{4p}\leq Kh^{2p}(1+|x|^{4p\varkappa}),$$

which together with (5.3.27) and (5.3.26) implies that (5.3.20) satisfies (5.2.11) with $q_2=1$. □

Lemmas 5.3.1 and 5.3.3 and Theorem 5.2.3 imply the following result.

Proposition 5.3.4 *Under the assumptions of Lemmas 5.3.1 and 5.3.3 the balanced Euler scheme* (5.3.1) *has mean-square order half, i.e., for it the inequality* (5.2.14) *holds with* $q=q_2-1/2=1/2$.

Remark 5.3.5 *In the additive noise case, the mean-square order of the balanced Euler scheme* (5.3.1) *does not improve since* q_1 *and* q_2 *remain* 3/2 *and* 1, *respectively.*

Remark 5.3.6 *One can consider the following scheme for* (5.2.1) *instead of* (5.3.1)*:*

$$X_{k+1} = X_k + \frac{a(t_k, X_k)h + \sum_{r=1}^m \sigma_r(t_k, X_k)\xi_{rk}\sqrt{h}}{1 + h|a(t_k, X_k)| + \sqrt{h}\sum_{r=1}^m |\sigma_r(t_k, X_k)|}. \tag{5.3.28}$$

This scheme is still a balanced scheme but has less restricted conditions on p*-th order moments when* p_0 *is finite. The proof is left as an exercise at the end of this chapter.*

5.4 Numerical examples

5.4.1 Some numerical schemes

In this section we list some numerical schemes for nonlinear stochastic differential equations. In the following schemes, $\xi_{rk} = (W_r(t_{k+1}) - W_r(t_k))/\sqrt{h}$ are i.i.d. $\mathcal{N}(0, 1)$ (Gaussian) random variables.

Explicit schemes

- the drift-tamed Euler scheme (a modified balanced method) [235]:

$$X_{k+1} = X_k + h\frac{a(X_k)}{1 + h\,|a(X_k)|} + \sum_{r=1}^m \sigma_r(t_k, X_k)\xi_{rk}\sqrt{h}. \tag{5.4.1}$$

- the fully tamed scheme [233]:

$$X_{k+1} = X_k + \frac{a(X_k)h + \sum_{r=1}^m \sigma_r(t_k, X_k)\xi_{rk}\sqrt{h}}{\max\left(1, h\left|ha(X_k) + \sum_{r=1}^m \sigma_r(t_k, X_k)\xi_{rk}\sqrt{h}\right|\right)}. \tag{5.4.2}$$

- the balanced Euler method (5.3.1)

$$X_{k+1} = X_k + \frac{a(t_k, X_k)h + \sum_{r=1}^m \sigma_r(t_k, X_k)\xi_{rk}\sqrt{h}}{1 + h|a(t_k, X_k)| + \sqrt{h}\sum_{r=1}^m |\sigma_r(t_k, X_k)\xi_{rk}|}. \tag{5.4.3}$$

Drift-implicit schemes

- the drift-implicit Euler scheme (5.2.15)

$$X_{k+1} = X_k + a(t_{k+1}, X_{k+1})h + \sum_{r=1}^m \sigma_r(t_k, X_k)\xi_{rk}\sqrt{h}. \tag{5.4.4}$$

- the trapezoidal scheme [358, p. 30]:

$$X_{k+1} = X_k + \frac{h}{2}\left[a(X_{k+1}) + a(X_k)\right] + \sum_{r=1}^{m} \sigma_r(t_k, X_k)\xi_{rk}\sqrt{h}. \qquad (5.4.5)$$

Fully implicit schemes

- the fully implicit Euler scheme ((3.2.27) with $\lambda = 1$)

$$X_{k+1} = X_k + a(t_{k+1}, X_{k+1})h - \sum_{r=1}^{m}\sum_{j=1}^{d} \frac{\partial \sigma_r}{\partial x^j}(t_{k+1}, X_{k+1})\sigma_r^j(t_{k+1}, X_{k+1})h$$
$$+ \sum_{r=1}^{m} \sigma_r(t_{k+\lambda}, (1-\lambda)X_k + \lambda X_{k+1})\,(\zeta_{rh})_k\,\sqrt{h}. \qquad (5.4.6)$$

- the midpoint method ((3.2.27) with $\lambda = 1/2$)

$$X_{k+1} = X_k + a(t_{k+\frac{1}{2}}, \frac{X_k + X_{k+1}}{2})h$$
$$+ \sum_{r=1}^{m} \sigma_r(t_{k+\frac{1}{2}}, \frac{X_k + X_{k+1}}{2})\,(\zeta_{rh})_k\,\sqrt{h} - \frac{1}{2}\sum_{r=1}^{m}\sum_{j=1}^{d}$$
$$\frac{\partial \sigma_r}{\partial x^j}(t_{k+\frac{1}{2}}, \frac{X_k + X_{k+1}}{2})\sigma_r^j(t_{k+\frac{1}{2}}, \frac{X_k + X_{k+1}}{2})h, \qquad (5.4.7)$$

where $(\zeta_{rh})_k$ are i.i.d. random variables defined in (3.2.28), which are truncations of $\xi_{rk} \sim \mathcal{N}(0,1)$ and $A_h = \sqrt{4|\ln h|}$.

Convergence order of these schemes. The drift-tamed Euler scheme (5.4.1) converges with strong convergence order half under Assumption 5.2.1 together with Lipschitz diffusion coefficients [235]. The fully tamed Euler scheme (5.4.2) is proved to have strong convergence but without order given under Assumption 5.2.1, see [233]. A half-order strong convergence of (5.4.3) is proved in Chapter 5.3.

The half-order convergence of the drift-implicit scheme (5.4.4) is proved in [336] and in Chapter 5.2.1. The trapezoidal scheme (5.4.5) can be shown to be of mean-square convergence with order half using Theorem 5.2.3. It only requires to show boundedness of higher moments of the scheme, which is similar to the proof of bounded moments in [335].

When $1/2 < \lambda \leq 1$, the fully implicit scheme (3.2.27) is expected to converge with order half. When diffusion coefficients and $\frac{\partial \sigma_r}{\partial x^j}\sigma_r^j$ are Lipschitz continuous, the half-order convergence is proved in [448]. For $\lambda = 1/2$, the midpoint method is proved in [448] to converge with a mean-square convergence order half when the diffusion coefficients are uniformly bounded. Moreover, the midpoint method is of mean-square convergence one for SDEs with commutative noises (see (3.2.7)).

5.4.2 Numerical results

In all the experiments with fully implicit schemes, where the truncated random variables ζ are used, we took $l = 2$ in (3.2.29). The experiments were performed using Matlab R2012a on a Macintosh desktop computer with Intel Xeon CPU E5462 (quad-core, 2.80 GHz). In simulations we used the Mersenne Twister random generator with seed 100. Newton's method was used to solve the nonlinear algebraic equations at each step of the implicit schemes.

We test the methods on two model problems. The first one has non-globally Lipschitz drift and global Lipschitz diffusion with two noncommutative noises. The second example satisfies Assumption 5.2.1 (non-globally Lipschitz both drift and diffusion). The aim of the tests is to compare the performance of the methods: their accuracy and computational costs.

Remark 5.4.1 *Experiments cannot prove or disprove boundedness of moments of the schemes since experiments rely on a finite sample of trajectories run over a finite time interval while blow-up of moments in divergent methods (e.g., explicit Euler scheme) is, in general, a result of large deviations [341, 359].*

To compute the mean-square error, we run M independent trajectories $X^{(i)}(t)$, $X_k^{(i)}$:

$$\left(\mathbb{E}[(X(T) - X_N)^2]\right)^{1/2} \doteq \left(\frac{1}{M}\sum_{i=1}^{M}[X^{(i)}(T) - X_N^{(i)}]^2\right)^{1/2}. \tag{5.4.8}$$

We took time $T = 50$ and $M = 10^4$. The reference solution was computed by the midpoint method with small time step $h = 10^{-4}$. It was verified that using a different implicit scheme for simulating a reference solution does not affect the outcome of the tests. We chose the midpoint scheme as a reference since in all the experiments it produced the most accurate results. The number of trajectories $M = 10^4$ was sufficiently large for the statistical errors (the Monte Carlo error with 95% confidence) not to significantly hinder the mean-square errors.

Example 5.4.2 *Consider the following Stratonovich SDE:*

$$dX = (1 - X^5)\,dt + X \circ dW_1 + dW_2, \quad X(0) = 0. \tag{5.4.9}$$

In Ito's sense, the drift of the equation becomes $a(t, x) = 1 - x^5 + x/2$. Here we tested the balanced Euler method (5.4.3), the drift-tamed scheme (5.4.1), the fully implicit Euler scheme (5.4.6), and the midpoint method (5.4.7).

Mean-square errors of these schemes are plotted in Figure 5.1. The observed rates of convergence of all the tested methods are close to the predicted 1/2. For a fixed time step h, the midpoint method is the most accurate scheme while the balanced Euler method (5.4.3) is the least accurate.

Fig. 5.1. Mean-square errors of the selected schemes for Example 5.4.2.

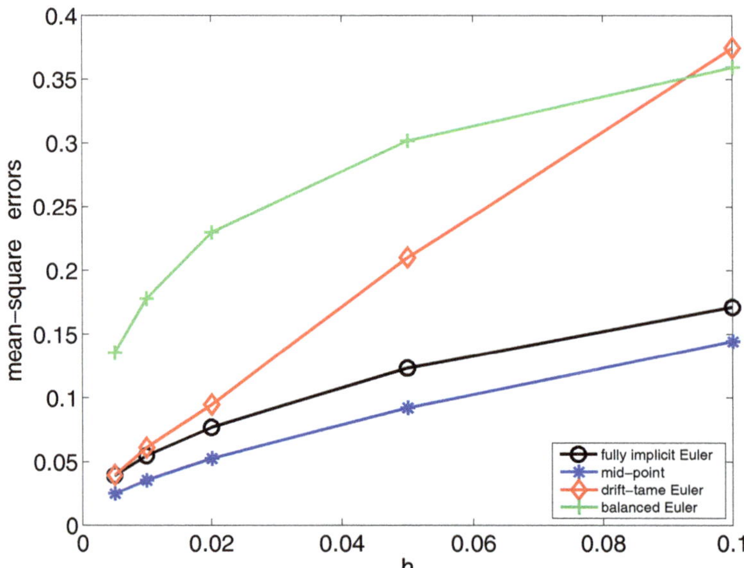

For large time step sizes, the convergence rates of balanced Euler scheme (5.4.3) are a bit off the predicted order 1/2. However, when time step sizes become smaller, the convergence rates are much closer to 1/2. See mean-square errors for the balanced Euler scheme (5.4.3) in Table 5.1.

Table 5.1. Mean-square errors of the balanced Euler scheme (5.4.3) for Example 5.4.2.

h	0.1	0.05	0.02	0.01	0.005	0.002	0.001
Errors	3.59e-01	3.02e-01	2.30e-01	1.78e-01	1.35e-01	9.27e-02	6.86e-02
Rate	–	0.25	0.30	0.37	0.39	0.41	0.44

To produce the result with accuracy $\sim 0.06 - 0.07$, in our experiment of running $M = 10^4$, the drift-tamed Euler (5.4.1) costs the least time and the balanced Euler scheme (5.4.3) costs the most. These numerical results confirmed the conclusion of [235] that the scheme (5.4.1) from [235] is highly competitive (Table 5.2).[3]

[3]However, (5.4.1) is not applicable when diffusion grows faster than a linear function.

Table 5.2. Comparison of mean-square errors at the magnitude of 0.06–0.07 of different schemes for Example 5.4.2 at $T = 50$.

Methods	h	Errors	Time (sec.)
The tamed Euler (5.4.1)	0.01	6.10e-02	170
The midpoint (5.4.7)	0.02	5.26e-02	329
The fully implicit Euler scheme (5.4.6)	0.01	5.48e-02	723
The balanced Euler (5.4.3)	0.001	6.86e-02	1870

Example 5.4.3 *Consider the SDE in the Stratonovich sense:*

$$dX = (1 - X^5)\, dt + X^2 \circ dW, \quad X(0) = 0. \tag{5.4.10}$$

In Ito's sense, the drift of the equation becomes $a(t, x) = 1 - x^5 + x^3$.

Here we tested the balanced Euler method (5.4.3), the fully tamed Euler scheme (5.4.2), the drift-implicit scheme (5.4.4), the fully implicit Euler scheme (5.4.6), the midpoint method (5.4.7), and the trapezoidal scheme (5.4.5).

It can be proved directly that implicit algebraic equations arising from application of the midpoint and fully implicit Euler schemes to (5.4.10) have unique solutions under a sufficiently small time step.

The fully tamed scheme (5.4.2) did not produce accurate results until the time step size is at least $h = 0.005$ and thus no errors presented here. A half-order convergence of this scheme is not expected, see [448, Remark 5.2] for detailed explanation.

Table 5.3 gives the mean-square errors and experimentally observed convergence rates for the corresponding methods. In addition to the data in the table, we evaluated errors for (5.3.1) for smaller time steps: $h = 0.002$ – the error is 3.70e-02 (rate 0.41), 0.001 – 2.73e-02 (0.44), 0.0005 – 2.00e-02 (0.45), i.e., for smaller h the observed convergence rate of (5.3.1) becomes closer to the theoretically predicted order 1/2. Since (5.4.10) is with single noise, the midpoint method demonstrates the first order of convergence. The other implicit schemes show the order half as expected.

Table 5.4 presents the time costs in seconds. Let us fix the tolerance level at 0.05–0.06. We highlighted in bold the corresponding values in both tables. In this example the midpoint scheme is the most efficient, due to its first order convergence in the commutative case. Among methods of half-order, the balanced Euler method (5.3.1) is the fastest and one can expect that for multi-dimensional SDEs the explicit scheme (5.3.1) can considerably outperform implicit methods (see a similar outcome for the drift-tamed method (5.4.1) supported by experiments in [235]. In comparison with the balanced Euler (5.4.3), the drift-tamed Euler scheme (5.4.1) is divergent when diffusion is growing faster than a linear function on infinity.

Table 5.3. Example 5.4.3. Mean-square errors of the selected schemes.

h	(5.4.4)	rate	(5.4.6)	rate	(5.4.7)	rate
0.2	3.449e-01	–	1.816e-01	–	1.378e-01	–
0.1	2.441e-01	0.50	1.331e-01	0.45	8.723e-02	0.66
0.05	1.592e-01	0.62	9.619e-02	0.47	**5.344e-02**	0.71
0.02	8.360e-02	0.70	6.599e-02	0.41	2.242e-02	0.95
0.01	**5.460e-02**	0.61	**4.919e-02**	0.42	1.145e-02	0.97
0.005	3.682e-02	0.57	3.522e-02	0.48	5.945e-03	0.95

h	(5.4.5)	rate	(5.4.3)	rate
0.2	4.920e-01	–	2.102e-01	–
0.1	3.526e-01	0.48	1.637e-01	0.36
0.05	2.230e-01	0.66	1.270e-01	0.37
0.02	1.048e-01	0.82	9.170e-02	0.36
0.01	**5.990e-02**	0.81	7.065e-02	0.38
0.005	3.784e-02	0.66	**5.393e-02**	0.39

Table 5.4. Example 5.4.3. Computational times for the selected schemes.

h	(5.4.4)	(5.4.6)	(5.4.7)	(5.4.5)	(5.4.3)
0.2	9.25e+00	1.10e+01	9.33e+00	1.20e+01	3.98e+00
0.1	1.77e+01	2.17e+01	1.80e+01	2.30e+01	7.49e+00
0.05	3.42e+01	4.26e+01	**3.51e+01**	4.48e+01	1.41e+01
0.02	8.33e+01	1.04e+02	8.69e+01	1.10e+02	3.37e+01
0.01	**1.64e+02**	**2.05e+02**	1.73e+02	**2.19e+02**	6.62e+01
0.005	3.25e+02	4.07e+02	3.47e+02	4.37e+02	**1.32e+02**

5.5 Summary and bibliographic notes

Under the assumption 5.2.1 (global monotone condition and polynomial growth of the coefficients), a solution to the equation (5.2.1) can be extremely large and then (5.2.1) becomes highly stiff. To deal with stiffness, several numerical methods have been proposed in the literature such as implicit schemes and balanced schemes. To show the convergence order of these methods, it is natural to investigate a basic relationship between local truncation error and global truncation error of numerical methods for SDEs with locally Lipschitz continuous and polynomially growing coefficients, see Theorem 5.2.3.

It is important to observe that numerical methods for SDEs with coefficients of nonlinear growth require the L^p ($p \geq 2$)-stability of numerical solutions for the mean-square convergence while for numerical methods of SDEs with Lipschitz coefficients they do not require this stability (though the L^p stability is naturally satisfied).

In Chapter 5.4, we show some comparisons among numerical results from balanced explicit schemes and implicit schemes. The balanced explicit scheme

(5.4.3) is competitive for SDEs with drift and diffusion coefficients of polynomial growth, see Example 5.4.3. However, the balanced scheme (5.4.3) exhibits very large errors for SDEs with Lipschitz diffusion coefficients. We observe that it requires small time step sizes to reach the asymptotic region of convergence even though it is proved to be of half order in the mean-square sense, see Examples 5.4.2 and 5.4.3.

Bibliographic notes. There have been many developments in numerical methods for SDEs with locally Lipschitz continuous coefficients since [448], see Refs. [232, 237, 411, 436, 503, 512]. In [512], a semi-tamed Euler is proposed for SDEs with non-Lipschitz continuous drift coefficients and Lipschitz continuous diffusion coefficients. The drift is decomposed into two parts: Lipschitz and non-Lipschitz parts, where only the non-Lipschitz part is tamed.

For SDEs with non-Lipschitz continuous drift and diffusion coefficients, Ref. [503] presents a class of first-order balanced schemes and proves L^p-stability of presented schemes using the fundamental theorem presented here. Ref. [411] proposes a slightly different tamed Euler scheme than (5.3.1)/(5.4.3) in this chapter:

$$X_{k+1} = X_k + \frac{a(t_k, X_k)h + \sum_{r=1}^{m} \sigma_r(t_k, X_k)\xi_{rk}\sqrt{h}}{1 + \sqrt{h}|a(t_k, X_k)| + \sqrt{h}\sum_{r=1}^{m} |\sigma_r(t_k, X_k)|}. \tag{5.5.1}$$

A general tamed scheme is proposed in [436] for Lyapunov stability rather than simply L^p-stability, one of which can be

$$X_{k+1} = X_k + \frac{a(t_k, X_k)h}{1 + \sqrt{h}|a(t_k, X_k)|} + \sum_{r=1}^{m} \frac{\sigma_r(t_k, X_k)}{1 + \sqrt{h}\sum_{r=1}^{m} |\sigma_r(t_k, X_k)|}\xi_{rk}\sqrt{h}. \tag{5.5.2}$$

Under even more general conditions, Refs. [232, 237] propose a tamed Euler scheme of a similar type for SDEs with *exponential moments* and prove Lyapunov stability and half-order convergence.

In [448], the authors present fully implicit (i.e., implicit both in drift and diffusion, (3.2.27)) mean-square schemes for one-sided Lipschitz drift coefficients, which grows superlinearly and not faster than polynomial growth at infinity. The fully implicit schemes (3.2.27) was proposed and motivated by geometric integration of stochastic Hamiltonian equations in [356] (see also [358]), where their convergence was proved under globally Lipschitz coefficients.

In this book, we will discuss numerical methods for SPDEs with coefficients of one-sided Lipschitz continuous in Chapters 9 and 10.

5.6 Suggested practice

Exercise 5.6.1 *Show that the scheme* (5.3.28) *has less restricted conditions on p-th order moments when* p_0 *is finite compared with the scheme* (5.3.1).

Exercise 5.6.2 *Consider the following stochastic Ginzburg-Landau equation*

$$dX(t) = \big(a(t)X(t) - b(t)X^3(t)\big)\, dt + \sigma(t)X(t)\, dW(t), \tag{5.6.1}$$
$$X(0) = x > 0. \tag{5.6.2}$$

Here $b(t) > 0$, $a(t)$, *and* $\sigma(t)$ *are bounded continuous functions on* $[0, \infty)$.

a) Show that the coefficients satisfy Assumption 5.2.1.
b) Show that the solution to stochastic Ginzburg-Landau equation is given by

$$X(t) = \frac{e^{\int_0^t a(s) - \frac{1}{2}\sigma^2(s)\, ds + \int_0^t \sigma(s)\, dW(s)}}{\left(x^{-2} + 2\int_0^t b(s)e^{2\int_0^s a(\theta) - \frac{1}{2}\sigma^2(\theta)\, d\theta + 2\int_0^s \sigma(\theta)\, dW(\theta)}\right)^{1/2}}.$$

c) Apply the schemes (5.4.1), (5.2.15), (5.4.5), (5.3.1), and (3.2.27) with $\lambda = 1/2$ *and* $\lambda = 1$ *to solve the stochastic Ginzburg-Landau equation and compare their mean-square convergence orders.*

Hint. Apply the integrating factor method. Let $Y(t)$ be the following exponential process $dY(t) = a(t)Y(t)\, dt + \sigma(t)Y(t)\, dW(t)$ and $X(t) = C(t)Y(t)$. By Ito formula, we have

$$dC(t) = -b(t)Y^2(t)C^3(t)\, dt,$$

from which we can find $C(t)$ and thus $X(t)$.

Exercise 5.6.3 *Consider the following SDE*

$$dX(t) = \big(X(t) - X^3(t)\big)\, dt + \sigma X^2(t)\, dW(t), \tag{5.6.3}$$
$$X(0) = 1. \tag{5.6.4}$$

Here $\sigma = 0.1$ *are bounded continuous functions on* $[0, \infty)$.

a) Show that the coefficients satisfy conditions in Assumption 5.2.1.
b) Apply the schemes (5.4.1), (5.2.15), (5.4.5), (5.3.1), and (3.2.27) with $\lambda = 1/2$ *and* $\lambda = 1$ *to solve the equation and compare their mean-square convergence orders. Note this time we have no analytical solution. Use a numerical solution by (3.2.27) with* $\lambda = 1/2$ *with a fine time step size as a reference solution ("exact solution").*

Part II

Temporal White Noise

The standard approach to constructing SPDE solvers starts with a space discretization of a SPDE, for which spectral methods (see, e.g., [78, 167, 249]), finite element methods (see, e.g., [6, 152, 491]) or spatial finite differences (see, e.g., [6, 189, 420, 495]) can be used. The result of such a space discretization is a large system of ordinary stochastic differential equations (SDEs), which requires time discretization to complete a numerical algorithm. In [101, 109] the SPDE is first discretized in time and then to this semi-discretization a finite-element or finite-difference method can be applied. Other numerical approaches include those making use of splitting techniques [33, 191, 293], quantization [161], or an approach based on the averaging-over-characteristic formula [361, 396]. In [315, 344] numerical algorithms based on the Wiener chaos expansion (WCE) were introduced for solving the nonlinear filtering problem for hidden Markov models. Since then, the WCE-based numerical methods have been successfully developed in a number of directions (see, e.g., [225, 489]).

In Part II of the book, we consider deterministic integration methods in random space for stochastic partial differential equations. In Chapter 6, we discuss Wiener chaos methods (WCE) and a multi-stage WCE for long time integration for linear advection-diffusion-reaction equations with multiplicative noise. In Chapter 7, we discuss stochastic collocation methods (precisely, sparse grid collocation methods) for linear parabolic equations with multiplicative noise. Subsequently, we compare the two methods for these linear equations in Chapter 8 while in Chapter 9 we apply stochastic collocation methods discussed in Chapters 7 and 8 to nonlinear equations, namely stochastic Euler equations for the one-dimensional piston problem.

6

Wiener chaos methods for linear stochastic advection-diffusion-reaction equations

In this chapter, we discuss numerical algorithms using Wiener chaos expansion (WCE) for solving second-order linear parabolic stochastic partial differential equations (SPDEs). The algorithm for computing moments of the SPDE solutions is deterministic, i.e., it does not involve any statistical errors from generating random numbers.

Although the Wiener chaos expansion (WCE) results in a triangular system of deterministic partial differential equations, WCE is only efficient for short time integration. Here, we present a recursive Wiener chaos expansion (WCE) method for longer time integration of linear parabolic equations with temporal white noise. We compare the deterministic algorithm with the Monte Carlo method and demonstrate that the new recursive WCE method is more efficient for highly accurate solutions.

This chapter is organized as follows. We first describe in Chapter 6.1 the multistage WCE for computing solutions and moments of solutions and discuss the complexity of the resulting algorithm. An illustration of the algorithm with a simple example is presented in Chapter 6.2. We compare the multistage WCE and Monte Carlo-type algorithms for one-dimensional problems in Chapter 6.3 and for a two-dimensional passive scalar equation in Chapter 6.4. In Chapter 6.5, we highlight the main points of this chapter and comment on WCE and the multistage WCE. We also provide two exercises focusing on time integration and the multistage WCE method at the end of the chapter.

6.1 Description of methods

In computing moments of SPDE solutions, the existing approaches to solving SPDEs are usually complemented by the Monte Carlo technique. Consequently, in these approaches numerical approximations of SPDEmoments

Z. Zhang, G.E. Karniadakis, *Numerical Methods for Stochastic Partial Differential Equations with White Noise*,
Applied Mathematical Sciences 196, DOI 10.1007/978-3-319-57511-7_6

have two errors: numerical integration (space-time discretization) error and Monte Carlo (statistical) error. To reach a high accuracy, we have to run a very large number of independent simulations of the SPDE to reduce the Monte Carlo error. In contrast, WCE methods for computing moments of the SPDE solutions are statistical-error free (no random number generators are used) but are only subject to the error from the truncation of the WCE.

6.1.1 Multistage WCE method

For the linear SPDE (3.3.23)–(3.3.24), we can use the propagator (3.3.16). In order to apply a truncation of the propagator, we introduce the following notation: the order of multi-index α:

$$d(\alpha) = \max\left\{l \geq 1 : \alpha_{k,l} > 0 \text{ for some } k \geq 1\right\},$$

and the truncated set of multi-indices:

$$\mathcal{J}_{\mathsf{N},\mathsf{n}} = \left\{\alpha \in \mathcal{J} : \ |\alpha| \leq \mathsf{N}, \ \ d(\alpha) \leq \mathsf{n}\right\}.$$

Recall that the multi-index length is $|\alpha| = \sum_{i,k=1}^{\infty} \alpha_{k,l}$. Here N is the highest Hermite polynomial order and n is the maximum number of Gaussian random variables for each Wiener process. Using (3.3.14), we introduce the truncated Wiener chaos solution

$$u_{\mathsf{N},\mathsf{n}}(t,x) = \sum_{\alpha \in \mathcal{J}_{\mathsf{N},\mathsf{n}}} \frac{1}{\sqrt{\alpha!}} \varphi_\alpha(t,x)\xi_\alpha, \tag{6.1.1}$$

with the basis $\{m_l(s)\}_{l\geq 1}$ given by

$$m_1(s) = \frac{1}{\sqrt{t}}, \quad m_l(s) = \sqrt{\frac{2}{t}}\cos(\frac{\pi(l-1)s}{t}), \quad l \geq 2, \quad 0 \leq s \leq t. \tag{6.1.2}$$

Letting $\alpha \in \mathcal{J}_{\mathsf{N},\mathsf{n}}$ in the propagator (3.3.16), we have that the coefficients $\varphi_\alpha(t,x;\phi)$ satisfy the propagator

$$\begin{aligned} \frac{\partial \varphi_\alpha(t,x;\phi)}{\partial t} &= \mathcal{L}\varphi_\alpha(t,x;\phi) + f(x)\mathbf{1}_{\{|\alpha|=0\}} \\ &\quad + \sum_{k=1}^{q}\sum_{l=1}^{\mathsf{n}} \alpha_{k,l} m_l(t) \left[\mathcal{M}_k \varphi_{\alpha^-(k,l)}(t,x;\phi) \right. \\ &\quad \left. + g_k(x)\mathbf{1}_{\{|\alpha|=1\}}\right], \quad t \in (0,T], \\ \varphi_\alpha(0,x) &= \phi(x)\mathbf{1}_{\{|\alpha|=0\}}, \end{aligned} \tag{6.1.3}$$

where $\alpha^-(k,l)$ is the multi-index with components

$$\left(\alpha^-(k,l)\right)_{i,j} = \begin{cases} \max(0, \alpha_{i,j} - 1), & \text{if } i = k \text{ and } j = l, \\ \alpha_{i,j}, & \text{otherwise.} \end{cases} \tag{6.1.4}$$

The truncated expansion (6.1.1) together with (6.1.3), (2.3.4), and (6.1.2) gives us a constructive approximation of the solution to (3.3.23), where implementation requires that we numerically solve the propagator (3.3.16).

It is proved in [315, Theorem 2.2] that when $b_i^k(t,x) = 0$, $c = 0$, $g_k = 0$ (reaction-diffusion equation) and the number of noises is finite there is a constant $C > 0$ such that for any $t \in (0,T]$

$$\mathbb{E}[\|u_{\mathsf{N},\mathsf{n}}(t,\cdot) - u(t,\cdot)\|_{L^2}^2] \le Ce^{Ct}\left(\frac{(Ct)^{\mathsf{N}+1}}{(\mathsf{N}+1)!} + \frac{t^3}{\mathsf{n}}\right). \tag{6.1.5}$$

It follows from the error estimates (6.1.5) that the error of the approximation $u_{\mathsf{N},\mathsf{n}}(t,\cdot)$ grows exponentially in time t, which severely limits its practical use. To overcome this difficulty, it was proposed in [315] to introduce a time discretization with step $\Delta > 0$ and view (6.1.1), (3.3.16), (2.3.4), (6.1.2) as the one-step approximation of the SPDE solution.

To this end, we introduce the multi-step basis for the WCE and its corresponding propagator. Let $0 = t_0 < t_1 < \cdots < t_{\mathsf{K}} = T$ be a uniform partition of the time interval $[0,T]$ with time step size Δ, see Figure 6.1. Let $\{m_k^{(i)}\} = \left\{m_k^{(i)}(s)\right\}_{k\ge 1}$ be the following CONS in $L^2([\mathsf{t}_i, \mathsf{t}_{i-1}])$:

$$\begin{aligned} m_l^{(i)} &= m_l(s - \mathsf{t}_i), \quad \mathsf{t}_i \le s \le \mathsf{t}_{i-1}, \\ m_l(s) &= \frac{1}{\sqrt{\Delta}}, \quad m_l(s) = \sqrt{\frac{2}{\Delta}}\cos\left(\frac{\pi(l-1)s}{\Delta}\right), \quad l \ge 2, \quad 0 \le s \le \Delta, \\ m_l(s) &= 0, \quad l \ge 1, \quad s \notin [0,\Delta]. \end{aligned} \tag{6.1.6}$$

Define the random variables $\xi_\alpha^{(i)}$, $i = 1,\ldots,K$, as

$$\xi_\alpha^{(i)} := \prod_\alpha^{(i)} \left(\frac{H_{\alpha_{k,l}}(\xi_{k,l}^{(i)})}{\sqrt{\alpha_{k,l}!}}\right), \; \alpha \in \mathcal{J}, \tag{6.1.7}$$

where $\xi_{k,l}^{(i)} = \int_{\mathsf{t}_i}^{\mathsf{t}_{i-1}} m_l^{(i)}(s)\, dW_k(s)$, and H_n are Hermite polynomials (2.3.3).

Let

$$u_{\Delta,\mathsf{N},\mathsf{n}}(0,x) = u_0(x) \tag{6.1.8}$$

and by induction for $i = 1,\ldots,K$:

$$u_{\Delta,\mathsf{N},\mathsf{n}}(\mathsf{t}_{i-1},x) = \sum_{\alpha\in\mathcal{J}_{\mathsf{N},\mathsf{n}}} \frac{1}{\sqrt{\alpha!}}\varphi_\alpha^{(i)}(\Delta,x)\xi_\alpha^{(i)}, \tag{6.1.9}$$

where $\varphi_\alpha^{(i)}(\Delta, x)$ solves the system

$$\begin{aligned}\frac{\partial \varphi_\alpha^{(i)}(s,x)}{\partial s} &= \mathcal{L}\varphi_\alpha^{(i)}(s,x) + f(x)\mathbf{1}_{\{|\alpha|=0\}} \\ &\quad + \sum_{k,l} \alpha_{k,l} m_l^{(i)}(s) \\ &\quad \left[\mathcal{M}_k \varphi^{(i)}_{\alpha^-(l,k)}(s,x) + g_k(x)\mathbf{1}_{\{|\alpha|=1\}}\right], \quad s \in (0,\Delta], \\ \varphi_\alpha^{(i)}(0,x) &= u_{\Delta,\mathsf{N},\mathsf{n}}(\mathsf{t}_i, x)\mathbf{1}_{\{|\alpha|=0\}}.\end{aligned} \tag{6.1.10}$$

Thus, (6.1.8)–(6.1.10) together with (6.1.6) and (6.1.7) give us a *recursive* method for solving the SPDE (3.3.23), where implementation requires to numerically solve the propagator (6.1.10) at every time step.

Based on the one-step error (6.1.5), the following global error estimate for the recursive WCE method is proved in [315, Theorem 2.4] (the case of $b_i^k(t,x) = 0$, $c = 0$, $g_k = 0$ and finite number of noises):

$$\mathbb{E}[\|u_{\Delta,\mathsf{N},\mathsf{n}}(\mathsf{t}_{i-1},\cdot) - u(\mathsf{t}_{i-1},\cdot)\|_{L^2}^2] \le Ce^{CT}\left(\frac{(C\Delta)^N}{(\mathsf{N}+1)!} + \frac{\Delta^2}{\mathsf{n}}\right), \quad i = 1,\ldots,K, \tag{6.1.11}$$

for some $C > 0$ independent of Δ, N, and n, i.e., this method is of global mean-square order $\mathcal{O}\left(\frac{\Delta^{\mathsf{N}/2}}{\sqrt{(\mathsf{N}+1)!}} + \frac{\Delta}{\sqrt{\mathsf{n}}}\right)$.

As mentioned above, the recursive WCE method requires to solve the propagator (6.1.10) at every time step, which is computationally rather expensive. To reduce the cost, we introduce a modification of this method following [315]. The idea is to expand the initial condition $u_0(x)$ in a basis $\{e_m\}$, present $u_{\Delta,\mathsf{N},\mathsf{n}}(\mathsf{t}_i,x)$ as $u_{\Delta,\mathsf{N},\mathsf{n}}(\mathsf{t}_i,x) = \sum_m c_m e_m(x)$ and note that $\varphi_\alpha(\Delta, x; u_{\Delta,\mathsf{N},\mathsf{n}}(\mathsf{t}_i,\cdot)) = \sum_m c_m \varphi_\alpha(\Delta, x; e_m)$, where $\varphi_\alpha(s,x;\phi)$ is the solution of the propagator (6.1.10) with the initial condition $\phi(x)$.

The idea is sketched in Figure 6.1. We can first compute the propagator (6.1.12) (see below) on $(0,\Delta]$ and obtain a problem-dependent basis $q_{\alpha,l,m}$ (6.1.13). This step is called "offline" as in [315]. Thus, one recursively computes the solution "online" by (6.1.14) and (6.1.15) only at time $i\Delta$ ($i = 2,\cdots,\mathsf{K}$) using the obtained basis $q_{\alpha,l,m}$. Specifically, we proceed as follows. Let $\{e_m\} = \{e_m(x)\}_{m\ge1}$ be a CONS in $L^2(\mathcal{D})$ with boundary conditions satisfied and $(\cdot,\cdot)$ be the inner product in that space.

Fig. 6.1. Illustration of the idea of multistage WCE. The dotted line denotes the "offline" computation, where we solve the propagator up to time Δ. The dashed line implies that one solves only the solution on certain time levels instead of on the entire time interval.

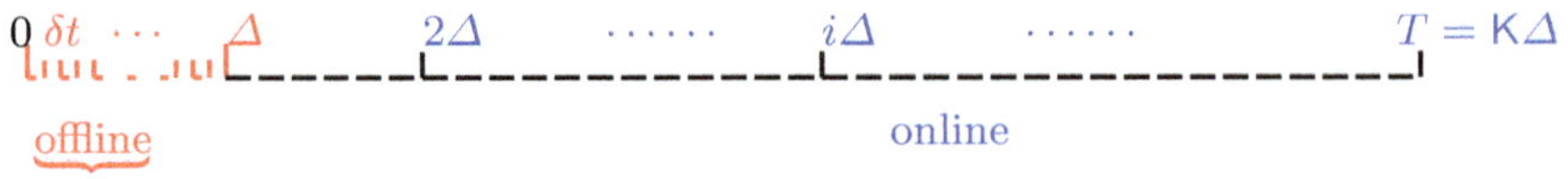

For simplicity, we assume that $f = g_k = 0$. Let $\varphi_\alpha(s,x;\phi)$ solve the following propagator:

$$\frac{\partial \varphi_\alpha(s,x;\phi)}{\partial s} = \mathcal{L}\varphi_\alpha(s,x;\phi) + \sum_{k,l} \alpha_{k,l} m_l(s) \mathcal{M}_k \varphi_{\alpha^-(l,k)}(s,x;\phi), \quad s \in (0,\Delta], \tag{6.1.12}$$
$$\varphi_\alpha(0,x) = \phi(x)\mathbf{1}_{\{|\alpha|=0\}},$$

where $m_l(s)$'s are the orthonormal cosine basis (6.1.2) on $L^2([0,\Delta])$. Define

$$q_{\alpha,l,m} = (\varphi_\alpha(\Delta,\cdot;e_l), e_m), \quad l,m \geq 1, \tag{6.1.13}$$

and then find by induction the coefficients

$$\psi_m(0;\mathsf{N},\mathsf{n}) := (u_0, e_m), \tag{6.1.14}$$
$$\psi_m(i;\mathsf{N},\mathsf{n}) := \sum_{\alpha\in\mathcal{J}_{\mathsf{N},\mathsf{n}}} \sum_l \frac{1}{\sqrt{\alpha!}} \psi_l(i-1;\mathsf{N},\mathsf{n}) q_{\alpha,l,m} \xi_\alpha^{(i)}, \; i = 1,\ldots,\mathsf{K}.$$

It can be readily shown that

$$u_{\Delta,\mathsf{N},\mathsf{n}}(\mathsf{t}_{i-1},x) = \sum_m \psi_m(i;\mathsf{N},\mathsf{n}) e_m(x), \; i = 0,\ldots,\mathsf{K}, \quad P\text{-a.s.} \; . \tag{6.1.15}$$

We refer to the numerical method (6.1.15), (6.1.12)–(6.1.14) together with (6.1.6)–(6.1.7) as the *multistage WCE method* for the SPDE (3.3.23).

In practice, if the equation (3.3.23) has an infinite number of Wiener processes, we truncate them to a finite number $\mathsf{r} \geq 1$ of noises. We introduce the correspondingly truncated set $\mathcal{J}_{\mathsf{N},\mathsf{n},\mathsf{r}}$ so that

$$\mathcal{J}_{\mathsf{N},\mathsf{n},\mathsf{r}} = \{\alpha \in \mathcal{J} : \; |\alpha| \leq \mathsf{N}, \;\; d_{\mathsf{r}}(\alpha) \leq \mathsf{n}\},$$

where $d_{\mathsf{r}}(\alpha) = \max\{l \geq 1 : \alpha_{k,l} > 0 \text{ for some } 1 \leq k \leq \mathsf{r}\}$. We have the following algorithm to compute the numerical solution.

Algorithm 6.1.1 *Choose a truncation of the number of noises* $r \geq 1$ *and the algorithm's parameters: a CONS* $\{e_m(x)\}_{m\geq 1}$ *and its truncation* $\{e_m(x)\}_{m=1}^{\mathsf{M}}$*; a time step* Δ*;* N *and* n *and* r *to determine the multi-index set* $\mathcal{J}_{\mathsf{N},\mathsf{n},\mathsf{r}}$*.*

Step 1. For each $m = 1,\ldots,\mathsf{M}$*, solve the propagator (6.1.12) for* $\alpha \in \mathcal{J}_{\mathsf{N},\mathsf{n},\mathsf{r}}$ *on the time interval* $[0,\Delta]$ *with the initial condition* $e_m(x)$ *and denote the obtained solution as* $\varphi_\alpha(\Delta,x;e_m)$*,* $\alpha \in \mathcal{J}_{\mathsf{N},\mathsf{n},r}$*,* $m = 1,\ldots,\mathsf{M}$*. We also need to choose a time step size* δt *to solve the equations in the propagator numerically.*

Step 2. Evaluate $\psi_m(0;\mathsf{N},\mathsf{n},\mathsf{M}) = (u_0,e_m)$*,* $m = 1,\ldots,\mathsf{M}$*, where* $u_0(x)$ *is the initial condition for (3.3.23), and* $q_{\alpha,l,m} = (\varphi_\alpha(\Delta,\cdot;e_l), e_m(\cdot))$*,* $l,m = 1,\ldots,\mathsf{M}$*.*

Step 3. On the i-th time step (at time $t = i\Delta$), generate the Gaussian random variables $\xi_\alpha^{(i)}$, $\alpha \in \mathcal{J}_{\mathsf{N},\mathsf{n},r}$, according to (6.1.7), compute the coefficients

$$\psi_m(i;\mathsf{N},\mathsf{n},\mathsf{M}) = \sum_{\alpha\in\mathcal{J}_{\mathsf{N},\mathsf{n},r}} \sum_{l=1}^{\mathsf{M}} \frac{1}{\sqrt{\alpha!}}\psi_l(i-1;\mathsf{N},\mathsf{n},\mathsf{M})q_{\alpha,l,m}\xi_\alpha^{(i)}, \quad m = 1,\dots,\mathsf{M},$$

and obtain the approximate solution of (3.3.23)

$$u_{\Delta,\mathsf{N},\mathsf{n}}^{\mathsf{M}}(\mathsf{t}_{i-1},x) = \sum_{m=1}^{\mathsf{M}} \psi_m(i;\mathsf{N},\mathsf{n},\mathsf{M})e_m(x).$$

In the next section we present an algorithm based on Algorithm 6.1.1, which allows us to compute moments of the solution to (3.3.23) without using the Monte Carlo technique.

Remark 6.1.2 *The cost of simulation of the random field $u(\mathsf{t}_i,x)$ by Algorithm 6.1.1 over K time steps is proportional to $\mathsf{K}\mathsf{M}^2\frac{(\mathsf{N}+\mathsf{n}r)!}{\mathsf{N}!(\mathsf{n}r)!}$.*

Remark 6.1.3 *Choosing an orthonormal basis is an important topic in the research of spectral methods, which can be found in [163] and many subsequent works. Here we choose the Fourier basis for Problem 3.3.23 because of periodic boundary conditions.*

6.1.2 Algorithm for computing moments

Implementation of Algorithm 6.1.1 requires the generation of the random variables $\xi_\alpha^{(i)}$ (see (6.1.7)). Then, for computing moments of the solution of the SPDE (3.3.23), we also need to make use of the Monte Carlo technique. In this section we present a deterministic algorithm (Algorithm 6.1.4) for computing moments, i.e., an algorithm which does not require any random numbers and does not have a statistical error. In Chapters 6.2, 6.3, and 6.4 we compare Algorithm 6.1.4 with some Monte Carlo-type methods and demonstrate that Algorithm 6.1.4 can be more computationally efficient when higher accuracy is required.

The mean solution $\mathbb{E}[u(t,x)]$ is equal to the solution $\varphi_{(0)}(t,x)$ of the propagator (6.1.12) with $\alpha = (0)$:

$$\mathbb{E}[u(t,x)] = \varphi_{(0)}(t,x).$$

Thus, evaluating the mean $\mathbb{E}[u(t,x)]$ is reduced to numerical solution of the linear deterministic PDE for $\varphi_{(0)}(t,x)$. We limit ourselves here to presenting an algorithm for computing the second moment of the solution, $\mathbb{E}[u^2(t,x)]$. Other moments of the solution $u(t,x)$ can be considered analogously.

According to Algorithm 6.1.1, we approximate the solution $u(\mathsf{t}_{i-1},x)$ of (3.3.23) by $u_{\Delta,\mathsf{N},\mathsf{n}}^{\mathsf{M}}(\mathsf{t}_{i-1},x)$ (when $f = g_k = 0$) as follows:

$$\psi_m(0;\mathsf{N},\mathsf{n},\mathsf{M}) = (u_0, e_m), \quad m = 1,\ldots,\mathsf{M},$$

$$\psi_m(\mathsf{t}_{i-1};\mathsf{N},\mathsf{n},\mathsf{M}) = \sum_{\alpha\in\mathcal{J}_{\mathsf{N},\mathsf{n},r}} \sum_{l=1}^{\mathsf{M}} \frac{1}{\sqrt{\alpha!}} \psi_l(\mathsf{t}_i;\mathsf{N},\mathsf{n},\mathsf{M}) q_{\alpha,l,m}\, \xi_\alpha^{(i)}, \quad m = 1,\ldots,\mathsf{M},$$

$$u^{\mathsf{M}}_{\Delta,\mathsf{N},\mathsf{n}}(\mathsf{t}_{i-1},x) = \sum_{m=1}^{\mathsf{M}} \psi_m(\mathsf{t}_{i-1};\mathsf{N},\mathsf{n},M) e_m(x), \quad i = 1,\ldots,\mathsf{K},$$

where $q_{\alpha,l,m}$ are from (6.1.13) and $\xi_\alpha^{(i)}$ are from (6.1.7). Then, we can evaluate the covariance matrices

$$\begin{aligned} Q_{lm}(0;\mathsf{N},\mathsf{n},\mathsf{M}) &:= \psi_l(0;\mathsf{N},\mathsf{n},\mathsf{M})\psi_m(0;\mathsf{N},\mathsf{n},\mathsf{M}), \; l,m = 1,\ldots,\mathsf{M}, \quad (6.1.16)\\ Q_{lm}(\mathsf{t}_{i-1};\mathsf{N},\mathsf{n},\mathsf{M}) &:= E[\psi_l(\mathsf{t}_{i-1};\mathsf{N},\mathsf{n},\mathsf{M})\psi_m(\mathsf{t}_{i-1};\mathsf{N},\mathsf{n},\mathsf{M})] \\ &= \sum_{j,k=1}^{M} Q_{jk}(\mathsf{t}_i;\mathsf{N},\mathsf{n},\mathsf{M}) \sum_{\alpha\in\mathcal{J}_{\mathsf{N},\mathsf{n},\mathsf{r}}} \frac{1}{\alpha!} q_{\alpha,j,l} q_{\alpha,k,m}, \\ & l,m = 1,\ldots,\mathsf{M}, \; i = 1,\ldots,\mathsf{K}, \end{aligned}$$

and, consequently, the second moment of the approximate solution

$$\mathbb{E}[(u^{\mathsf{M}}_{\Delta,\mathsf{N},\mathsf{n}}(\mathsf{t}_{i-1},x))^2] = \sum_{l,m=1}^{\mathsf{M}} Q_{lm}(\mathsf{t}_{i-1};\mathsf{N},\mathsf{n},\mathsf{M}) e_l(x) e_m(x), \; i = 1,\ldots,\mathsf{K}. \tag{6.1.17}$$

Implementation of (6.1.16)–(6.1.17) does not require generation of the random variables $\xi_\alpha^{(i)}$. Hence, we have constructed a deterministic algorithm for computing the second moments of the solution to the SPDE (3.3.23) when $f = g_k = 0$, which we formulate below.

Algorithm 6.1.4 (Recursive multistage Wiener chaos expansion, [505, Algorithm 2]) *Choose a truncation of the number of noises* $\mathsf{r} \geq 1$ *in (3.3.23) and the algorithm's parameters: a CONS* $\{e_m(x)\}_{m\geq 1}$ *and its truncation* $\{e_m(x)\}_{m=1}^{\mathsf{M}}$; *a time step* Δ; N *and* n *which together with* r *determine the size of the multi-index set* $\mathcal{J}_{\mathsf{N},\mathsf{n},\mathsf{r}}$.

Step 1. For each $m = 1,\ldots,\mathsf{M}$, *solve the propagator (6.1.12) for* $\alpha \in \mathcal{J}_{\mathsf{N},\mathsf{n},r}$ *on the time interval* $[0,\Delta]$ *with the initial condition* $\phi(x) = e_m(x)$ *and denote the obtained solution as* $\varphi_\alpha(\Delta,x;e_m)$, $\alpha \in \mathcal{J}_{\mathsf{N},\mathsf{n},r}$, $m = 1,\ldots,\mathsf{M}$. *Also, choose a time step size* δt *to solve the equations in the propagator numerically.*

Step 2. Evaluate $\psi_m(0;\mathsf{N},\mathsf{n},M) = (u_0,e_m)$, $m = 1,\ldots,\mathsf{M}$, *where* $u_0(x)$ *is the initial condition for (3.3.23), and* $q_{\alpha,l,m} = (\varphi_\alpha(\Delta,\cdot;e_l), e_m(\cdot))$, $l,m = 1,\ldots,\mathsf{M}$.

Step 3. Recursively compute the covariance matrices $Q_{lm}(\mathsf{t}_{i-1};\mathsf{N},\mathsf{n},\mathsf{M})$ *according to (6.1.16) and obtain the second moment* $\mathbb{E}[(u^{\mathsf{M}}_{\Delta,\mathsf{N},\mathsf{n}}(\mathsf{t}_{i-1},x))^2]$ *of the approximate solution to* (3.3.23) *by* (6.1.17).

The accuracy of Algorithm 6.1.4 for single noise ($\mathsf{r} = 1$) will be shown in Theorem 8.3.6 and its Corollary 8.3.2 in Chapter 8. The error estimates for approximation of the second moment $\mathbb{E}[u^2(\mathsf{t}_{i-1},x)]$ by $\mathbb{E}[u^2_{\Delta,\mathsf{N},\mathsf{n}}(\mathsf{t}_{i-1},x)]$ is the same as the errors given in (6.1.11). Due to the orthogonality of the random variables $\xi_\alpha^{(i)}$ in the sense that $\mathbb{E}[\xi_\alpha^{(i)}\xi_\beta^{(j)}]=0$ unless $i=j$ and $\alpha=\beta$, the following equality holds:

$$\mathbb{E}[u^2(t,x)]-\mathbb{E}[u^2_{\mathsf{N},\mathsf{n}}(t,x)]=\mathbb{E}[(u(t,x)-u_{\mathsf{N},\mathsf{n}}(t,x))^2]. \tag{6.1.18}$$

Here, we do not discuss errors arising from noise truncation and from truncation of the basis $\{e_m(x)\}_{m\geq 1}$.

Computational Cost. The computational costs of Steps 1 and 2 of Algorithm 6.1.4 are proportional to $\mathsf{M}^2\frac{(\mathsf{N}+\mathsf{nr})!}{\mathsf{N}!(\mathsf{nr})!}$ and the computational cost of Step 3 over K time steps is proportional to $\mathsf{KM}^4\frac{(\mathsf{N}+\mathsf{nr})!}{\mathsf{N}!(\mathsf{nr})!}$. Taking this into account together with the error estimates (6.1.11), it is computationally beneficial to choose $\mathsf{n}=1$ and $\mathsf{N}=2$ or 1.

The main computational cost of Algorithm 6.1.4 is due to the total number of basis functions M in physical space required for reaching a satisfactory accuracy. For a fixed accuracy, the number M of basis functions $\{e_m\}_{m=1}^{\mathsf{M}}$ is proportional to C^d, where C depends on a choice of the basis and on the problem. If the variance of $u^2(t,x_i)$ is relatively large and the problem considered does not require a very large number of basis functions M, then one expects Algorithm 6.1.4 to be computationally more efficient in evaluating second moments than the combination of Algorithm 6.1.1 with the Monte Carlo technique.

The efficiency of Algorithm 6.1.4 can often be improved by choosing an appropriate basis $\{e_m\}$ so that the majority of functions $q_{\alpha,l,m}$ are identically zero or negligible and hence can be dropped from computing the covariance matrix $\{Q_{lm}(\mathsf{t}_{i-1};\mathsf{N},\mathsf{M})\}_{l,m=1}^{\mathsf{M}}$, significantly decreasing the computational cost of Step 3. For instance, for the periodic passive scalar equation considered in Chapter 6.4 we choose the Fourier basis $\{e_m\}$. In this case the number of zero $q_{\alpha,l,m}$ is proportional just to M (the total number of $q_{\alpha,l,m}$ is proportional to M^2) and, consequently, the computational cost of Step 3 (and hence that of Algorithm 6.1.4) becomes proportional to M^2 instead of the original M^4. Moreover, computation of the covariance matrix according to (6.1.16) can be done in parallel. Clearly, the use of reduced-order methods with offline/online strategies [409] can greatly reduce the value of M and hence will make the recursive WCE method very efficient.

Remark 6.1.5 *It is more expensive to compute higher-order moments by a deterministic algorithm analogous to Algorithm 6.1.4. Since second moments give us such important, from the physical point of view, characteristics as energy and correlation functions, Algorithm 6.1.4 can be a competitive alternative to Monte Carlo-type methods in practice.*

6.2 Examples in one dimension

We consider two one-dimensional problems and illustrate the application of Algorithm 6.1.4 to these problems. We will test Algorithm 6.1.4 by evaluating the second moments $\mathbb{E}[u^2(t,x)]$ of the solutions to the stochastic advection-diffusion equation (3.3.18) and the stochastic reaction-diffusion equation (3.3.20).

Application of WCE algorithms to the model problem. The problems (3.3.18) and (3.3.20) are simpler than the general linear SPDE (3.3.23). Consequently, Algorithm 6.1.4 applied to them takes a simpler form, see Algorithm 6.2.1 below.

The model problems (3.3.18) and (3.3.20) have a single Wiener process and their solutions (3.3.19) and (3.3.21) have the form $u(t,x) = f(t,x,W(t))$, where $f(t,x,y)$ is a smooth function. Consequently, the solutions are expandable in the basis consisting just of $\xi_\alpha = H_k(W(t)/\sqrt{t})/\sqrt{k!} = H_k(\xi_1)/\sqrt{k!}$, $\alpha = (k,0,\ldots,0)$, $k = 0,1,\ldots$, i.e., we have

$$u(t,x) = \sum_{\alpha\in\mathcal{J}} \frac{\varphi_\alpha(t,x)}{\sqrt{\alpha!}}\xi_\alpha = \sum_{N=0}^{\infty}\sum_{\alpha\in\mathcal{J}_{N,1}} \frac{\varphi_\alpha(t,x)}{\sqrt{\alpha!}}\xi_\alpha = \sum_{k=0}^{\infty}\frac{\varphi_k(t,x)}{\sqrt{k!}}\eta_k, \quad (6.2.1)$$

where $\eta_k = \xi_\alpha$ with $\alpha = (k,0,\ldots,0)$, $k = 0,1,\ldots$. Hence

$$u_{\mathsf{N},1}(t,x) = :u_{\mathsf{N}}(t,x) = \sum_{k=0}^{\mathsf{N}}\frac{\varphi_k(t,x)}{\sqrt{k!}}\eta_k, \quad (6.2.2)$$

which corresponds to setting $\mathsf{n} = 1$ in (6.1.1). It is not difficult to show that applying Algorithm 6.1.4 to the model problems (3.3.18) and (3.3.20) is more accurate than in general cases of (3.3.23) (cf. (6.1.11) and (6.1.18)):

$$\left\|\mathbb{E}[u^2(t,\cdot)] - \mathbb{E}[u^2_{\Delta,\mathsf{N}}(t,\cdot)]\right\|_{L^2} \leq C\frac{(C\Delta)^{\mathsf{N}}}{(\mathsf{N}+1)!} \quad (6.2.3)$$

for all sufficiently small $\Delta > 0$ and a constant $C > 0$ independent of Δ and N (we assume the errors arising from truncation of the basis $\{e_m\}$ is negligible).

For the problems (3.3.18) and (3.3.20), the propagator (6.1.12) takes the form

$$\begin{aligned}\partial_t\varphi_0 &= a\partial^2_{xx}\varphi_0,\ \varphi_0(0,x;\phi) = \phi(x), \\ \partial_t\varphi_k &= a\partial^2_{xx}\varphi_k + \frac{1}{\sqrt{\Delta}}\sigma k\partial_x\varphi_{k-1},\quad \varphi_k(0,x;0) = 0,\ k>0,\end{aligned} \quad (6.2.4)$$

and

$$\begin{aligned}\partial_t\varphi_0 &= a\partial^2_{xx}\varphi_0,\ \varphi_0(0,x;\phi) = \phi(x), \\ \partial_t\varphi_k &= a\partial^2_{xx}\varphi_k + \frac{1}{\sqrt{\Delta}}\sigma k\varphi_{k-1},\quad \varphi_k(0,x;0) = 0,\ k>0,\end{aligned} \quad (6.2.5)$$

respectively. We solve these propagators numerically using the Fourier collocation method with M nodes in physical space and the Crank–Nicolson time discretization with step δt in time. Denote by $L_m(x)$, $m = 1, \ldots, \mathsf{M}$, the m-th Lagrangian trigonometric polynomials using M Fourier collocation nodes, i.e., $L_m(x)$ are m-th order trigonometric polynomials satisfying $L_m(x_l) = \delta_{m,l}$ and $x_l = \frac{2\pi}{\mathsf{M}}(l-1)$, $l = 1, \ldots, \mathsf{M}$. The m-th Lagrangian trigonometric polynomial can be written as

$$L_m(x) = \mathsf{M}\frac{\sin(\mathsf{M}x/2)}{\tan(x/2)}, \quad x \in [0, 2\pi]. \tag{6.2.6}$$

In fact, the Lagrangian trigonometric polynomials can be represented as

$$L_m(x) = \sum_{k=-\mathsf{M}/2+1}^{\mathsf{M}/2} \exp(-ikx)a_k, \text{ where } i = \sqrt{-1}.$$

Applying the conditions $L_m(x_l) = \delta_{m,l}$, we can find a_k's and rewrite $L_m(x)$ in the form of (6.2.6). More details of derivation of the formula (6.2.6) can be found in [447, Chapter 3].

Now we formulate the realization of Algorithm 6.1.4 in these two model problems.

Algorithm 6.2.1 *For given values of the model parameters a and σ, choose the algorithm parameters: a number of Fourier collocation nodes M, a time step δt for solving the propagator (6.2.4) (or (6.2.5)), a time step Δ, and the number of Hermite polynomials N.*

Step 1. Solve the propagator (6.2.4) (or (6.2.5)) on the time interval $[0, \Delta]$ with the initial condition $\phi(x) = L_k(x)$ using the Fourier collocation method with M nodes in physical space and the Crank–Nicolson scheme with step δt in time and denote the obtained numerical approximation of $\varphi_k(\Delta, x_l; L_m)$ as $\varphi_k^{\mathsf{M},\delta t}(\Delta, x_l; L_m)$, $l, m = 1, \ldots, \mathsf{M}$, $k = 0, 1, \ldots, \mathsf{N}$.

Step 2. Recursively compute the covariance matrices

$$Q_{lm}(\mathsf{t}_{i-1}; \mathsf{N}, \mathsf{M}) := \mathbb{E}[u_{\Delta,\mathsf{N}}^{\mathsf{M},\delta t}(\mathsf{t}_{i-1}, x_l)u_{\Delta,\mathsf{N}}^{\mathsf{M},\delta t}(\mathsf{t}_{i-1}, x_m)], \quad \mathsf{t}_{i-1} = i\Delta, \ i = 0, \ldots, \mathsf{K},$$

of the approximate solution to (3.3.18) *(or* (3.3.20)*):*

$$Q_{lm}(0; \mathsf{N}, \mathsf{M}) = u_0(x_l)u_0(x_m), \ l, m = 1, \ldots, \mathsf{M},$$

$$Q_{lm}(\mathsf{t}_{i-1}; \mathsf{N}, \mathsf{M}) = \sum_{k=1}^{\mathsf{N}}\sum_{q=1}^{\mathsf{M}}\sum_{r=1}^{\mathsf{M}} \frac{1}{k!} Q_{qr}(\mathsf{t}_i; \mathsf{N}, \mathsf{M})\varphi_k^{\mathsf{M},\delta t}(\Delta, x_l; L_q)\varphi_k^{\mathsf{M},\delta t}(\Delta, x_m; L_r),$$
$$l, m = 1, \ldots, \mathsf{M}, \quad i = 1, \ldots, \mathsf{K},$$

where $u_0(x)$ is the initial condition of (3.3.18) *(or* (3.3.20)*).*

In particular, we obtain the second moment of the approximate solution to (3.3.18) *(or* (3.3.20)*):*

$$\mathbb{E}[[u_{\Delta,\mathsf{N}}^{\mathsf{M},\delta t}(\mathsf{t}_{i-1}, x_j)]^2] = Q_{jj}(\mathsf{t}_{i-1}; \mathsf{N}, \mathsf{M}), \quad j = 1, \ldots, \mathsf{M}, \quad i = 1, \ldots, \mathsf{K}.$$

To approximate the solution of (3.3.18) (or (3.3.20)), one can use the truncated WCE $u_{\mathsf{N}}(t,x)$ from (6.2.2) and, in particular, evaluate the second moment $\mathbb{E}[u^2(t,x)]$ as

$$\mathbb{E}[u^2(t,x)] \approx \mathbb{E}[u_{\mathsf{N}}^2(t,x)] = \sum_{k=0}^{\mathsf{N}} \frac{\varphi_k^2(t,x)}{k!} \approx \sum_{k=0}^{\mathsf{N}} \frac{\left[\varphi_k^{\mathsf{M},\delta t}(t,x)\right]^2}{k!}, \tag{6.2.7}$$

where $\varphi_0(t,x) = \varphi_0(t,x;u_0(x))$ and $\varphi_k(t,x) = \varphi_k(t,x;0)$, $k > 0$, are solutions of the propagator (6.2.4) (or (6.2.5)) and $\varphi_k^{\mathsf{M},\delta t}(t,x)$ are their numerical approximations obtained, e.g., using the Fourier collocation method with M nodes in physical space and the Crank–Nicolson scheme with step δt in time. The approximation (6.2.7) can be viewed as a one-step approximation corresponding to Algorithm 6.2.1, i.e., the first step of Algorithm 6.2.1 with $\Delta = t$, and its error is estimated by

$$\left\|\mathbb{E}[u^2(t,\cdot)] - \mathbb{E}[u_{\mathsf{N}}^2(t,\cdot)]\right\|_{L^2} \leq Ce^{Ct}\frac{(Ct)^{\mathsf{N}+1}}{(\mathsf{N}+1)!}.$$

This error grows exponentially with t, which can be readily verified by numerical tests with (3.3.18). To reach a satisfactory accuracy of the approximation (6.2.7) for a fixed t, one has to take a sufficiently large N which is computationally expensive, even in the case of moderate values of t. In contrast, it is demonstrated in Chapter 6.2.1 that the error of Algorithm 6.2.1 grows linearly with time and it is relatively small even for $\mathsf{N} = 1$.

6.2.1 Numerical results for one-dimensional advection-diffusion-reaction equations

In this section we present some numerical results of Algorithm 6.2.1 on two model problems (3.3.18) and (3.3.20).

In approximating the propagators (6.2.4) and (6.2.5) we choose a sufficiently large number of Fourier collocation nodes M and a sufficiently small time step δt so that errors of numerical solutions to the propagators have a negligible influence on the overall accuracy of Algorithm 6.2.1 in our simulations. In all the numerical tests it was sufficient to take $\mathsf{M} = 20$; this choice of M was tested by running control tests with $\mathsf{M} = 80$.

We measure numerical errors in the following norms:

$$\rho_2(t) = \left(\frac{2\pi}{\mathsf{M}} \sum_{m=1}^{\mathsf{M}} (\mathbb{E}[u_{\Delta,\mathsf{N}}^{\mathsf{M},\delta t}(t,x_m))^2] - \mathbb{E}[u^2(t,x_m)])^2\right)^{1/2},$$

and

$$\rho_\infty(t) = \max_{1\leq m\leq \mathsf{M}} \left|\mathbb{E}[[u_{\Delta,\mathsf{N}}^{\mathsf{M},\delta t}(t,x_m)]^2] - \mathbb{E}[u^2(t,x_m)]\right|.$$

The results of our tests on the model problem (3.3.18) in the degenerate case (i.e., $\epsilon = 0$) and in the nondegenerate case (i.e., $\epsilon > 0$) are presented in Tables 6.1 and 6.2, respectively. Table 6.3 corresponds to the tests with the second model problem (3.3.20). Numerical tests with values of the parameters other than those used for Tables 6.1–6.3 were also performed and they gave similar results.

Analyzing the results in Tables 6.1, 6.2, and 6.3, we observe the convergence order of Δ^{N} for a fixed N in all the tests, which confirms our theoretical prediction (6.2.3). We also ran other cases (not presented here) to confirm the conclusion from Chapter 6.2 that the number n of random variables ξ_k used per step does not influence the accuracy of Algorithm 6.1.4 in the case of the model problems (3.3.18) and (3.3.20).

In Figure 6.2 we demonstrate the dependence of the relative numerical error

$$\rho_2^r(t) = \frac{\rho_2(t)}{\|\mathbb{E}[u^2(t,\cdot)]\|_{L^2}}$$

on integration time. These results were obtained in the degenerate case of the problem (3.3.18), but similar behavior of errors was observed in our tests with other parameters as well. One can conclude from Figure 6.2 that (after an initial fast growth) the error grows *linearly* with integration time. This is a remarkable feature of the recursive WCE algorithm, which implies that the algorithm can be used for long time integration of SPDEs.

Table 6.1. Performance of Algorithm 6.2.1 for Model (3.3.18). The parameters of the model (3.3.18) are $\sigma = 1$, $\epsilon = 0$, and the time $t = 10$. In Algorithm 6.2.1 we take $\mathsf{M} = 20$.

N	Δ	δt	$\rho_2(10)$	$\rho_\infty(10)$
1	0.1	1×10^{-3}	4.69×10^{-1}	1.87×10^{-1}
	0.01	1×10^{-4}	6.07×10^{-2}	2.42×10^{-2}
	0.001	1×10^{-5}	6.25×10^{-3}	2.49×10^{-3}
2	0.1	1×10^{-3}	1.92×10^{-2}	7.67×10^{-3}
	0.01	1×10^{-4}	2.07×10^{-4}	8.27×10^{-5}
	0.001	1×10^{-5}	2.09×10^{-6}	8.33×10^{-7}
3	0.1	1×10^{-3}	4.82×10^{-4}	1.99×10^{-4}
	0.01	1×10^{-4}	5.16×10^{-7}	2.06×10^{-7}
	0.001	1×10^{-5}	3.37×10^{-10}	1.81×10^{-10}
4	0.1	1×10^{-3}	9.36×10^{-6}	3.73×10^{-6}
	0.01	1×10^{-5}	9.35×10^{-10}	4.17×10^{-10}

Table 6.2. *Model* (3.3.18)*: performance of Algorithm 6.2.1.* The parameters of the model (3.3.18) are $\sigma = 1$, $\epsilon = 0.01$, and the time $t = 10$. In Algorithm 6.2.1 we take $\mathsf{M} = 20$.

N	Δ	δt	$\rho_2(10)$	$\rho_\infty(10)$
1	0.1	1×10^{-3}	3.84×10^{-1}	1.53×10^{-1}
	0.01	1×10^{-4}	4.97×10^{-2}	1.98×10^{-2}
	0.001	1×10^{-4}	5.11×10^{-3}	2.04×10^{-3}
2	0.1	1×10^{-3}	1.58×10^{-2}	6.28×10^{-3}
	0.01	1×10^{-4}	1.70×10^{-4}	6.77×10^{-5}
	0.001	1×10^{-4}	1.72×10^{-6}	6.88×10^{-7}
3	0.1	1×10^{-3}	3.95×10^{-4}	1.57×10^{-4}
	0.01	1×10^{-4}	4.22×10^{-7}	1.68×10^{-7}
	0.001	1×10^{-5}	3.65×10^{-10}	2.01×10^{-10}
4	0.1	1×10^{-3}	7.67×10^{-6}	3.06×10^{-6}
	0.01	1×10^{-5}	8.39×10^{-10}	3.90×10^{-10}

6.3 Comparison of the WCE algorithm and Monte Carlo type algorithms

In this section, we compare the recursive WCE algorithm with some Monte Carlo-type algorithms.

As discussed in Chapter 2, there are other approaches to solving SPDEs, which are usually complemented by the Monte Carlo technique when one is interested in computing moments of SPDE solutions. In this section, using the problem (3.3.18), we compare the performance of Algorithm 6.2.1 and two Monte Carlo-type algorithms, one of which is based on the method of characteristics [361] and another on the Fourier transform of the linear SPDE with subsequent simulation of SDEs and application of the Monte Carlo technique.

Table 6.3. Performance of Algorithm 6.2.1 for Model (3.3.20). The parameters of the model (3.3.20) are $\sigma = 1$, $a = 0.5$, and the time $t = 10$. In Algorithm 6.2.1 we take $\mathsf{M} = 20$.

N	Δ	δt	$\rho_2(10)$	$\rho_\infty(10)$
1	0.1	1×10^{-3}	5.75×10^{-1}	3.74×10^{-1}
	0.01	1×10^{-4}	7.44×10^{-2}	4.85×10^{-2}
	0.001	1×10^{-4}	7.65×10^{-3}	4.98×10^{-3}
2	0.1	1×10^{-3}	2.36×10^{-2}	1.53×10^{-2}
	0.01	1×10^{-4}	2.54×10^{-4}	1.65×10^{-4}
	0.001	1×10^{-4}	2.58×10^{-6}	1.68×10^{-6}
3	0.1	1×10^{-3}	5.90×10^{-4}	3.85×10^{-4}
	0.01	1×10^{-4}	6.32×10^{-7}	4.12×10^{-7}

Fig. 6.2. Dependence of the relative numerical error $\rho_2^r(t)$ on integration time. Model (3.3.18) is simulated by Algorithm 6.2.1 with $\mathsf{M} = 20$ and $\delta t = \Delta/100$ and various Δ and N. The parameters of (3.3.18) are $\sigma = 1$ and $\epsilon = 0$.

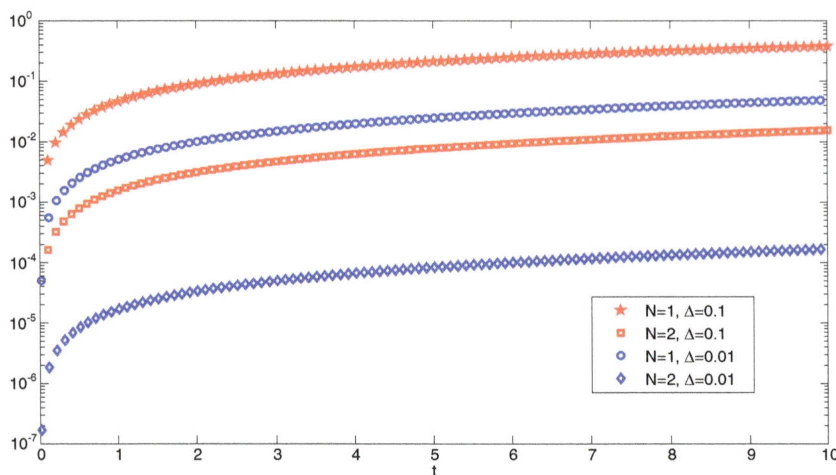

The solution of (3.3.18) with $\epsilon = 0$ (the degenerate case) can be represented via the method of characteristics [408]:

$$u(t,x) = \sin(X_{t,x}(0)), \tag{6.3.1}$$

where $X_{t,x}(s)$, $0 \le s \le t$, is the solution of the system of backward characteristics

$$dX_{t,x}(s) = \sigma \overleftarrow{dW}(s), \quad X_{t,x}(t) = x. \tag{6.3.2}$$

The notation "$\overleftarrow{dW}(s)$" means backward Itô integral (see, e.g., [408]). It follows from (6.3.2) that $X_{t,x}(0)$ has the same probability distribution as $x + \sigma\sqrt{t}\zeta$, where ζ is a standard Gaussian random variable (i.e., $\zeta \sim \mathcal{N}(0,1)$). Since we are interested only in computing statistical moments, it is assumed that

$$X_{t,x}(0) = x + \sigma\sqrt{t}\zeta. \tag{6.3.3}$$

Then, we can estimate the second moment $m_2(t,x) := \mathbb{E}[u^2(t,x)]$ as

$$m_2(t,x) \doteq \hat{m}_2(t,x) = \frac{1}{L}\sum_{l=1}^{L} \sin^2(x + \sigma\sqrt{t}\zeta^{(l)}), \tag{6.3.4}$$

where $\zeta^{(l)}$, $l = 1, \ldots, L$, are i.i.d. standard Gaussian random variables. The estimate $\hat{m}_2$ for m_2 is unbiased, and, hence, the numerical procedure for finding m_2 based on (6.3.4) has only the Monte Carlo (i.e., statistical) error which can be quantified via half of the length of the 95% confidence interval:

$$\rho_{MC}(t,x) = 2\frac{\sqrt{\mathrm{Var}(\sin^2(x+\sigma\sqrt{t}\zeta))}}{\sqrt{L}}.$$

Table 6.4 gives the statistical error for $\hat{m}_2(t,x)$ from (6.3.4) (no space-time discretization error in this algorithm), which is computed as

$$2 \cdot \max_j \frac{\sqrt{\frac{1}{L}\sum_{l=1}^{L} \sin^4(x_j+\sigma\sqrt{t}\zeta^{(l)}) - [\hat{m}_2(t,x_j)]^2}}{\sqrt{L}}, \tag{6.3.5}$$

where the set of x_j is the same as the one used in Table 6.5 by Algorithm 6.2.1 and $\zeta^{(l)}$ are as in (6.3.4).

All the tests were run using Matlab R2007b, on a Macintosh desktop computer with Intel Xeon CPU E5462 (quad-core, 2.80 GHz). Every effort was made to program and execute the different algorithms as much as possible in an identical way.

The cost of simulation due to (6.3.4) is directly proportional to L. The slower time increase for smaller L in Table 6.4 is due to the initialization time of the computer program in the time measurement.

Table 6.4. Performance of the method (6.3.4) for Model (3.3.18). The parameters of the model (3.3.18) are $\sigma = 1$, $\epsilon = 0$, and the time $t = 10$. The statistical error is computed according to (6.3.5).

L	Statistical error	CPU time (sec.)
10^2	8.87×10^{-2}	6×10^{-3}
10^4	7.40×10^{-3}	6.7×10^{-2}
10^6	7.09×10^{-4}	7.4×10^{0}
10^8	7.07×10^{-5}	7.4×10^{2}
10^{10}	7.07×10^{-6}	7.3×10^{4}

In Table 6.5 we repeat some of the results already presented in Table 6.1, which are now also accompanied by CPU time for comparison.

Comparing the results in Tables 6.4 and 6.5, we conclude that for a relatively large error the estimate $\hat{m}_2(t,x)$ from (6.3.4) is computationally more efficient than Algorithm 6.2.1; however, Algorithm 6.2.1 has lower costs in reaching a higher accuracy (errors of order equal to or smaller than 10^{-6}).

Now we compare Algorithm 6.2.1 with another approach exploiting the Monte Carlo technique for the problem (3.3.18) with $\epsilon = 0$. We can represent the solution of this periodic problem via the Fourier transform:

$$u(t,x) = \sum_{k\in\mathbb{Z}} e^{ikx} u_k(t) \tag{6.3.6}$$

Table 6.5. Performance of Algorithm 6.2.1 for Model (3.3.18). The parameters of the model (3.3.18) are $\sigma = 1$, $\epsilon = 0$, and the time $t = 10$. The parameters of Algorithm 6.2.1 are $\Delta = 0.1$, $\mathsf{M} = 20$, $\delta t = 0.001$.

N	$\rho_\infty(10)$	CPU time (sec.)
1	1.87×10^{-1}	5.7×10^0
2	7.67×10^{-3}	8.1×10^0
3	1.99×10^{-4}	1.1×10^1
4	3.73×10^{-6}	1.3×10^1

with $u_k(t)$, $t \geq 0$, $k \in \mathbb{Z}$, satisfying the system of SDEs:

$$du_k(t) = -k^2 \frac{1}{2}\sigma^2 u_k(t)dt + ik\sigma u_k(t)dW(t), \ \operatorname{Re} u_k(0) = 0, \qquad (6.3.7)$$
$$\operatorname{Im} u_k(0) = \frac{1}{2}(\delta_{1k} - \delta_{-1k}).$$

Noting that here $u_k(t) \equiv 0$ for all $|k| \neq 1$ and rewriting (6.3.6)-(6.3.7) in the trigonometric form, we get

$$u(t,x) = u^c(t)\cos x + u^s(t)\sin x, \qquad (6.3.8)$$

where

$$du^c(t) = -\frac{1}{2}\sigma^2 u^c(t)dt + \sigma u^s(t)dW(t), \quad u^c(0) = 0, \qquad (6.3.9)$$
$$du^s(t) = -\frac{1}{2}\sigma^2 u^s(t)dt - \sigma u^c(t)dW(t), \quad u^s(0) = 1.$$

The system (6.3.9) is a Hamiltonian system with multiplicative noise (see, e.g., [356, 358]). It is known [356, 358] that symplectic integrators have advantages in comparison with usual numerical methods in long time simulations of stochastic Hamiltonian systems. Here we use one of the symplectic methods – the midpoint scheme, to (6.3.9) and it takes the following form:

$$\bar{u}^c(t_{k+1}) = \bar{u}^c(t_k) + \frac{\sigma}{2}(\bar{u}^s(t_k) + \bar{u}^s(t_{k+1}))\sqrt{\Delta t}\zeta_{k+1}, \quad u^c(0) = 0, \qquad (6.3.10)$$
$$\bar{u}^s(t_{k+1}) = \bar{u}^s(t_k) - \frac{\sigma}{2}(\bar{u}^c(t_k) + \bar{u}^c(t_{k+1}))\sqrt{\Delta t}\zeta_{k+1}, \quad u^s(0) = 1,$$

where ζ_k are i.i.d. standard Gaussian random variables and $\Delta t > 0$ is a time step. The scheme (6.3.10) converges with the mean-square order 1/2 and weak order 1 [358]. It is implicit but can be resolved analytically since we are dealing with the linear system here.

Using (6.3.8) and (6.3.10), we evaluate the second moment of the solution to (3.3.18) with $\epsilon = 0$ as

$$m_2(t_k, x) := \mathbb{E}[u^2(t_k, x)] \doteq \mathbb{E}[(\bar{u}^c(t_k)\cos x + \bar{u}^s(t)\sin x)^2] \quad (6.3.11)$$
$$\doteq \hat{m}_2(t_k, x) = \frac{1}{L}\sum_{l=1}^{L}\left[\bar{u}^{c,(l)}(t_k)\cos x + \bar{u}^{s,(l)}(t_k)\sin x\right]^2,$$

where $\bar{u}^{c,(l)}(t_k)$, $\bar{u}^{s,(l)}(t_k)$ are independent realizations of the random variables $\bar{u}^c(t_k)$, $\bar{u}^s(t_k)$.

The estimate $\hat{m}_2(t_k, x)$ from (6.3.11) has two errors: the time discretization error due to the approximation of (6.3.9) by (6.3.10) and the Monte Carlo error. The errors presented in Table 6.6 are computed as $\max_j[\bar{m}_2(t_k, x_j) - \mathbb{E}[u^2(t_k, x_j)]]$ and are given together with the 95% confidence interval.

Table 6.6. Model (3.3.18): performance of the method (6.3.11). The parameters of the model (3.3.18) are $\sigma = 1$, $\epsilon = 0$, and the time $t = 10$.

Δt	L	Error	CPU time (sec.)
0.1	10^4	$8.06 \times 10^{-3} \pm 7.09 \times 10^{-3}$	4.72×10^{-1}
0.01	10^4	$6.55 \times 10^{-4} \pm 7.08 \times 10^{-4}$	3.90×10^{2}
0.001	10^6	$8.81 \times 10^{-5} \pm 7.07 \times 10^{-5}$	3.81×10^{5}

Comparing the results in Tables 6.6 and 6.5, we observe again that Algorithm 6.2.1 is computationally more efficient than the Monte Carlo-based algorithms in reaching a higher accuracy.

6.4 A two-dimensional passive scalar equation

We consider a special stochastic advection-diffusion equation (3.3.23)–(3.3.24) – a two-dimensional passive scalar equation. This equation is motivated by the study of the turbulent transport problem, see [146, 273, 314] and the references therein. Here we perform numerical tests on the *two-dimensional* ($d = 2$) passive scalar equation with periodic boundary conditions:

$$du(t,x) + \sum_{k=1}^{\infty}\sum_{i=1}^{d}\sigma_k^i(x)D_i u \circ dW_k(t) = 0,\ t > 0,\ x \in (0,\ell)^2, \quad (6.4.1)$$
$$u(t, x^1 + \ell, x^2) = u(t, x^1, x^2 + \ell) = u(t,x),$$
$$t > 0,\ x \in (0,\ell)^2,$$
$$u(0,x) = u_0(x),\ \ x \in (0,\ell)^2,$$

where '$\circ$' indicates the Stratonovich product, $\ell > 0$, the initial condition $u_0(x)$ is a periodic function with the period $(0,\ell)^2$, and $\sigma_k^i(x)$ are divergence-free periodic functions with the period $(0,\ell)^2$:

$$\operatorname{div}\sigma_k = 0. \quad (6.4.2)$$

In (6.4.1), we take a combination of such $\sigma_k(x)$ so that the corresponding spatial covariance C is symmetric and stationary: $C(x-y) = \sum_{k=1}^{\infty} \lambda_k \sigma_k(x)\sigma_k^\top(y)$, where λ_k are some nonnegative numbers. Namely, we consider

$$C(x-y) = \sum_{k=1}^{\infty} \lambda_k C(x-y; n_k, m_k), \tag{6.4.3}$$

where n_k, m_k is a sequence of positive integers, and

$$C(x-y;n,m) = \cos(2\pi \left(n[x^1-y^1] + m[x^2-y^2]\right)/\ell) \begin{bmatrix} m^2 & -nm \\ -nm & n^2 \end{bmatrix},$$

which can be decomposed as

$$\begin{aligned} C(x-y;n,m) &= \cos(2\pi[nx^1+mx^2]/\ell) \begin{bmatrix} -m \\ n \end{bmatrix} \cos(2\pi \left[ny^1+my^2\right]/\ell) \left[-m \; n\right] \\ &+ \sin(2\pi \left[nx^1+mx^2\right]/\ell) \begin{bmatrix} -m \\ n \end{bmatrix} \sin(2\pi[ny^1+my^2]/\ell) \left[-m \; n\right]. \end{aligned}$$

Hence, $\{\sigma_k(x)\}_{k\geq 1}$ in (6.4.1) is an appropriate combination of vector functions of the form

$$\cos(2\pi[nx^1+mx^2]/\ell) \begin{bmatrix} -m \\ n \end{bmatrix} \text{ and } \sin(2\pi \left[nx^1+mx^2\right]/\ell) \begin{bmatrix} -m \\ n \end{bmatrix}.$$

We rewrite (6.4.1) in the Ito's form:

$$\begin{aligned} du(t) + \frac{1}{2}\sum_{i,j=1}^{d} C_{ij}(0) D_i D_j u dt \quad &+ \sum_{k=1}^{\infty}\sum_{i=1}^{d} \sigma_k^i(x) D_i u dW_k(t) = 0, \\ u(t,x^1+\ell,x^2) &= u(t,x^1,x^2+\ell) = u(t,x), \; t>0, \; x\in(0,\ell)^2, \\ u(0,x) &= u_0(x), \quad x\in(0,\ell)^2. \end{aligned} \tag{6.4.4}$$

We present numerical results of Algorithm 6.1.4 applied to (6.4.4) and numerical results of the Monte Carlo-type algorithm from [361] based on the method of characteristics. We aim at computing the L^2-norm of the second moment of the SPDE solution

$$\left\|\mathbb{E}[u^2(T,\cdot)]\right\|_{L^2} = \left[\int_{[0,\ell]^2} (\mathbb{E}[u^2(T,x)])^2 dx\right]^{1/2}. \tag{6.4.5}$$

We considered the particular case of (6.4.1), (6.4.3) with $\ell = 2\pi$, the initial condition

$$u_0(x) = \sin(2x^1)\sin(x^2), \tag{6.4.6}$$

and with two noise terms:

$$\sigma_1(x) = \cos(x_1 + x_2) \begin{bmatrix} -1 \\ 1 \end{bmatrix}, \tag{6.4.7}$$

$$\sigma_2(x) = \sin(x_1 + x_2) \begin{bmatrix} -1 \\ 1 \end{bmatrix}, \quad \sigma_k(x) = 0 \ \text{ for } k > 2.$$

This example satisfies the so-called commutativity condition (3.3.29). The error estimate (6.1.11) holds in this case and is confirmed in the tests, see Chapter 8 for error estimates for single noise, one special case of commutative noises.

In Algorithm 6.1.4, we solve the propagator (6.1.12) corresponding to the SPDE (6.4.4) using a fourth-order explicit Runge–Kutta with step size δt in time and the Fourier spectral method with M modes in physical space.

The computational cost of Algorithm 6.1.4 is proportional to M^4 but with an appropriate choice of basis functions, this cost can be considerably reduced by exploiting the sparsity of the Wiener chaos expansion coefficients. Also, the use of Fourier basis for the problem (6.4.4) reduces the computational cost to being proportional to M^2. This significant reduction is based on the following observation. Since we consider finite number of noises with periodic $\sigma_k(x)$ and $\varphi_\alpha(\Delta, x; e_l)$ to be the solution of the propagator (6.1.12) with the initial condition equal to a single basis function $e_l(x)$, $\varphi_\alpha(\Delta, x; e_l)$ is expandable in a finite number of periodic functions $e_k(x)$ and this number does not depend on M. Hence, for fixed α and l the number of nonzero $q_{\alpha,l,m} = (\varphi_\alpha(\Delta, \cdot; e_l), e_m(\cdot))$ is finite and a small number. Therefore, the overall number of nonzero $q_{\alpha,l,m}$ is proportional to M instead of M^2.

The sparsity was tested and confirmed in our tests. We use the above fact in our computer realization of Algorithm 6.1.4 and reduce the computational cost of obtaining a single entry of the matrix $Q_{l,m}$ from the order of $\mathcal{O}(\mathsf{M}^2)$ to order $\mathcal{O}(1)$. Hence, the computational cost of Step 3 (and hence that of Algorithm 6.1.4) becomes proportional to M^2 instead of the original M^4.

We need a reference solution as we do not have an exact solution of the problem (6.4.1). To this end, the L_2-norm of the second moment of the SPDE solution at $T = 1$ was computed by Algorithm 6.1.4 with parameters $\mathsf{N} = 2$, $n = 1$, $\mathsf{M} = 900$ (i.e., 30 basis functions in each space direction), $\delta t = 1 \times 10^{-5}$, $\Delta = 1 \times 10^{-4}$ and which is equal to 1.57976 (5 d.p.). This result was also verified by the Monte Carlo-type method described below with $\Delta t = 1 \times 10^{-3}$, $\mathsf{M}_s = 10$ and $L = 8 \times 10^7$, which gave $1.579777 \pm 7.6 \times 10^{-5}$ where $\pm$ reflects the Monte Carlo error only.

For Algorithm 6.1.4, we measure the error of computing the L^2-norm of the second moment of the SPDE solution as follows

$$\rho(T) = \left\| \mathbb{E}[u_{\text{ref}}^2(T, \cdot)] \right\|_{l^2} - \left\| \mathbb{E}[(u_{\Delta,\mathsf{N}}^{\mathsf{M},\delta t}(T, \cdot))^2] \right\|_{l^2},$$

where $\|v(\cdot)\|_{l^2} = \frac{\ell}{\mathsf{M}_s} \left(\sum_{i,j=1}^{\mathsf{M}_s} v^2(x_i^1, x_j^2) \right)^{1/2}$, $x_i^1 = x_i^2 = (i-1)\ell/\mathsf{M}_s$, $i = 1, \ldots, \mathsf{M}_s$, and $\mathbb{E}[u_{\text{ref}}^2(T, \cdot)]$ is the reference solution computed as explained above. The results demonstrating second-order convergence (see (6.1.11) and the discussion after Algorithm 6.1.4) are given in Table 6.7. Some control tests with $\delta t = 1 \times 10^{-5}$ and $\mathsf{M} = 1600$ showed that the errors presented in this table are not essentially influenced by the errors caused by the choice of $\delta t = 1 \times 10^{-4}$ and cut-off of the basis at $\mathsf{M} = 900$.

Table 6.7. Performance of Algorithm 6.1.4 for passive scalar equation (6.4.4). The parameters of Algorithm 6.1.4 are $\mathsf{N} = 2$, $\mathsf{n} = 1$, $\mathsf{M} = 900$, $\delta t = 1 \times 10^{-4}$.

Δ	0.05	0.02	0.01	0.005	0.0025
$\rho(1)$	0.1539	0.0326	0.0089	0.0023	0.0006

6.4.1 A Monte Carlo method based on the method of characteristics

Let us now describe the Monte Carlo-type algorithm based on the method of characteristics. The solution $u(t, x)$ of (6.4.1) has the following (conditional) probabilistic representation (see [314, 408]):

$$u(t, x) = u_0(X_{t,x}(0)), \text{ a.s.,} \tag{6.4.8}$$

where $X_{t,x}(s)$, $0 \le s \le t$, is the solution of the system of (backward) characteristics

$$- dX = \sum_k \sigma_k(X) \overleftarrow{dW_k}(s), \quad X(t) = x. \tag{6.4.9}$$

Due to (6.4.2), it holds that

$$\sum_k \frac{\partial \sigma_k}{\partial x} \sigma_k = 0, \tag{6.4.10}$$

the phase flow of (6.4.9) preserves phase volume (see, e.g., [358, p. 247, Equation (5.5)]). We also recall that the Ito and Stratonovich forms of (6.4.9) coincide. As shown in [358], it is beneficial to approximate (6.4.9) using phase volume preserving schemes, e.g., by the midpoint method [358, Chapter 4], which for (6.4.9) takes the form (here we exploited that the Ito and Stratonovich forms of (6.4.9) coincide): for an integer $m \ge 1$,

$$\begin{aligned} X_m &= x, \\ X_l &= X_{l+1} + \sum_k \sigma_k \left(\frac{X_l + X_{l+1}}{2} \right) \left(\zeta_k^{\Delta t}\right)_l \sqrt{\Delta t}, \ l = n-1, \dots, 0, \end{aligned} \tag{6.4.11}$$

where $\left(\zeta_k^{\Delta t}\right)_l$ are, e.g., i.i.d. random variables with the law

$$\zeta_k^{\Delta t} = \begin{cases} \xi_k, \ |\xi_k| \leq A_{\Delta t}, \\ A_{\Delta t}, \ \xi_k > A_{\Delta t}, \\ -A_{\Delta t}, \ \xi_k < -A_{\Delta t}, \end{cases} \tag{6.4.12}$$

and ξ_k are i.i.d. standard Gaussian random variables, and $A_{\Delta t} = \sqrt{2c|\ln \Delta t|}$, $c \geq 1$. Its weak order is equal to one [358]. To solve the two-dimensional nonlinear equation at each step, we used the fixed-point method with the level of tolerance 10^{-13}. In this example, two fixed-point iterations were sufficient to reach this accuracy.

Using $\bar{X}_{t,x}(0) = X_0$ obtained by (6.4.11) with $\Delta t = T/m$, we simulate the L^2-norm of the second moment of the SPDE solution as follows:

$$\begin{aligned} \left\|\mathbb{E}[u^2(T,\cdot)]\right\|_{L^2} &= \left(\int_{[0,\ell]^2} (\mathbb{E}[u^2(T,x)])^2 \, dx\right)^{1/2} \approx \left\|\mathbb{E}[u^2(T,\cdot)]\right\|_{l^2} \\ &= \frac{\ell}{\mathsf{M}_s}\left[\sum_{i,j=1}^{\mathsf{M}_s} \left(\mathbb{E}[u_0^2(X_{T,x_i^1,x_j^2}(0))]\right)^2\right]^{1/2} \\ &\approx \frac{\ell}{\mathsf{M}_s}\left[\sum_{i,j=1}^{\mathsf{M}_s} (\mathbb{E}[u_0^2(\bar{X}_{T,x_i^1,x_j^2}(0))]^2\right]^{1/2} \\ &\approx \frac{\ell}{\mathsf{M}_s}\left[\sum_{i,j=1}^{\mathsf{M}_s} \left[\frac{1}{L}\sum_{l=1}^{L} u_0^2(\bar{X}_{T,x_i^1,x_j^2}^{(l)}(0))\right]^2\right]^{1/2}, \end{aligned} \tag{6.4.13}$$

where $x_i^1 = x_i^2 = (i-1)\ell/\mathsf{M}_s$, $i = 1, \ldots, \mathsf{M}_s$; $\bar{X}_{t,x_i^1,x_j^2}^{(l)}(0)$ are independent realizations of the random variables $\bar{X}_{t,x_i^1,x_j^2}(0)$. The approximation in (6.4.13) has three errors: (i) the error of discretization of the integral of the space domain $[0,\ell]^2$ which is negligible in our example even for $\mathsf{M}_s = 10$; (ii) the error of numerical integration due to replacement of $X_{t,x_i^1,x_j^2}(0)$ by $\bar{X}_{t,x_i^1,x_j^2}(0)$; (iii) the Monte Carlo error.

6.4.2 Comparison between recursive WCE and Monte Carlo methods

We compare Algorithm 6.1.4 and the Monte Carlo algorithm (6.4.13) by simulating the example (6.4.1), (6.4.6), (6.4) at $T = 1$. From Tables 6.8 and 6.9,[1] we observe similar effects as in one dimension that for lower accuracy the Monte Carlo algorithm (6.4.13) outperforms Algorithm 6.1.4; and that Algorithm 6.1.4 is more efficient for obtaining higher accuracy.

[1] Matlab R2010b was used for each test on a single core of two Intel Xeon 5540 (2.53 GHz) quad-core Nehalem processors.

Table 6.8. Performance of Algorithm 6.1.4 (recursive WCE) on passive scalar equation (6.4.4) at $T = 1$. The parameters of Algorithm 6.1.4 are $\mathsf{N} = 2$, $\mathsf{n} = 1$, $\mathsf{M} = 900$, $\delta t = 1 \times 10^{-4}$.

Δ	$\rho(1)$	CPU time
1×10^{-2}	8.89×10^{-3}	3.7×10^{4} (sec.)
1×10^{-3}	1.20×10^{-4}	$3.2 \times 10^{5} (sec.)$
5×10^{-4}	3.73×10^{-5}	$1.8 \times 10^{2} (hours)$

Algorithm 6.1.4 with the parameters $N = 2$, $n = 1$, $\Delta = 0.01$, $M = 900$, $\delta t = 1 \times 10^{-4}$ produced the result with the error 9×10^{-3}. It needed 3.7×10^{4} seconds of computer time. However, the Monte Carlo algorithm 6.4.11-6.4.13 required 12 seconds of computer time to produce the same level of accuracy (combined numerical integration and statistical error), with $\Delta t = 0.2$, $M_s = 10$, $L = 2.5 \times 10^{4}$.

Table 6.9. Performance of Algorithm 6.4.11-6.4.13 (a Monte Carlo method) for passive scalar equation (6.4.4) at $T = 1$. The parameter is $\mathsf{M} = 100$.

Δt	L	Error	CPU time
2×10^{-1}	2.5×10^{4}	$4.68 \times 10^{-3} \pm 4.38 \times 10^{-3}$	1.2×10^{1} (sec.)
1×10^{-2}	4×10^{7}	$1.46 \times 10^{-4} \pm 1.08 \times 10^{-4}$	3.5×10^{5} (sec.)
1×10^{-3}	4×10^{8}	$\sim \times 10^{-5} \pm 3.03 \times 10^{-5}$	9.7×10^{3} (hours)[2]

If we require higher accuracy, then Algorithm 6.1.4 outperforms the Monte Carlo algorithm 6.4.11-6.4.13. Algorithm 6.1.4 needs approximately 180 hours of computer time to obtain the error 3×10^{-5}, with the parameters $N = 2$, $n = 1$, $\Delta = 5 \times 10^{-4}$, $M = 900$, $\delta t = 1 \times 10^{-4}$. To achieve the same level of accuracy, the Monte Carlo algorithm 6.4.11-6.4.13 requires approximately 9700 hours of computer time (estimated, not experimentally measured), with $\Delta t = 0.001$, $M_s = 10$, $L = 4 \times 10^{8}$.

6.5 Summary and bibliographic notes

We have presented a multistage Wiener chaos expansion (WCE) method for advection-diffusion-reaction equations with multiplicative noise, and complemented this method by a deterministic algorithm for computing second moments of the SPDE solutions without any use of the Monte Carlo technique, see Algorithm 6.2.1.

[2]This is an estimated time according to the tests with smaller Δt, L and with $\mathsf{M} = 100$.

- The numerical tests demonstrated that the WCE-based deterministic algorithm can be more efficient than Monte Carlo-type methods in obtaining results of higher accuracy, scaling as Δ^{N}, where Δ is the time-step of the "online" integration and N is the order of Wiener chaos.
- For obtaining results of lower accuracy, Monte Carlo-type methods outperform the deterministic algorithm for computing moments even in the one-dimensional case. The recursive WCE algorithm is conceptually different from Monte Carlo-type methods and thus it can be used for independent verification of results obtained by Monte Carlo solvers.
- The efficiency of the algorithm can be greatly improved if it is combined with reduced-order methods so that only a handful of modes will be required to represent the solution accurately in physical space, i.e., a case with small M, see a discussion at the end of Chapter 6.1. We can also use the sparsity of the WCE solution in numerical simulation in Chapter 6.4 to reduce the computational cost.

We have shown the efficiency of the recursive WCE for linear stochastic parabolic equations with nonrandom coefficients, where WCE of linear equations leads to a lower triangular system of deterministic PDEs. While the coefficients are random, WCE will lead to a fully coupled system of PDEs and thus the computational convenience may be lost. In the next chapter, we will introduce stochastic collocation methods which lead to decoupled system of PDEs.

Bibliographic notes. The idea of the recursive WCE is originally from [315] for the Zakai equation of nonlinear filtering, where the operators $\mathcal{M}_k$ in (3.3.23) are just zeroth order differential operators. In Chapter 8, it will be shown that the technique for the convergence proof for the case of $\mathcal{M}_k$ being first-order differential operator is different from the one in [315]. The difference lies on the fact that the first-order differential operators $\mathcal{M}_k$'s require more regularity of the solution and stronger assumption on the coefficients in $\mathcal{M}_k$ even if a second-order convergence in random space is needed.

In this chapter, we require the solution to be square-integrable in the random space. In many cases, this requirement cannot be satisfied and we need to seek solutions in weaker sense. It is true that WCE can still be applied in this situation but solutions are only living in a weighted space, see, e.g., [34, 318, 319, 332, 384, 459, 469]. The WCE methods in these papers are all associated with the Ito product between two random fields, which can be seen as a zeroth order approximation of the classical product, see [346, 462, 470] and also Chapter 11.

Algorithm 6.1.1 coincides with the algorithm proposed in [315] for (3.3.23) in the case of $b_i^k(t,x) = 0$, $c = 0$, $g_k = 0$, and finite number of noises but generalizes it to a wider class of linear SPDEs of the form (3.3.23). In particular, the algorithm from [315] was applied to the nonlinear filtering problem for hidden Markov models in the case of *independent* noises in signal and observation, while Algorithm 6.1.1 is also applicable when noises in signal and observation are *dependent*.

Algorithm 6.1.1 allows us to simulate mean-square approximations of the solution to the SPDE (3.3.23). It can also be used together with the Monte Carlo technique for computing expectations of functionals of the solution to (3.3.23).

The recursive polynomial chaos has also been applied to linear parabolic equations driven by Gaussian color noise [321] (WCE) and Poisson noise [303] (generalized polynomial chaos).

It is possible to reduce the variance of the estimator on the right-hand side of (6.4.13) using *variance reduction techniques* to reduce the Monte Carlo error, see, e.g., [358, 361] and the references therein. However, for complex stochastic problems it is usually rather difficult to reduce the variance efficiently. In this chapter, we consider only direct Monte Carlo methods without variance reduction and give a comparison of computational costs for the WCE-based algorithm and these Monte Carlo methods.

One can recognize that (6.3.9) is a Kubo oscillator. A number of numerical tests with symplectic and non-symplectic integrators are done on a Kubo oscillator in [356, 358].

6.6 Suggested practice

Exercise 6.6.1 *Solve the Kubo oscillator* (6.3.9) *with the backward Euler scheme and numerically check the convergence rate.*

Hint. The exact solution is (3.3.18) when $\epsilon = 0$. The numerical convergence rate should be close to $1/2$ in the mean-square sense and 1 in the weak sense. The number of Monte Carlo sampling paths should be large enough to have the statistical error significantly smaller than the integration error.

Exercise 6.6.2 *Consider the following linear stochastic differential equation:*

$$\frac{dy}{dt} = \xi y, \quad y(0) = y_0, \tag{6.6.1}$$

where $y_0 = 1$, *and* $\xi \sim \mathcal{N}(0,1)$. *We can apply the multistage WCE to solve* (6.6.1) *in the following steps:*

a) Write down the propagator, i.e., equations for the WCE expansion coefficients y_α, $\alpha = 0, 1, \ldots$, *of Equation* (6.6.1).
b) Solve the propagator in a) on $[0, \Delta]$, *where* $\alpha = 0, 1, 2, \ldots, M$ *for a sufficient large* M, *say* 200. *Use the explicit fourth-order Runge-Kutta scheme in time. Specifically, for the equation* $\frac{dx}{dt} = f(t, x)$, $x(0) = x_0$, *the explicit fourth-order Runge-Kutta scheme reads*

$$x_{n+1} = x_n + \tfrac{h}{6}\left(k_1 + 2k_2 + 2k_3 + k_4\right), \quad n = 0, 1, 2, 3, \cdots, N, \; Nh = \Delta,$$

where h is the time step size and

$$
\begin{aligned}
k_1 &= f(t_n, x_n),\\
k_2 &= f(t_n + \tfrac{h}{2}, x_n + \tfrac{h}{2}k_1),\\
k_3 &= f(t_n + \tfrac{h}{2}, x_n + \tfrac{h}{2}k_2),\\
k_4 &= f(t_n + h, x_n + hk_3).
\end{aligned}
$$

c) *Compute the numerical solution at larger* $t \in (\Delta, T]$. *Denote by* $z^{(k)}$ *the solution in b) to the following problem on* $[0, \Delta]$.

$$
\frac{dz}{dt} = \xi z, \quad t \in [0, \Delta], \quad z(0) = H_k(\xi), \quad k = 1, \cdots, N. \tag{6.6.2}
$$

Then $z^{(k)} \approx \sum_{m=0}^{M} z_m^{(k)}(t) H_m(\xi), \quad t \in [0, \Delta]$. *Suppose* $y_0 = \sum_{k=0}^{N} y_k H_k(\xi)$. *Then a numerical solution to the problem* (6.6.1) *is*

$$
\begin{aligned}
y_{M,N}(t) &= \sum_{k=0}^{N} y_k \left(\sum_{m=0}^{M} z_m^{(k)}(t) H_m(\xi) \right)\\
&= \sum_{m=0}^{M} \left(\sum_{k=0}^{N} y_k z_m^{(k)}(t) \right) H_m(\xi), \quad t \in [0, \Delta].
\end{aligned} \tag{6.6.3}
$$

d) *Compute the moments of the numerical solution* $y_{M,N}(t)$ *for* $t = l\Delta$ *for* $l = 2, 3, \cdots$.

Use also a Monte Carlo method to compute moments of solutions approximately. Compare the computational time (for both low and high accuracy) with those for the above multistage WCE at $t = 1$ *and* $t = 10$.

Hint. The exact solution to (6.6.1) is $y_0 \exp(t\xi)$, where the moments can be computed explicitly. For comparison between multistage WCE and Monte Carlo methods, a similar conclusion as in this chapter is expected.

7

Stochastic collocation methods for differential equations with white noise

Stochastic collocation methods can lead to a fully decoupled system of PDEs, which can be readily implemented on parallel computers. However, stochastic collocation methods do not work when longer time integration is required. Though these methods are also cursed by the dimensionality, we apply the recursive strategy for longer time integration discussed in Chapter 6 to investigate the error behavior of stochastic collocation methods in one-time step approximation or short time integration.

Here, as an example, we consider one specific collocation method, namely, Smolyak's sparse grid collocation, which is a sparse grid collocation method, see, e.g., Chapter 2.5.4. If no confusion arises, we will still use the term stochastic collocation method (SCM) instead of sparse grid collocation method. We first analyze the error of Smolyak's sparse grid collocation used to evaluate expectations of functionals of solutions to stochastic differential equations discretized by the Euler scheme. We show theoretically and numerically that this algorithm can have satisfactory accuracy for small magnitude of noise or relatively short integration time. However, it does not converge, in general, neither with decrease of the Euler scheme's time step size nor with increase of Smolyak's sparse grid level. Subsequently, we use this method as a building block for presenting a new algorithm by combining sparse grid collocation with a recursive procedure. This approach allows us to numerically integrate linear stochastic partial differential equations over longer times. This is illustrated in numerical tests on a stochastic advection-diffusion equation.

7.1 Introduction

In a number of applications from physics, financial engineering, biology, and chemistry it is of interest to compute expectations of some functionals of solutions of ordinary stochastic differential equations (SDE) and stochastic

Z. Zhang, G.E. Karniadakis, *Numerical Methods for Stochastic Partial Differential Equations with White Noise*,
Applied Mathematical Sciences 196, DOI 10.1007/978-3-319-57511-7_7

partial differential equations (SPDE) driven by white noise. Usually, evaluation of such expectations requires to approximate solutions of stochastic equations and then to compute the corresponding averages with respect to the approximate trajectories. The most commonly used approach for computing the averages is the Monte Carlo technique, which is known for its slow rate of convergence and hence limiting computational efficiency of stochastic simulations. To speed up computation of the averages, variance reduction techniques (see, e.g., [358, 361] and the references therein), quasi-Monte Carlo algorithms [376, 423], and the multi-level Monte Carlo method [156, 157] have been proposed and used.

In this chapter, we consider a sparse grid collocation method accompanied by time discretization of differential equations perturbed by time-dependent noise. We obtain an error estimate for the SCM in conjunction with the Euler scheme for evaluating expectations of smooth functionals of solutions of a scalar linear SDE with additive noise. In particular, we conclude that the SCM can successfully work for a small magnitude of noise and relatively short integration time while in general it does not converge neither by decreasing the time discretization step used for SDE approximation nor by increasing the level of Smolyak's sparse grid; see Remark 7.2.4. Numerical tests in Chapter 7.2 confirm our theoretical conclusions and we also observe first-order convergence in time step size of the algorithm using the SCM as long as the SCM error is small relative to the error of time discretization of SDE. We note that our conclusion is, to some extent, similar to that for cubatures on Wiener space [70], for Wiener chaos method [225, 315, 316, 505] and some other functional expansion approaches [53, 54].

To achieve accurate longer time integration by numerical algorithms using the SCM, we exploit the idea of the recursive approach for a linear SPDE with *time-independent* coefficients presented in [315] and Chapter 6. The recursive approach works as follows. We first find an approximate solution of an SPDE at a relatively small time $t = h$, and subsequently take the approximation at $t = h$ as the initial value in order to compute the approximate solution at $t = 2h$, and so on, until we reach the final integration time $T = Nh$. To obtain second moments of the SPDE solution, we store a covariance matrix of the approximate solution at each time step kh and recursively compute the first two moments. Such an algorithm is presented in Chapter 7.3; in Chapter 7.4 we demonstrate numerically that this algorithm converges in time step h and works well on longer time intervals.

At the end of this chapter, we summarize and present a brief review on deterministic high-dimensional quadratures in random space including some disadvantage of the sparse grid collocation method. A restarting strategy for nonlinear SODEs is also presented for long-time integration. Two exercises are provided on applying stochastic collocation methods.

7.2 Isotropic sparse grid for weak integration of SDE

We consider application of the sparse grid rule (2.5.9) to the integral in (3.2.20). In this approach, the total error has two parts

$$|\mathbb{E}f(X(T)) - A(\mathsf{L},N)\varphi| \leq \left|\mathbb{E}f(X(T)) - \mathbb{E}f(\tilde{X}_N)\right| + \left|\mathbb{E}f(\tilde{X}_N) - A(\mathsf{L},N)\varphi\right|,$$

where $A(\mathsf{L},N)$ is defined in (2.5.9) and φ is from (3.2.19). The first part is controlled by the time step size h, see (3.2.15), and it converges to zero with order one in h. The second part is controlled by the sparse grid level L but it depends on h since decrease of h increases the dimension of the random space. Some illustrative examples will be presented in Chapter 7.2.2.

7.2.1 Probabilistic interpretation of SCM

It is not difficult to show that SCM admits a probabilistic interpretation, e.g., in the case of level $\mathsf{L} = 2$ we have

$$\begin{aligned} &A(2,N)\varphi(y_{1,1},\ldots,y_{r,1},\ldots,y_{1,N},\ldots,y_{r,N}) \qquad (7.2.1)\\ &= (Q_2 \otimes Q_1 \otimes \cdots \otimes Q_1)\,\varphi + (Q_1 \otimes Q_2 \otimes Q_1 \otimes \cdots \otimes Q_1)\,\varphi \\ &\quad + \cdots + (Q_1 \otimes Q_1 \otimes \cdots \otimes Q_2)\,\varphi - (Nr-1)\,(Q_1 \otimes Q_1 \otimes \cdots \otimes Q_1)\,\varphi \\ &= \sum_{i=1}^{N}\sum_{j=1}^{r} \mathbb{E}\varphi(0,\ldots,0,\zeta_{j,i},0,\ldots,0) - (Nr-1)\varphi(0,0,\ldots,0), \end{aligned}$$

where $\zeta_{j,i}$ are i.i.d. random variables with the law (3.2.17). Using (3.2.21), (7.2.1), Taylor's expansion and symmetry of $\zeta_{j,i}$, we obtain the relationship between the weak Euler scheme (3.2.16) and the SCM (2.5.9):

$$\begin{aligned} \mathbb{E}f(\tilde{X}_N) - A(2,N)\varphi &= \mathbb{E}\varphi(\zeta_{1,1},\ldots,\zeta_{r,1},\ldots,\zeta_{1,N},\ldots,\zeta_{r,N}) \qquad (7.2.2)\\ &\quad - \sum_{i=1}^{N}\sum_{j=1}^{r} \mathbb{E}\varphi(0,\ldots,0,\zeta_{j,i},0,\ldots,0) \\ &\qquad - (Nr-1)\varphi(0,0,\ldots,0) \\ &= \sum_{|\alpha|=4} \frac{4}{\alpha!}\mathbb{E}\left[\prod_{i=1}^{N}\prod_{j=1}^{r}(\zeta_{j,i})^{\alpha_{j,i}} \int_0^1 (1-z)^3 D^{\alpha}\right. \\ &\qquad \left. \varphi(z\zeta_{1,1},\ldots,z\zeta_{r,N})\,dz\right] \\ &\quad - \frac{1}{3!}\sum_{i=1}^{N}\sum_{j=1}^{r}\mathbb{E}\left[\zeta_{j,i}^4 \int_0^1 (1-z)^3 \frac{\partial^4}{(\partial y_{j,i})^4}\varphi(0,\ldots,0,\right. \\ &\qquad \left. z\zeta_{j,i},0,\ldots,0)dz\right], \end{aligned}$$

where the multi-index $\alpha = (\alpha_{1,1},\ldots,\alpha_{r,N}) \in \mathbb{N}_0^{rN}$, $|\alpha| = \sum_{i=1}^{N}\sum_{j=1}^{r}\alpha_{j,i}$, $\alpha! = \prod_{i=1}^{N}\prod_{j=1}^{r}\alpha_{j,i}!$ and $D^{\alpha} = \frac{\partial^{|\alpha|}}{(\partial y_{1,1})^{\alpha_{1,1}}\cdots(\partial y_{r,N})^{\alpha_{r,N}}}$. The error of the SCM applied to the weak approximation of SDE is further studied in Chapter 7.2.2.

In the SCM context, it is beneficial to exploit higher-order or higher-accuracy schemes for approximating the SDE (3.1.4) because they can allow us to reach a desired accuracy using larger time step sizes and therefore less random variables than the first-order Euler scheme (3.2.2) or (3.2.16). For example, we can use the second-order scheme (3.2.23). Roughly speaking, to achieve $O(h)$ accuracy with (3.2.23), we need only $\sqrt{2rN}$ ($\sqrt{rN}$ in the case of additive noise) random variables, while we need rN random variables for the Euler scheme (3.2.2). This reduces the dimension of the random space and hence can increase efficiency and the applicability of SCM (see, in particular Example 7.4.1 for a numerical illustration). We note that when noise intensity is relatively small, we can use high-accuracy low-order schemes designed for SDEs with small noise [357] (see also [358, Chapter 3]) in order to achieve a desired accuracy using less number of random variables than the Euler scheme (3.2.2).

Similarly, we can write down a probabilistic interpretation for any level L and derive a similar error representation. For example, we have for $\mathsf{L}=3$ that

$$
\begin{aligned}
&\mathbb{E}[\varphi(\zeta_{1,1}^{(3)},\cdots,\zeta_{r,N}^{(3)})]-A(3,N)\varphi\\
&=\sum_{|\alpha|=6}\frac{6}{\alpha!}\mathbb{E}[\prod_{i=1}^{N}\prod_{j=1}^{r}(\zeta_{j,i}^{(3)})^{\alpha_{j,i}}\int_0^1(1-z)^5D^\alpha\varphi(z\zeta_{1,1}^{(3)},\cdots,z\zeta_{r,N}^{(3)})\,dz]\\
&\quad-\sum_{\substack{|\alpha|=\alpha_{j,i}+\alpha_{l,k}=6\\(j-l)^2+(i-k)^2\neq 0}}\frac{6}{\alpha_{j,i}!\alpha_{k,l}!}\mathbb{E}[(\zeta_{j,i}^{(3)})^{\alpha_{j,i}}(\zeta_{l,k}^{(3)})^{\alpha_{l,k}}\int_0^1(1-z)^5D^\alpha\varphi(\cdots,z\zeta_{j,i}^{(3)},\\
&\qquad\qquad 0,\cdots,0,z\zeta_{l,k}^{(3)},\cdots)\,dz]\\
&\quad-\sum_{i=1}^{N}\sum_{j=1}^{r}\frac{6}{6!}\mathbb{E}[(\zeta_{j,i})^6\int_0^1(1-z)^5D^\alpha\varphi(0,\cdots,z\zeta_{j,i},\cdots,0)\,dz],
\end{aligned}
$$

where $\zeta_{j,i}$ are defined in (3.2.17) and $\zeta_{j,i}^{(3)}$ are i.i.d. random variables with the law $P(\zeta_{j,i}^{(3)}=\pm\sqrt{3})=1/6$, $P(\zeta_{j,i}^{(3)}=0)=2/3$.

7.2.2 Illustrative examples

In this section we show some limitations of the use of SCM in the weak approximation of SDEs. To this end, it is convenient and sufficient to consider the scalar linear SDE

$$dX=\lambda X dt+\varepsilon\, dW(t),\quad X_0=1, \tag{7.2.3}$$

where λ and ε are some constants.

We will compute expectations $\mathbb{E}f(X(T))$ for some $f(x)$ and $X(t)$ from (7.2.3) by applying the Euler scheme (3.2.2) and the SCM (2.5.9). This simple example provides us with a clear insight when algorithms of this type are able to produce accurate results and when they are likely to fail. Using direct calculations, we first (see Examples 7.2.1–7.2.2 below) derive an estimate for the error $|\mathbb{E}f(X_N) - A(2,N)\varphi|$ with X_N from (3.2.2) applied to (7.2.3) and for some particular $f(x)$. This will illustrate how the error of SCM with practical level (i.e., $\mathsf{L} \leq 6$) behaves. Then in Proposition 7.2.3, we obtain an estimate for the error $|\mathbb{E}f(X_N) - A(\mathsf{L},N)\varphi|$ for a smooth $f(x)$, which grows not faster than a polynomial function at infinity. We will observe that the considered algorithm is not convergent in time step h and the algorithm is not convergent in level L unless the noise intensity and integration time are small.

It follows from (3.2.15) and (3.2.18) that

$$\begin{aligned} |\mathbb{E}f(X_N) - A(\mathsf{L},N)\varphi| &\leq \left|\mathbb{E}f(\tilde{X}_N) - A(\mathsf{L},N)\varphi\right| + |\mathbb{E}f(X_N) - \mathbb{E}f(\tilde{X}_N)| \\ &\leq \left|\mathbb{E}f(\tilde{X}_N) - A(\mathsf{L},N)\varphi\right| + Kh, \end{aligned} \tag{7.2.4}$$

where $\tilde{X}_N$ is from the weak Euler scheme (3.2.16) applied to (7.2.3), which can be written as $\tilde{X}_N = (1+\lambda h)^N + \sum_{j=1}^{N}(1+\lambda h)^{N-j}\varepsilon\sqrt{h}\zeta_j$. Introducing the function

$$\bar{X}(N;y) = (1+\lambda h)^N + \sum_{j=1}^{N}(1+\lambda h)^{N-j}\varepsilon\sqrt{h}y_j, \tag{7.2.5}$$

we see that $\tilde{X}_N = \bar{X}(N;\zeta_1,\ldots,\zeta_N)$. We have

$$\frac{\partial}{\partial y_i}\bar{X}(N;y) = (1+\lambda h)^{N-i}\varepsilon\sqrt{h} \quad \text{and} \quad \frac{\partial^2}{\partial y_i \partial y_j}\bar{X}(N;y) = 0. \tag{7.2.6}$$

Then we obtain from (7.2.2):

$$\begin{aligned} R :&= \mathbb{E}f(\tilde{X}_N) - A(2,N)\varphi \\ &= \varepsilon^4 h^2 \sum_{|\alpha|=4} \frac{4}{\alpha!}\mathbb{E}\left[\prod_{i=1}^{N}(\zeta_i(1+\lambda h)^{N-i})^{\alpha_i} \int_0^1 (1-z)^3 z^4 \frac{d^4}{dx^4}\right. \\ &\qquad \left. f(\bar{X}(N, z\zeta_1,\ldots,z\zeta_N))\,dz\right] \\ &\quad -\frac{1}{3!}\varepsilon^4 h^2 \sum_{i=1}^{N}\mathbb{E}\left[\zeta_i^4 \int_0^1 (1-z)^3 z^4 \frac{d^4}{dx^4} f(\bar{X}(0,\ldots,0,z\zeta_i,0,\ldots,0))\right. \\ &\qquad \left.(1+\lambda h)^{4N-4i} dz\right]. \end{aligned} \tag{7.2.7}$$

Non-Convergence in time step h

We will illustrate no convergence in h for SCM for levels two and three through two examples, where sharp error estimates of $|\mathbb{E}f(X_N) - A(2,N)\varphi|$

are derived for SCM. Higher level SCM can be also considered but the conclusions do not change. In contrast, the algorithm of tensor-product integration in random space and the strong Euler scheme in time (i.e., the weak Euler scheme (3.2.16)–(3.2.17)) is convergent with order one in h. We also note that in practice, typically SCM with level no more than six are employed.

Example 7.2.1 *For $f(x) = x^p$ with $p = 1, 2, 3$, it follows from (7.2.7) that $R = 0$, i.e., SCM does not introduce any additional error, and hence by* (7.2.4)

$$|\mathbb{E}f(X_N) - A(2,N)\varphi| \leq Kh, \quad f(x) = x^p, \quad p = 1, 2, 3.$$

For $f(x) = x^4$, we get from (7.2.7):

$$R = \frac{6}{35}\varepsilon^4 h^2 \sum_{i=1}^{N} \sum_{j=i+1}^{N} (1+\lambda h)^{4N-2i-2j}$$

$$= \frac{6}{35}\varepsilon^4 \times \begin{cases} \frac{(1+\lambda h)^{2N}-1}{\lambda^2(2+\lambda h)^2}\left[\frac{(1+\lambda h)^{2N}+1}{1+(1+\lambda h)^2} - 1\right], & \lambda \neq 0, \quad 1+\lambda h \neq 0, \\ \frac{T^2}{2} - \frac{Th}{2}, & \lambda = 0. \end{cases}$$

We see that R does not go to zero when $h \to 0$ and that for sufficiently small $h > 0$

$$|\mathbb{E}f(X_N) - A(2,N)\varphi| \leq Kh + \frac{6}{35}\varepsilon^4 \times \begin{cases} \frac{1}{\lambda^2}(1+e^{4T\lambda}), & \lambda \neq 0, \\ \frac{T^2}{2}, & \lambda = 0. \end{cases}$$

We observe that the SCM algorithm does not converge with $h \to 0$ for higher moments. In the considered case of linear SDE, increasing the level L of SCM leads to the SCM error R being 0 for higher moments, e.g., for $\mathsf{L} = 3$ the error $R = 0$ for up to 5th moment but the algorithm will not converge in h for 6th moment and so on (see Proposition 7.2.3 below). Further (see the continuation of the illustration below), in the case of, e.g., $f(x) = \cos x$ for any L this error R does not converge in h, which is also the case for nonlinear SDE. We also note that one can expect that this error R is small when noise intensity is relatively small and either time T is small or SDE has, in some sense, stable behavior (in the linear case it corresponds to $\lambda < 0$).

Example 7.2.2 *Now consider $f(x) = \cos(x)$. It follows from* (7.2.7) *that*

$$R = \varepsilon^4 h^2 \sum_{|\alpha|=4} \frac{4}{\alpha!}\mathbb{E}\left[\prod_{i=1}^{N}(\zeta_i(1+\lambda h)^{N-i})^{\alpha_i} \int_0^1 (1-z)^3 z^4 \cos((1+\lambda h)^N \right.$$

$$\left. + z\sum_{j=1}^{N}(1+\lambda h)^{N-j}\varepsilon\sqrt{h}\zeta_j)\, dz\right]$$

$$-\frac{1}{3!}\varepsilon^4 h^2 \sum_{i=1}^{N}(1+\lambda h)^{4N-4i} \int_0^1 (1-z)^3 z^4 \mathbb{E}[\zeta_i^4 \cos((1+\lambda h)^N$$

$$+ z(1+\lambda h)^{N-i}\varepsilon\sqrt{h}\zeta_i)]\, dz$$

and after routine calculations we obtain

$$
\begin{aligned}
R = \varepsilon^4 h^2 \cos((1+\lambda h)^N) \Bigg[& \left(\frac{1}{6} \sum_{i=1}^{N} (1+\lambda h)^{4N-4i} + 2 \sum_{i=1}^{N} \sum_{j=i+1}^{N} (1+\lambda h)^{4N-2i-2j} \right) \\
& \times \int_0^1 (1-z)^3 z^4 \prod_{l=1}^{N} \cos(z(1+\lambda h)^{N-l} \varepsilon\sqrt{h}) dz \\
& + \left(\frac{2}{3} \sum_{i,j=1; i\neq j}^{N} (1+\lambda h)^{4N-3i-j} + 2 \sum_{\substack{k,i,j=1 \\ i\neq j, i\neq k, k\neq j}}^{N} (1+\lambda h)^{4N-2k-i-j} \right) \\
& \times \int_0^1 (1-z)^3 z^4 \prod_{l=i,j} \sin(z(1+\lambda h)^{N-l} \varepsilon\sqrt{h}) \prod_{\substack{l=1 \\ l\neq i, l\neq j}}^{N} \cos(z(1+\lambda h)^{N-l} \varepsilon\sqrt{h}) dz \\
& + 4 \sum_{\substack{i,j,k,m=1 \\ i\neq j, i\neq k, i\neq m, j\neq k, j\neq m, k\neq m}}^{N} (1+\lambda h)^{4N-i-j-k-m} \\
& \times \int_0^1 (1-z)^3 z^4 \prod_{l=i,j,k,m} \sin(z(1+\lambda h)^{N-l} \varepsilon\sqrt{h}) \\
& \prod_{l=1 l\neq i, l\neq j, l\neq k, l\neq m}^{N} \cos(z(1+\lambda h)^{N-l} \varepsilon\sqrt{h}) dz \\
& - \frac{1}{6} \sum_{i=1}^{N} (1+\lambda h)^{4N-4i} \int_0^1 (1-z)^3 z^4 \cos(z(1+\lambda h)^{N-i} \varepsilon\sqrt{h})] \, dz \Bigg].
\end{aligned}
$$

It is not difficult to see that R does not go to zero when $h \to 0$. In fact, taking into account that $|\sin(z(1+\lambda h)^{N-j}\varepsilon\sqrt{h})| \leq z(1+\lambda h)^{N-j}\varepsilon\sqrt{h}$, and that there are N^4 terms of order h^4 and N^3 terms of order h^3, we get for sufficiently small $h > 0$

$$
|R| \leq C\varepsilon^4(1 + e^{4T\lambda}),
$$

where $C > 0$ is independent of ε and h. Hence

$$
|\mathbb{E}f(X_N) - A(2,N)\varphi| \leq C\varepsilon^4(1 + e^{4T\lambda}) + Kh, \tag{7.2.8}
$$

and we have arrived at a similar conclusion for $f(x) = \cos x$ as for $f(x) = x^4$. Similarly, we can also have for $\mathsf{L} = 3$ that

$$
|\mathbb{E}f(X_N) - A(3,N)\varphi| \leq C\varepsilon^6(1 + e^{6T\lambda}) + Kh.
$$

This example shows for $\mathsf{L} = 3$, the error of SCM with the Euler scheme in time does not converge in h.

Error estimate for SCM with fixed level

Now we address the effect of the SCM level L. To this end, we will need the following error estimate of a Gauss-Hermite quadrature. Let $\psi(y)$, $y \in \mathbb{R}$, be a sufficiently smooth function which itself and its derivatives are growing not faster than a polynomial at infinity. Using the Peano kernel theorem (see, e.g., [96]) and that a Gauss-Hermite quadrature with n-nodes has the order of polynomial exactness $2n-1$, we obtain for the approximation error $R_{n,\gamma}\psi$ of the Gauss-Hermite quadrature $Q_n\psi$:

$$R_{n,\gamma}(\psi) := Q_n\psi - I_1\psi = \int_{\mathbb{R}} \frac{d^\gamma}{dy^\gamma}\varphi(y) R_{n,\gamma}(\Gamma_{y,\gamma})\, dy, \quad 1 \le \gamma \le 2n, \quad (7.2.9)$$

where $\Gamma_{y,\gamma}(z) = (z-y)^{\gamma-1}/(\gamma-1)!$ if $z \ge y$ and 0 otherwise. One can show (see, e.g., [339, Theorem 2]) that there is a constant $c > 0$ independent of n and y such that for any $0 < \beta < 1$

$$|R_{n,\gamma}(\Gamma_{y,\gamma})| \le \frac{c}{\sqrt{2\pi}} n^{-\gamma/2} \exp\left(-\frac{\beta y^2}{2}\right), \quad 1 \le \gamma \le 2n. \quad (7.2.10)$$

We also note that (7.2.10) and the triangle inequality imply, for $1 \le \gamma \le 2(n-1)$:

$$|R_{n,\gamma}(\Gamma_{y,\gamma}) - R_{n-1,\gamma}(\Gamma_{y,\gamma})| \le \frac{c}{\sqrt{2\pi}}[n^{-\gamma/2} + (n-1)^{-\gamma/2}] \exp\left(-\frac{\beta y^2}{2}\right). \quad (7.2.11)$$

Now we consider the error of the sparse grid rule (2.5.9) accompanied by the Euler scheme (3.2.2) for computing expectations of solutions to (7.2.3).

Proposition 7.2.3 *Assume that a function $f(x)$ and its derivatives up to $2\mathsf{L}$-th order satisfy the polynomial growth condition (3.2.14). Let X_N be obtained by the Euler scheme (3.2.2) applied to the linear SDE (7.2.3) and $A(L,N)\varphi$ be the sparse grid rule (2.5.9) with level L applied to the integral corresponding to $\mathbb{E}f(X_N)$ as in (3.2.20). Then for any L and sufficiently small $h > 0$*

$$|\mathbb{E}f(X_N) - A(\mathsf{L},N)\varphi| \le K\varepsilon^{2\mathsf{L}}(1 + e^{\lambda(2\mathsf{L}+\varkappa)T})\left(1 + (3c/2)^{\mathsf{L}\wedge N}\right)\beta^{-(\mathsf{L}\wedge N)/2}T^{\mathsf{L}}, \quad (7.2.12)$$

where $K > 0$ is independent of h, L and N; c and β are from (7.2.10); $\varkappa$ is from (3.2.14).

Proof. We recall (see (3.2.20)) that

$$\mathbb{E}f(X_N) = I_N\varphi = \frac{1}{(2\pi)^{N/2}} \int_{\mathbb{R}^{rN}} \varphi(y_1,\ldots,y_N) \exp\left(-\frac{1}{2}\sum_{i=1}^{N} y_i^2\right) dy.$$

Introduce the integrals

$$I_1^{(k)}\varphi = \frac{1}{\sqrt{2\pi}} \int_{\mathbb{R}} \varphi(y_1,\ldots,y_k,\ldots,y_N) \exp\left(-\frac{y_k^2}{2}\right) dy_k, \quad k = 1,\ldots,N, \quad (7.2.13)$$

and their approximations $Q_n^{(k)}$ by the corresponding one-dimensional Gauss-Hermite quadratures with n nodes. Also, let $\mathcal{U}_{i_k}^{(k)} = Q_{i_k}^{(k)} - Q_{i_k-1}^{(k)}$.

Using (2.5.9) and the recipe from the proof of Lemma 3.4 in [379], we obtain

$$I_N\varphi - A(\mathsf{L},N)\varphi = \sum_{l=2}^{N} S(\mathsf{L},l)\otimes_{k=l+1}^{N} I_1^{(k)}\varphi + (I_1^{(1)} - Q_{\mathsf{L}}^{(1)})\otimes_{k=2}^{N} I_1^{(k)}\varphi, \quad (7.2.14)$$

where

$$S(\mathsf{L},l) = \sum_{i_1+\cdots+i_{l-1}+i_l=\mathsf{L}+l-1} \otimes_{k=1}^{l-1}\mathcal{U}_{i_k}^{(k)} \otimes (I_1^{(l)} - Q_{i_l}^{(l)}). \quad (7.2.15)$$

Due to (7.2.9), we have for $n > 1$ and $1 \le \gamma \le 2(n-1)$

$$\begin{aligned} \mathcal{U}_n\psi &= Q_n\psi - Q_{n-1}\psi = [Q_n\psi - I_1(\psi)] - [Q_{n-1}\psi - I_1(\psi)] \quad (7.2.16)\\ &= \int_{\mathbb{R}} \frac{d^\gamma}{dy^\gamma}\psi(y)[R_{n,\gamma}(\Gamma_{y,\gamma}) - R_{n-1,\gamma}(\Gamma_{y,\gamma})]\,dy, \end{aligned}$$

and for $n = 1$

$$\mathcal{U}_n\psi = Q_1\psi - Q_0\psi = Q_1\psi = \psi(0). \quad (7.2.17)$$

By (7.2.15), (7.2.13), and (7.2.9), we obtain for the first term in the right-hand side of (7.2.14):

$$\begin{aligned} &S(\mathsf{L},l)\otimes_{n=l+1}^{N} I_1^{(n)}\varphi \\ &= \sum_{i_1+\cdots+i_l=\mathsf{L}+l-1} \otimes_{n=1}^{l-1}\mathcal{U}_{i_k}^{(k)} \otimes (I_1^{(l)} - Q_{i_l}^{(l)}) \otimes_{n=l+1}^{N} I_1^{(n)}\varphi \\ &= \sum_{i_1+\cdots+i_l=\mathsf{L}+l-1} \otimes_{n=1}^{l-1}\mathcal{U}_{i_k}^{(k)} \otimes (I_1^{(l)} - Q_{i_l}^{(l)}) \\ &\quad \otimes \int_{\mathbb{R}^{N-l}} \varphi(y)\frac{1}{(2\pi)^{(N-l)/2}}\exp(-\sum_{k=l+1}^{N}\frac{y_k^2}{2})\,dy_{l+1}\dots dy_N \\ &= -\sum_{i_1+\cdots+i_l=\mathsf{L}+l-1} \otimes_{n=1}^{l-1}\mathcal{U}_{i_k}^{(k)} \otimes \int_{\mathbb{R}^{N-l+1}} \frac{d^{2i_l}}{dy_l^{2i_l}}\varphi(y)R_{i_l,2i_l}(\Gamma_{y_l,2i_l}) \\ &\quad \times \frac{1}{(2\pi)^{(N-l)/2}}\exp(-\sum_{k=l+1}^{N}\frac{y_k^2}{2})\,dy_l\dots dy_N. \end{aligned}$$

Now consider two cases: if $i_{l-1} > 1$ then by (7.2.16):

$$\begin{aligned} S(\mathsf{L},l)\otimes_{n=l+1}^{N} I_1^{(n)}\varphi = &-\sum_{i_1+\cdots+i_l=\mathsf{L}+l-1} \\ &\otimes_{n=1}^{l-2}\mathcal{U}_{i_k}^{(k)} \otimes \int_{\mathbb{R}^{N-l+2}} \frac{d^{2i_{l-1}-2}}{dy_{l-1}^{2i_{l-1}-2}}\frac{d^{2i_l}}{dy_l^{2i_l}}\varphi(y)R_{i_l,2i_l}(\Gamma_{y_l,2i_l}) \\ &\times[R_{i_{l-1},2i_{l-1}-2}(\Gamma_{y_{l-1},2i_{l-1}-2}) - R_{i_{l-1}-1,2i_{l-1}-2} \\ &\qquad (\Gamma_{y_{i_{l-1}},2i_{l-1}-2})] \\ &\times\frac{1}{(2\pi)^{(N-l)/2}}\exp(-\sum_{k=l+1}^{N}\frac{y_k^2}{2})\,dy_{l-1}\dots dy_N \end{aligned}$$

otherwise (i.e., if $i_{l-1}=1$) by (7.2.17):

$$
\begin{aligned}
S(\mathsf{L},l)\otimes_{n=l+1}^{N} I_1^{(n)}\varphi = -\sum_{i_1+\cdots+i_l=\mathsf{L}+l-1} &\otimes_{n=1}^{l-2}\mathcal{U}_{i_k}^{(k)} \\
&\otimes\int_{\mathbb{R}^{N-l+1}} Q_1^{(l-1)}\frac{d^{2i_l}}{dy_l^{2i_l}}\varphi(y)R_{i_l,2i_l}(\Gamma_{y_l,2i_l}) \\
&\times\frac{1}{(2\pi)^{(N-l)/2}}\exp(-\sum_{k=l+1}^{N}\frac{y_k^2}{2})\,dy_l\ldots dy_N.
\end{aligned}
$$

Repeating the above process for $i_{l-2},\ldots,i_1$, we obtain

$$
\begin{aligned}
S(\mathsf{L},l)\otimes_{n=l+1}^{N} I_1^{(n)}\varphi = \sum_{i_1+\cdots+i_l=\mathsf{L}+l-1}\int_{\mathbb{R}^{N-\#F_{l-1}}}[\otimes_{m\in F_{l-1}}Q_1^{(m)}D^{2\alpha_l}\varphi(y)] \qquad (7.2.18) \\
\times\mathcal{R}_{l,\alpha_l}(y_1,\ldots,y_l)\frac{1}{(2\pi)^{(N-l)/2}}\exp(-\sum_{k=l+1}^{N}\frac{y_k^2}{2})\prod_{n\in G_{l-1}}dy_n\times dy_l\ldots dy_N,
\end{aligned}
$$

where the multi-index $\alpha_l=(i_1-1,\ldots,i_{l-1}-1,i_l,0,\ldots,0)$ with the m-th element α_l^m, the sets $F_{l-1}=F_{l-1}(\alpha_l)=\{m:\ \alpha_l^m=0,\ m=1,\ldots,l-1\}$ and $G_{l-1}=G_{l-1}(\alpha_l)=\{m:\ \alpha_l^m>0,\ m=1,\ldots,l-1\}$, the symbols $\#F_{l-1}$ and $\#G_{l-1}$ stand for the number of elements in the corresponding sets, and

$$
\begin{aligned}
\mathcal{R}_{l,\alpha_l}(y_1,\ldots,y_l) = -R_{i_l,2i_l}(\Gamma_{y_l,2i_l})\otimes_{n\in G_{l-1}}[&R_{i_n,2i_n-2}(\Gamma_{y_n,2i_n-2}) \\
&-R_{i_n-1,2i_n-2}(\Gamma_{y_n,2i_n-2})].
\end{aligned}
$$

Note that $\#G_{l-1}\le(\mathsf{L}-1)\wedge(l-1)$ and also recall that $i_j\ge1$, $j=1,\ldots,l$.
Using (7.2.10), (7.2.11) and the inequality

$$
\prod_{n\in G_{l-1}}[i_n^{-(i_n-1)}+(i_n-1)^{-(i_n-1)}]i_l^{-i_l}\le(3/2)^{\#G_{l-1}},
$$

we get

$$
\begin{aligned}
|\mathcal{R}_{l,\alpha}(y_1,\ldots,y_l)| &\le \prod_{n\in G_{l-1}}[i_n^{-(i_n-1)}+(i_n-1)^{-(i_n-1)}]i_l^{-i_l}\frac{c^{\#G_{l-1}+1}}{(2\pi)^{(\#G_{l-1}+1)/2}} \qquad (7.2.19) \\
&\quad\times\exp\left(-\sum_{n\in G_{l-1}}\frac{\beta y_n^2}{2}-\frac{\beta y_l^2}{2}\right) \\
&\le\frac{(3c/2)^{\#G_{l-1}+1}}{(2\pi)^{(\#G_{l-1}+1)/2}}\exp\left(-\sum_{n\in G_{l-1}}\frac{\beta y_n^2}{2}-\frac{\beta y_l^2}{2}\right).
\end{aligned}
$$

Substituting (7.2.19) in (7.2.18), we arrive at

$$\begin{aligned}
&\left|S(\mathsf{L},l)\otimes_{n=l+1}^{N} I_1^{(n)}\varphi\right| \qquad (7.2.20)\\
&\leq \sum_{i_1+\cdots+i_l=\mathsf{L}+l-1} \frac{(3c/2)^{\#G_{l-1}+1}}{(2\pi)^{(N-\#F_{l-1})/2}} \int_{\mathbb{R}^{N-\#F_{l-1}}} \left|\otimes_{m\in F_{l-1}} Q_1^{(m)} D^{2\alpha_l}\varphi(y)\right| \\
&\quad \times \exp\left(-\sum_{n\in G_{l-1}} \frac{\beta y_n^2}{2} - \frac{\beta y_l^2}{2} - \sum_{k=l+1}^{N} \frac{y_k^2}{2}\right) \prod_{n\in G_{l-1}} dy_n \times dy_l \ldots dy_N.
\end{aligned}$$

Using (7.2.6) and the assumption that $\left|\frac{d^{2\mathsf{L}}}{dx^{2\mathsf{L}}} f(x)\right| \leq K(1+|x|^{\varkappa})$ for some $K>0$ and $\varkappa \geq 1$, we get

$$\begin{aligned}
\left|D^{2\alpha_l}\varphi(y)\right| &= \varepsilon^{2L} h^{\mathsf{L}} \left|\frac{d^{2\mathsf{L}}}{dx^{2\mathsf{L}}} f(\bar{X}(N,y))\right| (1+\lambda h)^{2\mathsf{L}N-2\sum_{i=1}^{l} i\alpha_l^i} \qquad (7.2.21)\\
&\leq K\varepsilon^{2\mathsf{L}} h^{\mathsf{L}} (1+\lambda h)^{2\mathsf{L}N-2\sum_{i=1}^{l} i\alpha_l^i} (1+|\bar{X}(N,y)|^{\varkappa}).
\end{aligned}$$

Substituting (7.2.21) and (7.2.5) in (7.2.20) and doing further calculations, we obtain

$$\begin{aligned}
\left|S(\mathsf{L},l)\otimes_{n=l+1}^{N} I_1^{(n)}\varphi\right| &\leq K\varepsilon^{2\mathsf{L}} h^{\mathsf{L}} (1+e^{\lambda\varkappa T})(1+(3c/2)^{\mathsf{L}\wedge l})\beta^{-(\mathsf{L}\wedge l)/2} \\
&\qquad \sum_{i_1+\cdots+i_l=\mathsf{L}+l-1} (1+\lambda h)^{2\mathsf{L}N-2\sum_{i=1}^{l} i\alpha_l^i} \\
&\leq K\varepsilon^{2\mathsf{L}} h^{\mathsf{L}} (1+e^{\lambda(2\mathsf{L}+\varkappa)T})(1+(3c/2)^{\mathsf{L}\wedge l}) \\
&\qquad \beta^{-(L\wedge l)/2} \binom{\mathsf{L}+l-2}{\mathsf{L}-1} \\
&\leq K\varepsilon^{2\mathsf{L}} h^{\mathsf{L}} (1+e^{\lambda(2\mathsf{L}+\varkappa)T})(1+(3c/2)^{\mathsf{L}\wedge l})\beta^{-(\mathsf{L}\wedge l)/2} l^{\mathsf{L}-1}
\end{aligned} \qquad (7.2.22)$$

with a new $K>0$ which does not depend on h, ε, L, c, β, and l. In the last line of (7.2.22) we used

$$\binom{\mathsf{L}+l-2}{\mathsf{L}-1} = \prod_{i=1}^{L-1}(1+\frac{l-1}{i}) \leq \left[\frac{1}{\mathsf{L}-1}\sum_{i=1}^{\mathsf{L}-1}(1+\frac{l-1}{i})\right]^{\mathsf{L}-1} \leq l^{\mathsf{L}-1}.$$

Substituting (7.2.22) in (7.2.14) and observing that $\left|(I_1^{(1)} - Q_{\mathsf{L}}^{(1)})\otimes_{k=2}^{N} I_1^{(k)}\varphi\right|$ is of order $O(h^{\mathsf{L}})$, we arrive at (7.2.12).

Remark 7.2.4 *Due to Examples 7.2.1 and 7.2.2, the error estimate* (7.2.12) *proved in Proposition 7.2.3 is quite sharp and we conclude that in general the SCM algorithm for weak approximation of SDE does not converge by either decreasing the time step h or by increasing the level L. At the same time,* *the algorithm is convergent in* L *(when* $\mathsf{L} \leq N$*) if* $\varepsilon^2 T$ *is sufficiently small and SDE has some stable behavior (e.g.,* $\lambda \leq 0$*). Furthermore, the algorithm is sufficiently accurate when noise intensity* ε *and integration time* T *are relatively small.*

Remark 7.2.5 *It follows from the proof (see* (7.2.21)*) that if* $\frac{d^{2\mathsf{L}}}{dx^{2\mathsf{L}}} f(x) = 0$ *then the error* $I_N(\varphi) - A(\mathsf{L}, N)\varphi = 0$*. We emphasize that this is a feature of the linear SDE* (7.2.3) *thanks to* (7.2.6)*, while in the case of nonlinear SDE this error remains of the form* (7.2.12) *even if the* $2\mathsf{L}$*-th derivative of* f *is zero. See also the discussion at the end of Example 7.2.1 and numerical tests in Example 7.4.1.*

7.3 Recursive collocation algorithm for linear SPDEs

In the previous section we have demonstrated the limitations of SCM algorithms in application to SDEs that, in general, such an algorithm will not work unless the integration time T and the magnitude of noise are small. It is not difficult to understand that SCM algorithms have the same limitations in the case of SPDE as well, which, in particular, is demonstrated in Example 7.4.2, where a stochastic Burgers equation is considered. To treat this deficiency and achieve longer time integration in the case of linear SPDE, we will exploit the idea of the recursive approach presented in Chapter 6 and in [315, 505] in the case of a Wiener chaos expansion method. To this end, we apply the algorithm of SCM in conjunction with a time discretization of SPDE over a small interval $[(k-1)h, kh]$ instead of the whole interval $[0, T]$ as we did in the previous section, and build a recursive scheme to compute the second-order moments of the solutions to linear SPDE.

Consider the linear SPDE (3.3.23) with finite dimensional noises. We will continue to use the notation from the previous section: h is a step of uniform discretization of the interval $[0, T]$, $N = T/h$ and $t_k = kh$, $k = 0, \ldots, N$. We apply the midpoint rule in time to the SPDE (3.3.23):

$$\begin{aligned} u^{k+1}(x) &= u^k(x) + h[\tilde{\mathcal{L}} u^{k+1/2}(x) - \frac{1}{2}\sum_{l=1}^{r} \mathcal{M}_l g_l(x) + f(x)] \qquad (7.3.1) \\ &\quad + \sum_{l=1}^{r} \left[\mathcal{M}_l u^{k+1/2}(x) + g_l(x)\right] \sqrt{h}\, (\xi_{lh})_{k+1}, \; x \in \mathcal{D}, \\ u^0(x) &= u_0(x), \end{aligned}$$

where $u^k(x)$ approximates $u(t_k, x)$, $u^{k+1/2} = (u^{k+1} + u^k)/2$, and $(\xi_{lh})_k$ are i.i.d. random variables so that

$$\xi_h = \begin{cases} \xi, \; |\xi| \leq A_h, \\ A_h, \; \xi > A_h, \\ -A_h, \; \xi < -A_h, \end{cases} \qquad (7.3.2)$$

with $\xi \sim \mathcal{N}(0,1)$ and $A_h = \sqrt{2p|\ln h|}$ with $p \geq 1$. We note that the cut-off of the Gaussian random variables is needed in order to ensure that the implicitness of (7.3.1) does not lead to nonexistence of the second moment of $u^k(x)$ [356, 358]. Based on standard numerical methods for SDEs [358], it is natural to expect that under some regularity assumptions on the coefficients and the initial condition of (3.3.23), the approximation $u^k(x)$ from (7.3.1) converges with order 1/2 in the mean-square sense and with order 1 in the weak sense; in the latter case one can use discrete random variables $\zeta_{l,k+1}$ from (3.2.17) instead of $(\xi_{lh})_{k+1}$ (see also, e.g., [109, 167, 261] but we are not proving such a result here).

In the following, we denote for convenience that $u_H^k(x;\phi(\cdot)) = u_H^k(x;\phi(\cdot);(\xi_{lh})_k, l=1,\ldots,r)$ for the approximation (7.3.1) of the solution $u(t_k,x)$ to the SPDE (3.3.23) with $f(x)=0$ and $g_l(x)=0$ for all l (homogeneous SPDE) and with the initial condition $\phi(\cdot)$ prescribed at time $t=t_{k-1}$; $u_O^k(x) = u_O^k(x;(\xi_{lh})_k, l=1,\ldots,r)$ for the approximation (7.3.1) of the solution $u(t_k,x)$ to the SPDE (3.3.23) with the initial condition $\phi(x)=0$ prescribed at time $t=t_{k-1}$. Note that $u_O^k(x)=0$ if $f(x)=0$ and $g_l(x)=0$ for all l.

Let $\{e_i\} = \{e_i(x)\}_{i\geq 1}$ be a complete orthonormal system (CONS) in $L^2(\mathcal{D})$ with boundary conditions satisfied and $(\cdot,\cdot)$ be the inner product in that space. Then we can write

$$u^{k-1}(x) = \sum_{i=1}^{\infty} c_i^{k-1} e_i(x) \tag{7.3.3}$$

with $c_i^{k-1} = (u^{k-1}, e_i)$ and, due to the SPDE's linearity:

$$u^k(x) = u_O^k(x) + \sum_{i=1}^{\infty} c_i^{k-1} u_H^k(x; e_i(\cdot)).$$

We have

$$c_l^0 = (u_0, e_l), \quad c_l^k = q_{Ol}^k + \sum_{i=1}^{\infty} c_i^{k-1} q_{Hli}^k, \quad l=1,2,\ldots, \quad k=1,\ldots,N,$$

where $q_{Ol}^k = (u_O^k, e_l)$ and $q_{Hli}^k = (u_H^k(\cdot; e_i), e_l(\cdot))$.

Using (7.3.3), we represent the second moment of the approximation $u^k(x)$ from (7.3.1) of the solution $u(t_k,x)$ to the SPDE (3.3.23) as follows

$$\mathbb{E}[u^k(x)]^2 = \sum_{i,j=1}^{\infty} C_{ij}^k e_i(x) e_j(x), \tag{7.3.4}$$

where the covariance matrix $C_{ij}^k = \mathbb{E}[c_i^k c_j^k]$. Introducing also the means M_i^k, one can obtain the recurrent relations in k :

$$
\begin{aligned}
M_i^0 &= c_i^0 = (u_0, e_i), \quad C_{ij}^0 = c_i^0 c_j^0, \qquad (7.3.5)\\
M_i^k &= \mathbb{E}[q_{Oi}^k] + \sum_{l=1}^{\infty} M_l^{k-1}\mathbb{E}[q_{Hil}^k],\\
C_{ij}^k &= \mathbb{E}[q_{Oi}^k q_{Oj}^k] + \sum_{l=1}^{\infty} M_l^{k-1}\left(\mathbb{E}[q_{Oi}^k q_{Hjl}^k] + \mathbb{E}[q_{Oj}^k q_{Hil}^k]\right)\\
&\quad + \sum_{l,p=1}^{\infty} C_{lp}^{k-1}\mathbb{E}[q_{Hil}^k q_{Hjp}^k],\\
& i,j = 1,2,\ldots, \quad k = 1,\ldots,N.
\end{aligned}
$$

Since the coefficients of the SPDE (3.3.23) are time independent, none of the expectations involving the quantities q_{Oi}^k and q_{Hil}^k in (7.3.5) depend on k and hence it is sufficient to compute them just once, on a single step $k = 1$, and we get

$$
\begin{aligned}
M_i^0 &= c_i^0 = (u_0, e_i), \quad C_{ij}^0 = c_i^0 c_j^0, \qquad (7.3.6)\\
M_i^k &= \mathbb{E}[q_{Oi}^1] + \sum_{l=1}^{\infty} M_l^{k-1}\mathbb{E}[q_{Hil}^1],\\
C_{ij}^k &= \mathbb{E}[q_{Oi}^1 q_{Oj}^1] + \sum_{l=1}^{\infty} M_l^{k-1}\left(\mathbb{E}[q_{Oi}^1 q_{Hjl}^1] + \mathbb{E}[q_{Oj}^1 q_{Hil}^1]\right)\\
&\quad + \sum_{l,p=1}^{\infty} C_{lp}^{k-1}\mathbb{E}[q_{Hil}^1 q_{Hjp}^1],\\
& i,j = 1,2,\ldots, \quad k = 1,\ldots,N.
\end{aligned}
$$

These expectations can be approximated by quadrature rules from Chapter 2.5. If the number of noises r is small, then it is natural to use the tensor product rule (2.5.8) with one-dimensional Gauss–Hermite quadratures of order $n = 2$ or 3 (note that when $r = 1$, we can use just a one-dimensional Gauss–Hermite quadrature of order $n = 2$ or 3). If the number of noises r is large then it might be beneficial to use the sparse grid quadrature (2.5.9) of level $\mathsf{L} = 2$ or 3. More specifically,

$$
\begin{aligned}
\mathbb{E}[q_{Oi}^1] &\doteq \sum_{p=1}^{\eta}(u_O^1(\cdot;\mathsf{y}_p), e_i(\cdot))\mathsf{W}_p, \quad \mathbb{E}[q_{Hil}^1]\\
&\doteq \sum_{p=1}^{\eta}(u_H^1(\cdot;e_l;\mathsf{Y}_p), e_i(\cdot))\mathsf{W}_p, \qquad (7.3.7)\\
\mathbb{E}[q_{Oi}^1 q_{Oj}^1] &\doteq \sum_{p=1}^{\eta}(u_O^1(\cdot;\mathsf{Y}_p), e_i(\cdot))(u_O^1(\cdot;\mathsf{Y}_p), e_j(\cdot))\mathsf{W}_p,
\end{aligned}
$$

$$\mathbb{E}[q^1_{Oi}q^1_{Hjl}] \doteq \sum_{p=1}^{\eta}(u^1_O(\cdot;\mathsf{Y}_p), e_i(\cdot))(u^1_H(\cdot;e_l;\mathsf{Y}_p), e_j(\cdot))\mathsf{W}_p,$$

$$\mathbb{E}[q^1_{Hil}q^1_{Hjk}] \doteq \sum_{p=1}^{\eta}(u^1_H(\cdot;e_l;\mathsf{Y}_p), e_i(\cdot))(u^1_H(\cdot;e_k;\mathsf{Y}_p), e_j(\cdot))\mathsf{W}_p,$$

where $\mathsf{Y}_p \in \mathbb{R}^r$ are nodes of the quadrature, W_p are the corresponding quadrature weights, and $\eta = n^r$ in the case of the tensor product rule (2.5.8) with one-dimensional Gauss–Hermite quadratures of order n or η is the total number of nodes $\#S$ used by the sparse-grid quadrature (2.5.9) of level L. To find $u^1_O(x;\mathsf{Y}_p)$ and $u^1_H(x;e_l;\mathsf{Y}_p)$, we need to solve the corresponding elliptic PDE problems, which we do by using the spectral method in physical space, i.e., using a truncation of the CONS $\{e_l\}_{l=1}^{l_*}$ to represent the numerical solution.

To summarize, we formulate the following deterministic recursive algorithm for the second-order moments of the solution to the SPDE (3.3.23).

Algorithm 7.3.1 *Choose the algorithm's parameters: a complete orthonormal basis $\{e_l(x)\}_{l\geq 1}$ in $L^2(\mathcal{D})$ and its truncation $\{e_l(x)\}_{l=1}^{l_*}$; a time step size h; and a quadrature rule (i.e., nodes Y_p and the quadrature weights W_p, $p = 1,\ldots,\eta$).*

Step 1. For each $p = 1,\ldots,\eta$ and $l = 1,\ldots,l_*$, find approximations $\bar{u}^1_O(x;\mathsf{Y}_p) \approx u^1_O(x;\mathsf{y}_p)$ and $\bar{u}^1_H(x;e_l;\mathsf{Y}_p) \approx u^1_H(x;e_l;\mathsf{Y}_p)$ using the spectral method in physical space.

Step 2. Using the quadrature rule, approximately find the expectations as in (7.3.7) but with the approximate $\bar{u}^1_O(x;\mathsf{Y}_p)$ and $\bar{u}^1_H(x;e_l;\mathsf{Y}_p)$ instead of $u^1_O(x;\mathsf{Y}_p)$ and $u^1_H(x;e_l;\mathsf{Y}_p)$, respectively.

Step 3. Recursively compute the approximations of the means M_i^k, $i = 1,\ldots,l_*$, and covariance matrices $\{C_{ij}^k,\ i,j = 1,\ldots,l_*\}$ for $k = 1,\ldots,N$ according to (7.3.6) with the approximate expectations found in Step 2 instead of the exact ones.

Step 4. Compute the approximation of the second-order moment $\mathbb{E}[u^k(x)]^2$ using (7.3.4) with the approximate covariance matrix found in Step 3 instead of the exact one $\{C_{ij}^k\}$.

We emphasize that Algorithm 7.3.1 for computing moments does not have a statistical error. Error analysis of a slightly different version of this algorithm will be considered in Chapter 8.

Remark 7.3.2 *Algorithms analogous to Algorithm 7.3.1 can also be constructed based on other time-discretizations methods than the trapezoidal rule used here or based on other types of SPDE approximations, e.g., one can exploit the Wong-Zakai approximation as we will do in Chapter 8.*

Remark 7.3.3 *The cost of this algorithm is, similar to the algorithm in [505], $\frac{T}{\Delta}\eta l_*^4$ and the storage is ηl_*^2. The total cost can be reduced by employing some reduced order methods in physical space and be proportional to l_*^2 instead of l_*^4. The discussion on computational efficiency of the recursive Wiener chaos method is also valid here, see [505, Remark 4.1].*

7.4 Numerical results

In this section we illustrate via three examples how the SCM algorithms can be used for the weak-sense approximation of SDEs and SPDEs. The first example is a scalar SDE with multiplicative noise, where we show that the SCM algorithm's error is small when the noise magnitude is small. We also observe that when the noise magnitude is large, the SCM algorithm does not work well. In the second example we demonstrate that the SCM can be successfully used for simulating Burgers equation with additive noise when the integration time is relatively small. In the last example we show that the recursive algorithm from Chapter 7.3 works effectively for computing moments of the solution to an advection-diffusion equation with multiplicative noise over a longer integration time.

In all the tests we limit the dimension of random spaces by 40, which is an empirical limitation of the SCM of Smolyak on the dimensionality [393]. Also, we take the sparse grid level less than or equal to five in order to avoid an excessive number of sparse grid points. All the tests were run using Matlab R2012b on a Macintosh desktop computer with Intel Xeon CPU E5462 (quad-core, 2.80 GHz).

Example 7.4.1 (Modified Cox-Ingersoll-Ross (mCIR), see, e.g., [83]) *Consider the Ito SDE*

$$dX = -\theta_1 X \, dt + \theta_2 \sqrt{1 + X^2} \, dW(t), \quad X(0) = x_0. \tag{7.4.1}$$

For $\theta_2^2 - 2\theta_1 \neq 0$, the first two moments of $X(t)$ are equal to

$$\mathbb{E}X(t) = x_0 \exp(-\theta_1 t), \quad \mathbb{E}X^2(t) = -\frac{\theta_2^2}{\theta_2^2 - 2\theta_1} + (x_0^2 + \frac{\theta_F 2^2}{\theta_2^2 - 2\theta_1}) \exp((\theta_2^2 - 2\theta_1)t).$$

In this example we test the SCM algorithms based on the Euler scheme (3.2.1) and on the second-order weak scheme (3.2.23). We compute the first two moments of the SDE's solution and use the relative errors to measure accuracy of the algorithms as

$$\rho_1^r(T) = \frac{|\mathbb{E}X(T) - \mathbb{E}X_N|}{|\mathbb{E}X(T)|}, \quad \rho_2^r(T) = \frac{|\mathbb{E}X^2(T) - \mathbb{E}X_N^2|}{\mathbb{E}X^2(T)}. \tag{7.4.2}$$

Table 7.1 presents the errors for the SCM algorithms based on the Euler scheme (left) and on the second-order scheme (3.2.23) (right), when the noise magnitude is small. For the parameters given in the table's description, the exact values (up to 4 d.p.) of the first and second moments are 3.679×10^{-2} and 4.162×10^{-2}, respectively. We see that increase of the SCM level L above 2 in the Euler scheme case and above 3 in the case of the second-order scheme does not improve accuracy. When the SCM error is relatively small in comparison with the error due to time discretization, we observe decrease of the overall error of the algorithms in h: proportional to h for the Euler scheme and to h^2 for the second-order scheme. We underline that in this experiment the noise magnitude is small.

Table 7.1. Comparison of the SCM algorithms based on the Euler scheme (top) and on the second-order scheme (3.2.23) (bottom). The parameters of the model (7.4.1) are $x_0 = 0.1$, $\theta_1 = 1$, $\theta_2 = 0.3$, and $T = 1$.

h	L	$\rho_1^r(1)$	order	$\rho_2^r(1)$	order
5×10^{-1}	2	3.20×10^{-1}	–	3.72×10^{-1}	–
2.5×10^{-1}	2	1.40×10^{-1}	1.2	1.40×10^{-1}	1.4
1.25×10^{-1}	2	6.60×10^{-2}	1.1	4.87×10^{-2}	1.5
6.25×10^{-2}	2	3.21×10^{-2}	1.0	8.08×10^{-3}	2.6
3.125×10^{-2}	2	1.58×10^{-2}	1.0	1.12×10^{-2}	-0.5
2.5×10^{-2}	2	1.26×10^{-2}		1.49×10^{-2}	
2.5×10^{-2}	3	1.26×10^{-2}		1.48×10^{-2}	
2.5×10^{-2}	4	1.26×10^{-2}		1.55×10^{-2}	
2.5×10^{-2}	5	1.26×10^{-2}		1.56×10^{-2}	

h	L	$\rho_1^r(1)$	order	$\rho_2^r(1)$	order
5×10^{-1}	3	6.05×10^{-2}	–	8.52×10^{-2}	–
2.5×10^{-1}	3	1.14×10^{-2}	2.4	2.10×10^{-2}	2.0
1.25×10^{-1}	3	1.75×10^{-3}	2.7	6.73×10^{-3}	1.6
6.25×10^{-2}	4	3.64×10^{-4}	2.3	1.21×10^{-3}	2.5
3.125×10^{-2}	4	8.48×10^{-4}	-1.2	3.75×10^{-4}	1.7
2.5×10^{-2}	2	9.02×10^{-4}		5.72×10^{-2}	
2.5×10^{-2}	3	9.15×10^{-5}		2.84×10^{-3}	
2.5×10^{-2}	4	1.06×10^{-4}		2.77×10^{-4}	
2.5×10^{-2}	5	1.06×10^{-4}		1.81×10^{-4}	

In Table 7.2 we give results of the numerical experiment when the noise magnitude is not small. For the parameters given in the table's description, the exact values (up to 4 d.p.) of the first and second moments are 0.2718 and 272.3202, respectively. Although for the Euler scheme there is a proportional to h decrease of the error in computing the mean, there is almost no decrease of the error in the rest of this experiment. The large value of the second moment apparently affects the efficiency of the SCM here. For the Euler scheme, increasing L and decreasing h can slightly improve accuracy in computing the second moment, e.g., the smallest relative error for the second moment is 56.88% when $h = 0.03125$ and $\mathsf{L} = 5$ (this level requires 750337 sparse grid points) out of the considered cases of $h = 0.5$, 0.25, 0.125, 0.0625, and 0.03125 and $\mathsf{L} \leq 5$. For the mean, increase of the level L from 2 to 3, 4 or 5 does not improve accuracy. For the second-order scheme (3.2.23), relative errors for the mean can be decreased by increasing L for a fixed h: e.g., for $h = 0.25$, the relative errors are 0.5121 0.1753, 0.0316, and 0.0086 when $\mathsf{L} = 2$, 3, 4, and 5, respectively.

We also see in Table 7.2 that the SCM algorithm based on the second-order scheme may not admit higher accuracy than the one based on the Euler scheme, e.g., for $h = 0.5,\ 0,25,\ 0.125$ the second-order scheme yields higher accuracy while the Euler scheme demonstrates higher accuracy for smaller $h = 0.0625$ and 0.03125. Further decrease in h was not considered because this would lead to increase of the dimension of the random space beyond 40 when the sparse grid of Smolyak (2.5.9) may fail and the SCM algorithm may also lose its competitive edge with Monte Carlo-type techniques.

Table 7.2. Comparison of the SCM algorithms based on the Euler scheme (left) and on the second-order scheme (3.2.23) (right). The parameters of the model (7.4.1) are $x_0 = 0.08$, $\theta_1 = -1$, $\theta_2 = 2$, and $T = 1$. The sparse grid level $\mathsf{L} = 4$.

h	$\rho_1^r(1)$	Order	$\rho_2^r(1)$	$\rho_1^r(1)$	$\rho_2^r(1)$
5×10^{-1}	1.72×10^{-1}	–	9.61×10^{-1}	2.86×10^{-2}	7.69×10^{-1}
2.5×10^{-1}	1.02×10^{-1}	0.8	8.99×10^{-1}	8.62×10^{-3}	6.04×10^{-1}
1.25×10^{-1}	5.61×10^{-2}	0.9	7.87×10^{-1}	1.83×10^{-2}	7.30×10^{-1}
6.25×10^{-2}	2.96×10^{-2}	0.9	6.62×10^{-1}	3.26×10^{-2}	8.06×10^{-1}
3.125×10^{-2}	1.52×10^{-2}	1.0	5.64×10^{-1}	4.20×10^{-2}	8.40×10^{-1}

In this example, we have shown that the SCM algorithms based on first- and second-order schemes can produce sufficiently accurate results when the noise magnitude is small. The second-order scheme is preferable since for the same accuracy it uses random spaces of lower dimension than the first-order Euler scheme, compare, e.g., the error values highlighted by bold font in Table 7.1. When the noise magnitude is large (see Table 7.2), the SCM algorithms do not work well as it was predicted in Chapter 7.2.

Example 7.4.2 (Burgers equation with additive noise) *Consider the stochastic Burgers equation [93, 225]:*

$$du + u\partial_x u dt = \nu \partial_x^2 u dt + \sigma \cos(x) dW, \quad 0 \le x \le \ell, \quad \nu > 0 \tag{7.4.3}$$

with the initial condition $u_0(x) = 2\nu \frac{2\pi}{\ell} \frac{\sin(\frac{2\pi}{\ell}x)}{a+\cos(\frac{2\pi}{\ell}x)}$, $a > 1$, *and periodic boundary conditions. In the numerical tests the used values of the parameters are* $\ell = 2\pi$ *and* $a = 2$.

Apply the Fourier collocation method in physical space and the trapezoidal rule in time to (7.4.3):

$$\frac{\mathbf{u}_{j+1} - \mathbf{u}_j}{h} - \nu D^2 \frac{\mathbf{u}_{j+1} + \mathbf{u}_j}{2} = -\frac{1}{2} D (\frac{\mathbf{u}_{j+1} + \mathbf{u}_j}{2})^2 + \sigma \Gamma \sqrt{h} \xi_j, \tag{7.4.4}$$

where $\mathbf{u}_j = (u(t_j, x_1), \ldots, u(t_j, x_{\mathsf{M}}))^{\intercal}$, $t_j = jh$, D is the Fourier spectral differentiation matrix. The matrix D has entries $\partial_x L_i(x_j) = (-1)^{i-j} \cot((i-j)h/2)/2\delta_{i,j}$ and can be implemented in Matlab using a Toeplitz matrix as follows, see, e.g., [447, Chapter 3].

Code 7.1. Fourier spectral differentiation matrix

```
x_left=0; x_right=2*pi;  % l=2*pi
M=128;
h=(x_right-x_left)/M;
%set up  Fourier spectral differentiation matrix
column=[0  0.5*(-1).^(1:M-1).*cot((1:M-1)*h/2)];
D=toeplitz(column, column([1 M:-1:2]));
```

Also, ξ_j are i.i.d. $\mathcal{N}(0,1)$ random variables, and $\Gamma = (\cos(x_1),\ldots,\cos(x_{\mathsf{M}}))^{\mathsf{T}}$. The Fourier collocation points are $x_m = m\frac{\ell}{\mathsf{M}}$ $(m = 1,\ldots,\mathsf{M})$ in physical space and in the experiment we used $\mathsf{M} = 100$. We aim at computing moments of $\mathbf{u}_j$, which are integrals with respect to the Gaussian measure corresponding to the collection of ξ_j, and we approximate these integrals using the SCM from Chapter 2.5. The use of the SCM amounts to substituting ξ_j in (7.4.4) by sparse-grid nodes, which results in a system of (deterministic) nonlinear equations of the form (7.4.4). To solve the nonlinear equations, we used the fixed-point iteration method with tolerance $h^2/100$.

The errors in computing the first and second moments are measured as follows

$$\rho_1^{r,2}(T) = \frac{\|\mathbb{E}u_{\mathrm{ref}}(T,\cdot) - \mathbb{E}u_{\mathrm{num}}(T,\cdot)\|}{\|\mathbb{E}u_{\mathrm{ref}}(T,\cdot)\|}, \quad \rho_2^{r,2}(T) = \frac{\left\|\mathbb{E}u^2_{\mathrm{ref}}(T,\cdot) - \mathbb{E}u^2_{\mathrm{num}}(T,\cdot)\right\|}{\left\|\mathbb{E}u^2_{\mathrm{ref}}(T,\cdot)\right\|},$$
$$\rho_1^{r,\infty}(T) = \frac{\|\mathbb{E}u_{\mathrm{ref}}(T,\cdot) - \mathbb{E}u_{\mathrm{num}}(T,\cdot)\|_\infty}{\|\mathbb{E}u_{\mathrm{ref}}(T,\cdot)\|_\infty}, \quad \rho_2^{r,\infty}(T) = \frac{\left\|\mathbb{E}u^2_{\mathrm{ref}}(T,\cdot) - \mathbb{E}u^2_{\mathrm{num}}(T,\cdot)\right\|_\infty}{\left\|\mathbb{E}u^2_{\mathrm{ref}}(T,\cdot)\right\|_\infty}, \tag{7.4.5}$$

where $\|v(\cdot)\| = \left(\frac{2\pi}{\mathsf{M}}\sum_{m=1}^{\mathsf{M}} v^2(x_m)\right)^{1/2}$, $\|v(\cdot)\|_\infty = \max_{1\le m\le \mathsf{M}} |v(x_m)|$, x_m are the Fourier collocation points, and u_{num} and u_{ref} are the numerical solution obtained by the SCM algorithm and the reference solution, respectively. The first and second moments of the reference solution u_{ref} were computed by the same solver in space and time (7.4.4) but accompanied by the Monte Carlo method with a large number of realizations ensuring that the statistical errors were negligible.

First, we choose $\nu = 0.1$ and $\sigma = 1$. We obtain the reference solution with $h = 10^{-4}$ and 1.92×10^6 Monte Carlo realizations. The corresponding statistical error is 1.004×10^{-3} for the mean (maximum of the statistical error for $\mathbb{E}u_{\mathrm{ref}}(0.5, x_j)$) and 9.49×10^{-4} for the second moment (maximum of the statistical error for $\mathbb{E}u^2_{\mathrm{ref}}(0.5, x_j)$) with 95% confidence interval, and the corresponding estimates of L^2-norm of the moments are $\|\mathbb{E}u_{\mathrm{ref}}(0.5,\cdot)\| \doteq 0.18653$ and $\left\|\mathbb{E}u^2_{\mathrm{ref}}(0.5,\cdot)\right\| \doteq 0.72817$. We see from the results of the experiment presented in Table 7.3 that for $\mathsf{L} = 2$ the error in computing the mean decreases when h decreases up to $h = 0.05$ but the accuracy does not improve with further decrease of h. For the second moment, we observe no improvement in accuracy with decrease of h. For $\mathsf{L} = 4$, the error in computing the second moment decreases with h. When $h = 0.0125$, increasing the sparse grid level improves the accuracy for the mean: $\mathsf{L} = 3$ yields $\rho_1^{r,2}(0.5) \doteq 9.45 \times 10^{-3}$ and $\mathsf{L} = 4$ yields $\rho_1^{r,2}(0.5) \doteq 8.34 \times 10^{-3}$. As seen in Table 7.3, increase of the level L also improves the accuracy for the second moment when $h = 0.05$, 0.25, and 0.125.

Second, we choose $\nu = 1$ and $\sigma = 0.5$. We obtain the first two moments of the reference u_{ref} using $h = 10^{-4}$ and the Monte Carlo method with 3.84×10^6 realizations. The corresponding statistical error is 3.2578×10^{-4} for the mean and 2.2871×10^{-4} for the second moment with 95% confidence interval, and the corresponding estimates of L^2-norm of the moments are $\|\mathbb{E}u_{\text{ref}}(0.5, \cdot)\| \doteq 1.11198$ and $\|\mathbb{E}u_{\text{ref}}^2(0.5, \cdot)\| \doteq 0.66199$.

The results of the experiment are presented in Table 7.4. We see that accuracy is sufficiently high and there is some decrease of errors with decrease of time step h. However, as expected, no convergence in h is observed and further numerical tests (not presented here) showed that taking h smaller than 1.25×10^{-2} and level $\mathsf{L} = 2$ or 3 does not improve accuracy. In additional experiments we also noticed that there was no improvement of accuracy for the mean when we increased the level L up to 5. For the second moment, we observed some improvement in accuracy when L increases from 2 to 3 (see Table 7.4) but additional experiments (not presented here) showed that further increase of L (up to 5) did not reduce the errors.

For the errors measured in L^∞-norm (7.4.5) we had similar observations (not presented here) as in the case of L^2-norm.

In summary, this example has illustrated that SCM algorithms can produce accurate results in finding moments of solutions of nonlinear SPDE when the integration time is relatively small. Comparing Tables 7.3 and 7.4, we observe better accuracy for the first two moments when the magnitude of noise is smaller. In some situations, higher sparse grid levels L improve accuracy but dependence of errors on L is not monotone. No convergence in time step h or in level L was observed, which is consistent with the theoretical prediction in Chapter 7.2.

Table 7.3. Errors of the SCM algorithm to the stochastic Burgers equation (7.4.3) with parameters $T = 0.5$, $\nu = 0.1$ and $\sigma = 1$. Left: $\rho_1^{r,2}(0.5)$; Right: $\rho_2^{r,2}(0.5)$.

h	$\mathsf{L} = 2$	$\mathsf{L} = 3$	$\mathsf{L} = 2$	$\mathsf{L} = 3$	$\mathsf{L} = 4$
2.5×10^{-1}	1.28×10^{-1}	1.37×10^{-1}	4.01×10^{-2}	1.05×10^{-2}	1.25×10^{-2}
1.00×10^{-1}	4.70×10^{-2}	5.39×10^{-2}	4.48×10^{-2}	4.82×10^{-3}	4.69×10^{-3}
5.00×10^{-2}	2.75×10^{-2}	2.73×10^{-2}	4.73×10^{-2}	5.89×10^{-3}	2.82×10^{-3}
2.50×10^{-2}	2.51×10^{-2}	1.481×10^{-2}	4.87×10^{-2}	6.92×10^{-3}	2.34×10^{-3}
1.25×10^{-2}	2.67×10^{-2}	9.45×10^{-3}	4.95×10^{-2}	7.51×10^{-3}	2.29×10^{-3}

Table 7.4. Errors of the SCM algorithm applied to the stochastic Burgers equation (7.4.3) with parameters $\nu = 1$, $\sigma = 0.5$, and $T = 0.5$.

h	$\rho_1^{r,2}(0.5)$, $\mathsf{L} = 2$	$\rho_2^{r,2}(0.5)$, $\mathsf{L} = 2$	$\rho_2^{r,2}(0.5)$, $\mathsf{L} = 3$
2.5×10^{-1}	4.94×10^{-3}	8.75×10^{-3}	8.48×10^{-3}
1×10^{-1}	8.20×10^{-4}	1.65×10^{-3}	1.13×10^{-3}
5×10^{-2}	4.88×10^{-4}	1.18×10^{-3}	6.47×10^{-4}
2.5×10^{-2}	3.83×10^{-4}	1.08×10^{-3}	5.01×10^{-4}
1.25×10^{-2}	3.45×10^{-4}	1.07×10^{-3}	4.26×10^{-4}

Example 7.4.3 (Stochastic advection-diffusion equation) *Consider the stochastic advection-diffusion equation in the Ito sense:*

$$du = \left(\frac{\epsilon^2 + \sigma^2}{2} \partial_x^2 u + \beta \sin(x) \partial_x u \right) dt + \sigma \partial_x u \, dW(s), \quad (t,x) \in (0,T] \times (0, 2\pi), \tag{7.4.6}$$

$$u(0,x) = \phi(x), \quad x \in (0, 2\pi),$$

where $w(s)$ is a standard scalar Wiener process and $\epsilon \geq 0$, β, and σ are constants. In the tests we took $\phi(x) = \cos(x)$, $\beta = 0.1$, $\sigma = 0.5$, and $\epsilon = 0.2$.

We apply Algorithm 7.3.1 to (7.4.6) to compute the first two moments at a relatively large time $T = 5$. The Fourier basis was taken as CONS in physical space. Since (7.4.6) has a single noise only, we used one-dimensional Gauss–Hermite quadratures of order n. The implicitness due to the use of the trapezoidal rule was resolved by the fixed-point iteration with stopping criterion $h^2/100$.

As we have no exact solution of (7.4.6), we chose to find the reference solution by Algorithm 6.1.4 (multistage WCE method with the trapezoidal rule in time and Fourier collocation method in physical space) with the following parameters: the number of Fourier collocation points $\mathsf{M} = 30$, the length of time subintervals for the recursion procedure $h = 10^{-4}$, the highest order of Hermite polynomials $\mathsf{N} = 4$, the number of modes approximating the Wiener process $n = 4$, and the time step in the trapezoidal rule $h = 10^{-5}$. We obtain the second moment in the L^2-norm $\|\mathbb{E}u_{\text{ref}}^2(1,\cdot)\| \doteq 1.065195$. The errors are computed as follows

$$\varrho_2^2(T) = \left| \|\mathbb{E}u_{\text{ref}}^2(T,\cdot)\| - \|\mathbb{E}u_{\text{numer}}^2(T,\cdot)\| \right|, \quad \varrho_2^{r,2}(T) = \frac{\varrho_2^2(T)}{\|\mathbb{E}u_{\text{ref}}^2(T,\cdot)\|}, \tag{7.4.7}$$

where the norm is defined as in (7.4.5).

In Table 7.5, we observe first-order convergence in h for the second moments. We notice that increasing the quadrature order n from 2 to 3 does not improve accuracy which is expected. Indeed, the used trapezoidal rule is of weak order one in h in the case of multiplicative noise and more accurate quadrature rule cannot improve the order of convergence. This observation confirms that the total error should be expected to be $O(h)+O(h^{\mathsf{L}-1})$, which is proved in Chapter 8. In the case of additive noise, we expect to see the second order convergence in h when $n = 3$ due to the properties of the trapezoidal rule.

In conclusion, we showed that the recursive Algorithm 7.3.1 can work effectively for accurate computing of second moments of solutions to linear stochastic advection-diffusion equations at relatively large time. We observed convergence of order one in h.

Table 7.5. Errors in computing the second moment of the solution to the stochastic advection-diffusion equation (7.4.6) with $\sigma = 0.5$, $\beta = 0.1$, $\epsilon = 0.2$ at $T = 5$ by Algorithm 7.3.1 with $l_* = 20$ and the one-dimensional Gauss–Hermite quadrature of order $n = 2$ (left) and $n = 3$ (right).

h	$\varrho_2^{r,2}(5)$	Order	CPU time (sec.)	$\varrho_2^{r,2}(5)$	Order	CPU time (sec.)
5×10^{-2}	1.01×10^{-3}	–	7.41	1.06×10^{-3}	–	1.10×10
2×10^{-2}	4.07×10^{-4}	1.0	1.65×10	4.25×10^{-4}	1.0	2.43×10
1×10^{-2}	2.04×10^{-4}	1.0	3.43×10	2.12×10^{-4}	1.0	5.10×10
5×10^{-3}	1.02×10^{-4}	1.0	6.81×10	1.06×10^{-4}	1.0	1.00×10^{2}
2×10^{-3}	4.08×10^{-5}	1.0	1.70×10^{2}	4.25×10^{-5}	1.0	2.56×10^{2}
1×10^{-3}	2.04×10^{-5}	1.0	3.37×10^{2}	2.12×10^{-5}	1.0	5.12×10^{2}

7.5 Summary and bibliographic notes

We show when and how well Smolyak's sparse grid collocation works for a model problem in theory and for several problems in simulations. We also combine the collocation methods with the recursive strategy for longer-time integration of linear stochastic parabolic equations. The key points of this chapter are:

- We show that for the linear SODE (7.2.3) for computing expectations of solutions, the sparse grid collocation method of fixed level with the Euler scheme in time is convergent only when $\varepsilon^2 T$ is small, where ε is the noise intensity (diffusion coefficients) and T is the final integration time. See Proposition 7.2.3 and Chapter 7.2.2 for examples.
- The sparse grid collocation is not convergent in time step size. To have a convergent scheme in time, we apply the recursive strategy in [315] and in Chapter 6 and develop Algorithm 7.3.1 to compute the first two moments of solutions to linear stochastic advection-diffusion equations at relatively large time.

In the next chapter, we will compare theoretically and numerically the recursive WCE and the recursive SCM for linear problems.

Bibliographic notes. For SDE with time-dependent white noise, stochastic collocation methods, which are deterministic high-dimensional quadratures to evaluate integrals, have been discussed under different names: cubatures on Wiener space [322], de-randomization [373], optimal quantization [389, 390], and sparse grids of Smolyak type [153, 154, 172]. See a brief review in Chapter 2.5.4. The sparse grid of Smolyak type has been considered in [153, 154, 172, 398], where high accuracy was observed. However, the use of sparse grid in [153, 172] relies on exact sampling of geometric Brownian motion and of solutions of other simple SDE models.

While de-randomization and optimal quantization aim at finding quadrature rules which are in some sense optimal for computing a particular expectation under consideration, cubatures on Wiener space and a stochastic collocation method using Smolyak sparse grid quadratures use predetermined quadrature rules in a universal way without being tailored towards a specific expectation unless some adaptive strategies are applied. One of the major differences between SCM and other aforementioned methods is that SCM is endowed with negative weights. In practice, the difference leads to different numerical performance from those of cubatures on Wiener space, where only quadrature rules with positive weights are used.

For long-time simulation, deterministic methods for SDEs with white noise, e.g., stochastic collocation methods and functional expansion methods, do not work well unless some restarting strategies are employed. The fundamental obstacle is the increasing number of random variables induced by the discretization of Brownian motion, which requires significant reduction to compress history data. For ordinary SDEs, the approach proposed in [300] can be promising, compressing the history data via l_1 regression at each time step, which is summarized in the following two paragraphs.

Suppose at time t_k, we obtain a solution at cubature points x_i $(1 \leq i \leq n)$ and first m-th moments of the solution by the employed cubature rule. If we still use the same cubature rule at $(t_k, t_{k+1}]$, according to the definition of Brownian motion, we will have n^2 cubature points (the use of tensor product reflects Markovian property of the solution). To control the increase of the number of cubature points, we want to reduce the number of points at x_k by the following procedure:

- Choose a subset of $\{x_i\}_{i=1}^n$, say $\{\tilde{x}_k\}_{k=1}^{\tilde{n}}$ $(\tilde{n} < n)$ and some positive weights $\{\tilde{w}_k\}_{k=1}^n$ to form a new cubature rule such that the first m-moments obtained by the new cubature rule match the moments computed from the old cubature rule.

This procedure can be formulated as l_1 regression; see more details in [300]. As this requires compression at each time step, more analysis and reduction should be done to further reduce the computational cost. For SPDEs, it is not straightforward to extend the approach in [300] as we are dealing with a random field in both space and time. It seems that the recursive strategy proposed in [315] and in Chapter 6 may be the only feasible one. However, it is only efficient for the first two moments of solutions of linear equations. Hence, more effort should be put on long-time simulation with deterministic sampling methods.

7.6 Suggested practice

Exercise 7.6.1 *Apply the stochastic collocation method to solve the Kubo oscillator* (6.3.9) *and also its corresponding stochastic advection-diffusion equation* (3.3.18) *with periodic boundary condition. Use the midpoint scheme in time and Fourier spectral collocation method in space. Compare the accuracy when the levels of sparse grid are* $\mathsf{L} = 0, 1, 2, 3$.

Exercise 7.6.2 *Consider the following stochastic advection-diffusion-reaction equation*

$$du(t,x) = \epsilon u_{xx}(t,x)dt + f(u) + \sigma u_x(t,x) \circ dW(t), \quad t > 0, \;\; x \in (0, 2\pi), \tag{7.6.1}$$

$$u(0,x) = \sin(x),$$

where $f(u) = u - u^3$. *Use the midpoint scheme in time and Fourier spectral collocation method in space. Apply the sparse grid collocation in random space with the sparse grid level* $\mathsf{L} = 0, 1, 2, 3$. *Plot the mean and variance of the solution at different locations at* $t = k\delta$, $\delta = 0.1$, $k = 0, 1, 2, \cdots, 10$.

8

Comparison between Wiener chaos methods and stochastic collocation methods

In the last two chapters, we incorporated the recursive strategy into both Wiener chaos expansion (WCE) methods and stochastic collocation methods (SCM). In this chapter, we will compare both methods for *linear* stochastic advection-reaction-diffusion equations with commutative and noncommutative noises. To make a fair comparison, we develop a recursive multistage SCM using a spectral truncation of Brownian motion as in the case of WCE.

Numerical results demonstrate that the recursive multistage SCM is of order Δ (time step size) in the second-order moments while the recursive multistage WCE is of order $\Delta^{\mathsf{N}} + \Delta^2$ (N is the order of Wiener chaos) for advection-diffusion-reaction equations with commutative noises. These numerical results are in agreement with the theoretical error estimates. However, for noncommutative noises, both methods are of order one (Δ) in the second-order moments.

8.1 Introduction

In this chapter, we will show theoretically and through numerical examples that for white noise driven PDEs, WCE, and SCM have quite different performance when the noises are commutative. This is different than the case of color noise where WCE and SCM exhibit similar performance for smooth solutions.

To apply WCE and SCM and have a fair comparison between these two methods, we first discretize the Brownian motion with its truncated spectral expansion, see Chapter 2.2 and also, e.g., [391, Chapter IX] and [315], and subsequently we employ the corresponding functional expansion (WCE and SCM) to represent the solution in random space. In principle, we can employ any functional expansion, however, different expansions are preferred for different stochastic products because of computational efficiency.

Z. Zhang, G.E. Karniadakis, *Numerical Methods for Stochastic Partial Differential Equations with White Noise*,
Applied Mathematical Sciences 196, DOI 10.1007/978-3-319-57511-7_8

In practice, WCE is associated with the Ito-Wick product, see (3.3.15) and Chapter 11, as the product is defined with Wiener chaos modes yielding a weakly coupled system (lower-triangular system) of PDEs for linear equations. On the other hand, SCM is associated with the Stratonovich product, see (3.3.26), yielding a decoupled system of PDEs. These different formulations lead to different numerical performance as we demonstrate in Chapter 8.4; in particular, WCE can be of second-order convergence in time while SCM is only of first-order in time in the second-order moments for commutative noises. Further, when the noises serve as the advection coefficients, SCM can be more accurate than WCE when both methods are of first order convergence as the SCM (Stratonovich formulation) can lead to smaller diffusion coefficients than those for WCE (Ito-Wick formulation).

This chapter is organized as follows. After the Introduction section, we briefly revisit the WCE method and SCM for linear parabolic SPDEs, and present a new recursive SCM using a spectral truncation of Brownian motion, following the same recursive procedure as in Chapters 6 and 7. In Chapter 8.3, we present the error estimates for both methods for linear advection-diffusion-reaction equations and their proofs. In Chapter 8.4, we present numerical results of WCE and SCM for linear SPDEs with both commutative and non-commutative noises and verify the error estimates of WCE and SCM for commutative noises. We conclude in Chapter 8.5 and comments for multi-stage WCE and SCM. Four exercises are provided for readers to practice and compare multistage WCE and SCM for stochastic advection-diffusion-reaction equations in one dimension.

8.2 Review of Wiener chaos and stochastic collocation

In this section, we briefly revisit WCE and SCM for the linear SPDE (3.3.23). In both WCE and SCM, we discretize the Brownian motion using the spectral representations (2.2.1):

$$\lim_{\mathsf{n}\to\infty} \mathbb{E}[(W(t)-W^{(\mathsf{n})}(t))^2]=0, \quad W^{(\mathsf{n})}(t)=\sum_{i=1}^{\mathsf{n}}\int_0^t m_i(s)\,ds\xi_i,\ t\in[0,T], \tag{8.2.1}$$

where $\{m_i\}_{i=1}^{\infty}$ is a CONS in $L^2([0,T])$, and ξ_i are mutually independent standard Gaussian random variables.

8.2.1 Wiener chaos expansion (WCE)

The SPDE (3.3.23) with finite dimensional noises can be written in the following form using the Ito-Wick product

$$du(t,x) = [\mathcal{L}u(t,x) + f(x)]\,dt + \sum_{k=1}^{q} [\mathcal{M}_k u(t,x) + g_k(x)] \diamond \dot{W}_k\,dt,$$
$$(t,x) \in (0,T] \times \mathcal{D},$$
$$u(0,x) = u_0(x), \quad x \in \mathcal{D}, \tag{8.2.2}$$

where $\dot{W}_k$ is formally the first-order derivative of W_k in time, i.e., $\dot{W} = \frac{d}{dt}W$. To obtain the coefficients $\varphi_\alpha(t,x;\phi)$, we approximate W_k with the spectral truncation (8.2.1), $W_k^{(\mathsf{n})}$, and then we substitute the representation (3.3.14) into (3.3.23). By multiplying ξ_α on both sides of (3.3.23), and taking expectation with the properties of the Ito-Wick product $\xi_\alpha \diamond \xi_\beta = \sqrt{\frac{(\alpha+\beta)!}{\alpha!\beta!}}\xi_{\alpha+\beta}$ and $\mathbb{E}[\xi_\alpha\xi_\beta] = \delta_{\alpha=\beta}$, we then have that the coefficients $\varphi_\alpha(t,x;\phi)$ satisfy the propagator (6.1.3).

In practical computations, we are only interested in the truncated Wiener chaos solution (6.1.1). However, the error induced by the truncation of Wiener chaos expansion grows exponentially with time and thus WCE is not efficient for long-time integration. To control the error behavior, we can use the recursive WCE (see Algorithm 6.1.4) for computing the second moments, $\mathbb{E}[u^2(t,x)]$, of the solution of the SPDE (3.3.23). See Chapter 6 for more details.

Note that in Algorithm 6.1.4 we discretize the Brownian motion using the following spectral representation in a multi-element version, i.e., using K multi-elements:

$$w^{(\mathsf{n},\mathsf{K})}(t) = \sum_{k=1}^{\mathsf{K}} \sum_{i=1}^{\mathsf{n}} \int_{\mathsf{t}_{k-1}\wedge t}^{\mathsf{t}_k \wedge t} m_{i,k}(s)\,ds\xi_{i,k}, \ t \in [0,T], \tag{8.2.3}$$

where $0 = \mathsf{t}_0 < \mathsf{t}_1 < \cdots < \mathsf{t}_\mathsf{K} = T$, $\mathsf{t}_k \wedge t$ is the minimum of t_k and t, $\{m_{i,k}\}_{i=1}^{\infty}$ is a CONS in $L^2([\mathsf{t}_k, \mathsf{t}_{k+1}])$, and $\xi_{i,k}$ are mutually independent standard Gaussian random variables. This approximation of the Brownian motion will be also used in the stochastic collocation methods presented below.

8.2.2 Stochastic collocation method (SCM)

As we discussed in the last chapter, this method leads to a fully decoupled system instead of a weakly coupled system from the WCE. First, we rewrite the SPDE (3.3.23) with finite dimensional noises in Stratonovich form (3.3.26). Second, we replace the Brownian motion with its multi-element spectral expansion (8.2.3), and obtain the following partial differential equation with smooth random inputs:

$$d\tilde{u}(t,x) = [\tilde{\mathcal{L}}\tilde{u}(t,x) + f(x)]\,dt + \sum_{k=1}^{q} [\mathcal{M}_k \tilde{u}(t,x) + g_k(x)] \tag{8.2.4}$$
$$dW_k^{(\mathsf{n},\mathsf{K})}(t), \ (t,x) \in (0,T] \times \mathcal{D}, \tilde{u}(0,x) = u_0(x), \quad x \in \mathcal{D}.$$

Now we can apply standard numerical techniques of high integration to obtain p-th moments of the solution to (3.3.23)

$$\mathbb{E}[\tilde{u}^p_{\mathsf{n},\mathsf{K}}(x,t)] = \frac{1}{(2\pi)^{\mathsf{n}q\mathsf{K}/2}} \int_{\mathbb{R}^{\mathsf{n}q\mathsf{K}}} F(u_0(x),x,t,\mathbf{y}) e^{-\frac{\mathbf{y}^\top \mathbf{y}}{2}}\, d\mathbf{y}, \quad p=1,2,\cdots \tag{8.2.5}$$

where $\mathbf{y}=(y_{i,k,l})$, $i\leq \mathsf{n}, k\leq \mathsf{K}$, $l\leq q$ and the functional F represents the solution functional for (8.2.4). Here, we employ sparse grid collocation if the dimension nK is moderately large. As pointed out in [11, 486], we are led to a fully *decoupled* system of equations as in the case of Monte Carlo methods.

In practice, we use the sparse grid quadrature rule (2.5.9). Here again, the direct application of SCM is efficient only for short-time integration. To achieve long-time integration, we apply the recursive multistage idea used in Algorithm 6.1.4, i.e., we use SCM over small time interval $(\mathsf{t}_{i-1},\mathsf{t}_i]$ instead of over the whole interval $(0,T]$ and compute the second-order moments of the solution recursively in time. The derivation of such a recursive algorithm will make use of properties of the problem (3.3.23) and orthogonality of the basis both in physical space and in random space, as will be shown shortly.

We solve (8.2.4) with spectral methods in physical space, i.e., using a truncation of a CONS in physical space $\{e_m\}_{m=1}^{\mathsf{M}}$ to represent the numerical solution. The corresponding approximation of $\tilde{u}_{\Delta,\mathsf{n}}(t,x)$ is denoted by $\tilde{u}^{\mathsf{M}}_{\Delta,\mathsf{n}}(t,x)$. Further, let $v(t,x;s,v_0)$ be the approximation $\tilde{u}^{\mathsf{M}}_{\Delta,\mathsf{n}}(t,x)$ of $\tilde{u}_{\Delta,\mathsf{n}}(t,x)$ with the initial data v_0 prescribed at s: $\tilde{u}_{\Delta,\mathsf{n}}(s,x)=v_0(x)$. Note that

$$\tilde{u}^{\mathsf{M}}_{\Delta,\mathsf{n}}(\mathsf{t}_i,x) = v(\mathsf{t}_i,x;\mathsf{t}_{i-1},\tilde{u}^{\mathsf{M}}_{\Delta,\mathsf{n}}(\mathsf{t}_{i-1},\cdot)), \quad t_i = i\Delta. \tag{8.2.6}$$

Denote $\Phi_m(\mathsf{t}_i;\Delta,\mathsf{n},\mathsf{M}) = (\tilde{u}^{\mathsf{M}}_{\Delta,\mathsf{n}}(\mathsf{t}_i,\cdot),e_m)$. Then the second moments are computed by

$$\mathbb{E}[(\tilde{u}^{\mathsf{M}}_{\Delta,\mathsf{n}}(\mathsf{t}_i,x))^2] = \sum_{l,m=1}^{\mathsf{M}} H_{lm}(\mathsf{t}_i;\Delta,\mathsf{n},\mathsf{M}) e_l(x) e_m(x), \tag{8.2.7}$$

where $H_{lm}(\mathsf{t}_i;\Delta,\mathsf{n},\mathsf{M}) = \mathbb{E}[\Phi_l(\mathsf{t}_i;\Delta,\mathsf{n},\mathsf{M})\Phi_m(\mathsf{t}_i;\Delta,\mathsf{n},\mathsf{M})]$. Now we show how the matrix $H_{lm}(\mathsf{t}_i;\Delta,\mathsf{n},\mathsf{M})$ can be computed recursively. By the linearity of (8.2.4), we have

$$\tilde{u}^{\mathsf{M}}_{\Delta,\mathsf{n}}(\mathsf{t}_i,x) = \sum_{l=1}^{\mathsf{M}} \Phi_l(\mathsf{t}_{i-1};\Delta,\mathsf{n},\mathsf{M}) v(\mathsf{t}_i,x;\mathsf{t}_{i-1},e_l).$$

Denote $h_{l,m,i-1} = (v(\mathsf{t}_i,\cdot;\mathsf{t}_{i-1},e_l),e_m)$. Then by the orthonormality of e_m, we have

$$\Phi_m(\mathsf{t}_i;\Delta,\mathsf{n},\mathsf{M}) = \sum_{l=1}^{\mathsf{M}} \Phi_l(\mathsf{t}_{i-1};\Delta,\mathsf{n},\mathsf{M}) h_{l,m,i-1}.$$

The matrix $H_{lm}(\mathsf{t}_i; \Delta, \mathsf{n}, \mathsf{M})$ can be computed recursively as

$$H_{lm}(\mathsf{t}_i; \Delta, \mathsf{n}, \mathsf{M}) = \sum_{j=1}^{\mathsf{M}} \sum_{k=1}^{\mathsf{M}} H_{jk}(\mathsf{t}_{i-1}; \Delta, \mathsf{n}, \mathsf{M}) \mathbb{E}[h_{j,l,i-1} h_{k,m,i-1}].$$

We note that the expectation $\mathbb{E}[h_{j,l,i-1} h_{k,m,i-1}]$ does not depend on $i - 1$ because according to (8.2.4) and (3.3.24), $v(\mathsf{t}_i, x; \mathsf{t}_{i-1}, e_l)$ depend on the length of the time interval Δ and the random variables $\xi_{l,k,i}$ ($l \leq \mathsf{n}$, $k \leq q$) but is independent of time t_{i-1}. Denote $v(\mathsf{t}_i, \cdot; \mathsf{t}_{i-1}, e_l)$ with $\xi_{l,k,i}$ anchored at the sparse grid point $\mathsf{x}_\kappa \in \mathcal{H}_{\mathsf{L}}^{\mathsf{n}q}$ by $v_\kappa(\Delta, \cdot; e_l)$. Let $h_{\kappa,l,m} = (v_\kappa(\Delta, \cdot; e_l), e_m)$. Then, using the sparse grid quadrature rule (2.5.9), we obtain the recursive approximation of $H_{lm}(\mathsf{t}_i; \Delta, \mathsf{n}, \mathsf{M})$:

$$H_{lm}(\mathsf{t}_i; \Delta, \mathsf{n}, \mathsf{M}) \approx H_{lm}(\mathsf{t}_i; \Delta, \mathsf{L}, \mathsf{n}, \mathsf{M}) := \sum_{j=1}^{\mathsf{M}} \sum_{k=1}^{\mathsf{M}} H_{jk}(\mathsf{t}_{i-1}; \Delta, \mathsf{L}, \mathsf{n}, \mathsf{M}) \sum_{\kappa=1}^{\eta(\mathsf{L},\mathsf{n}q)} h_{\kappa,j,l} h_{\kappa,k,m} \mathsf{W}_\kappa. \tag{8.2.8}$$

Substituting (8.2.8) in (8.2.7), we obtain an approximation for the second moments of $u(t, x)$, denoted by $\mathbb{M}_{\Delta,\mathsf{L},\mathsf{n}}^{\mathsf{M}}(\mathsf{t}_i, x)$. When $\mathsf{M} = \infty$ (i.e., when the CONS $\{e_m\}$ is not truncated), we denote this approximation by $\mathbb{M}_{\Delta,\mathsf{L},\mathsf{n}}(\mathsf{t}_i, x)$.

Remark 8.2.1 *For nonhomogeneous equations, i.e., with forcing terms, we can have similar algorithms. Indeed, the same procedure applies once we can split the nonhomogeneous equations into two equations: nonhomogeneous equation with zero initial value and homogeneous equation with initial value. See Chapter 7.3 for a derivation of similar algorithms where only increments of Brownian motion are used, which is different from the spectral approximation of Brownian motion used here.*

Now we have the following algorithm for the second moments of the approximate solution when $f = g_k = 0$.

Algorithm 8.2.2 (Recursive multistage stochastic collocation method) *Choose a CONS $\{e_m(x)\}_{m\geq 1}$ and its truncation $\{e_m(x)\}_{m=1}^{\mathsf{M}}$; a time step Δ; the sparse grid level L and n, which together with the number of noises q determine the sparse grid $\mathcal{H}_{\mathsf{L}}^{\mathsf{n}q}$ which contains $\eta(\mathsf{L}, \mathsf{n}q)$ sparse grid points.*

Step 1. For each $m = 1, \ldots, \mathsf{M}$, solve the system of equations (8.2.4) on the sparse grid $\mathcal{H}_{\mathsf{L}}^{\mathsf{n}q}$ in the time interval $[0, \Delta]$ with the initial condition $\phi(x) = e_m(x)$ and denote the obtained solution as $v_\kappa(\Delta, x; e_m)$, $m = 1, \ldots, \mathsf{M}$, and $\kappa = 1, \cdots, \eta(\mathsf{L}, \mathsf{n}q)$. Also, choose a time step size δt to solve (8.2.4) numerically.

Step 2. Evaluate $h_{\kappa,l,m} = (v_\kappa(\Delta, \cdot; e_l), e_m)$, $l, m = 1, \ldots, \mathsf{M}$.

Step 3. Recursively compute the covariance matrices $H_{lm}(\mathsf{t}_i;\mathsf{L},\mathsf{n},\mathsf{M})$, $l,m=1,\ldots,\mathsf{M}$, *as follows:*

$$H_{lm}(0;\Delta,\mathsf{L},\mathsf{n},\mathsf{M}) = (u_0,e_l)(u_0,e_m),$$

$$H_{lm}(\mathsf{t}_i;\Delta,\mathsf{L},\mathsf{n},\mathsf{M}) = \sum_{j,k=1}^{\mathsf{M}} H_{jk}(\mathsf{t}_{i-1};\Delta,\mathsf{L},\mathsf{n},\mathsf{M}) \sum_{\kappa=1}^{\eta(\mathsf{L},\mathsf{n}q)} h_{\kappa,j,l}h_{\kappa,k,m}\mathsf{W}_\kappa,\ i=1,\ldots,\mathsf{K},$$

where $u_0(x)$ *is the initial condition for* (3.3.23) *and obtain the approximate second moments* $\mathbb{M}^{\mathsf{M}}_{\Delta,\mathsf{L},\mathsf{n}}(\mathsf{t}_i,x)$ *of the solution* $u(t,x)$ *to* (3.3.23) *as*

$$\mathbb{M}^{\mathsf{M}(\mathsf{t}_i,x)}_{\Delta,\mathsf{L},\mathsf{n}} = \sum_{l,m=1}^{\mathsf{M}} H_{lm}(\mathsf{t}_i;\Delta,\mathsf{L},\mathsf{n},\mathsf{M})e_l(x)e_m(x),\ i=1,\ldots,\mathsf{K}. \tag{8.2.9}$$

Remark 8.2.3 *Similar to Algorithm 6.1.4, the cost of this algorithm is* $\frac{T}{\Delta}\eta(\mathsf{L},\mathsf{n}q)\mathsf{M}^4$ *and the storage is* $\eta(\mathsf{L},\mathsf{n}q)\mathsf{M}^2$. *The total cost can be reduced to the order of* M^2 *by adopting some reduced order methods in physical space. The discussion on computational efficiency of the recursive WCE methods, see Chapter 6.1, is also valid for Algorithm 8.2.2.*

8.3 Error estimates

Although WCE and SCM use the same spectral truncation of Brownian motion, the former is associated with the Ito-Wick product while the latter is related to the Stratonovich product. Note that WCE employs orthogonal polynomials as a basis but SCM does not have such orthogonality. This difference enables WCE to have a better convergence rate than SCM in the second-order moments, see Corollary 8.3.2 and Theorem 8.3.6.

Assume that the operator $\mathcal{L}$ generates a semi-group $\{\mathcal{T}_t\}_{t\geq 0}$, which has the following properties: for $v\in H^k(\mathcal{D})$,

$$\|\mathcal{T}_t v\|^2_{H^k} \leq C(k,\mathcal{L})e^{2C_{\mathcal{L}}t}\|v\|^2_{H^k}, \tag{8.3.1}$$

where $C(0,\mathcal{L})=1$ and

$$\int_s^t e^{2C_{\mathcal{L}}(t-\theta)}\|\mathcal{T}_t v\|^2_{H^{k+1}}\,d\theta \leq \delta_{\mathcal{L}}^{-1}C(k,\mathcal{L})e^{2C_{\mathcal{L}}(t-s)}\|v\|^2_{H^k}. \tag{8.3.2}$$

Also, we assume that there exists a constant $\tilde{C}(r,\mathcal{M})$ such that

$$\|\mathcal{M}_l v\|^2_{H^k} \leq \tilde{C}(k,\mathcal{M})\|v\|^2_{H^{k+1}}, \quad \text{for } v\in H^{k+1},\ l=1,\ldots,q, \tag{8.3.3}$$

and that there exists a constant $\tilde{C}(k,\mathcal{L})$ such that

$$\|\mathcal{L}v\|_{H^k}^2 \le \tilde{C}(k,\mathcal{L}) \|v\|_{H^{k+2}}^2, \quad \text{for } v \in H^{k+2}. \tag{8.3.4}$$

The conditions (8.3.1) and (8.3.3) are satisfied with $k \le r$ and (8.3.4) is satisfied with $k \le r-1$ when the coefficients from (3.3.24) belong to the Hölder space $\mathcal{C}_b^{r+1}(\mathcal{D})$. Define also

$$C_k = \max_{1\le j\le k} \left\{ C(j,\mathcal{L})\tilde{C}(j-1,\mathcal{M}) \right\}. \tag{8.3.5}$$

8.3.1 Error estimates for WCE

For the WCE for the SPDE (3.3.23) with single noise ($q=1$), we have the convergence results stated below. In the general case, we have not succeeded in proving such theorems but we numerically check convergence orders using examples with commutative noises and noncommutative noises in Chapter 8.4.

Theorem 8.3.1 *Let $q = 1$ in (3.3.23). Assume that $\sigma_{i,1}, a_{i,j}, b_i, c, \nu_1$ in (3.3.24) belong to $\mathcal{C}_b^{r+1}(\mathcal{D})$ and $u_0 \in H^r(\mathcal{D})$, where $r \ge \mathsf{N}+2$ and N is the order of Wiener chaos. Also assume that (3.3.25) holds. Then for $C_1 < \delta_{\mathcal{L}}$, the error of the truncated Wiener chaos solution $u_{\mathsf{N},\mathsf{n}}(\mathsf{t}_i, x)$ from (6.1.1) is estimated as*

$$\begin{aligned}
&(\mathbb{E}[\|u_{\mathsf{N},\mathsf{n}}(\mathsf{t}_i,\cdot) - u(\mathsf{t}_i,\cdot)\|^2])^{1/2} \\
&\le (C_{\lfloor r\rfloor}\Delta)^{\mathsf{N}/2} e^{C_{\mathcal{L}}T} \left[\frac{e^{C_{\lfloor r\rfloor}T}}{(\mathsf{N}+1)!} + \frac{(C_{\lfloor r\rfloor}\Delta)^{\lfloor r\rfloor-\mathsf{N}-1}}{\lfloor r\rfloor!} \frac{\delta_{\mathcal{L}}}{\delta_{\mathcal{L}} - C_1} \right]^{1/2} \|u_0\|_{H^r} \\
&\quad + \sqrt{2C_{\mathsf{N}+2}C(\mathsf{N}+2,\mathcal{L})\tilde{C}(\mathsf{N},\mathcal{L})e^{C_{\mathsf{N}+2}T + C_{\mathcal{L}}T}} \frac{\Delta}{\sqrt{\mathsf{n}\pi}} \|u_0\|_{H^{\mathsf{N}+2}},
\end{aligned} \tag{8.3.6}$$

where $\mathsf{t}_i = i\Delta$, the constants $\delta_{\mathcal{L}}$ and $C_{\mathcal{L}}$ are from (3.3.25), $C_{\lfloor r\rfloor}$ is defined in (8.3.5), $\tilde{C}(\mathsf{N},\mathcal{L})$ is from (8.3.4), and $C(\mathsf{N}+2,\mathcal{L})$ is from (8.3.1).

From Theorem 8.3.1, we have that the mean-square error of the recursive multistage WCE is $\mathcal{O}(\Delta^{\mathsf{N}/2}) + \mathcal{O}(\Delta)$. The same result is proved for $q=1$ and $\sigma_{i,r} = 0$ in [315], where the condition $C_1 < \delta_{\mathcal{L}}$ is not required. Also, for the case of $\sigma_{i,r} \ne 0$, the mean-square convergence without order but not requiring the condition $C_1 < \delta_{\mathcal{L}}$ was proved in [314, 316].

Corollary 8.3.2 *Under the assumptions of Theorem 8.3.1, we have*

$$\begin{aligned}
&\left|\mathbb{E}[\|u_{\mathsf{N},\mathsf{n}}(\mathsf{t}_i,\cdot)\|^2] - \mathbb{E}[\|u(\mathsf{t}_i,\cdot)\|^2]\right| = \mathbb{E}[\|u_{\mathsf{N},\mathsf{n}}(\mathsf{t}_i,\cdot) - u(\mathsf{t}_i,\cdot)\|^2] \\
&\le (C_{\lfloor r\rfloor}\Delta)^{\mathsf{N}} e^{2C_{\mathcal{L}}T} \left[\frac{e^{C_{\lfloor r\rfloor}T}}{(\mathsf{N}+1)!} + \frac{(C_{\lfloor r\rfloor}\Delta)^{\lfloor r\rfloor-\mathsf{N}-1}}{\lfloor r\rfloor!} \frac{\delta_{\mathcal{L}}}{\delta_{\mathcal{L}} - C_1} \right] \|u_0\|_{H^r}^2 \\
&\quad + 2C_{\mathsf{N}+2}C(\mathsf{N}+2,\mathcal{L})\tilde{C}(\mathsf{N},\mathcal{L})e^{2C_{\mathsf{N}+2}T + 2C_{\mathcal{L}}T} \frac{\Delta^2}{\mathsf{n}\pi^2} \|u_0\|_{H^{\mathsf{N}+2}}^2.
\end{aligned} \tag{8.3.7}$$

This corollary states that the convergence rate of the error in second-order moments (8.3.7) is twice that of the mean-square error (8.3.6), i.e., $\mathcal{O}(\Delta^{\mathsf{N}}) + \mathcal{O}(\Delta^2)$. This corollary can be proved by the orthogonality of WCE. In fact, it holds that

$$\mathbb{E}[u^2(\mathsf{t}_i,x)] - \mathbb{E}[u^2_{\mathsf{N,n}}(\mathsf{t}_i,x)] = \mathbb{E}[(u(\mathsf{t}_i,x) - u_{\mathsf{N,n}}(\mathsf{t}_i,x))^2], \tag{8.3.8}$$

as the different terms in the Cameron-Martin basis are mutually orthogonal [57]. Then integration over the physical domain and by the Fubini Theorem (see Appendix D), we reach the conclusion in Theorem 8.3.1.

The idea for the proof Theorem 8.3.1 is to first establish an estimate for the one-step ($\Delta = T$) error where the global error can be readily derived from. We need the following two lemmas for the one-step errors. Introduce (cf. (3.3.14))

$$u_{\mathsf{N}}(t,x) = \sum_{|\alpha|\leq \mathsf{N},\, \alpha\in\mathcal{J}_q} \frac{1}{\sqrt{\alpha!}}\varphi_\alpha(t,x)\xi_\alpha. \tag{8.3.9}$$

Lemma 8.3.3 *Let $q = 1$ in (3.3.23). Assume that $\sigma_{i,1}, a_{i,j}, b_i, c, \nu_1$ belong to $C_b^{r+1}(\mathcal{D})$ and $u_0 \in H^r(\mathcal{D})$, where $r \geq \mathsf{N}+1$. Let u in (3.3.14) be the solution to (3.3.23) and u_{N} is from (8.3.9). For $C_1 < \delta_{\mathcal{L}}$, the following estimate holds*

$$\begin{aligned}
&\mathbb{E}[\|u(\Delta,\cdot) - u_{\mathsf{N}}(\Delta,\cdot)\|^2] \\
&\leq (C_{\lfloor r\rfloor}\Delta)^{\mathsf{N}+1} e^{2C_{\mathcal{L}}\Delta}\Big[\frac{e^{C_{\lfloor r\rfloor}\Delta}}{(\mathsf{N}+1)!} + \frac{(C_{\lfloor r\rfloor}\Delta)^{\lfloor r\rfloor-\mathsf{N}-1}}{\lfloor r\rfloor!}\frac{\delta_{\mathcal{L}}}{\delta_{\mathcal{L}}-C_1}\Big]\, \|u_0\|^2_{H^{\lfloor r\rfloor}},
\end{aligned}$$

where the constants $\delta_{\mathcal{L}}$ and $C_{\mathcal{L}}$ are from (3.3.25) and $C_{\lfloor r\rfloor}$ is from (8.3.5).

Lemma 8.3.4 *Under the assumptions of Lemma 8.3.3 and $r \geq \mathsf{N}+2$, we have*

$$\mathbb{E}[\|u_{\mathsf{N,n}}(\Delta,\cdot) - u_{\mathsf{N}}(\Delta,\cdot)\|^2] \leq \frac{2\Delta^3}{\mathsf{n}\pi^2} C(\mathsf{N}{+}2,\mathcal{L})\tilde{C}(\mathsf{N},\mathcal{L})C_{\mathsf{N}+2}e^{2C_{\mathsf{N}+2}\Delta+2C_{\mathcal{L}}\Delta}\, \|u_0\|^2_{H^{\mathsf{N}+2}},$$

where $C_{\mathcal{L}}$ is from (3.3.25), $C(\mathsf{N}{+}2,\mathcal{L})$ is from (8.3.1), $\tilde{C}(\mathsf{N},\mathcal{L})$ is from (8.3.4) and $C_{\mathsf{N}+2}$ is from (8.3.5).

Using Lemmas 8.3.3 and 8.3.4, we can establish the estimate of the global error stated in Theorem 8.3.1. Specifically, the one-step error is bounded by the sum of $\mathbb{E}[(u(\Delta){-}u_{\mathsf{N}}(\Delta))^2]$ and $\mathbb{E}[(u_{\mathsf{N}}(\Delta){-}u_{\mathsf{N,n}}(\Delta))^2]$, which are estimated in Lemmas 8.3.3 and 8.3.4. Then, the global error is estimated based on the recursion nature of Algorithm 6.1.4 as in the proof in [315, Theorem 2.4], which completes the proof of Theorem 8.3.1.

Now we proceed to proving Lemmas 8.3.3 and 8.3.4. Let us denote by $\mathbf{s}^k$ the ordered set $(s_1, \cdots, s_k)$ and for $k \geq 1$, denote $d\mathbf{s}^k := ds_1 \ldots ds_k$, and

$$\int^{(k)} (\cdots)\, d\mathbf{s}^k = \int_0^{\Delta}\int_0^{s_k}\cdots\int_0^{s_2} (\cdots)\, ds_1 \ldots ds_k,$$
$$\int_{(k)} (\cdots)\, d\mathbf{s}^k = \int_0^{\Delta}\int_{s_1}^{\Delta}\cdots\int_{s_{k-1}}^{\Delta} (\cdots)\, ds_k \ldots ds_2\, ds_1,$$

and $F(\Delta; \mathbf{s}^k; x) = \mathcal{T}_{\Delta - s_k}\mathcal{M}\cdots\mathcal{T}_{s_2-s_1}\mathcal{M}\mathcal{T}_{s_1}u_0(x)$, where $\mathcal{M} := \mathcal{M}_1$.

Proof of Lemma 8.3.3. It follows from (3.3.25) and the assumptions on the coefficients that (8.3.1) and (8.3.2) hold, cf. [125, Section 7.1.3]. Also, by the assumption that $\sigma_{i,1}, \nu_1$ belong to $\mathcal{C}_b^{r+1}(\mathcal{D})$, it can be readily checked that (8.3.3) holds.

By (3.3.14), (8.3.9), and orthogonality of ξ_α (see (6.1.7)), we have

$$\mathbb{E}[\|u(\Delta,\cdot) - u_{\mathsf{N}}(\Delta,\cdot)\|^2] = \sum_{k>N}\sum_{|\alpha|=k}\frac{\|\varphi_\alpha(\Delta,\cdot)\|^2}{\alpha!}.$$

Similar to the proof of Proposition A.1 in [315], we have

$$\sum_{|\alpha|=k}\frac{\varphi_\alpha^2(\Delta,x)}{\alpha!} = \int^{(k)} \left|F(\Delta;\mathbf{s}^k;x)\right|^2 d\mathbf{s}^k.$$

Then by the Fubini theorem (see Appendix D),

$$\sum_{|\alpha|=k}\frac{\|\varphi_\alpha(\Delta,\cdot)\|^2}{\alpha!} = \int^{(k)} \left\|F(\Delta;\mathbf{s}^k;\cdot)\right\|^2 d\mathbf{s}^k. \tag{8.3.10}$$

Assume that $r > 0$ is a integer. When $r > 0$ is not an integer, we use $\lfloor r \rfloor$ instead.

Denote $X_k = \mathcal{T}_{s_k - s_{k-1}}\mathcal{M}\cdots\mathcal{T}_{s_2-s_1}\mathcal{M}\mathcal{T}_{s_1}u_0$, $Y_k = \mathcal{M}X_k$, $k \geq 1$ and also $X = \mathcal{T}_{\Delta - s_k}Y_k$. Then $X_k = \mathcal{T}_{s_k-s_{k-1}}Y_{k-1}$ and $Y_{k-1} = \mathcal{M}X_{k-1}$.

By the definition of F, (8.3.1), (8.3.3) and (8.3.5), we have for $r \geq k$:

$$\begin{aligned}\left\|F(\Delta;\mathbf{s}^k;\cdot)\right\|^2 &\leq e^{2C_{\mathcal{L}}(\Delta - s_k)}\left\|Y_k\right\|_{H^0}^2 = e^{2C_{\mathcal{L}}(\Delta - s_k)}\left\|\mathcal{M}_{i_k}X_k\right\|_{H^0}^2\\ &\leq \tilde{C}(0,\mathcal{M})e^{2C_{\mathcal{L}}(\Delta - s_k)}\left\|X_k\right\|_{H^1}^2\\ &\leq C_1 e^{2C_{\mathcal{L}}(\Delta - s_{k-1})}\left\|Y_{k-1}\right\|_{H^1}^2 \leq \cdots \leq C_k^k e^{2C_{\mathcal{L}}\Delta}\left\|u_0\right\|_{H^k}^2,\end{aligned}$$

where $\tilde{C}(r-1,\mathcal{M})$ is from (8.3.3) and C_k is defined in (8.3.5). We then have

$$\int^{(k)} \left\|F(\Delta;\mathbf{s}^k;\cdot)\right\|^2 d\mathbf{s}^k \leq C_k^k e^{2C_{\mathcal{L}}\Delta}\left\|u_0\right\|_{H^k}^2 \int^{(k)} d\mathbf{s}^k. \tag{8.3.11}$$

If $r < k$, by changing the integration order and applying (8.3.1), (8.3.3), and (8.3.2), we get

$$\int^{(k)} \left\|F(\Delta;\mathbf{s}^k;\cdot)\right\|^2 d\mathbf{s}^k = \int^{(k)} \|X\|^2 d\mathbf{s}^k = \int_{(k)} \|X\|^2 d\mathbf{s}^k$$
$$\leq \int_{(k)} e^{2C_{\mathcal{L}}(\Delta - s_k)} \|Y_k\|^2 d\mathbf{s}^k = \int_{(k)} e^{2C_{\mathcal{L}}(\Delta - s_k)} \|\mathcal{M}_{i_k} X_k\|^2 d\mathbf{s}^k$$
$$\leq \tilde{C}(0,\mathcal{M}) \int_{(k)} e^{2C_{\mathcal{L}}(\Delta - s_k)} \|X_k\|^2_{H^1} d\mathbf{s}^k$$
$$= \tilde{C}(0,\mathcal{M}) \int_{(k-1)} \int_{s_{k-1}}^{\Delta} e^{2C_{\mathcal{L}}(\Delta - s_k)} \|X_k\|^2_{H^1} ds_k\, d\mathbf{s}^{k-1}$$
$$\leq \delta_{\mathcal{L}}^{-1} C_1 \int_{(k-1)} e^{2C_{\mathcal{L}}(\Delta - s_{k-1})} \|Y_{k-1}\|^2 d\mathbf{s}^{k-1},$$

where C_1 is from (8.3.5). Repeating this procedure and using (8.3.11), we obtain

$$\int^{(k)} \left\|F(t;s^k;\cdot)\right\|^2 d\mathbf{s}^k \leq \delta_{\mathcal{L}}^{r-k} C_1^{k-r} \int_{(r)} e^{2C_{\mathcal{L}}(\Delta - s_r)} \|Y_r\|^2 ds^r$$
$$\leq \delta_{\mathcal{L}}^{r-k} C_1^{k-r} C_r^r e^{2C_{\mathcal{L}}t} \|u_0\|^2_{H^r} \int^{(r)} d\mathbf{s}^r. \tag{8.3.12}$$

By (8.3.9), (8.3.10), (8.3.11), and (8.3.12), and $\int^{(k)} d\mathbf{s}^k = \frac{\Delta^k}{k!}$, we conclude that, for $r \geq \mathsf{N}+1$ and $C_1 < \delta_{\mathcal{L}}$,

$$\mathbb{E}[\|u(\Delta,\cdot) - u_{\mathsf{N}}(\Delta,\cdot)\|^2] = \sum_{\mathsf{N}<k\leq r} \int^{(k)} \left\|F(\Delta;\mathbf{s}^k;\cdot)\right\|^2 d\mathbf{s}^k$$
$$+ \sum_{k>r} \int^{(k)} \left\|F(\Delta;\mathbf{s}^k;\cdot)\right\|^2 d\mathbf{s}^k$$
$$\leq \sum_{\mathsf{N}<k\leq r} \frac{\Delta^k}{k!} C_r^k e^{2C_{\mathcal{L}}\Delta} \|u_0\|^2_{H^k}$$
$$+ \frac{\Delta^r}{r!} C_r^r e^{2C_{\mathcal{L}}\Delta} \|u_0\|^2_{H^r} \sum_{k>r} \delta_{\mathcal{L}}^{r-k} C_1^{k-r}$$
$$\leq (C_r\Delta)^{\mathsf{N}+1} e^{2C_{\mathcal{L}}\Delta} [\frac{e^{C_r\Delta}}{(\mathsf{N}+1)!}$$
$$+ \frac{(C_r\Delta)^{r-\mathsf{N}-1}}{r!} \frac{\delta_{\mathcal{L}}}{\delta_{\mathcal{L}} - C_1}] \|u_0\|^2_{H^r}. \qquad \square$$

Remark 8.3.5 *Lemma 8.3.3 holds for* $r = \infty$ *if* $C_\infty < \infty$. *Based on* (8.3.11), *we can prove that*

$$\mathbb{E}[\|u(\Delta,\cdot) - u_{\mathsf{N}}(\Delta,\cdot)\|^2] \le \sum_{k\ge \mathsf{N}} \frac{\Delta^k}{k!} C_\infty^k e^{2C_{\mathcal{L}}\Delta} \|u_0\|_{H^k}^2 \le (C_\infty \Delta)^{\mathsf{N}+1} e^{2C_{\mathcal{L}}\Delta} \frac{e^{C_\infty \Delta}}{(\mathsf{N}+1)!} \|u_0\|_{H^\infty}^2 .$$

If $r < \infty$, *we need to require that* $C_1 < \delta_{\mathcal{L}}$, *i.e.,* $\tilde{C}(0,\mathcal{M})C(1,\mathcal{L}) < \delta_{\mathcal{L}}$. *For example,* $\mathcal{L} = \triangle$, $\mathcal{M}_1 = \frac{1}{2}D_1$, *for which* $\tilde{C}(0,\mathcal{M})C(1,\mathcal{L}) = \frac{1}{2} < \delta_{\mathcal{L}} = 1$.

Proof of Lemma 8.3.4. It can be proved as in [315, p. 447] that

$$\mathbb{E}[|u_{\mathsf{N}}(\Delta,\cdot) - u_{\mathsf{N},\mathsf{n}}(\Delta,\cdot)|^2] = \sum_{l\ge \mathsf{n}+1} \sum_{k=1}^{\mathsf{N}} \sum_{|\alpha|=k, i_k^\alpha = l} \frac{\varphi_\alpha^2(\Delta,\cdot)}{\alpha!}, \quad (8.3.13)$$

where $i_{|\alpha|}^\alpha$ is the index of last nonzero element of α and the last summation in the right-hand side can be bounded by, see [315, (3.7)],

$$\sum_{|\alpha|=k, i_k^\alpha = l} \frac{\varphi_\alpha^2(\Delta,x)}{\alpha!} \le \int^{(k-1)} \left| \sum_{j=1}^{k} \int_{s_{j-1}}^{s_{j+1}} F_j(\Delta;\mathbf{s}^k;x) M_l(s_j)\, ds_j \right|^2 d\mathbf{s}_j^k,$$

where $d\mathbf{s}_j^k = ds_1 \cdots ds_{j-1}\, ds_{j+1} \cdots ds_k$, $s_0 := 0$, $s_{k+1} := \Delta$, $M_l(t) = \int_0^t m_l(s)\, ds$ and

$$\begin{aligned} F_j(\Delta;\mathbf{s}^k;x) = \frac{\partial F(\Delta;\mathbf{s}^k;x)}{\partial s_j} &= \mathcal{T}_{\Delta - s_k}\mathcal{M} \cdots \mathcal{T}_{s_{j+1}-s_j}\mathcal{M}\mathcal{L}\mathcal{T}_{s_j - s_{j-1}} \cdots \mathcal{T}_{s_1} u_0(x) \\ &\quad - \mathcal{T}_{\Delta - s_k}\mathcal{M} \cdots \mathcal{M}\mathcal{L}\mathcal{T}_{s_{j+1}-s_j} \cdots T_{s_1} u_0(x) \\ &=: F_j^1 + F_j^2. \end{aligned}$$

Then by the Fubini theorem (see Appendix D) and the Cauchy-Schwarz inequality, we have

$$\sum_{|\alpha|=k, i_k^\alpha = l} \frac{\|\varphi_\alpha(\Delta,\cdot)\|^2}{\alpha!} \le k \int^{(k-1)} \sum_{j=1}^{k} \int_{s_{j-1}}^{s_{j+1}} \left\| F_j(\Delta;\mathbf{s}^k;\cdot) \right\|^2 ds_j \int_{s_{j-1}}^{s_{j+1}} M_l^2(s_j)\, ds_j\, d\mathbf{s}_j^k.$$

We claim that

$$\begin{aligned} \left\| F_j(\Delta;\mathbf{s}^k;\cdot) \right\|^2 &\le 2 \max_{1\le j\le k} \left\| F_j^1 \right\|^2 \\ &\le 2C_{k+2}^k \tilde{C}(k,\mathcal{L}) C(k+2,\mathcal{L}) e^{2C_{\mathcal{L}}\Delta} \|u_0\|_{H^{k+2}}^2 . \end{aligned} \quad (8.3.14)$$

Thus, by (8.3.14) we have

$$\sum_{|\alpha|=k, i_k^\alpha=l} \frac{\|\varphi_\alpha(\Delta,\cdot)\|^2}{\alpha!} \leq 2k\Delta C_{k+2}^k C(k+2,\mathcal{L})\tilde{C}(k,\mathcal{L})e^{2C_{\mathcal{L}}\Delta} \|u_0\|_{H^{k+2}}^2$$
$$\int_0^\Delta M_l^2(s)\,ds \int^{(k-1)} ds_j^k. \quad (8.3.15)$$

Then by (8.3.13), (8.3.15), and $M_l(t) = \frac{\sqrt{2\Delta}}{(l-1)\pi}\sin(\frac{(l-1)\pi}{\Delta}t)$ (by (2.2.6)), we obtain that

$$\begin{aligned}
\mathbb{E}[\|u_{\mathsf{N}}(\Delta,\cdot) - u_{\mathsf{N},\mathsf{n}}(\Delta,\cdot)\|^2] &\leq \sum_{l\geq \mathsf{n}+1} \frac{\Delta^2}{(l-1)^2\pi^2} e^{2C_{\mathcal{L}}\Delta} \\
&\quad \sum_{k=1}^{\mathsf{N}} C_{k+2}^k C(k+2,\mathcal{L})\tilde{C}(k,\mathcal{L}) \|u_0\|_{H^{k+2}}^2 \frac{2k\Delta^k}{(k-1)!} \\
&\leq \frac{2\Delta^3}{\mathsf{n}\pi^2} e^{2C_{\mathcal{L}}\Delta} \sum_{k=1}^{\mathsf{N}} C_{k+2}^k \tilde{C}(k,\mathcal{L}) C(k+2,\mathcal{L}) \\
&\quad \|u_0\|_{H^{k+2}}^2 \frac{k\Delta^{k-1}}{(k-1)!} \\
&\leq \frac{2\Delta^3}{\mathsf{n}\pi^2} C_{\mathsf{N}+2} C(\mathsf{N}+2,\mathcal{L}) \\
&\quad \tilde{C}(\mathsf{N},\mathcal{L}) e^{2C_{\mathsf{N}+2}\Delta + 2C_{\mathcal{L}}\Delta} \|u_0\|_{H^{\mathsf{N}+2}}^2 .
\end{aligned}$$

It remains to prove (8.3.14). Note that it is sufficient to estimate $\|F_j^1\|$ due to the same structure of the two terms in $F_j(\Delta;\mathbf{s}^k;x)$. By the assumption that $a_{i,j}$ b_i and c belongs to $\mathcal{C}_b^{\mathsf{N}+3}(\mathcal{D})$, it can be readily checked that (8.3.4) holds with $l \leq \mathsf{N}+1$. Repeatedly using (8.3.1) and (8.3.3) gives

$$\begin{aligned}
\|F_j^1\|^2 &= \|\mathcal{T}_{\Delta-s_k}\mathcal{M}\cdots\mathcal{T}_{s_{j+1}-s_j}\mathcal{M}\mathcal{L}\mathcal{T}_{s_j-s_{j-1}}\cdots\mathcal{T}_{s_1}u_0\|^2 \\
&\leq e^{2C_{\mathcal{L}}(\Delta-s_k)} \|\mathcal{M}\cdots\mathcal{T}_{s_{j+1}-s_j}\mathcal{M}\mathcal{L}\mathcal{T}_{s_j-s_{j-1}}\cdots\mathcal{T}_{s_1}u_0\|^2 \\
&\leq \tilde{C}(0,\mathcal{M})e^{2C_{\mathcal{L}}(\Delta-s_k)} \|\mathcal{T}_{s_k-s_{k-1}}\cdots\mathcal{T}_{s_{j+1}-s_j}\mathcal{M}\mathcal{L}\mathcal{T}_{s_j-s_{j-1}}\cdots\mathcal{T}_{s_1}u_0\|_{H^1}^2 \\
&\leq C_1 e^{2C_{\mathcal{L}}(\Delta-s_{k-1})} \|\mathcal{M}\cdots\mathcal{T}_{s_{j+1}-s_j}\mathcal{M}\mathcal{L}\mathcal{T}_{s_j-s_{j-1}}\cdots\mathcal{T}_{s_1}u_0\|_{H^1}^2 \\
&\leq \cdots \leq C_{k-j}^{k-j} e^{2C_{\mathcal{L}}(\Delta-s_j)} \|\mathcal{M}\mathcal{L}\mathcal{T}_{s_j-s_{j-1}}\cdots\mathcal{T}_{s_1}u_0\|_{H^{k-j}}^2 \\
&\leq C_{k-j}^{k-j}\tilde{C}(k-j,\mathcal{M})e^{2C_{\mathcal{L}}(\Delta-s_j)} \|\mathcal{L}\mathcal{T}_{s_j-s_{j-1}}\cdots\mathcal{T}_{s_1}u_0\|_{H^{k-j+1}}^2 \\
&\leq C_{k-j}^{k-j}\tilde{C}(k-j,\mathcal{M})\tilde{C}(k-j+1,\mathcal{L})e^{2C_{\mathcal{L}}(\Delta-s_j)} \|\mathcal{T}_{s_j-s_{j-1}}\cdots\mathcal{T}_{s_1}u_0\|_{H^{k-j+3}}^2 \\
&\leq C_{k-j+1}^{k-j+1}\tilde{C}(k-j+1,\mathcal{L})e^{2C_{\mathcal{L}}(\Delta-s_j)} \|\mathcal{T}_{s_j-s_{j-1}}\mathcal{M}\cdots\mathcal{T}_{s_1}u_0\|_{H^{k-j+3}}^2 .
\end{aligned}$$

where we have used (8.3.4) in the last but one line and the fact that $C(k-j+1,\mathcal{L})\geq 1$. Similarly, we have

$$\left\|\mathcal{T}_{s_j-s_{j-1}}\mathcal{M}\cdots\mathcal{T}_{s_1}u_0\right\|_{H^{k-j+3}}^2 \leq C(k-j+3,\mathcal{L})C_{k+2}^{j-1}e^{2C_{\mathcal{L}}s_j}\left\|u_0\right\|_{H^{k+2}}^2.$$

Thus, we arrive at (8.3.14). This ends the proof of Lemma 8.3.4. □

8.3.2 Error estimate for SCM

For SCM for the SPDE (3.3.23), we have the following estimates: the first one is weak convergence of the Wong-Zakai type approximation $\tilde{u}_{\Delta,\mathsf{n}}(t,x)$ from (8.2.4) to $u(t,x)$ from (3.3.23), see Theorem 8.3.6; the second one is the convergence of SCM, i.e., the convergence of $\mathbb{M}_{\Delta,\mathsf{L},\mathsf{n}}(\mathsf{t}_i,x)$ to $\mathbb{E}[\tilde{u}^2_{\Delta,\mathsf{n}}(\mathsf{t}_i,x)]$, see Theorem 8.3.8. Here we prove the convergence rate when $\sigma_{i,r}=0$, which belongs to the case of commutative noises (3.3.29). Our proof for Theorem 8.3.6 is based on the mean-square of convergence of the Wong-Zakai type approximation (8.2.4) to (3.3.23). When $\sigma_{i,r}\neq 0$, we have not succeeded in proving this mean-square convergence and, as far as we know, only a rate of almost sure convergence of the Wong-Zakai type approximations to (3.3.23) has been proved [201].

According to Theorems 8.3.6 and 8.3.8, the error of the SCM is $\mathcal{O}(\Delta^{2\mathsf{L}-1})+\mathcal{O}(\Delta)$ in the second-order moments. Compared to Corollary 8.3.2, the SCM is of one order lower than WCE when $\mathsf{N}=2$ as the error of WCE is $\mathcal{O}(\Delta^{\mathsf{N}})+\mathcal{O}(\Delta^2)$.

To prove Theorems 8.3.6 and 8.3.8, we need a probabilistic representation of the solution to (3.3.23). Let $(\{B_k(s)\},1\leq k\leq d,\mathcal{F}_s^B)$ be a system of one-dimensional standard Wiener processes on a complete probability space $(\Omega^1,\mathcal{F}^1,Q)$ and independent of $w(s)$ on the space $(\Omega\otimes\Omega^1,\mathcal{F}\otimes\mathcal{F}^1,P\otimes Q)$. Consider the following backward stochastic differential equation on $(\Omega^1,\mathcal{F}^1,Q)$, for $0\leq s\leq t$,

$$\hat{d}\hat{X}_{t,x}(s)=b(\hat{X}_{t,x}(s))\,ds+\sum_{r=1}^{d}\alpha_r(\hat{X}_{t,x}(s))\,\hat{d}B_r(s),\quad \hat{X}_{t,x}(t)=x. \tag{8.3.16}$$

The symbol "$\hat{d}$" means backward integral, see, e.g., [280, 408] for treatment of backward stochastic integrals. The $d\times d$ matrix $\alpha(x)$ is defined by $\alpha(x)\alpha^\top(x)=2a(x)$. Here $a(x)$ and $b(x)$ are from (3.3.24). Consider the following backward stochastic differential equation on $(\Omega\otimes\Omega^1,\mathcal{F}\otimes\mathcal{F}^1,P\otimes Q)$ for $0\leq s\leq t$,

$$\hat{d}\hat{Y}_{t,1,x}(s)=c(\hat{X}_{t,x}(s))\hat{Y}_{t,1,x}(s)\,ds+\sum_{r=1}^{q}\nu_r(\hat{X}_{t,x}(s))\hat{Y}_{t,1,x}(s)]\,\hat{d}W_r,$$
$$\hat{Y}_{t,1,x}(t)=1. \tag{8.3.17}$$

Here $c(x)$ and $\nu_r(x)$ are from (3.3.24). When $u_0(x)\in C_b^2(\mathcal{D})$ and $\alpha(x)$, $b(x)$, $c(x)$, $\nu_r(x)\in C_b^0(\mathcal{D})$ and $\sigma_{i,r}=0$, the solution to (3.3.23)–(3.3.24) can be represented by, see, e.g., [280],

$$u(t,x) = \mathbb{E}_Q[u_0(\hat{X}_{t,x}(0)) \exp(\sum_{r=1}^{q} \int_0^t \nu_r(\hat{X}_{t,x}(s))\, \hat{d}W_r(s) + \int_0^t \bar{c}(\hat{X}_{t,x}(s))\, ds)], \tag{8.3.18}$$

where $\bar{c}(x) = c(x) - \frac{1}{2}\sum_{r=1}^{q} \nu_r^2(x)$.

Theorem 8.3.6 (Weak convergence of Wong-Zakai approximation) *Assume that $\sigma_{i,r} = 0$ and that the initial condition u_0 and the coefficients in (3.3.24) are in $C_b^2(\mathcal{D})$. Let $u(t,x)$ be the solution to (3.3.23) and $\tilde{u}_{\Delta,\mathsf{n}}(t,x)$ be the solution to (8.2.4). Then for any $\varepsilon > 0$, there exists a constant $C > 0$ such that the one-step error is estimated by*

$$\left|\mathbb{E}[u^2(\Delta,x)] - \mathbb{E}[\tilde{u}^2_{\Delta,\mathsf{n}}(\Delta,x)]\right| \le C\exp(C\Delta)(\Delta^6 + \Delta^2)\mathsf{n}^{-1+\varepsilon}, \tag{8.3.19}$$

and the global error is estimated by

$$\left|\mathbb{E}[u^2(\mathsf{t}_i,x)] - \mathbb{E}[\tilde{u}^2_{\Delta,\mathsf{n}}(\mathsf{t}_i,x)]\right| \le C\exp(CT)\Delta\mathsf{n}^{-1+\varepsilon}, \quad 1 \le i \le \mathsf{K}. \tag{8.3.20}$$

To prove Theorem 8.3.6, we first establish the one-step error (8.3.19) and then the global error (8.3.20). We follow the recipe of the proofs in [227, Theorem 3.1] and [54, Theorem 4.4] where $\mathsf{n} = 1$ and $\mathsf{K} > 1$.

We need the following mean-square convergence rate for the one-step error.

Proposition 8.3.7 (Mean-square convergence) *Assume that $\sigma_{i,r} = 0$ and that the initial condition u_0 and the coefficients in (3.3.24) are in $C_b^2(\mathcal{D})$. Let $u(t,x)$ be the solution to (3.3.23) and $\tilde{u}_{\Delta,\mathsf{n}}(t,x)$ the solution to (8.2.4). Then for any $\varepsilon > 0$,*

$$\mathbb{E}[|u(\Delta,x) - \tilde{u}_{\Delta,\mathsf{n}}(\Delta,x)|^2] \le C\exp(C\Delta)(\Delta^3 + \Delta^2)\mathsf{n}^{-1+\varepsilon}, \tag{8.3.21}$$

where the constant $C > 0$ is independent of n.

Proof. The solution to (8.2.4) using the spectral truncation of Brownian motion $W_r^{(\Delta,\mathsf{n})}$ from (8.2.3) can be represented by, see, e.g., [54, 227],

$$\tilde{u}_{\Delta,\mathsf{n}}(\Delta,x) = \mathbb{E}_Q[u_0(\hat{X}_{\Delta,x}(0)) \exp(\sum_{r=1}^{q} \int_0^\Delta \nu_r(\hat{X}_{\Delta,x}(s))\, dW_r^{(\Delta,\mathsf{n})}(s) + \int_0^\Delta \bar{c}(\hat{X}_{\Delta,x}(s))\, ds)]. \tag{8.3.22}$$

Using $e^x - e^y = e^{\theta x + (1-\theta)y}(x-y)$, $0 \le \theta \le 1$, boundedness of $\bar{c}(x)$ and $u_0(x)$ and the Cauchy-Schwarz inequality (twice), we have for some $C > 0$:

$$
\begin{aligned}
&\mathbb{E}[|\tilde{u}_{\Delta,\mathsf{n}}(\Delta,x)-u(\Delta,x)|^2] \qquad (8.3.23)\\
&=\mathbb{E}\Big[\Big(\mathbb{E}_Q[u_0(\hat{X}_{\Delta,x}(0))\exp(\int_0^{\Delta}\bar{c}(\hat{X}_{\Delta,x}(s))\,ds))\\
&\quad\exp(\sum_{r=1}^{q}\int_0^{\Delta}\nu_r(\hat{X}_{\Delta,x}(s))[\theta\,dW_r^{(\Delta,\mathsf{n})}(s)+(1-\theta)\,\hat{d}W_r(s)])\\
&\quad\times(\sum_{r=1}^{q}\int_0^{\Delta}\nu_r(\hat{X}_{\Delta,x}(s))[\,dW_r^{(\Delta,\mathsf{n})}(s)-\,\hat{d}W_r(s)])]\Big)^2\Big]\\
&\le C\exp(C\Delta)\mathbb{E}\Big[\Big(\mathbb{E}_Q[\exp(\sum_{r=1}^{q}\int_0^{\Delta}\nu_r(\hat{X}_{\Delta,x}(s))[\theta\,dW_r^{(\Delta,\mathsf{n})}(s)\\
&\quad+(1-\theta)\,\hat{d}W_r(s)])\times\Big|\sum_{r=1}^{q}\int_0^{\Delta}\nu_r(\hat{X}_{\Delta,x}(s))[\,dW_r^{(\Delta,\mathsf{n})}(s)-\,\hat{d}W_r(s)]\Big|]\Big)^2\Big]\\
&\le C\exp(C\Delta)\big(\mathbb{E}[\mathbb{E}_Q[\exp(\sum_{r=1}^{q}\int_0^{\Delta}4\nu_r(\hat{X}_{\Delta,x}(s))[\theta\,dW_r^{(\Delta,\mathsf{n})}(s)\\
&\quad+(1-\theta)\,\hat{d}W_r(s)])]]\big)^{1/2}\\
&\quad\times\big(\mathbb{E}[\mathbb{E}_Q[(\sum_{r=1}^{q}\int_0^{\Delta}\nu_r(\hat{X}_{\Delta,x}(s))[\,dW_r^{(\Delta,\mathsf{n})}(s)-\,\hat{d}W_r(s)])^4]]\big)^{1/2}.
\end{aligned}
$$

Recall that $\mathbb{E}[\cdot]=\mathbb{E}_P[\cdot]$ is the expectation with respect to P only. Hence, we need to estimate $\mathrm{I}_1=\big(\mathbb{E}[\mathbb{E}_Q[(\sum_{r=1}^{q}\int_0^{\Delta}\nu_r(\hat{X}_{\Delta,x}(s))[\,dW_r^{(\Delta,\mathsf{n})}(s)-\hat{d}\,W_r(s)])^4]]\big)^{1/2}$ and

$$
\mathrm{I}_2=\big(\mathbb{E}[(\mathbb{E}_Q[\exp(\sum_{r=1}^{q}\int_0^{\Delta}4\nu_r(\hat{X}_{\Delta,x}(s))[\theta\,dW_r^{(\Delta,\mathsf{n})}(s)+(1-\theta)\,\hat{d}W_r(s)])]]\big)^{1/2}.
$$

We first estimate I_1. Due to the independence of B_k and W_r, and according to [386] and (8.2.1), we have

$$
\begin{aligned}
\int_0^{\Delta}\nu_r(\hat{X}_{\Delta,x}(s))\,\hat{d}W_r(s)&=\int_0^{\Delta}\nu_r(\hat{X}_{\Delta,x}(s))\circ dW_r(s)\\
&=\sum_{i=0}^{\infty}\xi_{r,i}\int_0^{\Delta}\nu_r(\hat{X}_{\Delta,x}(s))m_{r,i}(s)\,ds.
\end{aligned}
$$

Thus by the Fubini theorem (see Appendix D), (8.2.1) and (8.2.3), we can represent I_1 as

$$
\begin{aligned}
\mathrm{I}_1 &= \Big(\mathbb{E}_Q[\mathbb{E}[\Big|\sum_{r=1}^{q}[\int_0^{\Delta} \nu_r(\hat{X}_{\Delta,x}(s))\, dW_r^{(\Delta,\mathsf{n})}(s) - \int_0^{\Delta} \nu_r(\hat{X}_{\Delta,x}(s)) \circ dW_r(s)]\Big|^4]]\Big)^{1/2} \\
&= \Big(\mathbb{E}_Q[\mathbb{E}[\Big|\sum_{r=1}^{q}\sum_{i=\mathsf{n}+1}^{\infty} \xi_{r,i}\int_0^{\Delta} \nu_r(\hat{X}_{\Delta,x}(s))m_{r,i}(s)\, ds\Big|^4]]\Big)^{1/2} \\
&\le \Big(3\mathbb{E}_Q[(\sum_{r=1}^{q}\sum_{i=\mathsf{n}+1}^{\infty}(\int_0^{\Delta} \nu_r(\hat{X}_{\Delta,x}(s))m_{r,i}(s)\, ds)^2)^2]\Big)^{1/2},
\end{aligned}
$$

where we have used twice the fact that $\hat{X}_{\Delta,x}$ are independent of W_r and $W_r^{(\Delta,\mathsf{n})}$. Then by standard estimates of L^2-projection error (cf. [58, (5.1.10)]), we have for $0 < \varepsilon < 1$,

$$
\sum_{i=\mathsf{n}+1}^{\infty}(\int_0^{\Delta} \nu_r(\hat{X}_{\Delta,x}(s))m_{r,i}(s)\, ds)^2 \le C\Delta^{1-\varepsilon}\mathsf{n}^{-1+\varepsilon}\left|\nu_r(\hat{X}_{\Delta,x}(\cdot))\right|^2_{\frac{1-\varepsilon}{2},2,[0,\Delta]}, \tag{8.3.24}
$$

where the Slobodeckij semi-norm $|f|_{\theta,p,[0,\Delta]}$ is defined by (3.3.10) and the constant $\Delta^{1-\varepsilon}$ appears due to the length of domain, see, e.g., [58, Chapter 5.4]. Thus, we obtain

$$
\mathrm{I}_1 \le C\Delta^{1-\varepsilon}\mathsf{n}^{-1+\varepsilon}\Big(\sum_{r=1}^{q}\mathbb{E}_Q[\left|\nu_r(\hat{X}_{\Delta,x}(\cdot))\right|^4_{\frac{1-\varepsilon}{2},2,[0,\Delta]}]\Big)^{1/2}, \; 0<\varepsilon<1. \tag{8.3.25}
$$

By (8.3.16) and the Ito formula, we have

$$
\begin{aligned}
\hat{X}_{\Delta,x}(s) - \hat{X}_{\Delta,x}(s_1) = \int_{s_1}^{s} b(\hat{X}_{\Delta,x}(s_2))\, ds_2 + \sum_{k=1}^{p}\alpha_k(\hat{X}_{\Delta,x}(s_1))[B_k(s) \\
- B_k(s_1)] + R(s_1,s),
\end{aligned}
$$

where $\mathbb{E}_Q[|R(s_1,s)|^{2l}] \le C\,|s_1 - s|^{2l}$ $(l \ge 1)$ when $b(x)$ and $\alpha_k(x)$ belong to $\mathcal{C}_b^2(\mathcal{D})$. By the Lipschitz continuity of ν_1, the definition of the Slobodeckij semi-norm, it is not difficult to show that

$$
\mathbb{E}_Q[\left|\nu_r(\hat{X}_{\Delta,x}(\cdot))\right|^4_{\frac{1-\varepsilon}{2},2,[0,\Delta]}] \le C(\Delta^{4+2\varepsilon} + \Delta^{2+2\varepsilon}). \tag{8.3.26}
$$

Thus, by (8.3.25) and (8.3.26), we have

$$
\mathrm{I}_1 \le C(\Delta^3 + \Delta^2)\mathsf{n}^{-1+\varepsilon}. \tag{8.3.27}
$$

Now we estimate I_2. Using the following facts (see, e.g., [227, Lemma 2.5]),

$$\mathbb{E}[\exp(\sum_{r=1}^{q}\int_0^{\Delta} 4\nu_r(\hat{X}_{\Delta,x}(s))\, dW_r)] = \exp(\sum_{r=1}^{q} 8\int_0^{\Delta} \nu_r^2(\hat{X}_{\Delta,x}(s))\, ds),$$

$$\mathbb{E}[\exp(\sum_{r=1}^{q}\int_0^{\Delta} 4\nu_r(\hat{X}_{\Delta,x}(s))\, dW_r^{(\Delta,\mathsf{n})}(s)] \leq 4\exp(\sum_{r=1}^{q} 8\int_0^{\Delta} \nu_r^2(\hat{X}_{\Delta,x}(s))\, ds),$$

we have $\mathrm{I}_2 \leq 4\exp(C\Delta)$. From here, (8.3.27) and (8.3.23), we reach (8.3.21). □

Now we are ready to prove Theorem 8.3.6, i.e., the convergence in the second moments.

Proof of Theorem 8.3.6. For simplicity of notation, we consider $q = 1$ while the case $q > 1$ can be proved similarly. Denote

$$U_{\Delta,\mathsf{n},\mathsf{m},\theta}(t,x,\mathbf{y}) =: u_0(\hat{X}_{t,x}(0))\exp(\sum_{i=1}^{\mathsf{n}} \nu_{1,i}y_i + \theta\sum_{j=\mathsf{n}+1}^{\mathsf{m}} \nu_{1,j}y_j$$
$$+ \int_0^t \bar{c}(\hat{X}_{t,x}(s))\, ds), \quad \mathsf{m} \geq \mathsf{n},$$

where $\nu_{1,i}(t,x) = \int_0^t \nu_1(\hat{X}_{t,x}(s))m_i(s)\, ds$ for $i \leq \mathsf{m}$ ($\hat{X}_{t,x}(s)$ is the solution to (8.3.16)) and $\mathbf{y} = (y_1,\dots,y_{\mathsf{n}},y_{\mathsf{n}+1},\dots,y_{\mathsf{m}})$. Let us write $\tilde{u}_{\Delta,\mathsf{n},\mathsf{m},\theta}(t,x,\Xi) = \mathbb{E}_Q[U_{\Delta,\mathsf{n},\mathsf{m},\theta}(t,x,\Xi)]$, where $\Xi = (\xi_1,\dots,\xi_{\mathsf{n}},\xi_{\mathsf{n}+1},\dots,\xi_{\mathsf{m}})$. With this notation, we have

$$\tilde{u}_{\Delta,\mathsf{m}}(t,x) = \tilde{u}_{\Delta,\mathsf{n},\mathsf{m},1}(t,x,\Xi), \quad \tilde{u}_{\Delta,\mathsf{n}}(t,x) = \tilde{u}_{\Delta,\mathsf{n},\mathsf{m},0}(t,x,\Xi).$$

For $\mathsf{m} > \mathsf{n}$, by the first-order Taylor expansion, we have

$$\begin{aligned}
&\left|\mathbb{E}[\tilde{u}^2_{\Delta,\mathsf{m}}(\Delta,x) - \tilde{u}^2_{\Delta,\mathsf{n}}(\Delta,x)]\right| \\
&= |2\sum_{i,j=\mathsf{n}+1}^{\mathsf{m}} \frac{1}{(\delta_{i,j}+1)}\int_0^1 \theta(1-\theta)\mathbb{E}[\tilde{u}_{\Delta,\mathsf{n},\mathsf{m},\theta}(\Delta,x,\Xi) \\
&\quad \mathbb{E}_Q[U_{\Delta,\mathsf{n},\mathsf{m},\theta}(\Delta,x,\Xi)\nu_{1,i}(t,x)\nu_{1,j}(t,x)]\xi_i\xi_j]\, d\theta \\
&\quad +2\sum_{i,j=\mathsf{n}+1}^{\mathsf{m}} \frac{1}{(\delta_{i,j}+1)}\int_0^1 \theta(1-\theta)\mathbb{E}[\mathbb{E}_Q[U_{\Delta,\mathsf{n},\mathsf{m},\theta}(\Delta,x,\Xi)\nu_{1,i}(t,x)] \\
&\quad \mathbb{E}_Q[U_{\Delta,\mathsf{n},\mathsf{m},\theta}(t,x,\Xi)\nu_{1,j}(t,x)]\xi_i\xi_j]\, d\theta| \\
&\leq 2\left|\int_0^1 (1-\theta)\theta\mathbb{E}[\tilde{u}_{\Delta,\mathsf{n},\mathsf{m},\theta}(\Delta,x,\Xi)\right. \\
&\quad \left.\mathbb{E}_Q[U_{\Delta,\mathsf{n},\mathsf{m},\theta}(\Delta,x,\Xi)\left(\sum_{i=\mathsf{n}+1}^{\mathsf{m}} \nu_{1,i}(\Delta,x)\xi_i\right)^2]]\, d\theta\right| \\
&\quad +2\left|\int_0^1 (1-\theta)\theta\mathbb{E}[\left(\sum_{i=\mathsf{n}+1}^{\mathsf{m}} \mathbb{E}_Q[U_{\Delta,\mathsf{n},\mathsf{m},\theta}(\Delta,x,\Xi)\nu_{1,i}(\Delta,x)]\xi_i\right)^2]\, d\theta\right|,
\end{aligned} \tag{8.3.28}$$

where $\delta_{i,j} = 1$ if $i = j$ and 0 otherwise and we have used the facts that ξ_i, $i > \mathsf{n}$, are independent of $\tilde{u}_{\mathsf{n}}(t, x)$ and $\mathbb{E}[\xi_i] = 0$.

By the Cauchy-Schwarz inequality (twice), we have for the first term in (8.3.28):

$$2\Big|\int_0^1 (1-\theta)\theta\mathbb{E}[\tilde{u}_{\Delta,\mathsf{n},\mathsf{m},\theta}(\Delta, x, \Xi)$$

$$\mathbb{E}_Q[U_{\Delta,\mathsf{n},\mathsf{m},\theta}(\Delta, x, \Xi)\left(\sum_{i=\mathsf{n}+1}^{\mathsf{m}} \nu_{1,i}(\Delta, x)\xi_i\right)^2]]\, d\theta\Big|$$

$$\leq C(\mathbb{E}[\mathbb{E}_Q[\left(\sum_{i=\mathsf{n}+1}^{\mathsf{m}} \nu_{1,i}(\Delta, x)\xi_i\right)^8]])^{1/4}. \qquad (8.3.29)$$

Here we also used that $\mathbb{E}[\tilde{u}^2_{\Delta,\mathsf{n},\mathsf{m},\theta}(\Delta, x, \Xi)], \mathbb{E}[(\mathbb{E}_Q[U^2_{\Delta,\mathsf{n},\mathsf{m},\theta}(\Delta, x, \Xi)])^2] \leq C$, which can be readily checked in the same way as in the proof of Proposition 8.3.7.

By the Taylor expansion for $U_{\Delta,\mathsf{n},\mathsf{m},\theta}(\Delta, x, \mathbf{y})$, we have

$$U_{\Delta,\mathsf{n},\mathsf{m},\theta}(\Delta, x, \mathbf{y}) = U_{\Delta,\mathsf{n},\mathsf{m},0}(\Delta, x, \mathbf{y}) + \sum_{i=\mathsf{n}+1}^{\mathsf{m}} \nu_{1,i}(\Delta, x)$$

$$\int_0^1 (1-\theta_1)\theta_1\theta U_{\Delta,\mathsf{n},\mathsf{m},\theta\theta_1}(\Delta, x, \mathbf{y})\, d\theta_1 y_i.$$

Then by the Cauchy-Schwarz inequality (several times) and the fact that ξ_i, $i > \mathsf{n}$ are independent of $U_{\Delta,\mathsf{n},\mathsf{m},0}(t, x, \Xi)$, we have for the second term in (8.3.28):

$$2\Big|\int_0^1 (1-\theta)\theta\mathbb{E}[\left(\sum_{i=\mathsf{n}+1}^{\mathsf{m}} \mathbb{E}_Q[U_{\Delta,\mathsf{n},\mathsf{m},\theta}(\Delta, x, \Xi)\nu_{1,i}(\Delta, x)]\xi_i\right)^2]\, d\theta\Big|$$

$$\leq 4\Big|\int_0^1 (1-\theta)\theta \sum_{i=\mathsf{n}+1}^{\mathsf{m}} \mathbb{E}[(\mathbb{E}_Q[U_{\Delta,\mathsf{n},\mathsf{m},0}((\Delta, x, \Xi)\nu_{1,i}(\Delta, x)])^2]\, d\theta\Big|$$

$$+4\Big|\int_0^1 (1-\theta)\theta^3\mathbb{E}[\Big(\mathbb{E}_Q[\int_0^1 (1-\theta_1)\theta_1 U_{\Delta,\mathsf{n},\mathsf{m},\theta\theta_1}(\Delta, x, \Xi)\, d\theta_1$$

$$(\sum_{i=\mathsf{n}+1}^{\mathsf{m}} \nu_{1,i}(\Delta, x)\xi_i))^2]\Big)^2]\, d\Big|$$

$$\leq \mathbb{E}[\mathbb{E}_Q[U^2_{\Delta,\mathsf{n},\mathsf{m},0}((\Delta, x, \Xi)]] \sum_{i=\mathsf{n}+1}^{\mathsf{m}} \mathbb{E}_Q[\nu^2_{1,i}(\Delta, x)]$$

$$+C\left|\int_0^1 (1-\theta)\theta^3(\mathbb{E}[\mathbb{E}_Q[\int_0^1 (1-\theta_1)\theta_1 U^4_{\Delta,\mathsf{n},\mathsf{m},\theta\theta_1}(\Delta,x,\Xi)\,d\theta_1])^{1/2}\,d\theta]\right|$$

$$(\mathbb{E}[\mathbb{E}_Q[(\sum_{i=\mathsf{n}+1}^{\mathsf{m}} \nu_{1,i}((\Delta,x)\xi_i)^8]])^{1/2}$$

$$\leq C\sum_{i=\mathsf{n}+1}^{\mathsf{m}} \mathbb{E}_Q[\nu^2_{1,i}(\Delta,x)] + C(\mathbb{E}[\mathbb{E}_Q[(\sum_{i=\mathsf{n}+1}^{\mathsf{m}} \nu_{1,i}(\Delta,x)\xi_i)^8]])^{1/2}. \tag{8.3.30}$$

Here we used that $\mathbb{E}[\mathbb{E}_Q[U^2_{\Delta,\mathsf{n},\mathsf{m},,0}((\Delta,x,\Xi)]]$, $\mathbb{E}[\mathbb{E}_Q[U^4_{\Delta,\mathsf{n},\mathsf{m},\theta\theta_1}(\Delta,x,\Xi)]] \leq C$, which can be readily checked in the same way as in the proof of Proposition 8.3.7.

By (8.3.28), (8.3.29), and (8.3.30), we have

$$\begin{aligned}&\left|\mathbb{E}[\tilde{u}^2_{\Delta,\mathsf{m}}(\Delta,x) - \tilde{u}^2_{\Delta,\mathsf{n}}(\Delta,x)]\right| \\ &\leq C\sum_{i=\mathsf{n}+1}^{\mathsf{m}} \mathbb{E}_Q[\nu^2_{1,i}(\Delta,x)] + C\mathbb{E}[\mathbb{E}_Q[(\sum_{i=\mathsf{n}+1}^{\mathsf{m}} \nu_{1,i}(\Delta,x)\xi_i)^8]])^{1/4} \\ &+C(\mathbb{E}[\mathbb{E}_Q[(\sum_{i=\mathsf{n}+1}^{\mathsf{m}} \nu_{1,i}(\Delta,x)\xi_i)^8]])^{1/2}.\end{aligned} \tag{8.3.31}$$

Similar to the proof of (8.3.25), we have

$$\begin{aligned}\mathbb{E}[\mathbb{E}_Q[(\sum_{i=\mathsf{n}+1}^{\mathsf{m}} \nu_{1,i}(\Delta,x)\xi_i)^8]] &\leq C\mathbb{E}_Q[\left(\sum_{i=\mathsf{n}+1}^{\mathsf{m}} \nu^2_{1,i}(\Delta,x)\right)^4] \\ &\leq C\Delta^{4(1-\varepsilon)}\mathsf{n}^{4(1-\varepsilon)}\mathbb{E}_Q[\left|\nu_1(\hat{X}_{\Delta,x}(\cdot))\right|^8_{\frac{1-\varepsilon}{2},2,[0,\Delta]}].\end{aligned}$$

Similar to the proof of (8.3.26), we can estimate $\mathbb{E}_Q[\left|\nu_1(\hat{X}_{\Delta,x}(\cdot))\right|^8_{\frac{1-\varepsilon}{2},2,[0,\Delta]}]$ as follows:

$$\mathbb{E}_Q[\left|\nu_1(\hat{X}_{\Delta,x}(\cdot))\right|^8_{\frac{1-\varepsilon}{2},2,[0,\Delta]}] \leq C(\Delta^{8+4\varepsilon} + C\Delta^{4+4\varepsilon}),$$

and thus

$$\mathbb{E}[\mathbb{E}_Q[(\sum_{i=\mathsf{n}+1}^{\mathsf{m}} \nu_{1,i}(\Delta,x)\xi_i)^8]] \leq C(\Delta^{12} + C\Delta^8). \tag{8.3.32}$$

Similarly, we have

$$\mathbb{E}[\mathbb{E}_Q[(\sum_{i=\mathsf{n}+1}^{\mathsf{m}} \nu_{1,i}(\Delta,x)\xi_i)^2]] = \sum_{i=\mathsf{n}+1}^{\mathsf{m}} \mathbb{E}_Q[\nu^2_{1,i}(\Delta,x)] \leq C(\Delta^3 + C\Delta^2). \tag{8.3.33}$$

By (8.3.31), (8.3.32), and (8.3.33), we have

$$\left|\mathbb{E}[\tilde{u}^2_{\Delta,\mathsf{m}}(\Delta,x)-\tilde{u}^2_{\Delta,\mathsf{n}}(\Delta,x)]\right| \leq C\exp(C\Delta)(\Delta^6+\Delta^2)\mathsf{n}^{-1+\varepsilon}. \tag{8.3.34}$$

By the triangle inequality and the Cauchy-Schwarz inequality, we obtain

$$\begin{aligned}\left|\mathbb{E}[u^2(\Delta,x)-\tilde{u}^2_{\Delta,\mathsf{n}}(\Delta,x)]\right| &\leq \left|\mathbb{E}[u^2(\Delta,x)-\tilde{u}^2_{\Delta,\mathsf{m}}(\Delta,x)]\right| \\ &\quad+\left|\mathbb{E}[\tilde{u}^2_{\Delta,\mathsf{m}}(\Delta,x)-\tilde{u}^2_{\Delta,\mathsf{n}}(\Delta,x)]\right|, \\ &\leq C(\mathbb{E}[|u(\Delta,x)-\tilde{u}_{\Delta,\mathsf{m}}(\Delta,x)|^2])^{1/2} \\ &\quad+\left|\mathbb{E}[\tilde{u}^2_{\Delta,\mathsf{m}}(\Delta,x)-\tilde{u}^2_{\Delta,\mathsf{n}}(\Delta,x)]\right|.\end{aligned}$$

The one-step error (8.3.19) then follows from (8.3.34), Proposition 8.3.7 and taking m to $+\infty$. The global error (8.3.20) is estimated from the recursion nature of Algorithm 7.3.1 as in the proof in [315, Theorem 2.4]. □

The following theorem is on the convergence of the second moments by SCM to those of the solution to (8.2.4).

Theorem 8.3.8 *Let $\tilde{u}_{\Delta,\mathsf{n}}(t,x)$ be the solution to* (8.2.4) *and $\mathbb{M}_{\Delta,\mathsf{L},\mathsf{n}}(\mathsf{t}_i,x)$ be the limit of $\mathbb{M}^{\mathsf{M}}_{\Delta,\mathsf{L},\mathsf{n}}(\mathsf{t}_i,x)$ from* (8.2.9) *when $\mathsf{M}\to\infty$. Under the assumptions of Theorem 8.3.6, for any $\varepsilon>0$, the one-step error is estimated by*

$$\begin{aligned}\left|\mathbb{M}_{\Delta,\mathsf{L},\mathsf{n}}(\Delta,x)-\mathbb{E}[\tilde{u}^2_{\Delta,\mathsf{n}}(\Delta,x)]\right| &\leq C\exp(C\Delta)(\Delta^{3\mathsf{L}}+\Delta^{2\mathsf{L}}) \\ &\quad(1+(3c/2)^{\mathsf{L}\wedge\mathsf{n}})\beta^{-(\mathsf{L}\wedge\mathsf{n})/2}\varepsilon^{-\mathsf{L}}\mathsf{L}^{-1}\mathsf{n}^{\mathsf{L}\varepsilon},\end{aligned}$$

and the global error is estimated by, for $1\leq i\leq \mathsf{K}$,

$$\begin{aligned}\left|\mathbb{M}_{\Delta,\mathsf{L},\mathsf{n}}(\mathsf{t}_i,x)-\mathbb{E}[\tilde{u}^2_{\Delta,\mathsf{n}}(\mathsf{t}_i,x)]\right| &\leq C\exp(CT)\Delta^{2\mathsf{L}-1} \\ &(1+(3c/2)^{\mathsf{L}\wedge\mathsf{n}})\beta^{-(\mathsf{L}\wedge\mathsf{n})/2}\varepsilon^{-\mathsf{L}}\mathsf{L}^{-1}\mathsf{n}^{\mathsf{L}\varepsilon}.\end{aligned}$$

Here the positive constants C, c, $\beta<1$ are independent of Δ, L, and n. The expression $\mathsf{L}\wedge\mathsf{n}$ means the minimum of L and n.

Proof. Setting $\varphi(y_1,\cdots,y_\mathsf{n}) = \tilde{u}^2_{\Delta,\mathsf{n}}(t,x,y_1,\cdots,y_\mathsf{n})$, we then have that $A(\mathsf{L},\mathsf{n})\varphi$ is an approximation of the second moment of the solution obtained by the sparse grid collocation methods. Recall from (8.3.22) that $\tilde{u}_{\Delta,\mathsf{n}}(t,x,\mathbf{y})=\mathbb{E}_Q[U_{\Delta,\mathsf{n}}(t,x,\mathbf{y})]$ where $U_{\Delta,\mathsf{n}}(t,x,\mathbf{y})=U_{\Delta,\mathsf{n},\mathsf{m},0}(t,x,\mathbf{y})$.

Now we estimate $D^{2\alpha_l}[\tilde{u}^2_{\Delta,\mathsf{n}}(\Delta,x,y_1,\cdots,y_\mathsf{n})]$. To this end, we need to first estimate $D^{\beta_l}[\tilde{u}_{\Delta,\mathsf{n}}(\Delta,x,y_1,\cdots,y_\mathsf{n})]$, where $\beta_l\leq 2\alpha_l$. By (8.3.24), we have for $0<\varepsilon<1$,

$$\nu^2_{1,k}(\Delta,x)\leq C(\Delta\max(k-1,1))^{\varepsilon-1}\left|\nu_1(\hat{X}_{\Delta,x}(\cdot))\right|^2_{\frac{1-\varepsilon}{2},2,[0,\Delta]},$$

we have, by the Cauchy-Schwarz inequality,

$$\left|D^{\beta_l}\tilde{u}_{\Delta,\mathsf{n}}(\Delta,x,\mathbf{y})\right| = \left|\mathbb{E}_Q[U_{\Delta,\mathsf{n}}(\Delta,x,\mathbf{y})\prod_{k=1}^{l}(\nu_{1,k}(\Delta,x))^{\beta_l^k}]\right| \tag{8.3.35}$$

$$\leq (\mathbb{E}_Q[U^2_{\Delta,\mathsf{n}}(\Delta,x,\mathbf{y})])^{1/2}(\mathbb{E}_Q[\prod_{k=1}^{l}(\nu_{1,k}(\Delta,x))^{2\beta_l^k}])^{1/2}$$

$$\leq (C\Delta^{1-\varepsilon})^{|\beta_l|/2}\prod_{k=2}^{l}(k-1)^{(\varepsilon-1)\beta_l^k/2}$$

$$(\mathbb{E}_Q[U^2_{\Delta,\mathsf{n}}(\Delta,x,\mathbf{y})])^{1/2}(\mathbb{E}_Q[\left|\nu_1(\hat{X}_{\Delta,x}(\cdot))\right|^{|2\beta_l|}_{\frac{1-\varepsilon}{2},2,[0,\Delta]}])^{1/2}.$$

By the chain rule for multivariate functions, we have

$$D^{2\alpha_l}[\tilde{u}^2_{\Delta,\mathsf{n}}(\Delta,x,\mathbf{y})] = \sum_{\beta_l+\gamma_l=2\alpha_l}(2\alpha_l)!\frac{D^{\beta_l}\tilde{u}_{\Delta,\mathsf{n}}(\Delta,x,\mathbf{y})}{\beta_l!}\frac{D^{\gamma_l}\tilde{u}_{\Delta,\mathsf{n}}(\Delta,x,\mathbf{y})}{\gamma_l!},$$

and thus by (8.3.35) and the fact that $\sum_{\beta_l+\gamma_l=2\alpha_l}\frac{(2\alpha_l)!}{\beta_l!\gamma_l!} = 2^{2|\alpha_l|-1}$, we have

$$\left|D^{2\alpha_l}[\tilde{u}^2_{\Delta,\mathsf{n}}(\Delta,x,\mathbf{y})]\right| \leq 2^{2|\alpha_l|-1}(C\Delta^{1-\epsilon})^{|\alpha_l|}\mathbb{E}_Q[U^2_{\Delta,\mathsf{n}}(\Delta,x,\mathbf{y})]\prod_{k=2}^{l}(k-1)^{(\varepsilon-1)\alpha_l^k}$$

$$\times \max_{\beta_l+\gamma_l=2\alpha_l}((\mathbb{E}_Q[\left|\nu_1(\hat{X}_{\Delta,x}(\cdot))\right|^{2|\beta_l|}_{\frac{1-\varepsilon}{2},2,[0,\Delta]}])^{1/2}$$

$$(\mathbb{E}_Q[\left|\nu_1(\hat{X}_{\Delta,x}(\cdot))\right|^{2|\gamma_l|}_{\frac{1-\varepsilon}{2},2,[0,\Delta]}])^{1/2}).$$

Similar to (8.3.32), we have $\mathbb{E}_Q[\left|\nu_1(\hat{X}_{\Delta,x}(\cdot))\right|^{2|\beta_l|}_{\frac{1-\varepsilon}{2},2,[0,\Delta]}] \leq C(\Delta^{|\beta_l|(2+\varepsilon)}+\Delta^{|\beta_l|(1+\varepsilon)})$ and

$$\left|D^{2\alpha_l}[\tilde{u}^2_{\Delta,\mathsf{n}}(\Delta,x,\mathbf{y})]\right| \leq C(\Delta^{3|\alpha_l|}+\Delta^{2|\alpha_l|})\prod_{k=2}^{l}(k-1)^{(\varepsilon-1)\alpha_l^k}\mathbb{E}_Q[U^2_{\Delta,\mathsf{n}}(\Delta,x,\mathbf{y})]. \tag{8.3.36}$$

Then by (7.2.20) and (8.3.36), we obtain

$$\left|S(\mathsf{L},l)\otimes_{n=l+1}^{\mathsf{n}}I_1^{(n)}\varphi\right|$$

$$\leq C(\Delta^{3\mathsf{L}}+\Delta^{2\mathsf{L}})(1+(3c/2)^{L\wedge l})\beta^{-(\mathsf{L}\wedge l)/2}$$

$$\mathbb{E}[\mathbb{E}_Q[U^2_{\Delta,\mathsf{n}}(\Delta,x,\mathbf{y})]]\sum_{i_1+\cdots+i_l=\mathsf{L}+l-1}\prod_{k=2}^{l}(k-1)^{(\varepsilon-1)\alpha_l^k}$$

$$\leq C(\Delta^{3\mathsf{L}}+\Delta^{2\mathsf{L}})(1+(3c/2)^{L\wedge l})\beta^{-(\mathsf{L}\wedge l)/2}\epsilon^{1-\mathsf{L}}(l-1)^{\mathsf{L}\varepsilon-1} \tag{8.3.37}$$

with the constant $C > 0$ which does not depend on n, ε, L, c, β, and l. In the last line we used the fact that $\mathbb{E}[\mathbb{E}_Q[U^2_{\Delta,\mathsf{n}}(\Delta, x, \mathbf{y})]]$ is bounded and that

$$\sum_{i_1+\cdots+i_l=\mathsf{L}+l-1} \prod_{k=2}^{l} (k-1)^{(\varepsilon-1)\alpha_l^k} = (l-1)^{\varepsilon-1}$$

$$\begin{aligned}\sum_{i_1+\cdots+i_l=\mathsf{L}+l-1} \prod_{k=2}^{l} (k-1)^{(\varepsilon-1)(i_k-1)} \\ \leq (l-1)^{\varepsilon-1}\Big(\sum_{k=2}^{l}(k-1)^{\varepsilon-1}\Big)^{\mathsf{L}-1} \\ \leq (l-1)^{\varepsilon-1}(\varepsilon^{-1}(l-1)^{\varepsilon})^{\mathsf{L}-1} = \varepsilon^{1-\mathsf{L}}(l-1)^{\mathsf{L}\varepsilon-1}.\end{aligned}$$

Then by (7.2.15) and (8.3.37), we have

$$\begin{aligned}|I_{\mathsf{n}}\varphi - A(\mathsf{L},\mathsf{n})\varphi| &\leq C(\Delta^{3\mathsf{L}} + \Delta^{2\mathsf{L}})(1+(3c/2)^{L\wedge \mathsf{n}})\beta^{-(\mathsf{L}\wedge\mathsf{n})/2}\varepsilon^{1-\mathsf{L}}\sum_{l=2}^{\mathsf{n}}(l-1)^{L\varepsilon-1} \\ &\quad + \left|(I_1^{(1)} - Q_{\mathsf{L}}^{(1)}) \otimes_{k=2}^{\mathsf{n}} I_1^{(k)}\varphi\right| \\ &\leq C(\Delta^{3\mathsf{L}} + \Delta^{2\mathsf{L}})(1+(3c/2)^{\mathsf{L}\wedge\mathsf{n}})\beta^{-(\mathsf{L}\wedge\mathsf{n})/2}\varepsilon^{-\mathsf{L}}\mathsf{L}^{-1}\mathsf{n}^{\mathsf{L}\varepsilon},\end{aligned}$$

where the term in the second line is estimated by the classical error estimate for the Gauss-Hermite quadrature Q, see, e.g., [339], and the estimation of derivatives (8.3.36).

The global error is estimated from the recursion nature of Algorithm 7.3.1 as in the proof in [315, Theorem 2.4].

8.4 Numerical results

In this section, we compare Algorithms 6.1.4 and 8.2.2 for linear stochastic advection-diffusion-reaction equations with commutative and noncommutative noises. We will test the computational performance of these two methods in terms of accuracy and computational cost. All the tests were run using Matlab R2012b, on a Macintosh desktop computer with Intel Xeon CPU E5462 (quad-core, 2.80 GHz). Every effort was made to program and execute the different algorithms as much as possible in an identical way.

We do not have exact solutions for all examples and hence we will evaluate the errors of the second-order moments using reference solutions, denoted by $\mathbb{E}[u^2_{\mathsf{ref}}(T,x)]$, which are obtained by either Algorithm 6.1.4 or Algorithm 8.2.2 with fine resolution. We do not use solutions obtained from Monte Carlo methods as reference solutions since Monte Carlo methods are of low accuracy and are less accurate than the recursive multistage WCE, see for comparison between WCE and Monte Carlo methods in Chapter 6 and also below.

The following error measures are used in the numerical examples below:

$$\varrho_2^2(T) = \left| \left\|\mathbb{E}[u_{\text{ref}}^2(T,\cdot)]\right\|_{l^2} - \left\|\mathbb{M}_\Delta^{\mathsf{M}}(T,\cdot)\right\|_{l^2} \right|, \quad \varrho_2^{r,2}(T) = \frac{\varrho_2^2(T)}{\left\|\mathbb{E}[u_{\text{ref}}^2(T,\cdot)]\right\|_{l^2}}, \tag{8.4.1}$$

$$\varrho_2^\infty(T) = \left| \left\|\mathbb{E}[u_{\text{ref}}^2(T,\cdot)]\right\|_{l^\infty} - \left\|\mathbb{M}_\Delta^{\mathsf{M}}(T,\cdot))\right\|_{l^\infty} \right|, \quad \varrho_2^{r,\infty}(T) = \frac{\varrho_2^\infty(T)}{\left\|\mathbb{E}[u_{\text{ref}}^2(T,\cdot)]\right\|_{l^\infty}}, \tag{8.4.2}$$

where $\mathbb{M}_\Delta^{\mathsf{M}}(T,x)$ is either $\mathbb{E}[(u_{\Delta,\mathsf{N},\mathsf{n}}^{\mathsf{M}}(T,x))^2]$ from Algorithm 6.1.4 or $\mathbb{M}_{\Delta,\mathsf{L},\mathsf{n}}^{\mathsf{M}}$ (T,x) from Algorithm 8.2.2, $\|v\|_{l^2} = \left(\frac{2\pi}{\mathsf{M}}\sum_{m=1}^{\mathsf{M}} v^2(x_m)\right)^{\frac{1}{2}}$, $\|v\|_{l^\infty} = \max_{1\le m\le \mathsf{M}}$ $|v(x_m)|$, x_m are the Fourier collocation points.

The computational complexity for Algorithm 6.1.4 is $\binom{\mathsf{N}+\mathsf{n}q}{\mathsf{N}}\frac{T}{\Delta}\mathsf{M}^4$ and that for Algorithm 8.2.2 is $\eta(\mathsf{L},\mathsf{n}q)\frac{T}{\Delta}\mathsf{M}^4$. The ratio of the computational cost of SCM over that of WCE is $\eta(\mathsf{L},\mathsf{n}q)/\binom{\mathsf{N}+\mathsf{n}q}{\mathsf{N}}$. For example, when $\mathsf{N}=1$ and $\mathsf{L}=2$, the ratio is $(1+2\mathsf{n}q)/(1+\mathsf{n}q)$, which will be used in the three numerical examples. The complexity increases exponentially with $\mathsf{n}q$ and L, see, e.g., [148], or N but increases linearly with $\frac{T}{\Delta}$. Hence, we only consider low values of L and N.

Example 8.4.1 (Single noise) *We consider the advection-diffusion equation* (3.3.27) *over the domain* $(0,T]\times(0,2\pi)$,

$$du = [(\epsilon + \frac{1}{2}\sigma^2)\partial_x^2 u + \beta(\sin(x)\partial_x u + u)]\,dt + \sigma\partial_x u\,dW(t),$$

with periodic boundary conditions and nonrandom initial condition $u(0,x) = \cos(x)$.

In this example, we compare Algorithms 6.1.4 and 8.2.2 for (3.3.27) with the parameters $\beta = 0.1$, $\sigma = 0.5$, and $\epsilon = 0.02$. We will show that the recursive multistage WCE is at most of order Δ^2 in the second-order moments and the recursive multistage SCM is of order Δ.

In Step 1, Algorithm 6.1.4, we employ the Crank-Nicolson scheme in time and Fourier collocation in physical space. We obtain the reference solution by Algorithm 6.1.4 with the same solver but finer resolution as a reference solution since we have no exact solution to (3.3.27). The reason for this choice of reference solution is as follows. For single noise, it is proved in Theorem 8.3.1 that the recursive multistage WCE is of second-order convergence in second-order moments. The second-order convergence is numerically verified in Chapter 6. For this specific example, a Monte Carlo method with 10^6 sampling paths (which costs 27.6 hours) gives $\left\|\mathbb{E}[u_{\text{MC}}^2]\right\| = 1.06517 \pm 6.1\times10^{-4}$ and $\left\|\mathbb{E}[u_{\text{MC}}^2]\right\|_\infty = 0.51746 \pm 6.1\times10^{-4}$, where the numbers after '±' are the statistical errors with the 95% confidence interval. We use Fourier collocation in space with $\mathsf{M}=20$ and Crank-Nicolson in time with $\delta t = 10^{-3}$ for the Monte Carlo method.

The reference solution is obtained via Algorithm 6.1.4 by $\mathsf{M} = 30$, $\Delta = 10^{-4}$, $\mathsf{N} = 4$, $\mathsf{n} = 4$, $\delta t = 10^{-5}$. It gives the second-order moments in l^2-norm $\left\|\mathbb{E}[u_{\text{ref}}^2]\right\|_{l^2} \doteq 1.065194550063$ and in the l^∞-norm $\left\|\mathbb{E}[u_{\text{ref}}^2]\right\|_{l^\infty} \doteq 0.5174746141105$.

From Table 8.1, we observe that the recursive WCE is $\mathcal{O}(\Delta^{\mathsf{N}}) + \mathcal{O}(\Delta^2)$ for the second-order moments. When $\mathsf{N} = 2$, the method is of second-order convergence in Δ and of first-order convergence when $\mathsf{N} = 1$. When $\mathsf{N} = 3$, the method is still second-order in Δ (not presented here). This verifies the estimate in Corollary 8.3.2.

Table 8.1. Algorithm 6.1.4: recursive multistage Wiener chaos method for (3.3.27) at $T = 5$: $\sigma = 0.5$, $\beta = 0.1$, $\epsilon = 0.02$, and $\mathsf{M} = 20$, $\mathsf{n} = 1$.

Δ	δt	N	$\varrho_2^{r,2}(T)$	Order	$\varrho_2^{r,\infty}(T)$	Order	CPU time (sec.)
1.0e-1	1.0e-2	1	1.5249e-2	–	8.8177e-3		3.57
1.0e-2	1.0e-3	1	1.5865e-3	$\Delta^{0.98}$	8.9310e-4	$\Delta^{0.99}$	33.22
1.0e-3	1.0e-4	1	1.5934e-4	$\Delta^{1.00}$	8.9429e-5	$\Delta^{1.00}$	348.03
1.0e-1	1.0e-2	2	1.9070e-4	–	4.1855e-5	–	5.14
1.0e-2	1.0e-3	2	2.0088e-6	$\Delta^{1.98}$	4.2889e-7	$\Delta^{1.99}$	51.75
1.0e-3	1.0e-4	2	2.0386e-8	$\Delta^{1.99}$	4.8703e-9	$\Delta^{1.94}$	490.04

In Step 1, Algorithm 8.2.2, we use the Crank-Nicolson scheme in time and Fourier collocation method in physical space. The errors are also measured as in (8.4.1) and (8.4.2). The reference solution is obtained by Algorithm 6.1.4 as in the case of WCE. We observe in Table 8.2 that the convergence order for second-order moments is one in Δ even when the sparse grid level L is 2, 3, and 4 (the last is not presented here). The errors for $\mathsf{L} = 3$ are more than half in magnitude smaller than those for $\mathsf{L} = 2$ while the time cost for $\mathsf{L} = 3$ is about 1.5 times of that for $\mathsf{L} = 2$.

Table 8.2. Algorithm 8.2.2: recursive multistage stochastic collocation method for (3.3.27) at $T = 5$: $\sigma = 0.5$, $\beta = 0.1$, $\epsilon = 0.02$, and $\mathsf{M} = 20$, $\mathsf{n} = 1$.

Δ	δt	L	$\varrho_2^{r,2}(T)$	Order	$\varrho_2^{r,\infty}(T)$	Order	CPU time (sec.)
1e-01	1e-02	2	3.4808e-04	–	3.0383e-03	–	3.71
1e-02	1e-03	2	3.4839e-05	$\Delta^{1.00}$	3.0130e-04	$\Delta^{1.00}$	33.88
1e-03	1e-04	2	3.4844e-06	$\Delta^{1.00}$	3.0106e-05	$\Delta^{1.00}$	325.06
1e-01	1e-02	3	1.6815e-04	–	3.4829e-04	–	5.16
1e-02	1e-03	3	1.6230e-05	$\Delta^{1.02}$	3.2283e-05	$\Delta^{1.03}$	50.59
1e-03	1e-04	3	1.6170e-06	$\Delta^{1.00}$	3.2026e-06	$\Delta^{1.00}$	486.08

In summary, from Tables 8.1 and 8.2, we observe that the recursive multistage WCE is $\mathcal{O}(\Delta^{\mathsf{N}}) + \mathcal{O}(\Delta^2)$ and the recursive multistage SCM is $\mathcal{O}(\Delta)$, as predicted by the error estimates in Chapter 8.3. While the SCM and the WCE are of the same order when $\mathsf{N} = 1$ and $\mathsf{L} \geq 2$, the former can be more accurate than the latter. In fact, when $\mathsf{N} = 1$ and $\mathsf{L} = 2$, the recursive multistage SCM error is almost two orders of magnitude smaller than the recursive multistage WCE while the computational cost for both is almost the same, as predicted ($\binom{\mathsf{N}+\mathsf{n}q}{\mathsf{N}} = \eta(\mathsf{L}, \mathsf{n}q) = 2$). The recursive multistage WCE with $\mathsf{N} = 2$ is of order Δ^2 and its errors are almost two orders of magnitude smaller than those by the recursive multistage SCM (with level 2 or 3) for the second-order moments.

In this example, the recursive multistage SCM outperforms the recursive multistage WCE with $\mathsf{N} = 1$. The reason can be as follows. In SCM, we solve an advection-dominated equation rather than a diffusion-dominated equation in WCE, as SCM is associated with the Stratonovich product which leads to the removal of the term $\frac{1}{2}\sigma^2\partial_x^2 u$ in the resulting equation, see (3.3.28). The larger σ is, the more dominant the diffusion is. In fact, results for $\sigma = 1$ and $\sigma = 0.1$ (not presented here) show that when $\sigma = 1$, the relative error of SCM with $\mathsf{L} = 2$ is almost three orders of magnitude smaller than WCE with $\mathsf{N} = 1$; when $\sigma = 0.1$, the relative error of SCM with $\mathsf{L} = 2$ is only less than one order of magnitude smaller than WCE with $\mathsf{N} = 1$. With the Crank-Nicolson scheme in time and Fourier collocation in physical space, we cannot achieve better accuracy for WCE with $\mathsf{N} = 1$ and Δ no less than 0.0005 when $\mathsf{M} \leq 40$.

Example 8.4.2 (Commutative noises) *We consider Equation* (3.3.30)

$$du = [(\epsilon + \frac{1}{2}\sigma_1^2 \cos^2(x))\partial_x^2 u + (\beta \sin(x) - \frac{1}{4}\sigma_1^2 \sin(2x))\partial_x u]\, dt$$
$$+\sigma_1 \cos(x)\partial_x u\, dW_1(t) + \sigma_2 u\, dW_2(t),$$

with two commutative noises over the domain $(0, T] \times (0, 2\pi)$, *with periodic boundary conditions and nonrandom initial condition* $u(0, x) = \cos(x)$. *The problem has commutative noises* (3.3.29)*:*

$$\sigma_1 \cos(x)\partial_x \sigma_2 \mathrm{Id} u = \sigma_2 \mathrm{Id} \sigma_1 \cos(x)\partial_x = \sigma_1\sigma_2 \cos(x)\partial_x.$$

Here Id is the identity operator.

In this example, we take $\sigma_1 = 0.5$, $\sigma_2 = 0.2$, $\beta = 0.1$, $\epsilon = 0.02$. We again observe first-order convergence for SCM and WCE with $\mathsf{N} = 1$, and second-order convergence for WCE with $\mathsf{N} = 2$ as in the last example with single noise.

We choose the same space-time solver for the recursive multistage WCE and SCM as in the last example. We compute the errors as in (8.4.1) and (8.4.2). For WCE, the reference second moments are $\left\|\mathbb{M}^{\mathsf{M}}_{\Delta=10^{-4},\mathsf{N},\mathsf{n}}(T,\cdot))\right\|_{l^2}$, and $\left\|\mathbb{M}^{\mathsf{M}}_{\Delta=10^{-4},\mathsf{N},\mathsf{n}}(T,\cdot)\right\|_{l^\infty}$ obtained by Algorithm 6.1.4 with $\delta t = 10^{-5}$ and all the other truncation parameters are the same as stated in the table. For SCM, the reference second moments are $\left\|\mathbb{M}^{\mathsf{M}}_{\Delta=10^{-4},\mathsf{L},\mathsf{n}}(T)\right\|_{l^2}$ and $\left\|\mathbb{M}^{\mathsf{M}}_{\Delta=10^{-4},\mathsf{L},\mathsf{n}}(T)\right\|_{l^\infty}$ obtained by Algorithm 8.2.2 with $\delta t = 10^{-5}$ while all the other truncation parameters are the same.

Here we do not compare the performance of Monte Carlo simulations with our algorithms as the main cost of Monte Carlo methods is to reduce the statistical errors. For the same parameters described above, when we used 10^6 Monte Carlo sampling paths, we could only reach the statistical error of 8.3×10^{-4}, in 3.9 hours. To obtain an error of 1×10^{-5}, seven thousand times more Monte Carlo sampling paths should be used, requiring three years of computational time and thus is not considered here. In the next example, we have similar situations and hence we will not consider Monte Carlo simulations. This also demonstrates the computational efficiency of Algorithms 6.1.4 and 8.2.2 in comparison with Monte Carlo methods.

For WCE, we observe in Figure 8.1 convergence of order Δ^{N} ($\mathsf{N} \leq 2$) in the second-order moments: first-order convergence when $\mathsf{N} = 1$, and second-order convergence when $\mathsf{N} = 2$. Numerical results for $\mathsf{N} = 3$ (not presented here) show that the convergence order is still two even though the accuracy is further improved when N increases from 2 to 3. This is consistent with our estimate $\mathcal{O}(\Delta^{\mathsf{N}}) + \mathcal{O}(\Delta^2)$ in Corollary 8.3.2.

We also tested the case $\mathsf{n} = 2$, which gives similar results and the same convergence order.

Fig. 8.1. Relative errors in second-order moments of recursive WCE and SCM for Example 8.4.2 with commutative noises at $T = 1$. The parameters are $\delta t = \Delta/10$, $M = 30$ and $\mathsf{n} = 1$. $\sigma_1 = 0.5$, $\sigma_2 = 0.2$, $\beta = 0.1$, $\epsilon = 0.02$. Left: l^2 errors; Right: l^∞ errors.

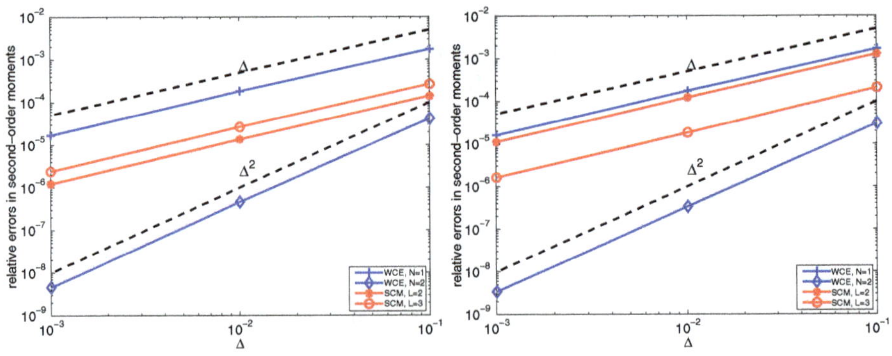

For SCM, we observe first-order convergence in Δ from Figure 8.1 when $\mathsf{L} = 2, 3$. We note that further refinement in truncation parameters in random space, i.e., increasing L and/or n do not change the convergence order nor improve the accuracy. The case $\mathsf{L} = 3$ actually leads to a bit worse accuracy, compared with the case $\mathsf{L} = 2$. We tested the case $\mathsf{L} = 4$, which leads to the same magnitudes of errors as $\mathsf{L} = 3$. We also tested $\mathsf{n} = 2$ and observed no improved accuracy for $\mathsf{L} = 2, 3, 4$. These numerical results are not presented here.

Table 8.3. Recursive multistage WCE (left) and SCM (right) for commutative noises (3.3.30) at $T = 1$: $\sigma_1 = 0.5$, $\sigma_2 = 0.2$, $\beta = 0.1$, $\epsilon = 0.02$, and $\mathsf{M} = 30$, $\mathsf{n} = 1$.

Δ	N	$\bar{\varrho}_2^{r,2}(T)$	Order	CPU time (sec.)	L	$\bar{\varrho}_2^{r,2}(T)$	Order	CPU time (sec.)
1e-01	1	1.6994e-03	–	3.19	2	1.2453e-03	–	5.18
1e-02	1	1.7838e-04	$\Delta^{0.98}$	32.74	2	1.2009e-04	$\Delta^{1.02}$	54.70
1e-03	1	1.6323e-05	$\Delta^{1.04}$	329.15	2	1.0889e-05	$\Delta^{1.04}$	545.20
1e-01	2	4.0658e-05	–	6.53	3	2.0482e-04	–	13.26
1e-02	2	4.4805e-07	$\Delta^{1.96}$	65.89	3	1.7897e-05	$\Delta^{1.06}$	142.23
1e-03	2	4.4682e-09	$\Delta^{2.00}$	657.55	3	1.6062e-06	$\Delta^{1.05}$	1420.24

For the two commutative noises, we conclude from this example that the recursive multistage WCE is of order $\Delta^{\mathsf{N}} + \Delta^2$ in the second-order moments and that the recursive multistage SCM is of order Δ in the second-order moments no matter what sparse grid level is used. The errors of recursive multistage SCM are one order of magnitude smaller than those of recursive multistage WCE with $\mathsf{N} = 1$ while the time cost of SCM is about 1.6 times of that cost of WCE, see Table 8.3. For large magnitude of noises ($\sigma_1 = \sigma_2 = 1$, numerical results are not presented), we observed that the SCM with $\mathsf{L} = 2$ and WCE with $\mathsf{N} = 1$ have the same order-of-magnitude accuracy. In this example, the use of SCM with $\mathsf{L} = 2$ for small magnitude of noises is competitive with the use of WCE with $\mathsf{N} = 1$.

Example 8.4.3 (Noncommutative noises) *We consider* (3.3.32)

$$du = [(\epsilon + \frac{1}{2}\sigma_1^2)\partial_x^2 u + \beta \sin(x)\partial_x u + \frac{1}{2}\sigma_2^2 \cos^2(x)u]\, dt + \sigma_1 \partial_x u\, dW_1(t) + \sigma_2 \cos(x)u\, dW_2(t),$$

with two noncommutative noises over the domain $(0, T] \times (0, 2\pi)$, *with periodic boundary conditions and nonrandom initial condition* $u(0, x) = \cos(x)$. *The problem has noncommutative noises as the coefficients do not satisfy* (3.3.29).

We take the same constants $\epsilon > 0$, β, σ_1, σ_2 as in the last example. We also take the same space-time solver as in the last example. In the current example, we observe only first-order convergence for SCM (level $\mathsf{L} = 2, 3, 4$) and WCE ($\mathsf{N} = 1, 2, 3$) when $\mathsf{n} = 1, 2$, see Table 8.4 for parts of the numerical results.

The errors are computed as in the last example. The reference solutions are obtained by Algorithm 6.1.4 for the recursive multistage WCE solutions and by Algorithm 8.2.2 for the recursive multistage SCM solutions, with $\Delta = 5 \times 10^{-4}$ and $\delta t = 5 \times 10^{-5}$ and all the other truncation parameters the same as stated in Tables 8.4 and 8.5.

Table 8.4. Algorithm 6.1.4 (recursive multistage Wiener chaos expansion, left) and Algorithm 8.2.2 (recursive multistage stochastic collocation method, right) for (3.3.32) at $T = 1$: $\sigma_1 = 0.5$, $\sigma_2 = 0.2$, $\beta = 0.1$, $\epsilon = 0.02$, and $\mathsf{M} = 20$, $\mathsf{n} = 1$. The time step size δt is $\Delta/10$. The reported CPU time is in seconds.

Δ	N	$\bar{\varrho}_2^{r,2}(T)$	Order	time (sec.)	L	$\bar{\varrho}_2^{r,2}(T)$	Order	time (sec.)
1.0e-1	1	3.7516e-03	–	1.04	2	6.4343e-04	–	1.65
5.0e-2	1	1.8938e-03	$\Delta^{0.99}$	2.11	2	3.1738e-04	$\Delta^{1.02}$	3.31
2.0e-2	1	7.5292e-04	$\Delta^{1.01}$	5.12	2	1.2440e-04	$\Delta^{1.02}$	8.64
1.0e-2	1	3.6796e-04	$\Delta^{1.03}$	10.19	2	**6.0502e-05**	$\Delta^{1.04}$	**17.12**
5.0e-3	1	1.7457e-04	$\Delta^{1.08}$	20.01	2	2.8635e-05	$\Delta^{1.08}$	33.82
2.0e-3	1	**5.8246e-05**	$\Delta^{1.20}$	**50.39**	2	9.5401e-06	$\Delta^{1.12}$	86.44
1.0e-1	2	9.4415e-05	–	2.16	3	1.5803e-04	–	4.03
5.0e-2	2	**3.7303e-05**	$\Delta^{1.81}$	**4.11**	3	**7.6548e-05**	$\Delta^{1.05}$	**8.68**
2.0e-2	2	1.2282e-05	$\Delta^{1.34}$	9.97	3	2.9673e-05	$\Delta^{1.03}$	22.08
1.0e-2	2	5.5807e-06	$\Delta^{1.21}$	20.03	3	1.4378e-05	$\Delta^{1.05}$	43.85
5.0e-3	2	2.5471e-06	$\Delta^{1.14}$	40.25	3	6.7925e-06	$\Delta^{1.08}$	88.35
2.0e-3	2	8.2965e-07	$\Delta^{1.22}$	101.34	3	2.2605e-06	$\Delta^{1.20}$	223.15

In this example, our error estimate for recursive multistage WCE is not valid any more and the numerical results suggest that the errors behave as $\Delta^{\mathsf{N}} + C\Delta/\mathsf{n}$. For $\mathsf{N} = 1$ and $\mathsf{n} = 10$ (not presented), the error is almost the same as $\mathsf{n} = 1$. While $\mathsf{N} = 2$ and $\mathsf{n} = 10$, the error first decreases as $\mathcal{O}(\Delta^2)$ for large time step size and then as $\mathcal{O}(\Delta)$ for small time step size; see Table 8.5. When $\mathsf{N} = 2$ and $\mathsf{n} = 10$, the errors with $\Delta = 0.005, 0.002, 0.001$ are ten percent $(1/\mathsf{n})$ of those with the same parameters but $\mathsf{n} = 1$ in Table 8.4. Here the constant in front of Δ, C/n, plays an important role: when Δ is large and this constant is small, then the order of two can be observed; when Δ is small, $C\Delta/\mathsf{n}$ is dominant so that only first-order convergence can be observed.

The recursive multistage SCM is of first-order convergence when $\mathsf{L} = 2, 3, 4$ and $\mathsf{n} = 1, 2, 10$ (only parts of the results presented). In contrast to Example 8.4.2, the errors from $\mathsf{L} = 3$ are one order of magnitude smaller those from $\mathsf{L} = 2$. Recalling that the number of sparse grid points is $\eta(2, 2) = 5$ and $\eta(3, 2) = 13$, we have the cost for $\mathsf{L} = 3$ is about 2.6 times of that for $\mathsf{L} = 2$. However, it is expected that in practice, a low level sparse grid is more efficient than a high level one when $\mathsf{n}q$ is large as the number of sparse grid points $\eta(\mathsf{L}, \mathsf{n}q)$ is increasing exponentially with $\mathsf{n}q$ and L. In other words, $\mathsf{L} = 2$ is preferred when SPDEs with many noises (large q) are considered.

As discussed in the beginning of this section, the ratio of time cost for SCM and WCE is $\eta(\mathsf{L}, \mathsf{n}q)/\binom{\mathsf{N}+\mathsf{n}q}{\mathsf{N}}$. The cost of recursive multistage SCM with $\mathsf{L} = 2$ is at most 1.8 times (1.6 predicted by the ratio above, $q = 2$ and $\mathsf{n} = 1$) of that of recursive multistage WCE with $\mathsf{N} = 1$. However, in this example, the accuracy of the recursive multistage SCM is one order of magnitude smaller than that of the recursive multistage WCE when $\mathsf{N} = 1$ and $\mathsf{L} = 2$. In Table 8.4, we present in bold the errors between 3.5×10^{-5} and 8.0×10^{-5}. Among the four cases listed in the table, the most efficient, for the given accuracy above, is WCE with $\mathsf{N} = 2$, which outperforms SCM with $\mathsf{L} = 3$ and $\mathsf{L} = 2$. WCE with $\mathsf{N} = 1$ is less efficient than the other three cases. We also observed that when $\sigma_1 = \sigma_2 = 1$, the error in SCM with $\mathsf{L} = 2$ is one order of magnitude smaller than WCE with $\mathsf{N} = 1$ (results not presented here).

For noncommutative noises in this example, we show that the error for WCE is $\Delta^2 + C\Delta/\mathsf{n}$ and the error for SCM is Δ. The numerical results suggest that SCM with $\mathsf{L} = 2$ is competitive with WCE with $\mathsf{N} = 1$ for both small and large magnitude of noises if $\mathsf{n} = 1$.

Table 8.5. Algorithm 6.1.4: recursive multistage Wiener chaos expansion for (3.3.32) at $T = 1$: $\sigma_1 = 0.5$, $\sigma_2 = 0.2$, $\beta = 0.1$, $\epsilon = 0.02$. The parameters are $\mathsf{M} = 20$, $\mathsf{N} = 2$, and $\mathsf{n} = 10$. The time step size δt is $\Delta/10$.

Δ	$\bar{\varrho}_2^{r,2}(T)$	Order	$\bar{\varrho}_2^{r,\infty}(T)$	Order	CPU time (sec.)
1.0e-1	4.9310e-05	–	2.6723e-05	–	84.00
5.0e-2	1.4031e-05	$\Delta^{1.81}$	7.3571e-06	$\Delta^{1.86}$	160.50
2.0e-2	2.9085e-06	$\Delta^{1.71}$	1.4171e-06	$\Delta^{1.80}$	391.40
1.0e-2	9.8015e-07	$\Delta^{1.57}$	4.4324e-07	$\Delta^{1.68}$	749.40
5.0e-3	3.5978e-07	$\Delta^{1.45}$	1.5082e-07	$\Delta^{1.56}$	1557.60
2.0e-3	9.8910e-08	$\Delta^{1.41}$	3.8369e-08	$\Delta^{1.49}$	3887.50

With these three examples, we observe that the convergence order of the recursive multistage SCM in the second-order moments is one for commutative and noncommutative noises. We verified that our error estimate for WCE, $\Delta^{\mathsf{N}} + \Delta^2$, is valid for commutative noises, see Examples 8.4.1 and 8.4.2; the numerical results for noncommutative noises, see Example 8.4.3, suggest that the errors are of order $\Delta^{\mathsf{N}} + C\Delta/\mathsf{n}$, where C is a constant depending on the coefficients of the noises.

For stochastic advection-diffusion-reaction equations, different formulations of stochastic products (Ito-Wick product for WCE, Stratonovich product for SCM) lead to different numerical performance. When the white noise is in the velocity, the Ito formulation will have stronger diffusion than that in the Stratonovich formulation in the resulting PDE. As stronger diffusion requires more resolution, the recursive multistage WCE with $\mathsf{N} = 1$ may produce less accurate results than those by the recursive multistage SCM with $\mathsf{L}=2$ with the same PDE solver under the same resolution, as shown in the first and the third examples.

To achieve convergence of approximations of second moments with first-order in time step Δ, we can use the recursive multistage SCM Algorithm 8.2.2 with $\mathsf{L}=2$, $\mathsf{n}=1$ and also the recursive multistage WCE Algorithm 6.1.4 with $\mathsf{N}=1$, $\mathsf{n}=1$, as both can outperform each other in certain cases. For commutative noises, Algorithm 6.1.4 with $\mathsf{N}=2$ is preferable when the number of noises, q, is small and hence the number of WCE modes is small so that the computational cost would grow slowly.

We also note that the errors of Algorithms 6.1.4 and 8.2.2 depend on the SPDE coefficients and integration time (cf. theoretical results of Chapter 8.3). For some SPDEs, the constants at powers of Δ in the errors can be very large and, to reach desired levels of accuracy, we need to use very small step size Δ or develop numerical algorithms further (e.g., higher-order or structure-preserving approximations, see such ideas for SODEs, e.g., in [358]). Further, in practice, we need to aim at balancing the three parts (truncation of Wiener processes, functional truncation of WCE/SCM, and space-time discretizations of the deterministic PDEs appearing in the algorithms) of the errors of Algorithms 6.1.4 and 8.2.2 for higher computational efficiency.

8.5 Summary and bibliographic notes

We have shown that both methods, WCE and SCM in conjunction with the recursive strategy, for linear advection-diffusion-reaction equations are comparable in computational performance, even though WCE can be formally of higher order:

- For commutative noises, the accuracy of the recursive WCE is $\mathcal{O}(\Delta^{\mathsf{N}})+\mathcal{O}(\Delta/\mathsf{n})$ (see Theorem 8.3.1 and Corollary 8.3.2) while the accuracy of the recursive SCM is $\mathcal{O}(\Delta)$ (see Theorem 8.3.8). Here N is the order of Wiener chaos and n is the truncation parameter of spectral expansion of Brownian motion.
- For noncommutative noises, the accuracy of both the recursive WCE and the recursive SCM is $\mathcal{O}(\Delta)$, see numerical results in Example 8.4.3.
- With truncation of spectral expansion of Brownian motions in use, WCE and SCM are associated with different stochastic products: WCE is associated with the Ito-Wick product and SCM is with the Stratonovich product. Hence, the WCE usually has more diffusion than the SCM does, see (8.2.2) and (3.3.26). This effect requires special attention of employing efficient numerical solvers in time and physical space.
- The numerical performance of these two methods depends on the intensity of the noises and the dynamics of the underlying equations. See numerical examples in Chapter 8.4.

However, when the underlying equations are nonlinear rather than linear, the recursive strategy is not applicable any more. Then SCM is preferable because of its resulting fully decoupled systems of equations while WCE results in fully coupled systems of equations. In the next chapter, we show how we can apply SCM to a classic practical nonlinear problem and evaluate the performance of SCM in stochastic flow simulations.

Bibliographic notes. Comparison between WCE and SCM is demonstrated in [13, 123] for elliptic equations with color noise. One big difference between color noise and white noise is that there is no issue of definition for stochastic products for color noise (to be more precise, the noise smoother than Brownian motion). Also, for color noise, it is shown in [13, 123] that there are only small differences in the numerical performance of generalized polynomial chaos expansion and SCM.

The convergence of WCE has been discussed in [314–316, 318] with white noise in the reaction rate and white noise in the advection velocity. However, the convergence rate of WCE is only known in the case of white noise in the reaction rate, see, e.g., [315]. In this case, the problem has commutative noises, (3.3.29). In this chapter, we discuss the convergence rate in the case of single white noise in the advection velocity, which is a special case of commutative noises (3.3.29). The case of noncommutative noises has not been discussed in literature while only numerical results are presented in this chapter.

The convergence rate of SCM has been discussed for color noise, see, e.g., [11, 379, 380, 500] for Smolyak's sparse grid collocation. For white noise, the convergence rate has been discussed in [506, 507]. Due to the low regularity of the solution in random space to PDEs with white noise, the convergence cannot be high as in the case of color noise.

The number of operations for the recursive multistage WCE is of order M^4, where M is the number of nodes employed in the discretization of physical space. As shown in Chapter 6.4, this computational complexity can be reduced to the order of M^2 using sparse representations (see, e.g., [418]) and many coefficients of WCE modes are negligible in computation. The complexity of the recursive multistage SCM is also of order M^4, but it is not clear that SCM admits sparse representations as WCE does.

8.6 Suggested practice

Exercise 8.6.1 *Consider the following stochastic advection-diffusion-reaction equation over the domain* $(0,T] \times (0,2\pi)$,

$$\begin{aligned} du = & [(\epsilon + \frac{1}{2}\sigma_1^2 \cos^2(x))\partial_x^2 u + (\beta \sin(x) - \frac{1}{4}\sigma_1^2 \sin(2x))\partial_x u + \sin(3x)]\, dt \\ & + [\sigma_1 \sin(x)\partial_x u + \cos(2x)]\, dW_1(t) + \sigma_2 u\, dW_2(t), \end{aligned}$$

where the initial condition is $u(0,x) = \cos(x)$ and periodic boundary conditions are imposed. The equation is similar to that in Example 8.4.2, with commutative noises but there are also forcing terms. Derive the recursive WCE and SCM algorithms to compute second moments of solutions as suggested in Remark 8.2.1.

Exercise 8.6.2 *Consider the following stochastic advection-diffusion-reaction equation over the domain $(0,T] \times (0,2\pi)$:*

$$du = [(\epsilon + \frac{1}{2}\sigma_1^2)\partial_x^2 u + \beta\sin(x)\partial_x u + \frac{1}{2}\sigma_2^2\cos^2(x)u + \sin(3x)]\,dt$$
$$+[\sigma_1\partial_x u + \cos(x)\,dW_1(t) + \sigma_2\cos(x)u\,dW_2(t).$$

The equation has two noncommutative noises, where periodic boundary conditions are imposed and nonrandom initial condition $u(0,x) = \cos(x)$. Derive the recursive WCE and SCM algorithms to compute second moments of solutions as suggested in Remark 8.2.1.

Exercise 8.6.3 *Write Matlab code for Exercises 8.6.1 and 8.6.2. Compare the performance of WCE and SCM with respect to the following aspects:*

- *convergence order in time*
- *accuracy when the order of WCE and the level of sparse grid changes.*

Hint. Use the multistage WCE with a fine resolution in physical space, time, and in random space to produce a reference solution.

Exercise 8.6.4 *Apply the stochastic collocation method (Hermite collocation method in random space) instead of WCE in Exercise 6.6.2.*

9

Application of collocation method to stochastic conservation laws

In this chapter we demonstrate how to apply the methods we presented in the previous chapters to stochastic nonlinear conservations laws. Specifically, since the problem is nonlinear we apply the stochastic collocation method (SCM) using Smolyak's sparse grid for a one-dimensional piston problem and test its computational performance. This is a classical problem in every aerodynamics textbook with an analytical solution if the piston velocity is fixed. However, here we consider a piston with a velocity perturbed by Brownian motion moving into a straight tube filled with a perfect gas at rest. The shock generated ahead of the piston can be located by solving the one-dimensional Euler equations driven by white noise using the Stratonovich or Ito formulations. We apply the Lie-Trotter splitting method before we approximate the Brownian motion with its spectral truncation and subsequently apply stochastic collocation using either sparse grid or the quasi-Monte Carlo (QMC) method. Numerical results verify the Stratonovich-Euler and Ito-Euler models against stochastic perturbation results, and demonstrate the efficiency of sparse grid collocation and QMC for small and large random piston motions, respectively.

9.1 Introduction

We consider the stochastic piston problem in [298], which defines a testbed for numerical solvers in both random and physical space. The piston driven by time-varying random motions moves into a straight tube filled with a perfect gas at rest. Of interest is to quantify the perturbation of the shock position ahead of the piston corresponding to the random motion. For the perturbed shock position, Lin et al. [298] obtained analytical solutions for small amplitudes of noises and numerical solutions for large amplitudes of

Z. Zhang, G.E. Karniadakis, *Numerical Methods for Stochastic Partial Differential Equations with White Noise*,
Applied Mathematical Sciences 196, DOI 10.1007/978-3-319-57511-7_9

noises, with the method of stochastic perturbation analysis and polynomial chaos, respectively. A specific random motion of the piston was studied, where the piston velocity was perturbed by a *correlated* stochastic process with zero mean and exponential covariance kernel. It was concluded that the variance of the shock location grows quadratically with time for small time and linearly for large time by both the perturbation analysis and numerical simulations of the corresponding Euler equations. Numerical results from the Monte Carlo method and the polynomial chaos method (e.g., [488]) for the stochastic Euler equations showed good agreement with the results from the perturbation analysis.

Here we consider the case of piston velocity perturbed by *Brownian motion*, which leads to the Euler equations subject to white noise rather than the Euler equations subject to color noise in [298]. We will use the Monte Carlo method and the stochastic collocation method presented in the previous two chapters for equations driven by white noise. Note that the method of perturbation analysis in [298] is independent of the type of noises when they have *continuous paths* in the random space so that the results by the perturbation analysis can be understood in a path-wise sense. Therefore, the stochastic piston problem defined in [298] can serve as a rigorous testbed of evaluating numerical stochastic solvers. So we will use the variances from perturbation analysis as reference solutions.

We use the stochastic collocation method (SCM) for time-dependent equations driven by white noise in time, presented in the previous chapters. Here we also adopt the quasi-Monte Carlo (QMC) method to compute up to larger time and/or for large amplitudes of noises. The QMC method is efficient and converges faster than the Monte Carlo method if relatively low dimensional integration is considered, see, e.g., [376, 424]; see also [165] for the application of the QMC method to elliptic equations in random porous media.

This chapter is organized as follows. In Chapter 9.2, we describe the piston problem driven by stochastic processes and review two different approaches to obtain the shock location: the perturbation analysis and the one-dimensional Euler equations. When the piston is driven by the Brownian motion, we introduce two types of Euler equations according to different interpretations of stochastic products for white noise, i.e., the Stratonovich-Euler equations and the Ito-Euler equations. In Chapter 9.3, we describe a splitting method for the Euler equations before comparing the variances from the two stochastic Euler equations with those from first-order perturbation analysis. We demonstrate that the Stratonovich-Euler equations are suitable for obtaining the variances of perturbations piston locations. We apply the stochastic collocation method in Chapter 9.4 to solve the Stratonovich-Euler equations in the splitting-method setting. We conclude in Chapter 9.5 with a summary and comments on computational efficiency, where we also compare different shock locations when the piston is driven by three different stochastic processes. Two exercises are presented for readers to practice the splitting method for stochastic parabolic equations.

9.2 Theoretical background

Suppose that the piston velocity is perturbed by a time-dependent stochastic process so that the piston velocity is $u_p = U_p + v_p(t, \omega)$, where ω is a point in random space; see Figure 9.1 for a sketch of shock tube driven by a piston perturbed with random motion. Here we write $v_p(t, \omega) = \epsilon U_p V(t, \omega)$ and denote the stochastic process $V(t, \omega)$ as $V(t)$ for brevity.

When $\epsilon = 0$, i.e., no perturbation is imposed on the piston, the piston moves into the tube with a constant velocity U_p, the shock speed S (and thus the shock location) can be determined analytically, see [297, 298]. When ϵ is very small, one can determine the perturbation process of the shock location using the first-order perturbation analysis [298], that is:

Fig. 9.1. A sketch of piston-driven shock tube with random piston motion.

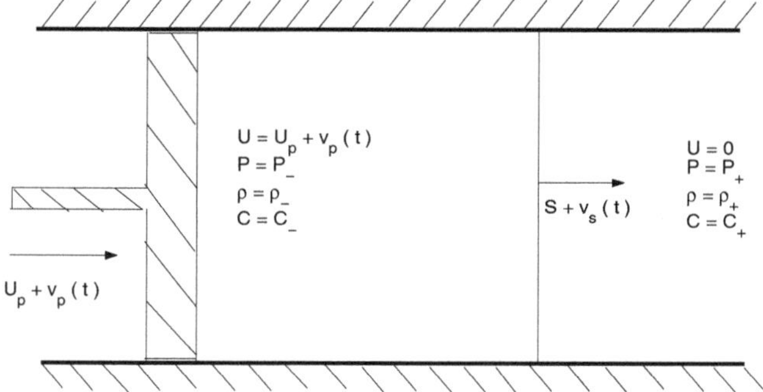

$$z(t) = \epsilon U_p q S' \sum_{n=0}^{\infty} (-r)^n \int_0^t V(\alpha \beta^n t_1)\, dt_1, \tag{9.2.1}$$

where $z(t) + tS$ is the shock location induced by the random motion of piston,

$$S' = \frac{\gamma+1}{4} \frac{S}{S - \frac{\gamma+1}{4} U_p},$$

$$q = \frac{2}{1+k}, \quad r = \frac{1-k}{1+k}, \quad k = C_- \frac{S + S'U_p}{1 + \gamma S U_p},$$

$$\alpha = \frac{C_- + U_p - S}{C_-}, \quad \beta = \frac{C_- + U_p - S}{C_- + S - U_p}.$$

Here γ is the ratio of the specific heats and C_- the sound speed behind the shock when the piston is unperturbed. The first two moments of the perturbation process $z(t)$ are

$$\mathbb{E}[z(t)] = 0,$$
$$\mathbb{E}[z^2(t)] = (\epsilon U_p q S')^2 \mathbb{E}[\Big(\sum_{n=0}^{\infty}(-r)^n \int_0^t V(\alpha\beta^n t_1,\cdot)\, dt_1\Big)^2].$$

We note that the perturbation analysis in [298] is independent of the perturbation process whenever the process is continuous such that the analysis can be understood in a path-wise way. By taking $V(t,\omega)$ as the Brownian motion $W(t)$ (omitting ω), we then have

$$\begin{aligned}
\mathbb{E}[z^2(t)] &= (\epsilon U_p q S')^2 \mathbb{E}[\Big(\sum_{n=0}^{\infty}(-r)^n \int_0^t W(\alpha\beta^n t_1)\, dt_1\Big)^2] \\
&= (\epsilon U_p q S')^2 \mathbb{E}[\Big(\sum_{n=0}^{\infty}(-r)^n \int_0^t \sqrt{\alpha\beta^n} W(t_1)\, dt_1\Big)^2] \\
&= (\epsilon U_p q S')^2 \Big(\sum_{n=0}^{\infty}(-r)^n \sqrt{\alpha\beta^n}\Big)^2 \mathbb{E}[\Big(\int_0^t W(t_1)\, dt_1\Big)^2] \\
&= \frac{\alpha t^3}{3}(\epsilon U_p q S')^2 \frac{1}{(1+r\beta^{\frac{1}{2}})^2},
\end{aligned} \tag{9.2.2}$$

where we use the scaling property of Brownian motion ($W(\alpha\beta^n t_1) = \sqrt{\alpha\beta^n}$ $W(t_1)$) and $\int_0^t W(t_1)\, dt_1$ is a Gaussian process with zero mean and variance $\frac{t^3}{3}$.

9.2.1 Stochastic Euler equations

The stochastic piston problem can be modeled by the following Euler equations with unsteady stochastic boundary:

$$\frac{\partial}{\partial t}\mathbf{U} + \frac{\partial}{\partial x}(f(\mathbf{U})) = 0, \tag{9.2.3}$$

where $\mathbf{U} = \begin{pmatrix} \rho \\ \rho u \\ E \end{pmatrix}$, $f(\mathbf{U}) = \begin{pmatrix} \rho u \\ \rho u^2 + P \\ u(P+E) \end{pmatrix}$, ρ is density, u is velocity, E is total energy, and P is pressure given by $(\gamma-1)(E - \frac{1}{2}\rho u^2)$ and $\gamma = 1.4$. The initial and boundary conditions are given by

$$u(x,0) = 0,\ P(x,0) = P_+,\ \rho(x,0) = \rho_+,\ x > X_p(t),$$
$$P(X_p(t),0) = P_-,\ \rho(X_p(t),0) = \rho_-,$$

and

$$u(X_p(t),t) = \frac{\partial}{\partial t}X_p(t) = u_p(t),\ t > 0,$$

where $X_p(t)$ is the position of the piston, and $u_p(t)$ is the velocity of the piston.

This problem is a moving boundary problem and can be transformed to a fixed boundary problem by defining a new coordinate (y, τ) from (x, t) via the following transform:

$$y = x - \int_0^\tau u_p(\tau_1, \omega)\, d\tau_1, \quad \tau = t. \tag{9.2.4}$$

Defining $v = u - u_p$, we then have the following Euler equations with a source term [298]:

$$\frac{\partial}{\partial \tau}\mathbf{V} + \frac{\partial}{\partial y}(f(\mathbf{V})) = g(\mathbf{V})\frac{\partial u_p}{\partial \tau}, \tag{9.2.5}$$

where $\mathbf{V} = \begin{pmatrix} \rho \\ \rho v \\ \tilde{E} \end{pmatrix}$, $\tilde{E} = \frac{P}{\gamma - 1} + \frac{1}{2}\rho v^2$ and $g(\mathbf{V}) = \begin{pmatrix} 0 \\ -\rho \\ -\rho v \end{pmatrix}$. The initial and boundary conditions are given by

$$\begin{aligned} v(y,0) = -U_p,\ P(y,0) = P_+,\ \rho(y,0) = \rho_+,\ y > 0, \\ P(0,0) = P_-,\ \rho(0,0) = \rho_-, \end{aligned} \tag{9.2.6}$$

and

$$v(0,\tau) = 0, \ \tau \geq 0.$$

Our goal here is to compute the variance of the shock location perturbation $z(\tau)$. The perturbation of the shock location is $z(\tau) = X_s(\tau) - \tau S = X_s(t) - tS$, where $X_s(\tau) = Y_s(\tau) + \int_0^\tau u_p(t_1)\, dt_1$ is the shock location while $Y_s(\tau)$ is the shock location under the new coordinate (y, τ).

If we take $u_p(t) = U_p(1+\epsilon W(t))$, where $W(t)$ is a scalar Brownian motion, we are led to the following Euler equations

$$\frac{\partial}{\partial \tau}\mathbf{V} + \frac{\partial}{\partial y}(f(\mathbf{V})) = \epsilon U_p g(\mathbf{V}) * \dot{W}, \tag{9.2.7}$$

where "$*$" denotes two different products as follows:

(1) Stratonovich-Euler equations

$$\frac{\partial}{\partial \tau}\mathbf{V} + \frac{\partial}{\partial y}(f(\mathbf{V})) = \epsilon U_p g(\mathbf{V}) \circ \dot{W}, \tag{9.2.8}$$

(2) Ito-Euler equations

$$\frac{\partial}{\partial \tau}\mathbf{V} + \frac{\partial}{\partial y}(f(\mathbf{V})) = \epsilon U_p g(\mathbf{V}) \dot{W}. \tag{9.2.9}$$

The initial and boundary conditions are imposed as above.

We will verify these two models (9.2.8) and (9.2.9) by solving them numerically with a splitting method in the next section.

9.3 Verification of the Stratonovich- and Ito-Euler equations

In the previous section, we introduced two approaches (perturbation analysis and stochastic Euler equations) to obtain the variances of the shock location. Here, we verify the correctness of the stochastic Euler equations by comparing the variances of the shock location obtained by two approaches, i.e., the first-order perturbation analysis and the numerical solution of the stochastic Euler equations, up to time $T = 5$.

For numerical simulations, we consider the piston velocity $U_p = 1.25$, where the Mach number of the shock is $M = 2$ and $\gamma = 1.4$. We normalize all velocities with C_+, the sound speed ahead of the shock, i.e., $C_+ = 1$. Then, the initial conditions are given through the unperturbed relations of state variables [298] as follows:

$$P_+ = 4.5, \quad P_- = 1.0, \quad \rho_+ = 3.73, \quad \rho_- = 1.4.$$

9.3.1 A splitting method for stochastic Euler equations

We use a source-term (noise-term) splitting method proposed in [224] for a scalar conservation law with time-dependent white noise source term. Holden and Risebro [224] considered a Cauchy problem on the whole line with multiplicative white noise in Ito's sense: $\frac{\partial}{\partial t}u + \frac{\partial}{\partial x}f(u) = g(u)\dot{W}(t)$ with deterministic essentially bounded initial condition where f, g are both Lipschitz, and g has bounded support. They proved the *almost-sure-convergence* of this splitting method to a weak solution of the Cauchy problem assuming an initial condition having bounded support and finitely many extrema while no convergence rate was provided. According to our discussion in Chapter 3.4.3 for parabolic equations with multiplicative white noise, the convergence order of splitting methods is usually half ($\sqrt{\Delta\tau}$, $\Delta\tau$ is a time step size for splitting methods) and at most one ($\Delta\tau$).

Here we extend this splitting method to the system (9.2.7). Specifically, given the solution at τ_n, $\mathbf{V}^n$, to obtain the solution at τ_{n+1}, we first solve, on the small time interval $[\tau_n, \tau_{n+1})$,

$$\frac{\partial}{\partial\tau}\mathbf{V}^{(1)} + \frac{\partial}{\partial y}(f(\mathbf{V}^{(1)})) = 0, \tag{9.3.1}$$

with the boundary conditions (9.2.6) and initial condition $\mathbf{V}^{(1)}(\tau_n) = \mathbf{V}^n$; then we solve the following Cauchy problem, again on $[\tau_n, \tau_{n+1})$,

$$\frac{\partial}{\partial\tau}\mathbf{V}^{(2)} = \epsilon U_p g(\mathbf{V}^{(2)}) * \dot{W}, \tag{9.3.2}$$

with the initial condition $\mathbf{V}^{(2)}(\tau_n) = \mathbf{V}^{(1)}(\tau_{n+1})$. Then the solution at time τ_{n+1}, $\mathbf{V}^{n+1}$, is set as $\mathbf{V}^{(2)}(\tau_{n+1})$ (subject to the error from the splitting).

Let us denote by $S(\tau, \tau_n)$ the operator which takes $\mathbf{V}(\tau_n)$ as initial condition at τ_n to the weak solution of (9.3.1) and by $R(\tau, \tau_n)$ the operator which takes the initial condition at time τ_n to the solution of the stochastic differential equation (9.3.2). Then the approximate solution at τ_{n+1} is defined by $\mathbf{V}^{n+1} = R(\tau_{n+1}, \tau_n)S(\tau_{n+1}, \tau_n)\mathbf{V}^n$. Thus, we define a sequence of approximate solution, $\{\mathbf{V}^n\}$, to (9.2.7) at time $\{\tau_n\}$.

The application of splitting technique requires numerical methods for (9.3.1) and (9.3.2). The splitting scheme allows us to employ efficient existing methods to solve them separately. To solve (9.3.1), we use a fifth-order WENO scheme in physical space and second-order strong-property-preserving (SPP) Runge-Kutta in time [253]. In solving (9.3.2), we will employ two different methods: the Monte Carlo method and the stochastic collocation method. We employ 1000 points for the fifth-order WENO scheme over the interval $[0, 5]$ and the time step size $d\tau = 0.0005$ so that the error from time discretization is negligible. As we mentioned before, our goal is to compute the variance of the perturbed shock location. Since there is always only one shock, we obtain $Y_s(\tau)$ by finding the biggest jump of pressure, where the error is of order $\mathcal{O}(dx)$ (dx is the mesh size in physical space).

9.3.2 Stratonovich-Euler equations versus first-order perturbation analysis

We first compare the results obtained by solving the Stratonovich-Euler equations with the Monte Carlo method and those obtained from first-order perturbation analysis.

To solve the Stratonovich-Euler equations (9.2.8) with the splitting method, we solve Equation (9.3.2) by the following Crank-Nicolson scheme

$$\mathbf{V}^{(2)}(\tau_{n+1}) = \mathbf{V}^{(2)}(\tau_n) + \epsilon U_p g(\mathbf{V}^{(2)}(\tau_{n+1/2}))\Delta W_n, \tag{9.3.3}$$

to accommodate the definition of the Stratonovich integral.

In our simulation, the values of function $g(\mathbf{V}^{(2)}(\tau))$ at $\tau_{n+1/2}$ are approximated by the average values $\frac{g(\mathbf{V}^{(2)}(\tau_n))+g(\mathbf{V}^{(2)}(\tau_{n+1}))}{2}$. Note that for the specific form of g, we do not have to invert the resulting matrix in (9.3.3).

Figure 9.2 verifies that the Stratonovich-Euler equations (9.2.8) can capture the variances of shock location for the stochastic piston problem driven by Brownian motion. Here we employ 10,000 realizations so that the statistical error can be neglected for noises with amplitude larger than 0.05 but smaller than 0.5. For noises with amplitude less than 0.05, the error of the adopted methods is dominated by the statistical error from the Monte Carlo method and also the space discretization error from WENO. Figure 9.2 presents the variances obtained by the Monte Carlo method (9.3.1)–(9.3.3) and those from variances estimates by the first-order perturbation analysis (9.2.2). We observe the agreement between the results from the Monte Carlo method and the perturbation analysis within small time and for small

Fig. 9.2. Comparison between the results from first-order perturbation analysis (9.2.2) and solving the Stratonovich-Euler equations (9.2.8) by the splitting method (9.3.1)–(9.3.3).

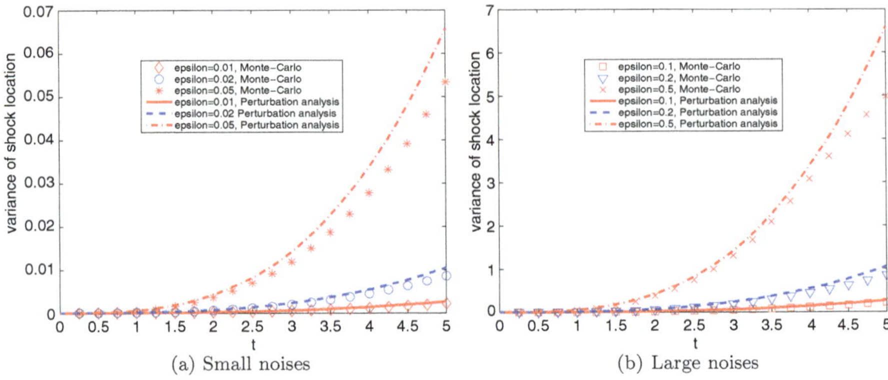

(a) Small noises (b) Large noises

noises. Figure 9.2(a) shows the results for small noises, i.e., $\epsilon \sim O(10^{-2})$ while Figure 9.2(b) for large noises, i.e., $\epsilon \sim O(10^{-1})$. The difference between the variances from the Monte Carlo method and the first-order perturbation analysis (9.2.2) is at most 12% – 13% of the variances (9.2.2), up to time $T = 5$, for all cases except for the case $\epsilon = 0.5$; for the latter, the difference between the variances is at most 19.3% of the variance (9.2.2). However, for small time ($t < 1$) the variances by Monte Carlo and perturbation analysis agree well, while they deviate much after $t = 2$. This effect can be explained as follows. For $t < 1$, the variance of the driving process (Brownian motion) has small value ($\sqrt{t}$) corresponding to a weak perturbation; while at later time it has larger value increasing substantially the perturbation. (We remind the reader that the perturbation process in [298] has unit variance.)

9.3.3 Stratonovich-Euler equations versus Ito-Euler equations

For the Ito-Euler equations (9.2.9), we solve (9.3.2) by the forward Euler scheme

$$\mathbf{V}^{(2)}(\tau_{n+1}) = \mathbf{V}^{(2)}(\tau_n) + \epsilon U_p g(\mathbf{V}^{(2)}(\tau_n))\Delta W_n. \tag{9.3.4}$$

Next we compare the numerical results for the Stratonovich-Euler equations and the Ito-Euler ones using the above discretization in time. We observe from Figure 9.3 that for both small and large noises, these two types of equations have almost the same variances for the perturbed shock location $\mathbb{E}[z^2(t)]$ up to time $T = 5$. Actually, the difference of variances by the Stratonovich-Euler and Ito-Euler equations for $\epsilon \leq 0.2$ is less than 10^{-3} up to time $t = 5$ which lies within the discretization errors. For $\epsilon = 0.5$, we present in Table 9.1 the difference of variances for these two approaches using the same sequence of Monte Carlo points. The Stratonovich-Euler equations exhibit larger variances in large time but the difference from those by the

Fig. 9.3. Comparison between solving Stratonovich-Euler equations (9.2.8) and Ito-Euler equations (9.2.9) by the splitting method (9.3.1)–(9.3.2).

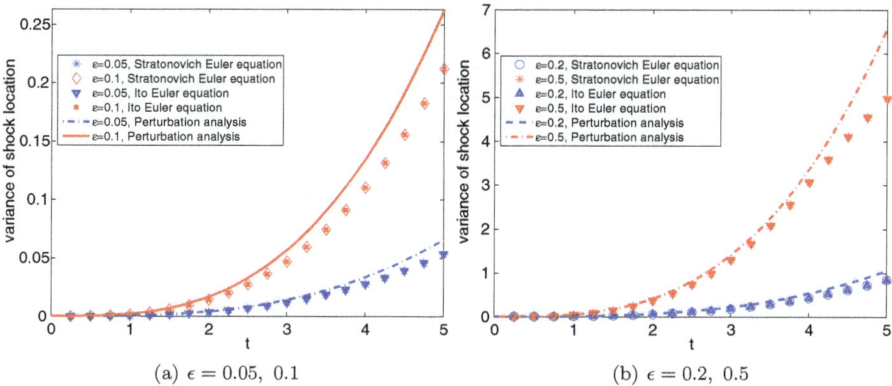

(a) $\epsilon = 0.05,\ 0.1$

(b) $\epsilon = 0.2,\ 0.5$

Ito-Euler equations is less than 10% of the variances by Ito-Euler equations. We then conclude that the Stratonovich-Euler equations are a suitable model for the piston problem driven by Brownian motion and we will consider only this approach hereafter.

Table 9.1. The difference of variances of shock location by Stratonovich-Euler and Ito-Euler equations for $\epsilon = 0.5$.

t	1.0	2.0	3.0	4.0	5.0
Error in variance	0.0007	0.0129	0.0742	0.2353	0.2421

9.4 Applying the stochastic collocation method

Next we test the stochastic collocation method versus the Monte Carlo method for the Stratonovich-Euler equations (9.2.8). To solve the Stratonovich-Euler equations (9.2.8), we again use the splitting method (9.3.1)–(9.3.2). In (9.3.2), we adopt the stochastic collocation method, where we first introduce a spectral approximation for the Brownian motions and subsequently apply the sparse grid method. Specifically, we first approximate Brownian motion with its spectral approximation, using K multi-elements:

$$W^{(\mathsf{n},\mathsf{K})}(\tau) = \sum_{k=0}^{\mathsf{K}-1} \sum_{i=1}^{n} \int_0^{\tau} \chi_{[\mathsf{t}_k,\mathsf{t}_{k+1})}(s) m_{k,i}(s)\, ds \xi_{k,i}, \quad \tau \in [0,T],$$

where $0 = \mathsf{t}_0 < \mathsf{t}_1 < \cdots < \mathsf{t}_\mathsf{K} = T$, $\chi_{[\mathsf{t}_k,\mathsf{t}_{k+1})}(\tau)$ is the indicator function of the interval $[\mathsf{t}_k, \mathsf{t}_{k+1})$, $\{m_{k,i}\}_{i=1}^{\infty}$ is a complete orthonormal basis in $L^2([\mathsf{t}_k, \mathsf{t}_{k+1}])$, and $\xi_{k,i}$ are mutually independent standard Gaussian random variables (with zero mean and variance one). Hence, we obtain the following partial differential equation with smooth inputs:

$$\frac{\partial}{\partial \tau}\mathbf{V}^{(2)} = \epsilon U_p g(\mathbf{V}^{(2)}) \sum_{k=0}^{\mathsf{K}} \sum_{i=1}^{\mathsf{n}} \chi_{[\mathsf{t}_{k-1},\mathsf{t}_k)}(\tau) m_{k,i}(\tau)\xi_{k,i}. \tag{9.4.1}$$

In (9.4.1) we apply the stochastic collocation method [11, 439, 486] for smooth noises. The stochastic collocation method we adopt here is the sparse grid of Smolyak type based on 1D Gaussian-Hermite quadrature, see, e.g., [148] and also Chapter 2.5.4. We use in this chapter the Matlab code('nwspgr.m') at `http://www.sparse-grids.de/`.

The first issue we have for the piston problem here is the discontinuity of the solution to (9.2.8), where the condition for spectral approximation to work may be invalid [434]. In practice, we solve the problem with the WENO scheme, which smears the shock somewhat, and thus we have higher regularity than that of the original problem. A second issue is that the use of the stochastic collocation method (Smolyak sparse grid) with Gaussian quadrature may not exhibit fast convergence because of the low regularity. Thus, we use $\mathsf{n} = 1$ or 2 with large K (small time step in $W^{(\mathsf{n},\mathsf{K})}$) instead of large n with small K. This choice of n is verified with control tests with $\mathsf{n} = 3,\ 4$ for different K, where the numerical results show large deviations from those of Monte Carlo method with high oscillations. We choose a low sparse grid level (i.e., $\mathsf{L} = 2$) to be consistent with the "available regularity" (numerical tests with high sparse grid level show an instability). The third issue is the so-called "curse-of-dimensionality." In practice, when the number of random variables, Kn, increases, the Smolyak sparse grid method will not work well and will be replaced by the QMC method.

Here we adopt a uniform partition of the time interval $[0,T]$, that is $\mathsf{t}_k = (k-1)\Delta$, $k = 1, \cdots, \mathsf{K}$. The complete orthonormal basis we employ in $L^2([\mathsf{t}_k, \mathsf{t}_{k+1}])$ is the cosine basis

$$m_{k,1}(t) = \frac{1}{\sqrt{\Delta}}, \quad m_{k,i}(t) = \sqrt{\frac{2}{\Delta}} \cos\left(\frac{(i-1)\pi}{\Delta}(t-\mathsf{t}_k)\right), \ i \geq 2.$$

Figure 9.4 compares the numerical results from the Monte Carlo method (9.3.1)–(9.3.3) and the stochastic collocation method for (9.3.1) and (9.4.1) with both small and large noises. For each ϵ, we use different Δ (the length of the uniform partition of time interval $[0,T]$), i.e., different size of elements K. We note that all the numerical solutions obtained by the stochastic collocation method agree with those from the Monte Carlo method (9.3.1)–(9.3.3) within small time. Here we do not observe convergence in n, recalling that such convergence requires smoothness in random space.

We note that smaller Δ and larger n may lead to a larger number of random variables and thus the break down of the sparse grid method [486].

So we first test the cases of small Δ such that we can apply the sparse grid method. Figure 9.4 shows that a low level sparse grid method works well for the piston problem with small perturbations. We note that our sparse grid level is two and thus the number of collocation points is $2\mathsf{n}\frac{T}{\Delta}+1$.

When $\mathsf{n} = 1$, we observe in Figure 9.4 good agreement of the results by the stochastic collocation method and the Monte Carlo method in small time ($t \leq 2$). Notice that when $\mathsf{n} = 1$, (9.4.1) is the classical Wong-Zakai approximation [481]

$$\frac{\partial}{\partial \tau}\mathbf{V}^{(2)} = g(\mathbf{V}^{(2)})\frac{1}{\sqrt{\Delta}}\sum_{k=0}^{\mathsf{K}-1}\chi_{[\mathsf{t}_k,\mathsf{t}_{k+1})}(\tau)\xi_{k,1}. \tag{9.4.2}$$

Fig. 9.4. Comparison between numerical results from Stratonovich-Euler equations (9.2.8) using the Monte Carlo method (9.3.1) and (9.3.3) and the stochastic collocation method (9.4.1). The sparse grid level is 2 and Δ is the size of element in time in the stochastic collocation method.

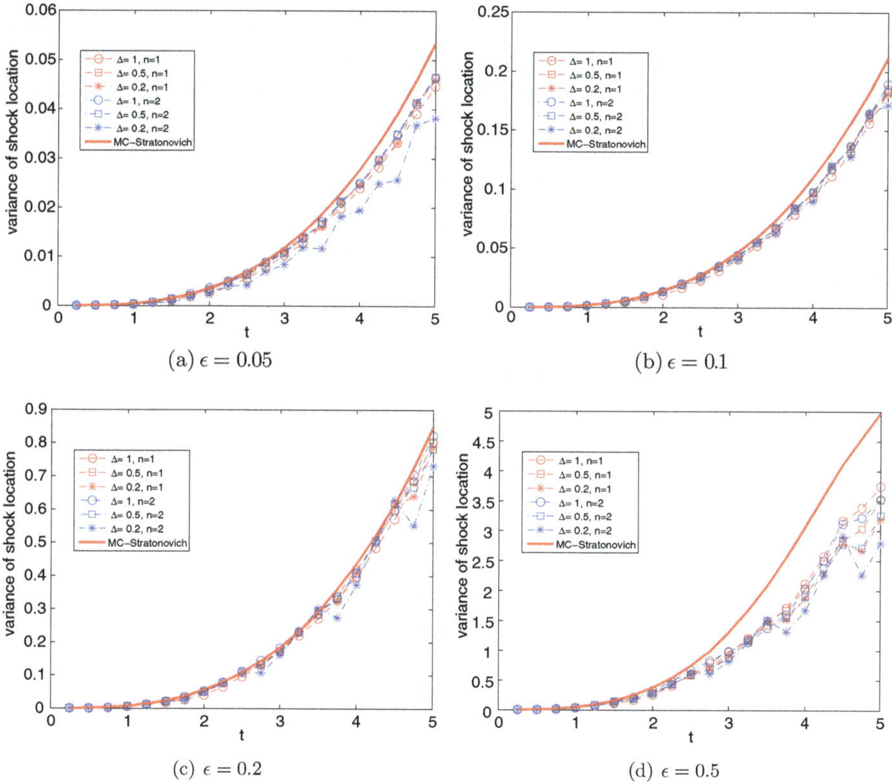

(a) $\epsilon = 0.05$

(b) $\epsilon = 0.1$

(c) $\epsilon = 0.2$

(d) $\epsilon = 0.5$

However, for $\mathsf{n} = 2$, there are some disagreements between the results. In Figure 9.4(a) and 9.4(c), the results of the case $\mathsf{n} = 2$ and $\Delta = 0.2$ (note that we have $\mathsf{nK} = 50$ random variables) underestimate those results from the Monte Carlo method and the stochastic collocation method with a smaller number of random variables ($\mathsf{n} = 1$). The larger number of random variables ($\mathsf{n} = 2$ here) does not result in convergence since we do not have a smooth solution as we mention above.

For the case with large perturbation, $\epsilon = 0.5$, we require smaller Δ and thus more random variables. This is why we observe the disagreement in Figure 9.4(d). For all cases in Figure 9.4, we observe a deviation of numerical results by stochastic collocation methods from those of Monte Carlo method over large time. Similar effects arise in the application of spectral methods in random space, e.g., in Wiener chaos methods. The interested reader may refer to [505] for a discussion of this effect.

To adapt to the high dimensionality (large number of random variables), we employ the QMC method instead of sparse grid methods. We consider two popular QMC sequences: one is a scrambled Halton with the method RR2 proposed in [262]; and the other is a scrambled Sobol sequence suggested in [340]. Both sequences lie in hypercube and thus an inverse transformation is adopted to generate sequences in the entire space based on these two sequences. The Matlab code for generating a scrambled Halton sequence is listed in Code 2.4, and the code for a scrambled Sobol sequence is listed in Code 2.5. In Figure 9.5, we test the large noise case, i.e., $\epsilon = 0.5$. Both Halton and Sobol sequences work if a moderately large sample of the sequences is adopted. For 1000 sample points, variances from both sequences are closer to those from Monte Carlo method (9.3.1)–(9.3.3) than those from 500 sample points of both sequences.

9.5 Summary and bibliographic notes

In this chapter we demonstrated how to apply the stochastic collocation method to nonlinear conservation laws driven by white noise. Specifically, we simulated a stochastic piston problem, where a piston is pushed with velocity being time-varying Brownian motion and ahead of the piston there is an adiabatic tube of constant area.

- This one-dimensional problem is modeled with the Euler equations driven by white noise, and we verified the simulation results with the first-order stochastic perturbation analysis presented in [298]. The stochastic products can be either an Ito product or a Stratonovich product. We showed numerically that the Stratonovich-Euler equations are a suitable model for the piston problem driven by Brownian motion.
- By splitting the Euler equations into two parts – a "deterministic part" and a "stochastic part" – we solved the "stochastic part" by the Monte Carlo method and the stochastic collocation method.

Fig. 9.5. Comparison between numerical results from Stratonovich-Euler equations (9.2.8) using direct Monte Carlo method (9.3.1)–(9.3.3) and the QMC method for (9.4.1) with a large noise: $\epsilon = 0.5$.

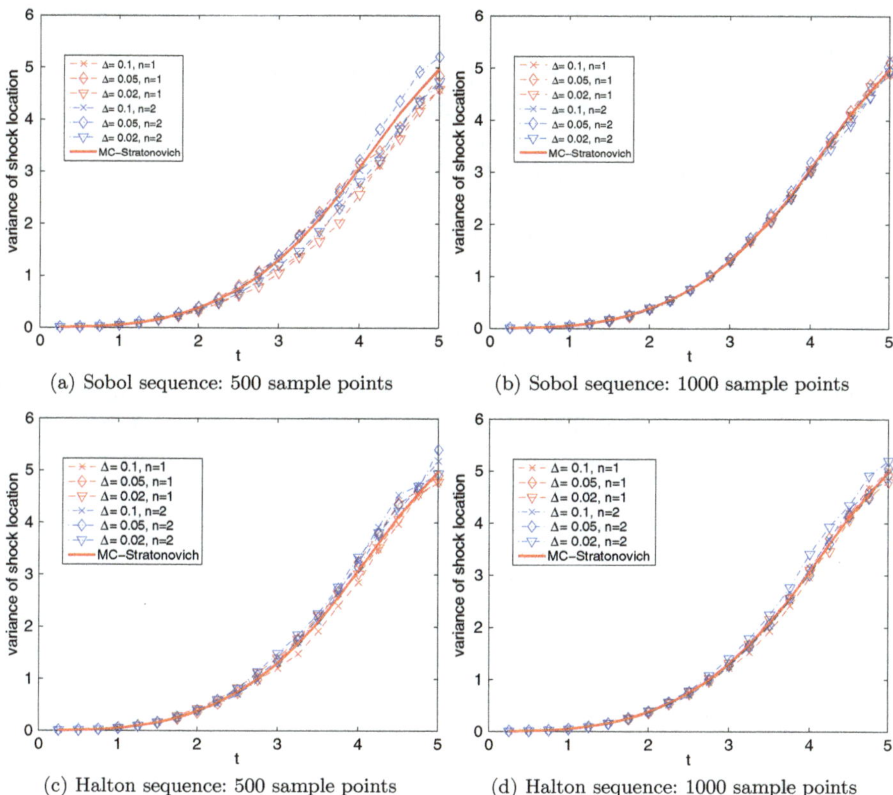

(a) Sobol sequence: 500 sample points

(b) Sobol sequence: 1000 sample points

(c) Halton sequence: 500 sample points

(d) Halton sequence: 1000 sample points

- QMC is employed for more accurate longer-time integration. The sparse grid collocation method is only efficient for short time integration (with large magnitudes of noises) and for a relative long time integration (with small magnitudes of noises). For large noises, we need small time-interval Δ for the stochastic collocation method to converge. For smaller time-interval Δ which leads to larger number of random variables, the QMC method leads to accurate solutions.
- We tested two types of QMC sequences for the "stochastic part" using a multi-element spectral approximation of the Brownian motion when the noise is large. The stochastic collocation and QMC methods are superior to the Monte Carlo method in the sense that they can achieve faster convergence than the classic Monte Carlo method.

The low accuracy of the stochastic collocation method, especially for long times, is caused by the discontinuity of the solution. Due to the deterministic solver, we have that the accuracy for the numerical shock location is only first-order in the spatial step size, i.e., $\mathcal{O}(dx)$ where dx is the mesh size in physical space.

With regards to computational efficiency, the stochastic collocation method is more efficient than Monte Carlo simulation when a small number of random variables is involved, where the number of collocation points is far less than Monte Carlo sampling points. As time becomes larger, we introduce more random variables and thus we need to employ the more efficient QMC method. In other applications involving long-time integration, it may be possible to use all three different ways of sampling, i.e., starting with sparse grid for early time, continuing with the QMC for moderate time and even switching to the Monte Carlo method for long time.

Bibliographic notes. The concept of a moving piston was used by the English physicist James Joule to demonstrate the mechanical equivalent of heat in his pioneering studies, almost two centuries ago. In the last century, the moving piston has also been used extensively in fundamental studies of fluid mechanics and shock discontinuities. This now classical problem has been solved analytically in one dimension and also in higher space dimensions see, e.g., [84, 478].

However, many variants of the moving piston problems cannot be solved analytically. In gas dynamics, in particular, in the context of normal shock waves, the one-dimensional classical problem describes a piston moving at constant speed in a tube of constant area and adiabatic walls; the shock wave is created ahead of the piston. Closed-form analytical solutions of this flow problem with general time-dependent piston speeds are difficult to obtain; see semi-analytical solutions in [292] for accelerating and decelerating pistons that are valid only for short times.

In [298], a moving piston with random velocity was investigated numerically, where the effect of random velocity was measured by variances of the shock location. Also, a stochastic perturbation analysis was also performed as reference solutions. When the random velocity is a Gaussian random field with zero mean and exponential covariance kernel $\exp(-|t_1 - t_2|)$, it was shown that the mean of the shock location is the same as the shock location under a constant velocity while the variance of the shock location is growing quadratically with the time t.

In this chapter, we show that the variances of the shock location grow *cubically* with time when the random velocity is the standard Brownian motion. These variances are significantly different from those from a piston driven by color noise. In Figure 9.6 we compare the variances of shock positions induced by three different Gaussian noises: Brownian motion, stochastic process with zero mean and exponential covariance kernel $\exp(-|t_1 - t_2|)$ (see [298]), and standard Gaussian random variable, where the noise amplitude is $\epsilon = 0.1$. The results are obtained via the stochastic perturbation analysis in [298].

Fig. 9.6. Comparison among variances of shock positions induced by three different Gaussian noises: Brownian motion, stochastic process with zero mean and exponential covariance kernel $\exp(-|t_1 - t_2|)$, and standard Gaussian random variable. The noise amplitude is $\epsilon = 0.1$.

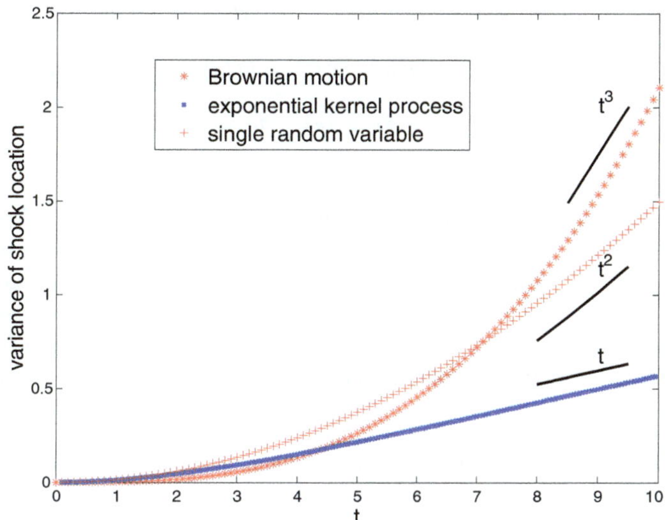

The case of Brownian motion induces smaller values of variances than the other two cases for short times and greater values of variances for longer times.

Polynomial chaos methods have been applied to solve stochastic hyperbolic problems with shocks, see, e.g., [343, 395]. Non-polynomial basis based chaos methods, e.g., wavelet based chaos methods proposed for smooth solutions in [294, Chapter 8], are used in [394] for solutions with strong discontinuities.

Some other statistical-error-free methods other than function expansion methods have also been proposed, e.g., equations of probability density function (PDF) methods [77, 461] and equations of cumulative distribution function (CDF) methods [471]. In these methods, some integro-differential equations for PDF or CDF methods are derived where numerical methods for deterministic equations are then applied to obtain PDF/CDF numerically.

The stochastic shock problem is listed as an open computational problem in [294] and it remains open today. An analytical methodology for wave propagation in random media has been presented in [134].

9.6 Suggested practice

Exercise 9.6.1 *Consider the following stochastic Burgers equation* $(0, T] \times (0, 1)$*:*

$$\partial_t u + (u + \sigma \dot{W}(t)) \circ \partial_x u = \mu \partial_x^2 u, \quad u(0,x) = u_0(x),\, u(t,0) = u(t,1). \quad (9.6.1)$$

Here $u_0(x) = \sin(2\pi x)$ *is a deterministic function.*

- *Write down the equation in Ito form.*
- *Solve the problem using the following splitting method*

$$\partial_t \bar{u} + \bar{u}\partial_x \bar{u} = \mu \partial_x^2 \bar{u}, \quad \bar{u}(t_n) = \tilde{u}(t_n), \quad t \in (t_n, t_{n+1}], \quad (9.6.2)$$

$$\partial_t \tilde{u} = \sigma \dot{W}(t)) \circ \partial_x \tilde{u}, \quad \tilde{u}(t_n) = \bar{u}(t_{n+1}), \quad t \in (t_n, t_{n+1}]. \quad (9.6.3)$$

Apply the explicit fourth-order Runge-Kutta scheme for the first equation and the midpoint scheme to solve the second equation. In random space, use the Monte Carlo method. The total error of this splitting scheme is dominated by the splitting error, which is expected to be of half-order in the mean-square sense. Check numerically that the convergence order is indeed half. Make sure the statistical error is much smaller than the integration error.

Hint. A solution to this problem can be obtained semi-analytically, see Appendix B.

Exercise 9.6.2 *Apply Ito product in Equation* (9.6.1) *instead of Stratonovich product and redo the last exercise.*

Part III

Spatial White Noise

In this part, we discuss numerical methods for spatial white noise. Specifically, we consider a semilinear elliptic equation with additive noise in Chapter 10 and an elliptic equation with multiplicative noise in Chapter 11.

After the truncation of Brownian motion (or white noise), the resulting equations are usually solved with standard numerical methods such as finite elements methods, finite difference methods, and spectral methods. We present some error estimates of finite element approximation of the resulting equations as well as the error of truncating the Brownian motion. It is shown that the convergence rate of the finite elements methods is the same as that for deterministic elliptic equations when one- or two-dimensional problems are considered. For additive noise, the convergence rate in three-dimensional space is lower than in the deterministic case because of the low regularity of spatial white noise.

For elliptic equations with coefficients that is spatial noise, the solutions may not be square-integrable in random space and instead they may lie in some weighted stochastic Sobolev space. However, when the solutions are square-integrable, the solutions can be numerically obtained with high accuracy using Wiener chaos expansion methods (WCE). Moreover, a perturbation technique called Wick-Malliavin approximation can be applied to reduce the computational cost of WCE.

10

Semilinear elliptic equations with additive noise

With temporal white noise, the solutions of stochastic parabolic equations have low regularity in time (Hölder continuous with exponent $1/2 - \epsilon$ and $\epsilon > 0$ is arbitrarily small) and thus the spectral approximation of Brownian motion leads to only half order convergence in its truncation mode n. With spatial white noise, however, the solution can be smoother and we can expect higher-order convergence with spectral approximation of Brownian motion.

We investigate in this chapter the strong and weak convergence order of piecewise linear finite element methods for a class of semilinear elliptic equations with additive spatial white noise using a spectral truncation of white noise. We show that the strong convergence order of the finite element approximation is $h^{2-d/2}$, where h is the element size and $d \leq 3$ is the dimension. We also show that the weak convergence order is $h^{\min(4-d,2)}$. Moreover, we consider a fourth-order equation and show that a spectral approximation of the white noise can lead to higher convergence order in both strong and weak sense when the solutions are smooth. Numerical results confirm our prediction for one- and two-dimensional elliptic problems.

10.1 Introduction

Let us first consider strong and weak convergence orders for spectral truncation of white noise in a linear elliptic equation with additive noise.

Z. Zhang, G.E. Karniadakis, *Numerical Methods for Stochastic Partial Differential Equations with White Noise*,
Applied Mathematical Sciences 196, DOI 10.1007/978-3-319-57511-7_10

Example 10.1.1 (Linear elliptic equation, see, e.g., [6, 118]) *Consider the following linear problem:*

$$-\partial_x^2 u + bu = g(x) + \partial_x W(x), \quad x \in (0,1),$$
$$u(0) = u(1) = 0,$$

where $b > -\pi^2$ and $g \in L^2([0,1])$.

The solution can be represented by $u = \sum_{k=1}^{\infty} \frac{\xi_k + g_k}{b + k^2\pi^2} e_k$, where $g_k = \int_0^1 g e_k \, dx$, $\frac{\partial}{\partial x} W(x) = \sum_{k=1}^{\infty} e_k \xi_k$, and ξ_k's are i.i.d. standard Gaussian random variables. The basis $\{e_k\}_{k=1}^{\infty}$ can be any orthonormal basis in $L^2([0,1])$, e.g., $e_k = \sqrt{2}\sin(k\pi x)$, which are the corresponding eigenfunctions of $-\Delta u = \lambda u$ over $[0,1]$ with $u(0) = u(1) = 0$. The first two moments of u are

$$\mathbb{E}[u] = \sum_{k=1}^{\infty} \frac{g_k}{b + k^2\pi^2} e_k \quad \mathbb{E}[\|u\|^2] = \sum_{k=1}^{\infty} \frac{1 + g_k^2}{(b + k^2\pi^2)^2} = \sum_{k=1}^{\infty} \frac{1}{(b + k^2\pi^2)^2} + \|\mathbb{E}[u]\|^2 .$$

It can be readily checked that there exists $C > 0$ independent of n such that

$$\mathbb{E}[\|u - u_{\mathsf{n}}\|^2] = \mathbb{E}[\|u\|^2 - \|u_{\mathsf{n}}\|^2] \le C \frac{1}{\mathsf{n}^3},$$

where $u_{\mathsf{n}} = \sum_{k=1}^{\mathsf{n}} \frac{\xi_k + g_k}{b + k^2\pi^2} e_k$. In this example, *the weak convergence order is twice of the mean-square convergence order.* This conclusion holds for nonlinear equations as well when some mild assumptions on nonlinear terms are made.

In this chapter, we study the numerical approximation of the following semilinear elliptic equation with additive white noise using a spectral approximation of spatial white noise:

$$-\Delta u(x) + f(u(x)) = g(x) + \frac{\partial^d}{\partial x_1 \partial x_2 \cdots \partial x_d} W(x), \quad x = (x_1, \cdots, x_d) \in \mathcal{D}, \tag{10.1.1}$$

with Dirichlet boundary condition

$$u(x) = 0, \quad x \in \partial\mathcal{D}, \tag{10.1.2}$$

where $\mathcal{D} = (0,1)^d$, $W(x)$ is a Brownian sheet on $\bar{\mathcal{D}} = [0,1]^d$, $f, g \in L^2(\mathcal{D})$ so that (10.1.1) is well posed, see Chapter 10.2 for details.

The piecewise linear approximation for Brownian motion leads to a piecewise constant approximation of white noise, which has been widely used in approximating temporal noise for solving SDEs (see, e.g., [259] and [358]) as well as in approximating spatial noises (see, e.g., [6, 64, 118, 194]). Using piecewise constant approximation, Gyongy and Martinez [194] considered a finite difference scheme in physical space for the problem (10.1.1) and obtained dimension-dependent convergence order in the mean-square sense: h if $d = 1$; and $h^{2-\frac{d}{2}-\epsilon}$ if $d = 2, 3$, where h is the finite difference step size. Here and throughout this chapter, $\epsilon > 0$ is an arbitrary small constant. For finite element methods for (10.1.1) in physical space, [6] obtained the mean-square convergence order h for a one-dimensional linear problem; also [64] considered a two-dimensional problem (10.1.1) over a general bounded convex domain and established the mean-square convergence order $h^{1-\epsilon}$. In other words, the finite element methods basically yield the same mean-square convergence order as the finite difference methods do for $d = 1$, 2, when the piecewise constant approximation of white noise is used.

With a spectral approximation of the spatial additive noise, we will show that the mean-square convergence order is $h^{2-d/2}$, where we use piecewise linear finite element approximation in physical space, see Theorem 10.4.1. Specifically, in the one-dimensional case, we obtain the mean-square convergence order $h^{3/2}$ instead of h from the piecewise constant approximation of white noise [6]. We note that for $d = 1$, the solution is actually in $H^{3/2-\epsilon}(\mathcal{D})$ and the spectral approximation benefits from the smoothness of the solution as will be shown in Chapter 10.2, where we also show similar effects for fourth-order equations.

We show that the weak convergence order of the spectral approximation of white noise is twice its mean-square convergence order when only the white noise is discretized in (10.1.1), see Theorem 10.3.2. While further discretizing (10.1.1) with a piecewise linear finite element method, we show that the weak error is $h^{\min(4-d,2)}$ for $d \leq 3$, see Theorem 10.4.3. We present some numerical results for one- and two-dimensional semilinear elliptic equations in Chapter 10.5. At the end of this chapter, we summarize the main points of the chapter and present a review on piecewise constant approximation of white noise and discuss some disadvantage of spectral approximation of white noise. Some exercises are provided.

10.2 Assumptions and schemes

For (10.1.1) to be well posed, we require the following assumption as in [48, 64, 194].

Assumption 10.2.1 *The function f satisfies the following conditions:*

- *There exists a constant $L < C_p$ such that*

$$[f(s) - f(t)](s - t) \geq -L\,|s - t|^2\,, \quad \forall s, t \in \mathbb{R}. \tag{10.2.1}$$

- *There exist constants $M \geq 0$ and $R \geq 0$ such that*

$$|f(s) - f(t)| \leq M + R\,|s - t|\,, \quad \forall s, t \in \mathbb{R}. \tag{10.2.2}$$

Here C_p is the constant in the Poincare inequality:

$$\|\nabla v\|^2 \geq C_p \|v\|^2, \quad v \in H_0^1(\mathcal{D}).$$

Under Assumption 10.2.1, the solution to (10.1.1) is proved to exist and be unique in $\mathbb{L}^p(\Omega, L^2(\mathcal{D}))$ when $d \leq 3$ [48]. For $d > 4$, for Equation (10.1.1) to be well posed in $\mathbb{L}^p(\Omega, L^2(\mathcal{D}))$, [337] considered additive color noise instead of white noise. The solution to (10.1.1) is understood in the sense of a mild solution:

$$u(x) + \int_{\mathcal{D}} \mathcal{K}(x,y) f(u(y))\,dy = \int_{\mathcal{D}} \mathcal{K}(x,y) g(y)\,dy + \int_{\mathcal{D}} \mathcal{K}(x,y)\,dW(y), \tag{10.2.3}$$

where $\mathcal{K}(x,y)$ is Green's function of the Poisson equation.

Remark 10.2.2 *The assumption* (10.2.1) *can allow the nonlinear function f of the following form $f = f_1 + f_2$, where f_1 is Lipschitz continuous with a small Lipschitz constant and f_2 can be a sum of nondecreasing bounded functions and a Lipschitz continuous function. For example, f_2 can be the sign function* $\operatorname{sgn}(x)$ *or Heaviside function or a summation of both functions.*

Here we represent the spatial white noise $\dot{W}(x)$ with an orthogonal series expansion

$$\frac{\partial^d}{\partial x_1 \partial x_2 \cdots \partial x_d} W(x) = \sum_{\alpha \in \mathcal{J}} e_\alpha(x) \xi_\alpha, \tag{10.2.4}$$

or for the spatial Brownian motion (Brownian sheet)

$$W(x) = \sum_{\alpha \in \mathbb{N}^d} \int_0^{x_d} \int_0^{x_{d-1}} \cdots \int_0^{x_1} e_\alpha(y)\,dy_1 \cdots dy_d \xi_\alpha, \tag{10.2.5}$$

where $\{e_\alpha(x)\}_{|\alpha|=1}^{\infty}$ is a complete orthonormal basis in $L^2(\mathcal{D})$; ξ_α, $\alpha = (\alpha_1, \alpha_2, \alpha_3)$ are independent standard Gaussian random variables. In practice, we can take any orthonormal basis in $L^2(\mathcal{D})$. Here we take the eigenfunctions of the elliptic equation

$$-\Delta \psi = \lambda \psi, \ x \in \mathcal{D} \quad \psi = 0, x \in \partial \mathcal{D}, \tag{10.2.6}$$

which can form an orthonormal basis in $L^2(\mathcal{D})$. We denote the truncation of $W(x)$ (10.2.5) by W_{n}:

$$W_{\mathsf{n}}(x) = \sum_{|\alpha| \leq \mathsf{n},\ \alpha \in \mathbb{N}^d} \int_0^{x_d} \int_0^{x_{d-1}} \cdots \int_0^{x_1} e_\alpha(y)\,dy_1 \cdots dy_d \xi_\alpha. \tag{10.2.7}$$

A semi-discrete scheme of (10.1.1) (to be precise, (10.2.3)) is then as follows

$$u_{\mathsf{n}}(x) + \int_{\mathcal{D}} \mathcal{K}(x,y) f(u_{\mathsf{n}}(y))\,dy = \int_{\mathcal{D}} \mathcal{K}(x,y) g(y)\,dy + \int_{\mathcal{D}} \mathcal{K}(x,y)\,dW_{\mathsf{n}}(y), \tag{10.2.8}$$

which is equivalent to

$$- \Delta u_{\mathsf{n}}(x) + f(u_{\mathsf{n}}(x)) = g(x) + \frac{\partial^d}{\partial x_1 \partial x_2 \cdots \partial x_d} W_{\mathsf{n}}(x). \tag{10.2.9}$$

The semi-discrete scheme (10.2.9) requires further discretization in physical space. Here we consider a finite element approximation. Let V_h be a linear finite element subspace of $H_0^1(\mathcal{D})$ with quasi-uniform triangulation T_h. Then, the linear finite element approximation of u_{n} in (10.2.9) is to find $u_{\mathsf{n}}^h \in V_h$ such that

$$(\nabla u_{\mathsf{n}}^h, \nabla v) + (f(u_{\mathsf{n}}^h), v) = (g + \frac{\partial^d}{\partial x_1 \cdots \partial x_d} W_{\mathsf{n}}, v), \quad \forall v \in V_h. \tag{10.2.10}$$

Before ending this section, we remark that this approach can be further extended as follows:

- The domain $\mathcal{D}$ can be a bounded domain with a smooth boundary $\partial \mathcal{D}$, while conclusions in this chapter remain true. For example, one can consider the problem in [64] where the domain $\mathcal{D}$ is bounded convex.
- The operator Δ can be replaced by general self-adjoint, positive-definite, linear operators, say $\mathcal{A}$, with compact inverse.

We emphasize again that any orthonormal basis in $L^2(\mathcal{D})$ can be used in the spectral expansion (10.2.5), though it may be convenient to use the eigenfunctions of $\mathcal{A}$ as a basis if they can be explicitly obtained.

10.3 Error estimates for strong and weak convergence order

In this section, we will discuss strong and weak convergence orders.

Theorem 10.3.1 (Strong error) *Let u be the solution to* (10.1.1) *and u_{n} the solution to* (10.2.9). *Under Assumption 10.2.1, we have*

$$\mathbb{E}[\|u - u_{\mathsf{n}}\|^2] \leq C\big(M(\sum_{|\alpha|=\mathsf{n}+1}^{\infty} \lambda_\alpha^{-2})^{1/2} + \sum_{|\alpha|=\mathsf{n}+1}^{\infty} \lambda_\alpha^{-2}\big) \leq C\big(M\mathsf{n}^{-(2-d/2)}$$
$$+(C_p + R)^2 \mathsf{n}^{-(4-d)}\big),$$

where the constant C depends only on d, C_p, L, M, R and λ_α are eigenvalues of the problem (10.2.6). *The constants L, M, R are from Assumption 10.2.1.*

Under further smoothness of the nonlinear function f, we show that the weak convergence order is twice of the mean-square convergence order.

Theorem 10.3.2 (Weak error) *Let u be the solution to* (10.1.1) *and u_{n} the solution to* (10.2.9)*. In addition to Assumption 10.2.1, assume also that f and F and their derivatives up to fourth-order are of at most polynomial growth at infinity:*

$$\left|\frac{d^k}{dx^k}G(x)\right| \leq c(1+|x|^{\varkappa}), \quad \varkappa < \infty, k = 1,2,3,4 \text{ and } G = f,\ F. \tag{10.3.1}$$

Furthermore, we assume that $M = 0$ in Assumption 10.2.1. Then we have

$$\|\mathbb{E}[F(u)] - \mathbb{E}[F(u_{\mathsf{n}})]\|_{L^q} \leq C \sum_{|\alpha|=\mathsf{n}+1}^{\infty} \lambda_\alpha^{-2} \leq C\mathsf{n}^{-(4-d)}, \quad 1 \leq q < \infty. \tag{10.3.2}$$

The constant C depends on d, C_p, L, M, R as in Theorem 10.3.1 and also the constant in (10.3.1).

10.3.1 Examples of other PDEs

It seems that the fact that the weak convergence order is twice that of the strong-order convergence is quite general. In the following example, we show that it is true for an advection equation with multiplicative noises and a fourth-order equation with additive noise using the spectral approximation (10.2.7).

Example 10.3.3 (Advection-reaction, see, e.g., [438])

$$\partial_t u + \partial_x u = \sigma(u-1) \circ \dot{W}(x), \qquad x \in [0, L] \tag{10.3.3}$$

with initial condition $u_0(x)$ and zero inflow. The stochastic product $u \circ \dot{W}$ is the Stratonovich product. Equation (10.3.3) *can be written in Ito's form*

$$\partial_t u + \partial_x u = \frac{\sigma^2}{2}(u-1) + \sigma(u-1) \diamond W(x), \qquad x \in [0, L], \tag{10.3.4}$$

where "$\diamond$" represents the Ito-Wick product. The exact solution of (10.3.3) *is*

$$u = 1 + [u_0(x-t) - 1]\exp[\sigma W(x) - \sigma W(x-t)]. \tag{10.3.5}$$

Applying the truncated spectral expansion (10.2.7) *in one-dimensional physical space, we then have the following approximation to Equation* (10.3.3)*:*

$$\partial_t u_{\mathsf{n}} + \partial_x u_{\mathsf{n}} = \sigma(u_{\mathsf{n}} - 1)\frac{d}{dx}W_{\mathsf{n}}(x), \qquad x \in [0, L] \tag{10.3.6}$$

whose solution is

$$u_{\mathsf{n}} = 1 + [u_0(x-t) - 1]\exp[\sigma W_{\mathsf{n}}(x) - \sigma W_{\mathsf{n}}(x-t)]. \tag{10.3.7}$$

Theorem 10.3.4 *Let u be the solution to (10.3.3) and u_{n} the solution to (10.3.6). Then we have*

$$\mathbb{E}[|u-u_{\mathsf{n}}|^2] \le C_1\frac{1}{\mathsf{n}}, \quad \left|\mathbb{E}[u^k - u_{\mathsf{n}}^k]\right| \le C_2\frac{1}{\mathsf{n}}, \quad \forall k>0, \tag{10.3.8}$$

where C_1 and C_2 depend only on t, x, and σ in the former inequality and C_2 depends also on k in the latter.

Proof. We first prove the strong convergence. Note that

$$\mathbb{E}[(u-u_{\mathsf{n}})^2] = (u_0(x-t)-1)^2\mathbb{E}[(\exp(\sigma W(x)-\sigma W(x-t)) \\ - \exp(\sigma W_{\mathsf{n}}(x)-\sigma W_{\mathsf{n}}(x-t)))^2]. \tag{10.3.9}$$

By the fact that $\exp(a)-\exp(b) = \exp(\theta a + (1-\theta)b)(a-b)$ where $0\le\theta\le 1$ and the Cauchy-Schwarz inequality, we have

$$\begin{aligned}
&\mathbb{E}[(u-u_{\mathsf{n}})^2] \\
&= (u_0(x-t)-1)^2\mathbb{E}[\big(\exp(\sigma W(x)-\sigma W(x-t)) - \exp(\sigma W_{\mathsf{n}}(x)-\sigma W_{\mathsf{n}}(x-t))\big)^2] \\
&\le (u_0(x-t)-1)^2\big(\mathbb{E}[\exp(4\sigma\theta(W(x)-W(x-t))+4\sigma(1-\theta)(W_{\mathsf{n}}(x)-W_{\mathsf{n}}(x-t)))]\big)^{1/2} \\
&\quad \times\sigma^2\big(\mathbb{E}[((W(x)-W(x-t))-(W_{\mathsf{n}}(x)-W_{\mathsf{n}}(x-t)))^4]\big)^{1/2}.
\end{aligned}$$

It requires to estimate the two expectations in this inequality. The first one is bounded as follows:

$$\begin{aligned}
&(\mathbb{E}[\exp(4\sigma\theta(W(x)-W(x-t))+4(1-\theta)\sigma(W_{\mathsf{n}}(x)-W_{\mathsf{n}}(x-t)))])^{1/2} \\
&\le (\mathbb{E}[\exp(8\sigma\theta(W(x)-W(x-t)))])^{1/4}(\mathbb{E}[\exp(8(1-\theta)\sigma(W_{\mathsf{n}}(x)-W_{\mathsf{n}}(x-t)))])^{1/4} \\
&\le \exp(8\sigma^2\theta^2 t)\exp(8(1-\theta)^2\sigma^2 t) \le \exp(8\sigma^2 t).
\end{aligned} \tag{10.3.10}$$

Now we estimate the second expectation $\mathbb{E}[(W(x)-W(x-t)-(W_{\mathsf{n}}(x)-W_{\mathsf{n}}(x-t)))^4]$. In fact,

$$\begin{aligned}
&\mathbb{E}[(W(x)-W(x-t)-(W_{\mathsf{n}}(x)-W_{\mathsf{n}}(x-t)))^4] \\
&= \mathbb{E}[(\sum_{k=\mathsf{n}+1}^{\infty}[M_k(x)-M_k(x-t)]\xi_k)^4] \\
&= \mathbb{E}[(\sum_{k=\mathsf{n}+1}^{\infty}\sum_{l=\mathsf{n}+1}^{\infty}[M_k(x)-M_k(x-t)]^2[M_l(x)-M_l(x-t)]^2\xi_k^2\xi_l^2] \\
&\le 3\sum_{k=\mathsf{n}+1}^{\infty}\sum_{l=\mathsf{n}+1}^{\infty}[M_k(x)-M_k(x-t)]^2[M_l(x)-M_l(x-t)]^2 \\
&= 3(\sum_{k=\mathsf{n}+1}^{\infty}[M_k(x)-M_k(x-t)]^2)^2 \le C\frac{1}{\mathsf{n}^2},
\end{aligned} \tag{10.3.11}$$

where $M_k = \int_0^x m_k(y)\,dy$ with $m_1(x) = 1/\sqrt{L}$, $m_k(x) = \sqrt{2/L}\cos(\pi(k-1)$ $x/L)$ and C depends only on t and x.

By (10.3.10) and (10.3.11), we have the first estimate in (10.3.8).

Now we prove the weak convergence. It suffices to check $\mathbb{E}[(u-1)^k] - \mathbb{E}[(u_n-1)^k]$. By (10.3.5) and (10.3.7), we have

$$
\begin{aligned}
&\left|\mathbb{E}[(u-1)^k] - \mathbb{E}[(u_n-1)^k]\right| \\
&= \Big|(u_0(x-t)-1)^k \exp(\frac{k^2}{2}\sigma^2\mathbb{E}[(W(x)-W(x-t))^2]) \\
&\quad -(u_0(x-t)-1)^k \exp(\frac{k^2}{2}\sigma^2\mathbb{E}[(W_n(x)-W_n(x-t))^2])\Big| \\
&\le |u_0(x-t)-1|^k \exp(\frac{k^2}{2}\sigma^2\mathbb{E}[(W(x)-W(x-t))^2]) \\
&\quad \times \frac{k^2}{2}\sigma^2\left(\mathbb{E}[(W(x)-W(x-t))^2] - \mathbb{E}[(W_n(x)-W_n(x-t))^2]\right),
\end{aligned}
$$

where we have used the fact $e^x - e^y = e^{\theta x+(1-\theta)y}(x-y)$ $(0\le\theta\le 1)$ and that $\mathbb{E}[(W_n(x)-W_n(x-t))^2] \le \mathbb{E}[(W(x)-W(x-t))^2]$. By $\mathbb{E}[(W(x)-W(x-t))^2] - \mathbb{E}[(W_n(x)-W_n(x-t))^2] = \sum_{k=n+1}^{\infty}\left(M_k(x)-M_k(x-t)\right)^2$, we then have

$$
\left|\mathbb{E}[(u-1)^k] - \mathbb{E}[(u_n-1)^k]\right| \le \frac{k^2}{2}\sigma^2\left|(u_0(x-t)-1)^k\right| \exp(\frac{k^2}{2}\sigma^2 t)\frac{C}{n}.
$$

Hence, the estimate of the weak convergence order follows.

For one-dimensional advection equations with multiplicative noise, we have the order of $1/\sqrt{n}$ for strong convergence and $1/n$ for weak convergence. We do not expect better convergence order as in the case of elliptic equation, where the smoothing of the inverse of Laplacian operator is involved. The following example shows that when better smoothing effects appear, e.g., for biharmonic equations, the strong convergence order can be even higher than that in the case of the elliptic operators.

Example 10.3.5 (Linear biharmonic equations with additive noise) *Consider the following linear biharmonic equation with additive noise*

$$
\Delta^2 u + bu = g(x) + \frac{\partial^d}{\partial x_1 \partial x_2 \cdots \partial x_d} W(x), \quad x = (x_1, \cdots, x_d) \in \mathcal{D} = [0,1]^d, \tag{10.3.12}
$$

with $u=0$ and $\Delta u = 0$ on $\partial\mathcal{D}$, $g \in L^2(\mathcal{D})$. Suppose the operator Δ^2 has eigenvalues λ_α and eigenfunctions e_α. Then, λ_α is proportional to $\pi^4(\alpha_1^4 + \cdots + \alpha_d^4)$. We approximate (10.3.12) *by truncating the white noise using the spectral representation* (10.2.4)*:*

$$
\Delta^2 u_n + bu_n = g(x) + \sum_{|\alpha|\le n} e_\alpha(x)\xi_\alpha. \tag{10.3.13}
$$

Then we have

$$u = \sum_{\alpha} \frac{g_\alpha + \xi_\alpha}{b + \lambda_\alpha} e_\alpha, \quad u_{\mathsf{n}} = \sum_{|\alpha| \le \mathsf{n}} \frac{g_\alpha + \xi_\alpha}{b + \lambda_\alpha} e_\alpha. \tag{10.3.14}$$

Similar to Theorem 10.3.1, we can conclude that

$$\mathbb{E}[\|u\|^2 - \|u_{\mathsf{n}}\|^2] = \mathbb{E}[\|u - u_{\mathsf{n}}\|^2] \le C\mathsf{n}^{-(8-d)}. \tag{10.3.15}$$

10.3.2 Proofs of the strong convergence order

The eigenvalue problem (10.2.6) has the following orthonormal eigenfunctions:

$$e_\alpha(x) := e_\alpha(x_1, x_2, \cdots, x_d) = \prod_{i=1}^{d} \sqrt{2}\sin(\pi \alpha_i x_i), \quad \alpha_i \ge 1 \tag{10.3.16}$$

and the corresponding eigenvalues $\lambda_\alpha = \pi^2 |\alpha|^2$. We also use single-indexed eigenvalues λ_i and eigenfunctions $e_i(x)$ if no confusion arises as the single-indexed system can be always achieved by a proper arrangement of multi-indices.

To prove the strong and weak convergence order, we need the following space

$$\dot{H}^s = \dot{H}^s(\mathcal{D}) = \mathcal{D}((-\Delta)^{s/2}) = \left\{ v | \left\|v\right\|_s = \left\|(-\Delta)^{s/2} v\right\| = (\sum_{k=1}^{\infty} \lambda_k^s (v, e_k))^{1/2} < \infty \right\}, \quad s \in \mathbb{R}.$$

It is known that this space is equivalent to the classical Sobolev-Hilbert space H^s, i.e., $\dot{H}^s = H^s$, see, e.g., [446].

The Green function $\mathcal{K}(x, y)$ can be represented by

$$\mathcal{K}(x, y) = \sum_{\alpha \in \mathbb{N}^d} \frac{1}{\pi^2 |\alpha|^2} e_\alpha(x) e_\alpha(y). \tag{10.3.17}$$

We first consider the regularity of solutions to (10.1.1).

Lemma 10.3.6 *There exists a constant C depending only on d that*

$$\int_{\mathcal{D}} |\mathcal{K}(x, y)|^2 \, dy \le C, \quad \int_{\mathcal{D}} \|\mathcal{K}(\cdot, y)\|_{2-d/2-\epsilon}^2 \, dy \le C. \tag{10.3.18}$$

Proof. By (10.3.17) and orthonormality of $\{e_\alpha\}$, we have

$$\int_{\mathcal{D}} |\mathcal{K}(x, y)|^2 \, dy = \sum_{\alpha} \frac{1}{\pi^4 |\alpha|^4} e_\alpha^2(x) \le C \sum_{\alpha} \frac{1}{\pi^2 |\alpha|^4} \le C.$$

By the fact that $\lambda_\alpha \le C |\alpha|^2$, we then have

$$\int_{\mathcal{D}} \|\mathcal{K}(\cdot, y)\|_{2-d/2-\epsilon}^2 \, dy = \sum_{\alpha} \lambda_\alpha^{2-d/2-\epsilon} \frac{1}{\pi^4 |\alpha|^4} \le C \sum_{\alpha} \frac{1}{|\alpha|^{d+\epsilon}} \le C\epsilon^{-1},$$

where we use the fact that the series $\displaystyle\sum_{\alpha \in N^d} \frac{1}{|\alpha|^s}$ converges if and only if $s > d + \epsilon$ with $\epsilon > 0$.

Lemma 10.3.7 *For any $\epsilon > 0$, we have*

$$\mathbb{E}[\left\|\frac{\partial^d}{\partial x_1 \partial x_2 \cdots \partial x_d} W_n\right\|^2_{-d/2-\epsilon}] \leq \mathbb{E}[\left\|\frac{\partial^d}{\partial x_1 \partial x_2 \cdots \partial x_d} W\right\|^2_{-d/2-\epsilon}] < C(d)\epsilon^{-1}. \tag{10.3.19}$$

Proof. By the definition of norms in $\dot{H}^{-\beta}$, where $\beta > 0$, we have

$$\begin{aligned}\mathbb{E}[\left\|\frac{\partial^d}{\partial x_1 \partial x_2 \cdots \partial x_d} W_n\right\|^2_{-\beta}] &\leq \mathbb{E}[\left\|\frac{\partial^d}{\partial x_1 \partial x_2 \cdots \partial x_d} W\right\|^2_{-\beta}] \\ &= \sum_{\alpha \in \mathbb{N}^d} \lambda_\alpha^{-\beta} = \frac{1}{\pi^{2\beta}} \sum_{\alpha \in \mathbb{N}^d} \frac{1}{|\alpha|^{2\beta}}.\end{aligned}$$

Then we have for $2\beta > d + \epsilon$, $\mathbb{E}[\left\|\frac{\partial^d}{\partial x_1 \partial x_2 \cdots \partial x_d} W\right\|^2_{-\beta}] \leq C(d)\epsilon^{-1} < \infty$.

From Lemmas 10.3.6 and 10.3.7, we have the following regularity for (10.1.1).

Theorem 10.3.8 (Regularity) *Under Assumption 10.2.1, we have for the solution to* (10.1.1)*:*

$$\mathbb{E}[\|u\|^q_{L^p}] < \infty, \quad 1 \leq p,\, q < \infty. \tag{10.3.20}$$

Furthermore, if $M = 0$ in (10.2.2) *of Assumption 10.2.1,*

$$\mathbb{E}[\|u\|^2_{2-d/2-\epsilon}] < C\epsilon^{-1}. \tag{10.3.21}$$

Proof. We prove the L^p-stability for Equation (10.1.1). For $p = 2$, the L^2 regularity can be found in [194]. By (10.2.3) and (10.2.2), taking L^p-norm over both sides, we then have

$$\mathbb{E}[\|u\|^q_{L^p}] \leq C\mathbb{E}[\left\|\int_{\mathcal{D}} |\mathcal{K}(\cdot,y)|\,(1+|u|)\,dy\right\|^q_{L^p}] + C\mathbb{E}[\left\|\sum_{|\alpha|=1}^{\infty} \frac{1}{\pi^2 |\alpha|^2} e_\alpha(x)\xi_\alpha\right\|^q_{L^p}]. \tag{10.3.22}$$

By the Cauchy-Schwarz inequality and by Lemma 10.3.6, we have

$$\mathbb{E}[\left\|\int_{\mathcal{D}} |\mathcal{K}(\cdot,y)|\,|u|\,dy\right\|^q_{L^p}] \leq \left\|(\int_{\mathcal{D}} \mathcal{K}^2(\cdot,y)\,dy)^{1/2}\right\|^q_{L^p} \mathbb{E}[\|u\|^q] \leq C\mathbb{E}[\|u\|^q].$$

By the Littlewood-Paley inequality (see Appendix D and [301]), we have for any $1 < p < \infty$,

$$\left\|\sum_{|\alpha|=1}^{\infty} \frac{1}{\pi^2 |\alpha|^2} e_\alpha(x)\xi_\alpha\right\|_{L^p} \leq C\left\|(\sum_{|\alpha|=1}^{\infty} \frac{1}{\pi^4 |\alpha|^4} e_\alpha^2(x)\xi_\alpha^2)^{\frac{1}{2}}\right\|_{L^p} \leq C(\sum_{|\alpha|=1}^{\infty} \frac{1}{\pi^4 |\alpha|^4} \xi_\alpha^2)^{\frac{1}{2}}.$$

Then by the Cauchy-Schwarz inequality and the triangle inequality, we have

$$\begin{aligned}
\mathbb{E}[\left\| \sum_{|\alpha|=1}^{\infty} \frac{1}{\pi^2 |\alpha|^2} e_\alpha(x)\xi_\alpha \right\|_{L^p}^q] &\leq C\mathbb{E}[(\sum_{|\alpha|=1}^{\infty} \frac{1}{\pi^4 |\alpha|^4} \xi_\alpha^2)^{\frac{q}{2}}] \\
&\leq C(\mathbb{E}[(\sum_{|\alpha|=1}^{\infty} \frac{1}{\pi^4 |\alpha|^4} \xi_\alpha^2)^q])^{\frac{1}{2}} \\
&\leq C(\sum_{|\alpha|=1}^{\infty} \frac{1}{\pi^4 |\alpha|^4} (\mathbb{E}[\xi_\alpha^{2q}])^{\frac{1}{q}})^{\frac{q}{2}} \\
&\leq C(\sum_{|\alpha|=1}^{\infty} \frac{1}{\pi^4 |\alpha|^4})^{\frac{q}{2}} < \infty.
\end{aligned}$$

Then by (10.3.22), we have the inequality in (10.3.20) when $p > 1$. The inequality for $p = 1$ follows readily from Cauchy-Schwarz inequality and the case $p = 2$.

With Lemma 10.3.7, the estimate (10.3.21) can be proved similarly.

Now, we can discuss the strong convergence order for the spectral truncation of white noise.

Lemma 10.3.9 *For* $\eta = \int_{\mathcal{D}} \mathcal{K}(x, y)\, d[W(y) - W_{\mathsf{n}}(y)]$, *we have*

$$\mathbb{E}[\|\eta\|^2] \leq C(d)\epsilon^{-1}(\mathsf{n} + 1)^{-(4-d)}. \tag{10.3.23}$$

Proof. By (10.3.17), (10.2.5), and (10.2.7), we have

$$\begin{aligned}
\mathbb{E}[\|\eta\|^2] &= \mathbb{E}[\left\| \left(\int_{\mathcal{D}} \mathcal{K}(x, y)\, d[W(y) - W_{\mathsf{n}}(y)] \right) \right\|^2] \\
&= \mathbb{E}[\int_{\mathcal{D}} \left(\int_{\mathcal{D}} \sum_{\alpha \in \mathbb{N}^d} \frac{1}{\pi^2 |\alpha|^2} e_\alpha(x) e_\alpha(y) \sum_{|\alpha| \geq \mathsf{n}+1} e_\alpha(y)\, dy \xi_\alpha \right)^2 dx] \\
&= \mathbb{E}[\int_{\mathcal{D}} \left(\sum_{|\alpha| \geq \mathsf{n}+1} \frac{1}{\pi^2 |\alpha|^2} e_\alpha(x)\xi_\alpha \right)^2 dx] \\
&= \sum_{|\alpha| \geq \mathsf{n}+1} \int_{\mathcal{D}} \frac{1}{\pi^4 |\alpha|^4} e_\alpha^2(x)\, dx \\
&= \sum_{|\alpha| \geq \mathsf{n}+1} \frac{1}{\pi^4 |\alpha|^4} \leq \int_{|\alpha| \geq \mathsf{n}+1} \frac{1}{\pi^4 |\alpha|^4}\, d\alpha \leq C(d)\mathsf{n}^{-(4-d)}.
\end{aligned}$$

Here we also used the mutual independence of ξ_α's and the orthonormality of the basis $\{e_\alpha\}$.

Lemma 10.3.10 (Cf. [194, Theorem 2.3]) *Under Assumption 10.2.1, then we have*

$$\mathbb{E}[\|u - u_{\mathsf{n}}\|^2] \leq C(M \left(\mathbb{E}[\|\eta\|^2]\right)^{\frac{1}{2}} + (C_p + R)^2 \mathbb{E}[\|\eta\|^2]),$$

where the constant C depends only on C_p, L, and M, R.

Proof. By (10.2.3) and (10.2.8), the error equation reads

$$u(x) - u_{\mathsf{n}}(x) = - \int_{\mathcal{D}} \mathcal{K}(x, y)[f(u(y)) - f(u_{\mathsf{n}}(y))] \, dy + \eta(x). \tag{10.3.24}$$

Multiplying $[f(u(y)) - f(u_{\mathsf{n}}(y))]$ over both sides of (10.1.1), applying the inequality (10.2.1) and integrating over the domain $\mathcal{D}$, we have

$$\begin{aligned} -L \|u - u_{\mathsf{n}}\|^2 \leq & \int_{\mathcal{D}} \eta(x)[f(u(x)) - f(u_{\mathsf{n}}(x))] \, dx \\ & - \int_{\mathcal{D}} \int_{\mathcal{D}} \mathcal{K}(x, y)[f(u(y)) - f(u_{\mathsf{n}}(y))] \, dy [f(u(x)) - f(u_{\mathsf{n}}(x))] \, dx \\ \leq & \int_{\mathcal{D}} \eta(x)[f(u(x)) - f(u_{\mathsf{n}}(x))] \, dx - C_p \\ & \int_{\mathcal{D}} \left(\int_{\mathcal{D}} \mathcal{K}(x, y)[f(u(y)) - f(u_{\mathsf{n}}(y))] \, dy \right)^2 dx \\ = & \int_{\mathcal{D}} \eta(x)[f(u(x)) - f(u_{\mathsf{n}}(x))] \, dx - C_p \int_{\mathcal{D}} [u - u_{\mathsf{n}} - \eta]^2 \, dx. \end{aligned}$$

Then, by (10.2.2) and the Cauchy-Schwarz inequality, we have

$$(C_p - L) \|u - u_{\mathsf{n}}\|^2 \leq (C_p + R) \|u - u_{\mathsf{n}}\| \|\eta\| + 2M \|\eta\| .$$

By Assumption 10.2.1 ($C_p - L > 0$), we have

$$\|u - u_{\mathsf{n}}\|^2 \leq C(M \|\eta\| + (C_p + R)^2 \|\eta\|^2),$$

where the constant C depends only on C_p, L and M, R.

Theorem 10.3.1 follows from the triangle inequality and Lemmas 10.3.9 and 10.3.10.

10.3.3 Weak convergence order

To prove Theorem 10.3.2, we need the following lemmas. We introduce the following equation

$$- \Delta u(x; z_1, \cdots, z_n, \cdots) + f(u(x; z_1, \cdots, z_n, \cdots)) = g(x) + \sum_{i=1}^{\infty} e_i(x) z_i, \tag{10.3.25}$$

where z_i are parameters in the real line. Note that $u(x; \xi_1, \cdots)$ is the solution to the problem (10.1.1). Here ξ_i's are also single-indexed in the same way as the eigenvalues λ_i's.

Lemma 10.3.11 *In addition to Assumption 10.2.1, assume also that f satisfies the polynomial growth condition* (10.3.1). *Then, there exists a constant $C > 0$ depending only on d, κ, β and those constants in Assumption 10.2.1 such that*

$$\mathbb{E}[\left\|D^\beta u\right\|_{L^q}^2] \le C \prod_i \lambda_i^{-2\beta_i}, \quad 1 \le |\beta| \le 4, \quad 1 \le q \le \infty.$$

Lemma 10.3.12 *Suppose that F satisfies the polynomial growth condition* (10.3.1). *Under the conditions of Lemma 10.3.11, we then have for some constant $C > 0$ depending only on d, κ, β and those constants in Assumption 10.2.1 that*

$$\mathbb{E}[\left\|D^\beta (F(u))\right\|_{L^q}^2] \le C \prod_i \lambda_i^{-2\beta_i}, \quad |\beta| \le 4, \quad 1 \le q < \infty.$$

Proof of Lemma 10.3.11. To estimate the derivatives of solution with respect to parameters, we need the following auxiliary equation: for $\tilde{g} \in L^2(\mathcal{D})$,

$$-\Delta v + f'(u)v = \tilde{g}(x), \quad x \in \mathcal{D}, \quad v = 0 \text{ on } \partial\mathcal{D}. \tag{10.3.26}$$

By Assumption 10.2.1, we claim the following estimate

$$\|v\|_{L^q} \le C \left\| \int_{\mathcal{D}} \mathcal{K}(x,y)\tilde{g}(y)\, dy \right\|_{L^\infty}, \quad 1 \le q \le \infty. \tag{10.3.27}$$

We first establish the case $q = 2$. Equation (10.3.26) can be written in the integral form as

$$v(x) + \int_{\mathcal{D}} \mathcal{K}(x,y) f'(u) v\, dy = \int_{\mathcal{D}} \mathcal{K}(x,y)\tilde{g}(y)\, dy. \tag{10.3.28}$$

Multiplying $f'(u)v$ over both sides of (10.3.28), by the Poincare inequality (see Appendix D) and (10.3.28), we have

$$\begin{aligned} 0 &= (f'(u)v, v) + (\int_{\mathcal{D}} \mathcal{K}(\cdot,y) f'(u(y))v(y)\, dy, f'(u)v) - (\int_{\mathcal{D}} \mathcal{K}(\cdot,y)\tilde{g}(y)\, dy, f'(u)v) \\ &\ge (f'(u)v, v) + C_p \left\| \int_{\mathcal{D}} \mathcal{K}(\cdot,y) f'(u(y))v(y)\, dy \right\|^2 - (\int_{\mathcal{D}} \mathcal{K}(\cdot,y)\tilde{g}(y)\, dy, f'(u)v) \\ &\ge -L\left\|v\right\|^2 + C_p \left\| v - \int_{\mathcal{D}} \mathcal{K}(\cdot,y)\tilde{g}(y)\, dy \right\|^2 - (\int_{\mathcal{D}} \mathcal{K}(\cdot,y)\tilde{g}(y)\, dy, f'(u)v). \end{aligned}$$

Then, by the fact that $f' \ge -L > -C_p$ and $|f'| \le R$, we have (10.3.27) when $q = 2$. After taking L^q-norm over both side of (10.3.28) and by $\int_{\mathcal{D}} \mathcal{K}^2(x,y)\, dy \le C$ (Lemma 10.3.6), we have

$$\|v\|_{L^q} \le RC\left\|v\right\| + \left\| \int_{\mathcal{D}} \mathcal{K}(\cdot,y)\tilde{g}(y)\, dy \right\|_{L^q}, \quad 1 \le q \le \infty,$$

and thus by Lemma 10.3.6, we reach (10.3.27).

Taking the derivative with respect to z_i in Equation (10.3.25), we have

$$-\Delta D^{\epsilon_i}u(x;z_1,\cdots,z_n,\cdots)+f'(u(x;z_1,\cdots,z_n,\cdots))D^{\epsilon_i}u(x;z_1,\cdots,z_n,\cdots)=e_i(x),$$

Thus, by (10.3.27) and (10.3.17), we have

$$\|D^{\epsilon_i}u\|_{L^q} \le C\left\|\int_{\mathcal{D}} \mathcal{K}(\cdot,y)e_i(y)\,dy\right\|_{L^\infty} = C\left\|\lambda_i^{-1}e_i\right\|_{L^\infty} \le C\lambda_i^{-1}, \quad 1\le q\le\infty. \tag{10.3.29}$$

Taking the derivatives with respect to z_i and z_j in Equation (10.3.25), we have the following equation:

$$-\Delta D^{\epsilon_i+\epsilon_j}u + f'(u)D^{\epsilon_i+\epsilon_j}u = -f''(u)D^{\epsilon_i}uD^{\epsilon_j}u.$$

Then by (10.3.27), Lemma 10.3.6, and (10.3.29), we have

$$\begin{aligned}\left\|D^{\epsilon_i+\epsilon_j}u\right\|_{L^q} &\le C\left\|\int_{\mathcal{D}} \mathcal{K}(\cdot,y)f''(u)D^{\epsilon_i}uD^{\epsilon_j}u\,dy\right\|_{L^\infty}\\ &\le C\lambda_i^{-1}\lambda_j^{-1}\left\|f''(u)\right\|, \quad 1\le q\le\infty.\end{aligned}$$

Similarly, we have

$$\begin{aligned}\left\|D^{\epsilon_i+\epsilon_j+\epsilon_k}u\right\|_{L^q} &\le C\lambda_i^{-1}\lambda_j^{-1}\lambda_k^{-1}(\left\|f^{(3)}(u)\right\| + \|f''(u)\|),\\ \left\|D^{\epsilon_i+\epsilon_j+\epsilon_k+\epsilon_l}u\right\|_{L^q} &\le C\lambda_i^{-1}\lambda_j^{-1}\lambda_k^{-1}\lambda_l^{-1}(\left\|f^{(4)}(u)\right\| + \left\|f^{(3)}(u)\right\| + \|f''(u)\|).\end{aligned}$$

By the assumption of polynomial growth at infinity for f and its derivatives and the L^p-stability (Theorem 10.3.8), we reach the conclusion. □

Proof of Lemma 10.3.12. By the multivariate chain rule (also known as multivariate Faa di Bruno formula), we have $D^{\epsilon_i+\epsilon_j}F(u) = F'(u)D^{\epsilon_i+\epsilon_j}u + F''(u)D^{\epsilon_i}uD^{\epsilon_j}u$, and thus by Lemma 10.3.11,

$$\left\|D^{\epsilon_i+\epsilon_j}F(u)\right\|_{L^q} \le C(\|F'(u)\|_{L^q} + \|F''(u)\|_{L^q})\lambda_i^{-1}\lambda_j^{-1}, \quad 1\le q<\infty.$$

Similarly, we have

$$\begin{aligned}\left\|D^{\epsilon_i+\epsilon_j+\epsilon_k+\epsilon_l}F(u)\right\|_{L^q} &\le C\lambda_i^{-1}\lambda_j^{-1}\lambda_k^{-1}\lambda_l^{-1}(\|F'(u)\|_{L^q}\\ &\quad + \|F''(u)\|_{L^q} + \left\|F^{(3)}(u)\right\|_{L^q} + \left\|F^{(4)}(u)\right\|_{L^q}),\end{aligned}$$

and then the conclusion follows from the assumption of polynomial growth of F and its derivatives at infinity (10.3.1) and Lemma 10.3.8. □

Proof of Theorem 10.3.2. By the first-order Taylor's expansion, we have, for $\mathsf{m} > \mathsf{n}$,

$$\begin{aligned}
&\mathbb{E}[F(u_{\mathsf{m}}) - F(u_{\mathsf{n}})] \\
&= \mathbb{E}[F\left(u_{\mathsf{m}}(\xi_1, \cdots, \xi_{\mathsf{n}'}, \cdots, \xi_{\mathsf{m}'})\right) - F\left(u_{\mathsf{n}}(\xi_1, \cdots, \xi_{\mathsf{n}})\right)] \\
&= \mathbb{E}[\sum_{i=\mathsf{n}'+1}^{\mathsf{m}'} D^{\epsilon_i}(F(u_{\mathsf{m}}(\xi_1, \cdots, \xi_{\mathsf{n}'}, 0, \cdots, 0)))\xi_i] \\
&+ \sum_{i,j=\mathsf{n}'+1}^{\mathsf{m}'} \mathbb{E}[\int_0^1 (1-t) D^{\epsilon_i+\epsilon_j}(F(u_{\mathsf{m}}(\xi_1, \cdots, \xi_{\mathsf{n}'}, t\xi_{\mathsf{n}'+1}, \cdots, t\xi_{\mathsf{m}'})))\xi_i\xi_j] \\
&= \sum_{i,j=\mathsf{n}'+1}^{\mathsf{m}} \int_0^1 (1-t)\mathbb{E}[D^{\epsilon_i+\epsilon_j}(F(u_{\mathsf{m}}(\xi_1, \cdots, \xi_{\mathsf{n}'}, t\xi_{\mathsf{n}'+1}, \cdots, t\xi_{\mathsf{m}'})))\xi_i\xi_j],
\end{aligned} \tag{10.3.30}$$

where $\mathsf{n}' = \mathsf{n}!/d!/(\mathsf{n}-d)!$ and $\mathsf{m}' = \mathsf{m}!/d!/(\mathsf{m}-d)!$ and we used the fact ξ_i $(i \geq \mathsf{n}'+1)$ is independent of $F(u_{\mathsf{m}}(\xi_1, \cdots, \xi_{\mathsf{n}'}, 0, \cdots, 0))$ and $\mathbb{E}[\xi_i] = 0$.

To estimate (10.3.30), we split the term into two parts:

$$\mathrm{I} = \sum_{i=\mathsf{n}'+1}^{\mathsf{m}'} \int_0^1 (1-t)\mathbb{E}[D^{2\epsilon_i}(F(u_{\mathsf{m}}(\xi_1, \cdots, \xi_{\mathsf{n}'}, t\xi_{\mathsf{n}'+1}, \cdots, t\xi_{\mathsf{m}'})))\xi_i^2],$$

$$\mathrm{II} = 2 \sum_{i<j,\ i,j=\mathsf{n}'+1}^{\mathsf{m}'} \int_0^1 (1-t)\mathbb{E}[D^{\epsilon_i+\epsilon_j}(F(u_{\mathsf{m}}(\xi_1, \cdots, \xi_{\mathsf{n}'}, t\xi_{\mathsf{n}'+1}, \cdots, t\xi_{\mathsf{m}'})))\xi_i\xi_j].$$

By Lemma 10.3.12, we have, for $1 \leq q < \infty$,

$$\begin{aligned}
\|I\|_{L^q} &= \left\| \sum_{i=\mathsf{n}'+1}^{\mathsf{m}'} \int_0^1 (1-t)\mathbb{E}[D^{2\epsilon_i}(F(u_{\mathsf{m}}(\xi_1, \cdots, \xi_{\mathsf{n}'}, t\xi_{\mathsf{n}'+1}, \cdots, t\xi_{\mathsf{m}'})))\xi_i^2]\, dt \right\|_{L^q} \\
&\leq C \sum_{i=\mathsf{n}'+1}^{\mathsf{m}'} \lambda_i^{-2}.
\end{aligned} \tag{10.3.31}$$

For II, we use the recipe of the proof of Theorem 2.8 in [74]. For simplicity, we define that $X_{i.j}^{t,r,s} = (\xi_1, \cdots, \xi_{\mathsf{n}'}, t\xi_{\mathsf{n}'+1}, \cdots, tr\xi_i, \cdots, ts\xi_j, \cdots, t\xi_{\mathsf{m}'})$.

Noticing that $\mathbb{E}[D^{\epsilon_i+\epsilon_j}(F(u_{\mathsf{m}}(X_{i,j}^{t,0,1})\xi_i\xi_j] = 0$ $(i<j)$, we have

$$\begin{aligned}
&\int_0^1 (1-t)\mathbb{E}[D^{\epsilon_i+\epsilon_j}(F(u_{\mathsf{m}}(X_{i,j}^{t,1,1}))\xi_i\xi_j]\, dt \\
&= \int_0^1 (1-t)\mathbb{E}[D^{\epsilon_i+\epsilon_j}(F(u_{\mathsf{m}}(X_{i,j}^{t,1,1}))\xi_i\xi_j]\, dt \\
&\qquad - \int_0^1 (1-t)\mathbb{E}[D^{\epsilon_i+\epsilon_j}(F(u_{\mathsf{m}}(X_{i,j}^{t,0,1}))\xi_i\xi_j]\, dt \\
&= \int_0^1 \int_0^1 (1-t)t\mathbb{E}[D^{2\epsilon_i+\epsilon_j}(F(u_{\mathsf{m}}(X_{i,j}^{t,r,1}))\xi_i\xi_j]\, dt\, dr.
\end{aligned}$$

With $\mathbb{E}[D^{2\epsilon_i+\epsilon_j}(F(u_{\mathsf{m}}(X_{i,j}^{t,r,0})\xi_i\xi_j] = 0$ $(i < j)$, we have similarly

$$\begin{aligned}&\int_0^1\int_0^1 (1-t)t\mathbb{E}[D^{2\epsilon_i+\epsilon_j}(F(u_{\mathsf{m}}(X_{i,j}^{t,r,1}))\xi_i\xi_j]\,dt\,dr,\\ &=\int_0^1\int_0^1\int_0^1 (1-t)t^2\mathbb{E}[D^{2\epsilon_i+2\epsilon_j}(F(u_{\mathsf{m}}(X_{i,j}^{t,r,s}))\xi_i^2\xi_j^2]\,dt\,dr\,ds,\end{aligned}$$

and thus for $i < j$,

$$\begin{aligned}&\int_0^1 (1-t)\mathbb{E}[D^{\epsilon_i+\epsilon_j}(F(u_{\mathsf{m}}(X_{i,j}^{t,1,1}))\xi_i\xi_j]\,dt\\ &=\int_0^1\int_0^1\int_0^1 (1-t)t^2\mathbb{E}[D^{2\epsilon_i+2\epsilon_j}(F(u_{\mathsf{m}}(X_{i,j}^{t,r,s}))\xi_i^2\xi_j^2]\,dt\,dr\,ds.\end{aligned}$$

Now we can bound II as, with Lemma 10.3.12,

$$\begin{aligned}\|\mathrm{II}\|_{L^q} &\leq \left\|2\sum_{i<j,\ i,j=\mathsf{n}'+1}^{\mathsf{m}'}\int_0^1 (1-t)\mathbb{E}[D^{2\epsilon_i+2\epsilon_j}(F(u_{\mathsf{m}}(X_{i,j}^{t,1,1}))\xi_i\xi_j]\,dt\right\|_{L^q} \\ &\leq c\sum_{i<j,\ i,j=\mathsf{n}'+1}^{\mathsf{m}'}\lambda_i^{-2}\lambda_j^{-2}.\end{aligned} \tag{10.3.32}$$

Thus, we have by (10.3.30), (10.3.31), and (10.3.32),

$$\|\mathbb{E}[F(u_{\mathsf{m}})-F(u_{\mathsf{n}})]\|_{L^q} \leq c\sum_{|\alpha|=\mathsf{n}+1}^{\mathsf{m}}\lambda_\alpha^{-2} + c\Big(\sum_{|\alpha|=\mathsf{n}+1}^{\mathsf{m}}\lambda_\alpha^{-2}\Big)^2. \tag{10.3.33}$$

Then by $\lambda_\alpha = \pi^2|\alpha|^2$, we arrive at the conclusion.

10.4 Error estimates for finite element approximation

In this section, we show that the strong convergence order of finite element approximation can be the same as the strong order of the spectral truncation of white noise (Theorem 10.4.1) while the weak convergence order is constrained by the convergence order of the piecewise linear finite element approximation in one dimension (Theorem 10.4.3). It then follows that for one-dimensional problems, a piecewise quadratic finite element approximation should be used to obtain higher convergence order.

Theorem 10.4.1 (Finite element approximation, strong error) *Let u be the solution to* (10.1.1) *and u_{n}^h the solution to* (10.2.10). *Under Assumption 10.2.1, we have the following estimate for piecewise linear finite element approximation of* (10.1.1),

$$\begin{aligned}\mathbb{E}[\left\|u-u_{\mathsf{n}}^h\right\|^2] &\leq 2\mathbb{E}[\|u-u_{\mathsf{n}}\|^2]+2\mathbb{E}[\left\|u_{\mathsf{n}}-u_{\mathsf{n}}^h\right\|^2]\\ &\leq C(\mathsf{n}^{-(4-d)}+M\mathsf{n}^{-(2-\frac{d}{2})})+C(h^4\mathsf{n}^d+Mh^2\mathsf{n}^{d/2}).\end{aligned}$$

When taking $\mathsf{n}=1/h$, *we have*

$$\mathbb{E}[\left\|u-u_{\mathsf{n}}^h\right\|^2]\leq C(h^{4-d}+Mh^{2-\frac{d}{2}}). \tag{10.4.1}$$

Remark 10.4.2 *The convergence order in Theorem 10.4.1 is optimal as the solution* u *to* (10.1.1) *belongs to* $H^{2-d/2-\epsilon}(\mathcal{D})$ *when* $M=0$ *in Assumption 10.2.1, see Theorem 10.3.8. Compared to the methods of finite difference and finite element in [6, 64, 194], the convergence order is half order higher than the convergence order presented in [6] for the one-dimensional problem and is the same for higher dimensional problems [64, 194].*

Define the Ritz projection $R_h: H_0^1(\mathcal{D})\to V_h$ by

$$(\nabla R_h w,\nabla v)=(\nabla w,\nabla v),\quad \forall v\in V_h,\quad w\in H_0^1(\mathcal{D}).$$

Then it holds that, see, e.g., [446], there is a constant C independent of h such that for $0\leq l<r\leq 2$

$$\|w-R_h w\|_l\leq Ch^{r-l}\|w\|_r,\quad w\in H^2(\mathcal{D})\cap H_0^1(\mathcal{D}). \tag{10.4.2}$$

Proof. It can be readily checked from (10.2.9) and (10.2.10) that

$$(\nabla(R_h u_{\mathsf{n}}-u_{\mathsf{n}}^h),\nabla v)+(f(u_{\mathsf{n}})-f(u_{\mathsf{n}}^h),v)=0,\quad v\in V_h. \tag{10.4.3}$$

Taking $v=R_h u_{\mathsf{n}}-u_{\mathsf{n}}^h$ and by (10.2.1) and (10.2.2), the Cauchy-Schwarz inequality, we have

$$\begin{aligned}\left\|\nabla(R_h u_{\mathsf{n}}-u_{\mathsf{n}}^h)\right\|^2 &= -(f(u_{\mathsf{n}})-f(u_{\mathsf{n}}^h),u_{\mathsf{n}}-u_{\mathsf{n}}^h)+(f(u_{\mathsf{n}})-f(u_{\mathsf{n}}^h),R_h u_{\mathsf{n}}-u_{\mathsf{n}})\\ &\leq L\left\|u_{\mathsf{n}}-u_{\mathsf{n}}^h\right\|^2+c(M+R\left\|u_{\mathsf{n}}-u_{\mathsf{n}}^h\right\|)\|R_h u_{\mathsf{n}}-u_{\mathsf{n}}\|)\\ &\leq \frac{L+C_p}{2}\left\|u_{\mathsf{n}}-u_{\mathsf{n}}^h\right\|^2+CM\|R_h u_{\mathsf{n}}-u_{\mathsf{n}}\|+C\|R_h u_{\mathsf{n}}-u_{\mathsf{n}}\|^2.\end{aligned}$$

Then by the Poincare inequality $\left\|\nabla(R_h u_{\mathsf{n}}-u_{\mathsf{n}}^h)\right\|^2\geq C_p\left\|R_h u_{\mathsf{n}}-u_{\mathsf{n}}^h\right\|^2$, the triangle inequality and $L<C_p$, there exists a constant C independent of h but dependent of C_p,R,L:

$$\left\|R_h u_{\mathsf{n}}-u_{\mathsf{n}}^h\right\|_1^2+\left\|u_{\mathsf{n}}-u_{\mathsf{n}}^h\right\|^2\leq C(M\|R_h u_{\mathsf{n}}-u_{\mathsf{n}}\|+\|R_h u_{\mathsf{n}}-u_{\mathsf{n}}\|^2). \tag{10.4.4}$$

Then by (10.4.2), we have

$$\left\|R_h u_{\mathsf{n}}-u_{\mathsf{n}}^h\right\|_1^2+\left\|u_{\mathsf{n}}-u_{\mathsf{n}}^h\right\|^2\leq C(Mh^2\|u_{\mathsf{n}}\|_2+h^4\|u_{\mathsf{n}}\|_2^2). \tag{10.4.5}$$

Similar to the proof of Theorem 10.3.8, we have

$$\mathbb{E}[\|u_{\mathsf{n}}\|_2^2] \leq C\mathbb{E}[\left\|g + \frac{\partial^d}{\partial x_1 \cdots \partial x_d} W_{\mathsf{n}}\right\|^2] \leq C\mathsf{n}^d, \tag{10.4.6}$$

where we have used the fact that

$$\mathbb{E}[\left\|\frac{\partial^d}{\partial x_1 \cdots \partial x_d} W_{\mathsf{n}}\right\|^2] = \sum_{|\alpha| \leq \mathsf{n}} \|e_\alpha\|^2 \leq C\mathsf{n}^d.$$

By (10.4.4) and (10.4.6), we obtain that

$$\left\|u_{\mathsf{n}} - u_{\mathsf{n}}^h\right\|^2 \leq C(Mh^2 \|u_{\mathsf{n}}\|_2 + h^4 \|u_{\mathsf{n}}\|_2^2),$$

whence we can reach the conclusion by setting $h = \mathsf{n}^{-1}$.

Theorem 10.4.3 (Finite element approximation, weak error) *Let u be the solution to (10.1.1) and u_{n}^h the solution to (10.2.10). Under the conditions of Theorem 10.3.2, we have, for $d \leq 3$,*

$$\left|\mathbb{E}[\|u\|^2 - \left\|u_{\mathsf{n}}^h\right\|^2]\right| \leq C[\mathsf{n}^{-(4-d)} + h^4\mathsf{n}^d + h^3\mathsf{n}^{d/2} + h^2\mathsf{n}^{\min(d-2,0)}].$$

To get optimal convergence in h, we take n at the order of $1/h$ and have

$$\left|\mathbb{E}[\|u\|^2 - \left\|u_{\mathsf{n}}^h\right\|^2]\right| \leq Ch^{\min(4-d,2)}.$$

The constant C depends on d, C_p, L, R as in Theorem 10.3.1 and also the constant c in (10.3.1).

Lemma 10.4.4 *Let u_{n} be the solution to (10.2.8) or (10.2.9) and u_{n}^h be the finite element solution in (10.2.10). Then we have*

$$\mathbb{E}[\|u_{\mathsf{n}}\|_{L^p}^q] < C < \infty, \quad p, q \geq 1. \tag{10.4.7}$$

and

$$\mathbb{E}[\|u_{\mathsf{n}}\|_1^2] \leq C\mathsf{n}^{\min(d-2,0)}, \tag{10.4.8}$$

where C does not depend on n. For u_{n}^h, we have

$$\mathbb{E}[\left\|u_{\mathsf{n}}^h\right\|_{L^p}^q] < \infty, \quad p, q \geq 1. \tag{10.4.9}$$

Proof. The proof of (10.4.7) is similar to the proof of (10.3.20). In particular, we have

$$\mathbb{E}[\|u_{\mathsf{n}}\|^2] \leq C.$$

Then multiplying u_{n} over both sides of (10.2.9) and using integration-by-parts, we have

$$\|\nabla u_{\mathsf{n}}\|^2 = -(f(u_{\mathsf{n}}, u_{\mathsf{n}})) + (g, u_{\mathsf{n}}) + (\frac{\partial^{d-1}}{\partial x_2 \cdots \partial x_d} W_{\mathsf{n}}, \partial_{x_1} u_{\mathsf{n}}).$$

By (10.3.16) and the definition of Brownian sheet (10.2.5), we have

$$\mathbb{E}[\left\|\frac{\partial^{d-1}}{\partial x_2\cdots\partial x_d}W_{\mathsf{n}}\right\|^2] = \mathbb{E}[(\int_{\mathcal{D}}(\sum_{|\alpha|\leq \mathsf{n}+1}\int_0^{x_1} e_\alpha(y_1,x_2,\cdots,x_d)\xi_\alpha\,dy_1)^2\,dx]$$
$$\leq C\mathsf{n}^{\min(d-2,0)}.$$

Then by Assumption 10.2.1 and Cauchy inequality, we have

$$\|\nabla u_{\mathsf{n}}\|^2 \leq C(1+\|u_{\mathsf{n}}\|^2+\|g\|^2)+\frac{1}{2}\left\|\frac{\partial^{d-1}}{\partial x_2\cdots\partial x_d}W_{\mathsf{n}}\right\|^2+\frac{1}{2}\|\partial_{x_1}u_{\mathsf{n}}\|^2,$$

and thus we reach (10.4.8) from the fact that $\|\partial_{x_1}u_{\mathsf{n}}\|^2 \leq \|\nabla u_{\mathsf{n}}\|^2$. In fact,

$$\mathbb{E}[\|u_{\mathsf{n}}\|_1^2] = \mathbb{E}[\|u_{\mathsf{n}}\|^2]+\mathbb{E}[\|\nabla u_{\mathsf{n}}\|^2] \leq C\mathsf{n}^{\min(d-2,0)}. \tag{10.4.10}$$

Now we prove (10.4.9). When $p=2$, (10.4.9) follows from (10.2.10) if we take $v=u_{\mathsf{n}}^h$ and apply Assumption 10.2.1. According to Chapter 2 in [446], we have

$$u_{\mathsf{n}}^h+\int_{\mathcal{D}}R_h\mathcal{K}(x,y)f(u_{\mathsf{n}}^h)\,dy = \int_{\mathcal{D}}R_h\mathcal{K}(x,y)g(y)\,dy+\int_{\mathcal{D}}R_h\mathcal{K}(x,y)\,dW_{\mathsf{n}}(y). \tag{10.4.11}$$

To prove (10.4.11), it is key to show that the inverse of $-\Delta_h$, a discrete Laplacian from S_h to S_h, defined by

$$-(\Delta_h\phi_h,v) := (\nabla\phi,v) = (g,v), \quad \phi_h, v\in S_h.$$

is $R_h(-\Delta)^{-1}$. Denote that $T=(-\Delta)^{-1}: L^2\to H_0^1(\mathcal{D})$. Then for the elliptic problem $-\Delta\phi=g$ over $\mathcal{D}$ with homogeneous Dirichlet boundary conditions, the solution is $\phi=Tg$. Denote the inverse of $-\Delta_h$ by T_h. Then we have, see, e.g., [446, (2.16)]

$$T_h = R_hT.$$

This gives that

$$T_hg = R_hTg = \int_{\mathcal{D}}R_h\mathcal{K}(x,y)g(y)\,dy. \tag{10.4.12}$$

The finite element approximation (10.2.10) can be rewritten as

$$-\Delta_hu_{\mathsf{n}}^h+P_hf(u_{\mathsf{n}}^h) = P_h(g+\frac{\partial^d}{\partial x_1\cdots\partial x_d}W_{\mathsf{n}}),$$

where P_h is the L^2 projection into S_h. Then by $T_hP_h=T_h$ (see [446, (2.24)]), we have

$$u_{\mathsf{n}}^h = -T_hf(u_{\mathsf{n}}^h)+T_h(g+\frac{\partial^d}{\partial x_1\cdots\partial x_d}W_{\mathsf{n}}).$$

Thus, by (10.4.12), we reach (10.4.11).

For $p \neq 2$, we follow the same idea as in the proof of Theorem 10.3.8. Taking L^p-norm over both sides of (10.4.11) and by (10.2.2), we then have

$$\mathbb{E}[\left\|u_{\mathsf{n}}^h\right\|_{L^p}^q] \leq C\mathbb{E}[\left\|\int_{\mathcal{D}} |R_h\mathcal{K}(\cdot,y)|\,(1+\left|u_{\mathsf{n}}^h\right|)\,dy\right\|_{L^p}^q] + C\mathbb{E}[\left\|\sum_{|\alpha|=1}^{\infty} \frac{1}{\pi^2\,|\alpha|^2} R_h e_\alpha(x)\xi_\alpha\right\|_{L^p}^q]. \tag{10.4.13}$$

By the Cauchy-Schwarz inequality, we have

$$\mathbb{E}[\left\|\int_{\mathcal{D}} |R_h\mathcal{K}(\cdot,y)|\left|u_{\mathsf{n}}^h\right| dy\right\|_{L^p}^q] \leq \left\|(\int_{\mathcal{D}} (R_h\mathcal{K}(\cdot,y))^2\,dy)^{1/2}\right\|_{L^p}^q \mathbb{E}[\left\|u_{\mathsf{n}}^h\right\|^q] \leq C\mathbb{E}[\left\|u_{\mathsf{n}}^h\right\|^q].$$

Similar to the proof of (10.3.20), we conclude that (10.4.9) holds.

Proof. We will use the duality argument and Theorem 10.3.2 to prove Theorem 10.4.3.

By Theorem 10.3.2 (taking $q = 1$ in (10.3.2)), we have

$$\left|\mathbb{E}[\|u\|^2 - \|u_{\mathsf{n}}\|^2]\right| \leq C\mathsf{n}^{-(4-d)}.$$

By the standard estimate of the Ritz operator in negative norms, (see, e.g., [446, Theorem 5.1]),

$$\|u_{\mathsf{n}} - R_h u_{\mathsf{n}}\|_{-r} \leq Ch^{q+r}\,\|u_{\mathsf{n}}\|_q\,, \quad 1 \leq q \leq s,\ 0 \leq r \leq s-2, \tag{10.4.14}$$

we have, taking $q = r = 1$ and by the fact $\|R_h u_{\mathsf{n}}\|_1 \leq C\,\|u_{\mathsf{n}}\|_1$,

$$\left|\mathbb{E}[\|u_{\mathsf{n}}\|^2 - \|R_h u_{\mathsf{n}}\|^2]\right| \leq \mathbb{E}[\|u_{\mathsf{n}} - R_h u_{\mathsf{n}}\|_{-1}\,\|u_{\mathsf{n}} + R_h u_{\mathsf{n}}\|_1] \leq Ch^2(\mathbb{E}[\|u_{\mathsf{n}}\|_1^2]). \tag{10.4.15}$$

Similar to the proof of Theorem 10.3.8, we have

$$\mathbb{E}[\|u_{\mathsf{n}}\|_1^2] \leq C,\ (d=1,2) \quad \text{and} \quad \mathbb{E}[\|u_{\mathsf{n}}\|_1^2] \leq C\mathsf{n},\ (d=3).$$

From here and (10.4.15), we have

$$\left|\mathbb{E}[\|u_{\mathsf{n}}\|^2 - \|R_h u_{\mathsf{n}}\|^2]\right| \leq Ch^2\mathsf{n}^{\min(d-2,0)}. \tag{10.4.16}$$

In order to estimate $\left|\mathbb{E}[\|u\|^2 - \left\|u_{\mathsf{n}}^h\right\|^2]\right|$, we only need to estimate $\left|\mathbb{E}[\|R_h u_{\mathsf{n}}\|]^2 - \mathbb{E}[\left\|u_{\mathsf{n}}^h\right\|^2]\right|$. We use the duality argument to obtain such an estimate. To this end, we introduce the following linear adjoint problem over the domain $\mathcal{D}$:

$$-\Delta\psi + f'(u_{\mathsf{n}})\psi = \phi, \quad \psi|_{\partial\mathcal{D}} = 0. \tag{10.4.17}$$

It holds that $\|\psi\|_2 \leq C\,\|\phi\|$ since $f'(u_{\mathsf{n}}) \geq -L > -C_p$ is bounded. Introducing $e = R_h u_{\mathsf{n}} - u_{\mathsf{n}}^h$, $e_1 = u_{\mathsf{n}} - u_{\mathsf{n}}^h$, $e_2 = R_h u_{\mathsf{n}} - u_{\mathsf{n}}$, we then have, by (10.4.17), the definition of the Ritz projection and the error equation (10.4.3),

$$
\begin{aligned}
(e, \phi) &= (\nabla e, \nabla \psi) + (f'(u_{\mathsf{n}})e, \psi) = (\nabla e, \nabla R_h \psi) + (f'(u_{\mathsf{n}})e, \psi) \\
&= -(f(u_{\mathsf{n}}) - f(u_{\mathsf{n}}^h), R_h \psi) + (f'(u_{\mathsf{n}})e, \psi) \\
&= (f(u_{\mathsf{n}}) - f(u_{\mathsf{n}}^h), \psi - R_h \psi) + (f'(u_{\mathsf{n}})e - f(u_{\mathsf{n}}) - f(u_{\mathsf{n}}^h), \psi)
\end{aligned}
$$

Thus we have, by (10.4.2), $|f'(u_{\mathsf{n}})| \leq R$ and Taylor's expansion,

$$
\begin{aligned}
|(e, \phi)| &\leq C \|e_1\| h^2 \|\psi\|_2 - \frac{1}{2}(f''(\theta u_{\mathsf{n}} + (1-\theta)u_{\mathsf{n}}^h)e^2, \psi) \\
&\leq C \|e_1\| h^2 \|\psi\|_2 + C \|e\|_{L^4}^2 \|\psi\|_\infty (1 + \|u_{\mathsf{n}}\|_{L^{2\kappa}}^{\kappa} + \left\|u_{\mathsf{n}}^h\right\|_{L^{2\kappa}}^{\kappa}),
\end{aligned} \tag{10.4.18}
$$

where $0 \leq \theta \leq 1$ and we used the polynomial growth condition (10.3.1) for f''. Then we have, by the embedding $\|v\|_\infty \leq C \|v\|_2$ and $\|e\|_{L^4} \leq C \|e\|_1$, (10.4.18), and $\|\psi\|_2 \leq C \|\phi\|$ that

$$
|(e, \phi)| \leq C\big(h^2 \|e_1\| + (1 + \|u_{\mathsf{n}}\|_{L^{2\kappa}}^{\kappa} + \left\|u_{\mathsf{n}}^h\right\|_{L^{2\kappa}}^{\kappa}) \|e\|_1^2\big) \|\phi\| .
$$

Thus, by the definition of negative norm and the Hölder inequality, we have, for any $0 \leq r \leq 1$,

$$
\begin{aligned}
\mathbb{E}[\|e\|_{-r}^2] &\leq Ch^4 \mathbb{E}[\|e_1\|^2] + \mathbb{E}[\|e\|_1^4 (1 + \|u_{\mathsf{n}}\|_{L^{2\kappa}}^{2\kappa} + \left\|u_{\mathsf{n}}^h\right\|_{L^{2\kappa}}^{2\kappa})] \\
&\leq Ch^4 \mathbb{E}[\|e_1\|^2] \\
&\quad + C(\mathbb{E}[\|e\|_1^{4(1+\epsilon)}])^{1/(1+\epsilon)} \Big(1 + (\mathbb{E}[\|u_{\mathsf{n}}\|_{L^{2\kappa}}^{2\kappa(1+1/\epsilon)}])^{\epsilon/(1+\epsilon)} \\
&\quad + (\mathbb{E}[\left\|u_{\mathsf{n}}^h\right\|_{L^{2\kappa}}^{2\kappa(1+1/\epsilon)}])^{\epsilon/(1+\epsilon)}\Big)
\end{aligned} \tag{10.4.19}
$$

Then by Lemma 10.4.4, (10.4.14) and (10.4.5), we have, for $d \leq 3$ and any $0 \leq r \leq 1$,

$$
(\mathbb{E}[\|e\|_{-r}^2])^{1/2} \leq Ch^2 (\mathbb{E}[\|e_1\|^2])^{1/2} + C(\mathbb{E}[\|e\|_1^{4(1+\epsilon)}])^{1/(2+2\epsilon)} \leq C(h^3 \mathsf{n}^{d/2} + h^4 \mathsf{n}^d),
$$

whence we obtain, by (10.4.14), (10.4.4), and (10.4.6),

$$
\begin{aligned}
\left|\mathbb{E}[\left\|u_{\mathsf{n}}^h\right\|^2 - \|R_h u_{\mathsf{n}}\|^2]\right| &\leq \mathbb{E}[\left\|u_{\mathsf{n}}^h - R_h u_{\mathsf{n}}\right\| \left\|u_{\mathsf{n}}^h + R_h u_{\mathsf{n}}\right\|] \\
&\leq (\mathbb{E}[\|e\|^2])^{1/2} ((\mathbb{E}[\|e\|^2])^{1/2} + 2(\mathbb{E}[\|R_h u_{\mathsf{n}}\|^2])^{1/2}) \\
&\leq Ch^4 \mathsf{n}^d + h^3 \mathsf{n}^{d/2}.
\end{aligned} \tag{10.4.20}
$$

where we applied (10.4.5) and (10.4.16).

Then by the triangle inequality, Theorem 10.3.2, (10.4.16), and (10.4.20), we reach the conclusion.

Remark 10.4.5 *When f is linear, we can improve the error for $\left|\mathbb{E}[\left\|u_{\mathsf{n}}^h\right\|^2 - \|R_h u_{\mathsf{n}}\|^2]\right|$, which can be checked from the proof. However, the conclusion on convergence order will not change since the error* (10.4.16) *is dominant in the total error.*

10.5 Numerical results

In this section, we present some numerical results of piecewise linear finite element approximation of one- and two- dimensional elliptic equations (10.1.1) with spatial Brownian motion approximated by its spectral truncation (10.2.7).

To compute the expectations, we use Monte Carlo sampling for both problems with the Mersenne Twister random generator (seed 100) to compute expectations. The experiments were performed using Matlab R2012a on a Macintosh desktop computer with Intel Xeon CPU E5462 (quad-core, 2.80 GHz). A fixed-point iteration method with tolerance $h^2/100$ was used to solve the nonlinear algebraic equations at each step of the implicit schemes.

Example 10.5.1 (One-dimensional elliptic)

$$-\partial_x^2 u = \frac{1}{2} u + \sigma \partial_x W(x), \quad x \in \mathcal{D} = (0,2), \tag{10.5.1}$$

with zero Dirichlet boundary condition.

In this example, we will truncate the Brownian motion as follows $W(x) = \sum_{k=1}^{\mathsf{n}} \int_0^x m_k(y)\, dy \xi_k$, where we use the cosine basis in $L^2(\mathcal{D})$:

$$m_1(x) = \frac{1}{\sqrt{|\mathcal{D}|}}, \quad m_k(x) = \sqrt{\frac{2}{|\mathcal{D}|}} \cos(\frac{(k-1)\pi}{|\mathcal{D}|} x), \qquad k \geq 2.$$

The errors are measured in the weak sense:

$$\varrho_2^r = \frac{\left| \mathbb{E}[\|u_{\mathsf{n}}^h\|^2] - \mathbb{E}[\|u_{\mathsf{ref}}\|^2] \right|}{\mathbb{E}[\|u_{\mathsf{ref}}^2\|]}, \tag{10.5.2}$$

where $\|v\|$ is the L^2 norm in physical space and $\mathsf{n} = 2/h$ in this example. We take $\sigma = 1$ and obtained the value of $\mathbb{E}[\|u_{\mathsf{ref}}\|^2] = 0.2731183$ (up to 7 digit) analytically as in Example 10.1.1.

In Table 10.1, we observe that the weak convergence of finite element methods is of second-order, which is in agreement of Theorem 10.4.3. We use 4×10^8 Monte Carlo sample paths to obtain the numerical solution. The numbers after "$\pm$" are the statistical errors with the 95% confidence interval.

Example 10.5.2 (Two-dimensional elliptic equation)

$$-\Delta u + \sin(u) = \sigma \frac{\partial^2}{\partial x_1 \partial x_2} W(x), \quad x \in \mathcal{D} = (0,1) \times (0,1), \tag{10.5.3}$$

with zero Dirichlet boundary conditions.

Table 10.1. Weak convergence of piecewise linear finite element methods for the one-dimensional problem (10.5.1) with a spectral approximation of white noise (10.2.7) using $\mathsf{n} = 2/h$.

# Element	n	ϱ_2^r	Order
4	4	$3.4237 \times 10^{-2} \pm 4.08 \times 10^{-5}$	–
8	8	$1.0658 \times 10^{-2} \pm 3.77 \times 10^{-5}$	1.68
16	16	$2.7521 \times 10^{-3} \pm 3.69 \times 10^{-5}$	1.95
32	32	$7.2822 \times 10^{-4} \pm 3.67 \times 10^{-5}$	1.92

In this example, we test the weak convergence of piecewise linear finite element (rectangular element) approximation of (10.5.3) with different noise magnitudes. The errors are measured in the following weak sense:

$$\rho_1^r = \frac{\left|\left\|(\mathbb{E}[u_{\mathsf{n}}^h])^2\right\| - \left\|(\mathbb{E}[u_{2\mathsf{n}}^{h/2}])^2\right\|\right|}{\left\|(\mathbb{E}[u_{2\mathsf{n}}^{h/2}])^2\right\|}, \quad \rho_2^r = \frac{\left|\mathbb{E}[\left\|u_{\mathsf{n}}^h\right\|^2] - \mathbb{E}[\left\|u_{2\mathsf{n}}^{h/2}\right\|^2]\right|}{\mathbb{E}[\left\|u_{2\mathsf{n}}^{h/2}\right\|^2]}.$$

When 32×32 elements are used, we employ 2×10^5 Monte Carlo sample paths and obtain

$\sigma = 0.5$, $\left\|(\mathbb{E}[u_{32}^{\sqrt{2}/32}])^2\right\| = 0.22861 \pm 2.3 \times 10^{-4}$ and $\mathbb{E}[\left\|u_{32}^{\sqrt{2}/32}\right\|^2] = 0.22965 \pm 4.5 \times 10^{-4}$;

$\sigma = 1.0$, $\left\|(\mathbb{E}[u_{32}^{\sqrt{2}/32}])^2\right\| = 0.22861 \pm 4.7 \times 10^{-4}$ and $\mathbb{E}[\left\|u_{32}^{\sqrt{2}/32}\right\|^2] = 0.23278 \pm 9.0 \times 10^{-4}$.

In Table 10.2, we observe a second-order convergence of piecewise approximation (10.2.10) for the two-dimensional semilinear problem (10.5.3), which is consistent with our theoretical prediction in Theorem 10.4.3.

10.6 Summary and bibliographic notes

For spatial noise, we can expect higher-order convergence from the spectral approximation of Brownian motion. When an explicit form of spectral representation of Brownian motion is available, we can use the spectral truncation: we have better convergence from it rather than from piecewise linear approximation of Brownian motion and have the same convergence as the piecewise linear approximation. With the spectral approximation of Brownian motion, we observe the following:

- For semilinear elliptic equations with additive noise, the weak convergence rate is twice the strong convergence rate if only white noise (Brownian motion) is truncated, see Theorems 10.3.1 and 10.3.2. This is also true for other PDEs, in Chapter 10.3.1.

Table 10.2. Weak convergence of piecewise linear finite element approximation of the two-dimensional semilinear problem (10.5.3) with a spectral approximation of white noise (10.2.7) using $\mathsf{n} = \sqrt{2}/h$.

σ	# MC	# element	ρ_1^r	order	CPU time(s.)
0.5	1×10^3	4×4	$1.831 \times 10^{-2} \pm 3.2 \times 10^{-3}$	–	0.1
0.5	4×10^4	8×8	$4.201 \times 10^{-3} \pm 5.3 \times 10^{-4}$	$h^{2.12}$	2.2
0.5	8×10^4	16×16	$1.113 \times 10^{-3} \pm 3.7 \times 10^{-4}$	$h^{1.92}$	80.1
1.0	2×10^3	4×4	$1.779 \times 10^{-2} \pm 4.5 \times 10^{-3}$	–	0.1
1.0	1×10^5	8×8	$4.281 \times 10^{-3} \pm 6.7 \times 10^{-4}$	$h^{2.05}$	71.6
1.0	2×10^5	16×16	$1.231 \times 10^{-3} \pm 4.7 \times 10^{-4}$	$h^{1.80}$	191.2

σ	# MC	# element	ρ_2^r	order
0.5	1×10^3	4×4	$1.800 \times 10^{-2} \pm 6.5 \times 10^{-3}$	–
0.5	4×10^4	8×8	$4.172 \times 10^{-3} \pm 1.0 \times 10^{-3}$	$h^{2.11}$
0.5	8×10^4	16×16	$1.121 \times 10^{-3} \pm 7.1 \times 10^{-4}$	$h^{1.90}$
1.0	2×10^3	4×4	$1.662 \times 10^{-2} \pm 9.2 \times 10^{-3}$	–
1.0	1×10^5	8×8	$4.177 \times 10^{-3} \pm 1.3 \times 10^{-3}$	$h^{1.99}$
1.0	2×10^5	16×16	$1.255 \times 10^{-3} \pm 9.0 \times 10^{-4}$	$h^{1.73}$

- For finite element discretization of semilinear elliptic equations with additive noise, the strong convergence order of the finite element approximation is $h^{2-d/2}$ (Theorem 10.4.1) and the weak convergence order is $h^{\min(4-d,2)}$ (Theorem 10.4.3).
- When solutions are smooth in random space, e.g., a fourth-order equation with additive noise, a spectral approximation of the white noise can lead to higher convergence order in both strong and weak sense when the solutions are smooth.

In this chapter, we consider nonlinear elliptic equations with additive noise where no stochastic products are involved. In the next chapter, we will consider elliptic equations with multiplicative noise, where the stochastic products have to be carefully defined.

Bibliographic notes. Investigating the benchmark problem (10.1.1) is helpful to better understand the influence of discretizing Brownian motion/sheet as well as more complex noises in the context of approximating stochastic partial differential equations. For example, when higher dimensional white noise is considered, which is the case for space-time white noise (see, e.g., [249]), we can combine one of the above approximation methods in each dimension and thus have different approximation of white noise. It is then crucial to understand the performance of different approximation methods in a simple case such as the problem (10.1.1).

Piecewise constant approximation of white noise has been used for many problems with white noise since its use in linear elliptic equations [6], see, e.g., [64, 66] for nonlinear elliptic problems, [67, 68] for Helmholtz equations, [118] for linear elliptic equations with additive color noise, [490] heat equation with additive space-time noise, [464] for reaction-diffusion equation with space-time white noise as the coefficient of nonlinear reaction term, [491] for space-time noise (color in space and white in time), and [256] for Allen-Cahn equations with additive space-time white noise, etc.

Spectral approximation can be and has been considered for elliptic equations with multiplicative noise, see, e.g., linear elliptic equation with lognormal diffusivity [73, 74, 140, 141] and with white noise diffusivity [469].

Disadvantage of spectral approximation. The spectral approximation may not be explicitly known, e.g., the eigenfunctions of the leading operator are not easily found, L^2-CONS in the spectral approximation may not be explicitly expressed even when domains have arbitrary but smooth boundary curves, cf. the last paragraph in Chapter 10.2.

It is crucial to assume that the nonlinear term f satisfies Assumption (10.2.1) as in [48, 64, 194], especially the monotone condition (10.2.1). The monotone condition (10.2.1) allows a large class of nonlinearity terms such as $f(x) = x(1 - x^2)$ (e.g., in Allen-Cahn equation).

10.7 Suggested practice

In the following problems, $\{e_k\}_{k=1}^{\infty}$ is a CONS on $L^2(\mathcal{D})$, where $\mathcal{D}$ is the domain considered and ξ_k's are i.i.d. standard Gaussian random variables. Let

$$\partial_x W_Q(x) = \sum_{k=1}^{\infty} \sqrt{q_k} e_k(x) \xi_k, \; q_k \geq 0.$$

Assume that a) $q_k = 1/k^2$ for any integer $k \geq 1$ or b) $q_k = 1/k^4$ for all positive integer k.

Exercise 10.7.1 *Consider the following stochastic elliptic equation with additive noise*

$$\begin{aligned} -\partial_x^2 u + bu(x) &= g(x) + \partial_x W_Q(x), \quad x \in (0,1), \\ u(0) &= u(1) = 0. \end{aligned}$$

Here g is smooth enough, say, $g(x) = \sin(x)$ and $b > 0$. Derive a regularity estimate for u that is similar as in Example 10.1.1, where $q_k = 1$ for any integer $k \geq 1$.

Exercise 10.7.2 *Consider the following nonlinear stochastic elliptic equation with additive noise*

$$-\partial_x^2 u + f(u) = g(x) + \partial_x W_Q(x), \quad x \in (0,1),$$
$$u(0) = u(1) = 0.$$

Assume that f satisfies Assumption (10.2.1)*. Derive a regularity estimate for u as in Theorem 10.3.8.*

Exercise 10.7.3 *Consider the following one-dimensional linear stochastic elliptic equation*

$$-\partial_x^2 u = \frac{1}{2}u + \partial_x W_Q(x), \quad x \in \mathcal{D} = (0,2) \tag{10.7.1}$$

with zero Dirichlet boundary conditions. Numerically check the mean-square convergence order when $q_k = 1/k^2$ for any integer $k \geq 1$.

Hint. See Example 10.5.1.

Exercise 10.7.4 *Consider the following one-dimensional elliptic equation*

$$-\partial_x^2 u = \sin(u) + \partial_x W_Q(x), \quad x \in \mathcal{D} = (0,2) \tag{10.7.2}$$

with zero Dirichlet boundary conditions. Numerically check the mean-square convergence order when $q_k = 1/k^2$ for any integer $k \geq 1$.

11

Multiplicative white noise: The Wick-Malliavin approximation

In this chapter, we consider Wiener chaos expansion (WCE) for elliptic equations with multiplicative noise. Unlike the stochastic collocation methods (SCM), a direct application of WCE will lead to a fully coupled linear system. To sparsify the resulting linear system, we present WCE with the use of Ito-Wick product and an approximation/reduction technique called Wick-Malliavin approximation. Specifically, we consider Wick-Malliavin approximation for elliptic equations with lognormal coefficients and use the Wick product for elliptic equations with spatial white noise as coefficients. Numerical results demonstrate that high-order Wick-Malliavin approximation is efficient even when the noise intensity is relatively large.

The Wick-Malliavin approximation can be used as reduction methods for WCE of nonlinear problems. It can significantly reduce the computational cost while maintaining the high accuracy of WCE. Moreover, the Wick-Malliavin approximation can be applied for SPDEs with non-Gaussian white noise. For stochastic collocation methods (SCM), such an approximation is not available since the Wick-Malliavin approximation is based on Wiener chaos expansion or more generally polynomial chaos expansion.

11.1 Introduction

Consider WCE with Wick-Malliavin approximation for the following stochastic elliptic equation with multiplicative noise:

$$-\operatorname{div}(a(x,\omega)\nabla u(x,\omega)) = f(x), \quad x \in \mathcal{D}, \qquad u(x) = 0, \quad x \in \partial\mathcal{D}, \tag{11.1.1}$$

where $\mathcal{D}$ is a domain with Lipschitz boundary, $a(x,\omega)$ is a random field, and f is either random or deterministic. Many methods have been proposed for (11.1.1) with different assumptions on $a(x,\omega)$, see the bibliographic note

Z. Zhang, G.E. Karniadakis, *Numerical Methods for Stochastic Partial Differential Equations with White Noise*,
Applied Mathematical Sciences 196, DOI 10.1007/978-3-319-57511-7_11

at the end of this chapter. Here we consider WCE for lognormal $a(x,\omega)$ and white noise $a(x,\omega)$.

WCE for (11.1.1) leads to a fully coupled linear system of deterministic elliptic equations. To reduce the computational cost of WCE, the product in (11.1.1) between $a(x,\omega)$ and ∇u has been replaced with the Ito-Wick product, see, e.g., [320, 466]:

$$-\operatorname{div}(a \diamond \nabla u) = f(x,\omega), \quad x \in \mathcal{D}, \qquad u(x) = 0, \quad x \in \partial\mathcal{D}. \tag{11.1.2}$$

The definition of the Ito-Wick product "$\diamond$" can be found in Chapter 2.3. The application of WCE to (11.1.2) results in a weakly coupled linear system which is lower triangular. One feature of (11.1.2) is that the mean field of the solution, $\mathbb{E}[u]$, satisfies the deterministic equations

$$-\operatorname{div}(\mathbb{E}[a]\nabla\mathbb{E}[u]) = f(x), \quad \mathbb{E}[u](x)|_{\partial\mathcal{D}} = 0.$$

However, the use of Wick product can deteriorate the existence of the solution in $\mathbb{L}^2(\Omega,\mathcal{F},\mathbb{P})$ and some weighted space has to be used, see Chapter 11.4.

The solution to (11.1.2) can be treated as an approximation of (11.1.1) of order two (in the intensity of the noise) in the mean-square sense. In Chapter 11.3, we consider a higher-order approximation using the recently developed Wick-Malliavin approximation in [346, 470],

$$-\nabla\cdot(a\diamond\nabla u+\sum_{q=1}^{\mathsf{Q}}\frac{1}{q!}\mathfrak{D}^q a\diamond\mathfrak{D}^q\nabla u) = f(x), \quad x \in \mathcal{D}, \qquad u(x) = 0, \quad x \in \partial\mathcal{D}, \tag{11.1.3}$$

where $\mathfrak{D}$ is the Malliavin derivative which will be defined shortly. When $\mathsf{Q} = 0$, (11.1.3) becomes (11.1.2).

This chapter is organized as follows. First, we introduce the Wick-Malliavin approximation in Chapter 11.2 and apply this approximation to elliptic equations with lognormal coefficients in Chapter 11.3. In Chapter 11.4, we discuss the case when a is a spatial white noise and $\mathsf{Q} = 0$ and its finite element approximation. We then present the Wick-Malliavin approximation for nonlinear equations with a Gaussian random forcing in Chapter 11.5 and for nonlinear equations with a non-Gaussian random forcing in Chapter 11.6. Numerical results are presented in most of the sections. At the end of the chapter, we summarize the conclusions in this chapter and present a brief review on numerical methods for elliptic equations with random coefficients, including different treatments in stochastic products between the random coefficient and the gradient of the solution. A review on convergence rates of WCE is also presented. Some exercises are provided to enhance the readers' understanding of generalized Wiener chaos methods and the Wick-Malliavin approximation.

11.2 Approximation using the Wick-Malliavin expansion

The Wick-Malliavin approximation states that the product of two square-integrable random variables can be approximated as in Taylor's expansion, see Theorem 11.2.1.

Let us introduce the Malliavin derivative. Consider a spatial white noise $\dot{W}_\phi = \sum_{k=1}^\infty \phi_k(x)\xi_k$ on the probability space $(\Omega, \mathcal{F}, \mathbb{P})$ taking values on the Hilbert space $\mathfrak{U} = L^2(\mathcal{D})$, where $\{\phi_k\}_{k=1}^\infty$ is a CONS of $\mathfrak{U}$. Here $\xi_k = \dot{W}_{\phi_k} = \int_{\mathcal{D}} \phi_k(x)\, dW(x)$ are i.i.d. standard Gaussian random variables and $\mathbb{E}[\dot{W}_\phi \dot{W}_\psi] = (\phi, \psi)_{\mathfrak{U}}$ with $(\cdot,\cdot)_{\mathfrak{U}}$ being the inner product associated with the Hilbert space $\mathfrak{U}$.

Denote by $\mathfrak{D}$ the Malliavin derivative with respect to $\dot{W}_\phi$ on $\mathbb{L}^2(\Omega, \mathcal{F}, \mathbb{P})$, which can be defined as a directional derivative

$$\mathfrak{D}\xi_\alpha = \sum_{k\geq 1} \sqrt{\alpha_k}\phi_k \xi_{\alpha-\epsilon_k} = \sum_{\beta\in\mathcal{J}} \Big(\sum_{|\gamma|=1} \mathbf{1}_{\beta+\gamma=\alpha} \frac{\sqrt{\alpha!}}{\sqrt{\beta!}} \prod_k (\phi_k)^{\gamma_k} \Big) \xi_\beta \in L^2((\Omega, \mathcal{F}, \mathbb{P}); \mathfrak{U}), \tag{11.2.1}$$

where $\varepsilon_k \in \mathcal{J}$ has only one nonzero component: $(\varepsilon_k)_k = 1$ and $(\varepsilon_k)_j = 0$ for any $j \neq k$. The high order Malliavin derivative can be defined through reduction as

$$\mathfrak{D}^n \xi_\alpha = \sigma^n \sum_{\beta\in\mathcal{J}} \Big(\sum_{|\gamma|=n} \mathbf{1}_{\beta+\gamma=\alpha} \frac{\sqrt{\alpha!}}{\sqrt{\beta!}} \phi_\gamma \Big) \xi_\beta \in L^2((\Omega, \mathcal{F}, \mathbb{P}); \mathfrak{U}^{\otimes n}) \tag{11.2.2}$$

where $\phi_\gamma = \sum_{k_1,\cdots,k_n} \sum_{\varepsilon_{k1}+\cdots+\varepsilon_{k_n}=\gamma} \phi_{k_1} \otimes \cdots \otimes \phi_{k_n} \in \mathcal{U}^{\otimes n}$. For example, $n = 2$,

$$\begin{aligned}
\mathfrak{D}^2\xi_\alpha = \mathfrak{D}(\mathfrak{D}\xi_\alpha) &= \sum_{|\gamma|=1} \sum_{\beta=\alpha-\gamma} \frac{\sqrt{\alpha!}}{\sqrt{\beta!}} \mathfrak{D}\xi_\beta \prod_k (\phi_k)^{\gamma_k} \\
&= \sum_{|\gamma|=1} \sum_{\beta=\alpha-\gamma} \frac{\sqrt{\alpha!}}{\sqrt{\beta!}} \Big(\sum_{|\eta|=1} \sum_{\theta=\beta-\eta} \frac{\sqrt{\beta!}}{\sqrt{\theta!}} \xi_\theta \prod_\eta (\phi_l)^{\eta_l} \Big) \otimes \prod_\gamma (\phi_k)^{\gamma_k} \\
&= \sum_{|\gamma|=1} \sum_{|\eta|=1} \sum_{\theta=\alpha-\gamma-\eta} \frac{\sqrt{\alpha!}}{\sqrt{\theta!}} \xi_\theta \prod_\eta (\phi_l)^{\eta_l} \otimes \prod_\gamma (\phi_k)^{\gamma_k}.
\end{aligned}$$

This can be written in the form of (11.2.2) as

$$\mathfrak{D}^2\xi_\alpha = \sum_{\theta\in\mathcal{J}} \Big(\sum_{|\zeta|=2} \mathbf{1}_{\zeta+\theta=\alpha} \frac{\sqrt{\alpha!}}{\sqrt{\theta!}} \phi_\zeta \Big) \xi_\theta.$$

By the definition of Malliavin derivatives for Cameron-Martin basis, it can be readily checked that for two square-integrable random variables u and v,

$$\mathfrak{D}(u \diamond v) = \mathfrak{D}u \diamond v + u \diamond \mathfrak{D}v. \tag{11.2.3}$$

Theorem 11.2.1 (Mikulevicius-Rozovsky formula, [346]) *For elements of the Cameron-Martin basis ξ_α and ξ_β, the following relation holds with probability 1:*

$$\xi_\alpha \xi_\beta = \xi_\alpha \diamond \xi_\beta + \sum_{q=1}^{\infty} \frac{\mathfrak{D}^q \xi_\alpha \diamond \mathfrak{D}^q \xi_\beta}{q!}. \tag{11.2.4}$$

Moreover, the relation can be extended to any square-integrable random variables X and Y, i.e.,

$$XY = X \diamond Y + \sum_{q=1}^{\infty} \frac{\mathfrak{D}^q X \diamond \mathfrak{D}^q Y}{q!}. \tag{11.2.5}$$

The theorem is a combination of Proposition 4 and Remark 11 in [346]. The proof of this theorem is based on the following two observations:

- the linearization coefficients of the product

$$\xi_\alpha \xi_\beta = \sum_{\gamma \le \alpha \wedge \beta} B(\alpha, \beta, \gamma) \xi_{\alpha+\beta-2\gamma}, \tag{11.2.6}$$

-

$$\mathfrak{D}^n \xi_\alpha \diamond \mathfrak{D}^n \xi_\beta = n! \sum_{|\gamma|=n,\ \gamma \le \alpha \wedge \beta} B(\alpha, \beta, \gamma) \xi_{\alpha+\beta-2\gamma}, \tag{11.2.7}$$

or its equivalent form

$$\mathfrak{D}^n \xi_\alpha \diamond \mathfrak{D}^n \xi_\beta = n! \sum_{\gamma} \sum_{\theta \le \gamma} \sum_{|\gamma|=n} 1_{\gamma+\gamma-\theta=\beta} 1_{\gamma+\theta=\alpha} B(\alpha, \beta, \gamma) \xi_{\alpha+\beta-2\gamma}, \tag{11.2.8}$$

where

$$B(\alpha, \beta, \gamma) = \left(\binom{\alpha}{\gamma} \binom{\beta}{\gamma} \binom{\alpha+\beta-2\gamma}{k-\gamma} \right)^{\frac{1}{2}}. \tag{11.2.9}$$

Proof. Assume that $u = \sum_{\kappa \in \mathcal{J}} u_\kappa \xi_\kappa$ and $v = \sum_{\kappa \in \mathcal{J}} u_\kappa \xi_\kappa$, we have

$$\begin{aligned} uv &= \sum_{\alpha \in \mathcal{J}} \sum_{\theta, \kappa \in \mathcal{J}} u_\theta v_\kappa \mathbb{E}[\xi_\theta \xi_\kappa \xi_\alpha] \xi_\alpha \\ &= \sum_{\alpha \in \mathcal{J}} \sum_{\theta, \kappa \in \mathcal{J}} u_\theta v_\kappa \sum_{\gamma \le \theta \wedge \kappa} B(\theta, \kappa, \gamma) \mathbb{E}[\xi_{\theta+\kappa-2\gamma} \xi_\alpha] \xi_\alpha \\ &= \sum_{\alpha \in \mathcal{J}} \sum_{\gamma \in \mathcal{J}} \sum_{(0) \le \beta \le \alpha} u_{\beta+\gamma} v_{\alpha-\beta+\gamma} \Phi(\alpha, \beta, \gamma) \xi_\alpha \end{aligned}$$

where $\Phi(\alpha, \beta, \gamma) = B(\beta+\gamma, \alpha-\beta+\gamma, \gamma)$ and

$$\Phi(\alpha, \beta, \gamma) = \left[\binom{\alpha}{\beta} \binom{\beta+\gamma}{\gamma} \binom{\alpha-\beta+\gamma}{\gamma} \right]^{1/2}. \tag{11.2.10}$$

The product of u and v then can be represented as

$$uv = u \diamond v + \sum_{Q=1}^{\infty} \sum_{\alpha \in \mathcal{J}} \sum_{\gamma \in \mathcal{J}, |\gamma|=Q} \sum_{(0) \le \beta \le \alpha} u_{\beta+\gamma} v_{\alpha-\beta+\gamma} \Phi(\alpha, \beta, \gamma) \xi_\alpha.$$

Then by the definition of Malliavin derivatives, (11.2.5) can be derived.

By the Mikulevicius-Rozovsky formula (Theorem 11.2.1) and the fact that $\mathfrak{D}^n \xi_\alpha = 0$ if $|\alpha| < n$, we have the following conclusion.

Corollary 11.2.2 *For elements of the Cameron-Martin basis ξ_θ and ξ_κ, the following relation holds with probability 1:*

$$\xi_\theta \xi_\kappa = \xi_\theta \diamond \xi_\kappa + \sum_{q=1}^{\mathrm{Q}} \frac{\mathfrak{D}^q \xi_\theta \diamond \mathfrak{D}^q \xi_\kappa}{q!}, \tag{11.2.11}$$

where $\mathrm{Q} = \min(|\theta|, |\kappa|)$.

Now let us look at Equation (11.1.1) again. Assume that $a(x,\omega)$ and ∇u are square-integrable in random space. Then by Theorem 11.2.1, Equation (11.1.1) can be rewritten as

$$-\nabla \cdot (a \diamond \nabla u + \sum_{q=1}^{\infty} \frac{1}{q!} \mathfrak{D}^q a \diamond \mathfrak{D}^q \nabla u) = f(x), \quad x \in \mathcal{D}, \qquad u(x) = 0, \quad x \in \partial\mathcal{D}. \tag{11.2.12}$$

Here the above assumption can be satisfied when $a(x,\omega)$ is smooth enough, see, e.g., [73, 74] and Chapter 11.3.

11.3 Lognormal coefficient

Consider the equation (11.1.1) where the coefficient $a(x,\omega)$ is lognormal:

$$a(x,\omega) = \exp\big(\sigma \sum_{k=1}^{\infty} \lambda_k^{1/2} \phi_k(x) \xi_k(\omega)\big), \tag{11.3.1}$$

where λ_k are nonnegative real numbers, $\phi_k(x)$ is a CONS in $L^2(\mathcal{D})$,and ξ_k's are i.i.d. standard Gaussian random variables. This representation can be obtained from the Karhunen-Loève expansion (Theorem 2.1.5) of $\ln(a(x,\omega))$, which is a Gaussian field with zero mean and covariance kernel $K(x-y)$, e.g., kernels in Table 2.1. When $K(z)$ is in $\mathcal{C}^{0,1}(\mathbb{R}^+)$, a and $\ln(a)$ belong to $\mathcal{C}^{0,\mu}(\bar{\mathcal{D}})$ a.s. for $\mu < 1/2$. By Mercer's theorem, the correlation function $K(x-y)$ can be expressed as

$$K(x,y) = \sigma^2 \sum_{i=1}^{\infty} \lambda_k \phi_k(x) \phi_k(y), \tag{11.3.2}$$

where $\{\phi_k(x)\}_{k=1}^{\infty}$ is a CONS of $L^2(\mathcal{D})$ and

$$\xi_k(\omega) = \sigma^{-1} \lambda_k^{-1/2} \int_{\mathcal{D}} \ln(a(x,\omega)) \phi_k(x) dx, \quad k = 1, 2, \cdots$$

Theorem 11.3.1 (Existence and uniqueness, [74, Proposition 1]) *Assume that $\mathcal{D}$ is an open bounded domain in $\mathbb{R}^d$ with $\mathcal{C}^2$ boundary. When $f \in L^2(\mathcal{D})$ and $K(z)$ is in $\mathcal{C}^{0,1}(\mathbb{R}^+)$, Equation* (11.1.1) *has a unique solution in $\mathbb{L}^q(\Omega, H_0^1(\mathcal{D}))$.*

By the Cameron-Martin theorem, $\xi_\alpha = \prod_k \frac{H_{\alpha_k}(\xi_k)}{\sqrt{\alpha_k!}}$ is a CONS in $\mathbb{L}^2(\Omega, \mathcal{F}, \mathbb{P})$. Then the lognormal process $a(x,\omega)$ in (11.3.1) can be represented by the following WCE

$$a(x,\omega) = e^{\sigma^2/2} \sum_{\alpha\in\mathcal{J}} \frac{\sigma^{|\alpha|} \prod_k \lambda_k^{\alpha_k/2} \phi_k^{\alpha_k}(x)}{\sqrt{\alpha!}} \xi_\alpha. \tag{11.3.3}$$

Now we truncate a with a_{n} as

$$a_{\mathsf{n}} = e^{\sigma^2/2} \exp(\sigma \sum_{k=1}^{\mathsf{n}} \lambda^{1/2} \phi_k(x)\xi_k) = e^{\sigma^2/2} \sum_{\substack{\alpha\in\mathcal{J} \\ \alpha_i=0,\, i\geq \mathsf{n}}} \frac{\sigma^{|\alpha|} \prod_k \lambda_k^{\alpha_k/2} \phi_k^{\alpha_k}(x)}{\sqrt{\alpha!}} \xi_\alpha.$$

We have the following stochastic elliptic equations with finite dimensional random variables.

$$-\operatorname{div}(a_{\mathsf{n}}(x)\nabla u_{\mathsf{n}}) = f(x), \quad x \in \mathcal{D}, \qquad u_{\mathsf{n}}(x) = 0, \quad x \in \partial\mathcal{D}. \tag{11.3.4}$$

For any μ, ν with $0 \leq \nu < \mu < 1/2$ and $q \geq 1$ then $\|a_{\mathsf{n}} - a\|_{\mathbb{L}^q(\Omega, \mathcal{C}^{0,\nu}(\bar{\mathcal{D}}))} \leq C(R_{\mathsf{n}}^\mu)^{1/2}$ where the positive C depends only on μ, ν, q and

$$R_{\mathsf{n}}^\mu = \sum_{k=\mathsf{n}+1}^{\infty} \lambda_k \|\phi_k\|_{\mathcal{C}^{0,\mu}}^2$$

is well defined and convergent for all $0 \leq \mu \leq 1/2$.

Theorem 11.3.2 ([74, Theorem 2.8]) *For $f \in L^p(\mathcal{D})$, $p \geq d$, $0 < \nu < \min(\frac{1}{2}, 1 - \frac{d}{p})$, the strong convergence order is*

$$\mathbb{E}[\|u - u_{\mathsf{n}}\|_{\mathcal{C}^{1,\nu}(\bar{\mathcal{D}}))}^q] \leq C(R_{\mathsf{n}}^\nu)^{q/2}, \qquad q \geq 1. \tag{11.3.5}$$

Assume also that $\psi \in \mathcal{C}^6(\mathbb{R})$ and ψ and its derivatives have at most polynomial growth, then the weak convergence holds

$$\|\mathbb{E}[\psi(u)] - \mathbb{E}[\psi(u_{\mathsf{n}})]\|_{\mathcal{C}^{1,\nu}(\bar{\mathcal{D}})} \leq C R_{\mathsf{n}}^\nu,$$

where the positive constant C depends only on β, p and f, ψ.

Remark 11.3.3 *Many important processes admit a well-defined R_{n}^μ, such as Brownian motion and Gaussian process with exponential kernel. For Brownian motion (spatial) over $(0, L)$,*

$$W(x) = \sum_{k=1}^{\infty} \int_0^x e_k(y)\, dy \xi_k, \qquad e_1(x) = \sqrt{\frac{1}{L}},\ e_k(x) = \sqrt{\frac{2}{L}} \cos(\frac{k\pi x}{\sqrt{L}}),\ k \geq 2,$$

$\lambda_k \sim \mathcal{O}(\frac{1}{k^2})$ and $\|\phi_k\|_{\mathcal{C}^{0,\nu}([0,L])} \sim \mathcal{O}(k^\nu)$ and thus $R_{\mathsf{n}}^\nu \sim \mathcal{O}(N^{2\nu-1})$. A similar conclusion holds for Gaussian process in one-dimension with exponential kernel $K(z) = \exp(-\frac{|z|}{l_c})$, where l_c is called correlation length, see [74].

The Wick-Malliavin approximation of Q-th order to (11.3.4) is

$$-\operatorname{div}(a_{\mathsf{n}} \diamond \nabla u_{\mathsf{n},\mathsf{Q}} + \sum_{q=1}^{\mathsf{Q}} \frac{1}{q!} \mathfrak{D}^q a_{\mathsf{n}} \diamond \mathfrak{D}^q \nabla u_{\mathsf{n},\mathsf{Q}}) = f(x), \quad x \in \mathcal{D}, \qquad (11.3.6)$$

where the boundary condition is $u_{\mathsf{n},\mathsf{Q}}|_{\partial\mathcal{D}} = 0$.

Theorem 11.3.4 *Let u_{n} be the solution to* (11.3.4) *and $u_{\mathsf{n},\mathsf{Q}}$ be the solution to* (11.3.6). *Then there exists a positive constant C such that*

$$\mathbb{E}[\|u_{\mathsf{n},\mathsf{Q}} - u_{\mathsf{n}}\|^2_{L^2(\mathcal{D})}] \leq C(C_1\sigma)^{2(\mathsf{Q}+1)}, \qquad (11.3.7)$$

where C depends on n and proper norms of u_{n}, $u_{\mathsf{n},\mathsf{Q}}$, a, a_{n} and f, $\lambda_k^{1/2}\phi_k$ ($k = 1, 2, \cdots, \mathsf{n}$) but is independent of σ and Q. Here C_1 is a constant depending only on f and a, a_n.

The conclusion can be proved similarly as in [470, Lemma 4.2]. We note that the constant C depends on n in (11.3.7) and may blow up when $\mathsf{n} \to \infty$. To have an estimate (11.3.7) where the constant does not depend on n, it is required that the λ_k decays fast enough, e.g., $\lambda_1 = 1$ and $\lambda_k = 0$ for $k \geq 2$.

11.3.1 One-dimensional example

Consider the following test model:

$$-\partial_x(a(\xi)\partial_x u) = 1, \qquad (11.3.8)$$

where $a(\xi) = e^{\sigma\xi - \sigma^2/2} = \sum_{k=0}^{\infty} \sigma^k \frac{H_k(\xi)}{k!}$ and ξ is a standard Gaussian random variable.

Suppose that $u^{\mathsf{N}} = \sum_{k=0}^{\mathsf{N}} u_k(x) \frac{H_k(\xi)}{\sqrt{k!}}$. The fully discretization of WCE yields a solution u^{N} satisfying that

$$-\sum_{k=0}^{\mathsf{N}} u_k(x)\mathbb{E}[a(\xi)\frac{H_k(\xi)}{\sqrt{k!}}\frac{H_l(\xi)}{\sqrt{l!}}] = \mathbb{E}[\frac{H_l(\xi)}{\sqrt{l!}}], \quad l \geq 0.$$

In matrix form, we have

$$S\mathbf{u}(x) = (1, 0, \cdots, 0)^\top,$$

where $\mathbf{u}(x) = (u_0(x), \cdots, u_{\mathsf{N}}(x))^\top$ and S is a full matrix of size $(\mathsf{N}+1) \times (\mathsf{N}+1)$. In fact,

$$\begin{aligned} S_{l,k} = \mathbb{E}[a(\xi)\frac{H_k(\xi)}{\sqrt{k!}}\frac{H_l(\xi)}{\sqrt{l!}}] &= \sum_{n=0}^{\infty} \frac{\sigma^n}{\sqrt{n!}}\mathbb{E}[\frac{H_n(\xi)}{\sqrt{n!}}\frac{H_k(\xi)}{\sqrt{k!}}\frac{H_l(\xi)}{\sqrt{l!}}] \\ &= \sum_{n=0}^{\infty} \frac{\sigma^n}{\sqrt{n!}}\mathbb{E}[\sum_{q \leq k \wedge n} B(k,n,q)\frac{H_{n+k-2q}(\xi)}{\sqrt{(n+k-2q)!}}\frac{H_l(\xi)}{\sqrt{l!}}] \\ &= \sum_{n=0}^{\infty} \frac{\sigma^n}{\sqrt{n!}} \sum_{q \leq k \wedge n} B(k,n,q)\delta_{n+k-2q,l}. \end{aligned}$$

The Q-th order Wick-Malliavin approximation (11.3.6) leads to

$$S^{(\mathsf{Q})}\mathbf{u}(x) = (1, 0, \cdots, 0)^\top,$$

where $S^{(\mathsf{Q})}_{l,k} = 0$ if $l - k < 0$ is odd or $l < k - 2\mathsf{Q}$ and can be calculated from the Mikulevicius-Rozovsky formula (Theorem 11.2.1) and

$$\begin{aligned} S^{(\mathsf{Q})}_{l,k} &= \mathbb{E}[(a(\xi) \diamond \frac{H_k(\xi)}{\sqrt{k!}} + \sum_{q=1}^{\mathsf{Q}} \frac{\mathfrak{D}^q a \diamond \mathfrak{D}^p(\frac{H_k(\xi)}{\sqrt{k!}})}{q!}) \frac{H_l(\xi)}{\sqrt{l!}}] \\ &= \sum_{n=0}^{\infty} \frac{\sigma^n}{\sqrt{n!}} \sum_{q \le k \wedge n,\, q \le Q} B(k, n, q)\delta_{n+k-2q,l}. \end{aligned}$$

It can be readily seen that $S = S^{(\mathsf{N})}$. Compared to S, $S^{(\mathsf{Q})}$ can be sparse, especially when $\mathsf{Q} \ll \mathsf{N}$. For example, $S^{(0)}$ is *lower-triangular*, as $S^{(0)}_{l,k} = 0$ if $0 \le l < k \le \mathsf{N}$.

Now we present some numerical results for different Q when N is fixed. Here we take $\mathsf{N} = 30$ ($\mathsf{N} = 40$ leads to similar results) and we measure the errors in the following way

$$\Big(\sum_{n=0}^{\mathsf{N}} \|u_n - u_n^{\mathsf{N},\mathsf{Q}})\|^2\Big)^{1/2},$$

which is a discrete analogue to $(\mathbb{E}[\|u - u^{(\mathsf{Q})}\|^2])^{1/2}$.

In Figure 11.1, we observe that the errors of the Wick-Malliavin approximation decrease with the increase of level Q. When the noise magnitude is small, e.g., $\sigma = 0.1$, level $\mathsf{Q} = 4$ can lead to accuracy of 10^{-7}. However, when the noise magnitude is large, e.g., $\sigma = 0.5$, we only achieve an accuracy of 10^{-2} when $\mathsf{Q} = 4$. Moreover, we note that when the noise magnitude is even larger, e.g., $\sigma = 0.65$, we need a large level Q: when $\mathsf{Q} = 1$, the mean-square error is large than 1. To have a reasonable accuracy, we need $\mathsf{Q} = 5$ where the error is around 10^{-1}.

Numerical results in this section verify the claim in Theorem 11.3.4. It seems likely that the Wick-Malliavin approximation is similar to perturbation methods: they are efficient when the noise magnitude is small and not working efficiently when the noise magnitude is large. However, the Wick-Malliavin approximation can be of high order for Q and we only need small Q for noises with small magnitude. When Q is large, the method can work for large magnitude noises such as $\sigma = 0.5$ which improves the results by a perturbation analysis in [39], where σ is less than 0.4.

11.4 White noise as coefficient

We discuss WCE with finite element methods for stochastic elliptic equations with *spatial white noise*, i.e., (11.1.1) with

$$a(x, \omega) = \mathbb{E}[a(x, \cdot)] + \dot{W}(x),$$

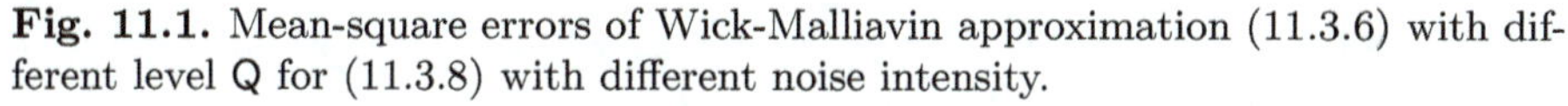

Fig. 11.1. Mean-square errors of Wick-Malliavin approximation (11.3.6) with different level Q for (11.3.8) with different noise intensity.

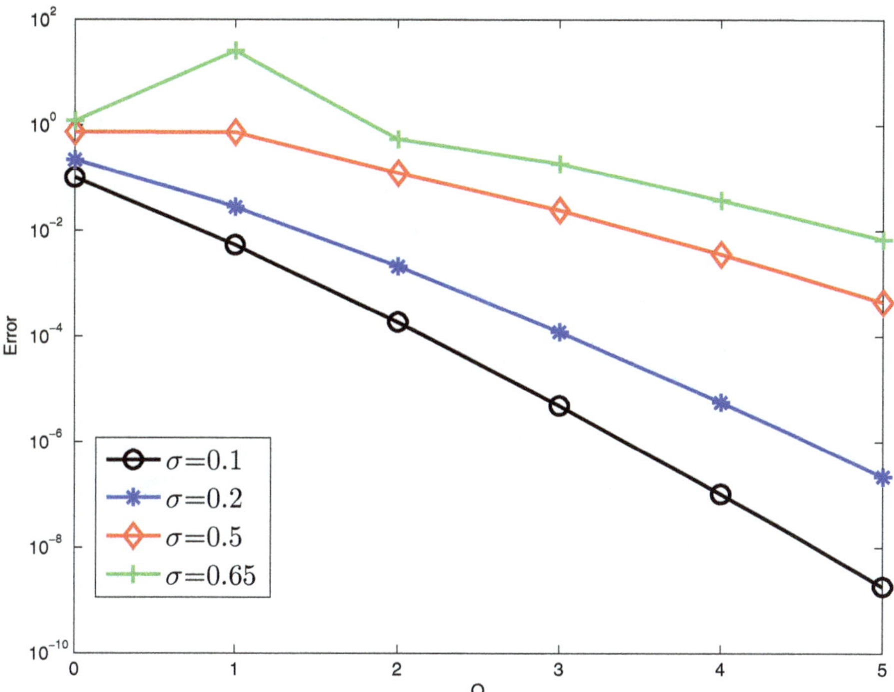

where $\dot{W} = \dot{W}(x)$ is a centered (zero mean) Gaussian white noise process in the spatial domain $\mathcal{D}$. Specifically, $\dot{W}(x) = \sum_{k=1}^{\infty} m_k \xi_k$ where $\{m_k(x)\}_{k\geq 1}$ is a complete orthonormal basis in $L^2(\mathcal{D})$ and ξ_k are mutually independent standard Gaussian random variables. Consider that the interaction of u and $\dot{W}$ is through the Wick product "$\diamond$," we then can write (11.1.1) as

$$\mathcal{A}u = \mathcal{M}u \diamond \dot{W} + f, \quad \text{in } \mathcal{D}, \quad u|_{\partial\mathcal{D}} = 0, \tag{11.4.1}$$

where $\mathcal{A} = \mathcal{M} = \Delta$. The operators $\mathcal{A}$, $\mathcal{M}$ can be extended to second-order differential operators of more general forms where $\mathcal{A}$ is uniformly elliptic and

$$\mathcal{M}v \diamond \dot{W} = \sum_{k\geq 1} (\mathcal{M}v) \diamond (\sum_{k\geq 1} m_k \xi_k) =: \sum_{k\geq 1} (\mathcal{M}_k v) \diamond \xi_k,$$

and $\mathcal{M}_k$ satisfies that for all $u, v \in H_0^1(\mathcal{D})$, $(\mathcal{M}_k u, v) \leq M_k \left\|\nabla u\right\|_1 \left\|v\right\|_1$.

Given the Cameron-Martin basis $\{\xi_\alpha\}_{\alpha\in\mathcal{J}}$, where $\mathcal{J}$ is the collection of multi-indices with only finitely many nonzeros, any solution in $\mathbb{L}^2(\Omega, \mathcal{F}, \mathbb{P})$ can be represented in the form of $\sum_{\alpha\in\mathcal{J}} u_\alpha \xi_\alpha$. However, the solution u for (11.4.1) belongs to the weighted space [319]

$$R\mathbb{L}^2(\mathcal{F}; H^1(\mathcal{D})) = \left\{ u = \sum_{\alpha\in\mathcal{J}} u_\alpha \xi_\alpha \middle| \sum_{\alpha\in\mathcal{J}} \left\|u_\alpha\right\|_{H^1}^2 r_\alpha^2 < \infty \right\},$$

instead of lying in $\mathbb{L}^2(\mathcal{F}; H^1(\mathcal{D})) = \left\{u = \sum_{\alpha\in\mathcal{J}} u_\alpha\xi_\alpha \middle| \sum_{\alpha\in\mathcal{J}} \|u_\alpha\|_{H^1}^2 < \infty\right\}$. Here, the weights r_α are [319, Theorem 3.1.1]

$$r_\alpha = \frac{q^\alpha}{\sqrt{|\alpha|!}}, \text{ where } q^\alpha = \prod_{k=1}^{\infty} q_k^{\alpha_k}, \tag{11.4.2}$$

and q_k, $k \geq 1$ are chosen such that $\sum_{k\geq 1} q_k^2 C_k^2 < 1$, with $\left\|\mathcal{A}^{-1}\mathcal{M}_k v\right\|_1 \leq C_k \left\|v\right\|_1$.

With an interpretation of solutions in weighted spaces, we still can plug the representation $\sum_{\alpha\in\mathcal{J}} u_\alpha\xi_\alpha$ into (11.4.1) and take expectation after multiplying ξ_α over both sides of the equations to obtain the so-called *propagator*, for each $\alpha \in \mathcal{J}$,

$$\mathcal{A}u_\alpha = \sum_{k=1}^{\infty} \sqrt{\alpha_k}\mathcal{M}_k u_{\alpha-\varepsilon_k} + f\mathbf{1}_{|\alpha|=0}, \quad \text{in } \mathcal{D}, \quad u_\alpha|_{\partial\mathcal{D}} = 0. \tag{11.4.3}$$

Here $\varepsilon_k \in \mathcal{J}$ and $|\varepsilon_k| = 1$ with $(\varepsilon_k)_k = 1$, and we also use the convention that $(\alpha - \varepsilon_k)_k = 0$ if $\alpha_k - \varepsilon_k \leq 0$. We observe that we have transformed the stochastic problem into a weakly coupled system of deterministic equations with the use of Wick product; otherwise, we are led to a fully coupled system of deterministic equations.

In numerical methods, we are only interested in some truncated propagator, e.g., those equations with $\alpha \in \mathcal{J}_{\mathsf{N},\mathsf{n}} =: \{\alpha = (\alpha_1, \cdots, \alpha_{\mathsf{n}}) | \, |\alpha| := \sum_{k=1}^{\mathsf{n}} \alpha_k\} \leq \mathsf{N}$ for the propagator (11.4.3) of the elliptic equation.

To facilitate the finite element approximation of the truncated propagator in physical space, we first state the propagator (11.4.3) in its variational form,

$$A(u_\alpha, v) = \sum_{k=1}^{\infty} \sqrt{\alpha_k} M_k(u_{\alpha-\varepsilon_k}, v) + (f\mathbf{1}_{|\alpha|=0}, v), \quad \forall\, v \in H_0^1(\mathcal{D}), \tag{11.4.4}$$

where $A(u, v)$, and $M_k(u, v)$ are bilinear form associated with $\mathcal{A}$ and $\mathcal{M}_k$, and $(\cdot, \cdot)$ is the inner product in $L^2(\mathcal{D})$. Suppose that we use finite element space $\{S_h\}$, which is finite-dimensional subspaces of $H_0^1(\mathcal{D})$ contains piecewise polynomials of at most $r-1$ order ($r \geq 2$) and the partition $\mathbb{T}_h$ of the domain $\mathcal{D}$ is quasi-uniform and h is the maximal length. The FEA of (11.4.1) (to be precise, FEA of the truncated propagator of (11.4.1)) is:

To find $u_{\mathsf{N},\mathsf{n}}^h = \sum_{\alpha\in\mathcal{J}_{\mathsf{N},\mathsf{n}}} u_\alpha^h \xi_\alpha$ where the Wiener chaos expansion coefficient $u_\alpha^h \in S_h$ satisfies

$$A^h(u_\alpha^h, v) = \sum_{k=1}^{\mathsf{n}} \sqrt{\alpha_k} M_k^h(u_{\alpha-\varepsilon_k}, v) + (f\mathbf{1}_{|\alpha|=0}, v), \quad \forall\, v \in S_h, \tag{11.4.5}$$

where $A^h(\cdot, \cdot)$ and $M_k^h(\cdot, \cdot)$ are approximations of $A(\cdot, \cdot)$ and $M_k(\cdot, \cdot)$ by numerical integrations, respectively. For simplicity of presentation, we assume that $A^h = A$ and $M_k^h = M$ over $S_h \times S_h$.

11.4.1 Error Estimates

In this section, we study how the finite element and Wiener Chaos truncation errors for (11.4.1), especially how errors from equations in the propagators grow in the weakly coupled systems of deterministic equations.

For the FEA of the problems (11.4.1) and (11.4.5), we have the following error estimate.

Theorem 11.4.1 *Assume the domain $\mathcal{D}$ is convex and open bounded with smooth boundary condition. Suppose that $f \in H^{m-1}(\mathcal{D})$, then $u \in R\mathbb{L}^2(\mathcal{F}, H^{m+1}(\mathcal{D}) \cap H_0^1(\mathcal{D}))$ solves (11.4.1). Suppose that $u^h_{\mathsf{N},\mathsf{n}} \in S_h$ is the FEA of (11.4.5). Under assumptions on S_h, we then have*

$$\begin{aligned}\left\|u - u^h_{\mathsf{N},\mathsf{n}}\right\|^2_{R\mathbb{L}^2(\mathcal{F};H^1(\mathcal{D}))} &= \sum_{\alpha\in\mathcal{J}_{\mathsf{N},\mathsf{n}}} \left\|u_\alpha - u^h_\alpha\right\|_1^2 r_\alpha^2 + \sum_{\alpha\notin\mathcal{J}_{\mathsf{N},\mathsf{n}}} \|u_\alpha\|_1^2 r_\alpha^2 \\ &\leq C\Big(\frac{1-(\hat{q}-\hat{q}_W)^{\mathsf{N}+1}}{1-(\hat{q}-\hat{q}_W)}\Big(\frac{1-\bar{q}_{\mathsf{n}}^{\mathsf{N}+1}}{1-\bar{q}_{\mathsf{n}}}\Big)^2 + \frac{1-\tilde{q}_{\mathsf{n}}^{\mathsf{N}+1}}{1-\tilde{q}_{\mathsf{n}}}\Big)h^{2m}\|f\|^2_{m-1} \\ &\quad + C\Big(\frac{\hat{q}_W}{(1-\hat{q})^2} + \frac{(\hat{q}-\hat{q}_W)^{\mathsf{N}+1}}{1-\hat{q}}\Big)\|f\|^2_{H^{-1}},\end{aligned}$$

where C is a constant depending solely on m, A_1, A_2 and $\mathcal{D}$, $\hat{q} = \sum_{k\geq1} C_k^2 q_k^2 < 1$ and $\hat{q}_W = \sum_{k>N} C_k^2 q_k^2$. The constants $\tilde{C}_k$'s come from (11.4.18) and $\tilde{q}_{\mathsf{n}} = \sum_{k=1}^{\mathsf{n}} \tilde{C}_k^2 q_k^2 < \bar{q}_{\mathsf{n}} = \sum_{k=1}^{\mathsf{n}} \bar{C}_k q_k < 1$ and $\bar{C}_k = \max(C_k, \tilde{C}_k)$.

Remark 11.4.2 *We have required that $\bar{q} = \sum_{k=1}^{\mathsf{n}} \bar{C}_k q_k < 1$, which leads to $\hat{q} < \bar{q} < 1$ and $\tilde{q} < \bar{q} < 1$. In this case, the constant in front of h^{2m} can be small. Error estimates for FEA of (11.4.1) have been investigated in [295, 469]. In [469], the authors present a total error estimate of FEA error and truncation of WCE, showing how the numerical solution method behaves with truncation parameters in random space. In a subsequent work [295], the authors show an optimal convergence in L^2-norm of some FEA in physical space, but there is a large constant in front of FEA error in physical space, $\binom{N+p}{p}$, i.e., the number of equations one solves in the truncated propagator. Theorem 11.4.1 implies that discretization in random space and physical space has very weak effects on each other.*

As we bound $\left\|u - u^h_{\mathsf{N},\mathsf{n}}\right\|^2_{R\mathbb{L}^2(\mathcal{F};H^1(\mathcal{D}))}$ by the two parts

$$\sum_{\alpha\in\mathcal{J}_{\mathsf{N},\mathsf{n}}} \left\|u_\alpha - u^h_\alpha\right\|_1^2 r_\alpha^2 \text{ and } \sum_{\alpha\notin\mathcal{J}_{\mathsf{N},\mathsf{n}}} \|u_\alpha\|_1^2 r_\alpha^2,$$

we then present the error estimates in two lemmas which readily lead to Theorem 11.4.1.

Lemma 11.4.3 (Error estimate of WCE truncation error, cf. [469]) *Assume the domain $\mathcal{D}$ is convex and open bounded with smooth boundary condition. Suppose that $f \in H^{-1}(\mathcal{D})$, then $u \in R\mathbb{L}^2(\mathcal{F}, H_0^1(\mathcal{D}))$ solves (11.4.1). We then have*

$$\sum_{\alpha \notin \mathcal{J}_{\mathsf{N},\mathsf{n}}} \|u_\alpha\|_1^2 r_\alpha^2 \leq C(\frac{\hat{q}_W}{(1-\hat{q})^2} + \frac{(\hat{q}-\hat{q}_W)^{\mathsf{N}+1}}{1-\hat{q}}) \|f\|_{H^{-1}}^2, \quad (11.4.6)$$

where C is a constant depending solely on m, A_1, A_2 and $\mathcal{D}$, $\hat{q} = \sum_{k\geq 1} C_k^2 q_k^2 < 1$ and $\hat{q}_W = \sum_{k>N} C_k^2 q_k^2$.

Proof. We observe that

$$\alpha \notin \mathcal{J}_{\mathsf{N},\mathsf{n}} = \{\alpha |\, |\alpha| > \mathsf{N}\} \oplus \{\alpha |\, |\alpha| \leq \mathsf{N} \text{ and } a_k > 0 \text{ for some } k > \mathsf{n}\} =: \mathcal{J}_1 \oplus \mathcal{J}_2.$$

Then we can write

$$\sum_{\alpha \notin \mathcal{J}_{\mathsf{N},\mathsf{n}}} \|u_\alpha\|_1^2 r_\alpha^2 = \sum_{\alpha \in \mathcal{J}_1} \|u_\alpha\|_1^2 r_\alpha^2 + \sum_{\alpha \in \mathcal{J}_2} \|u_\alpha\|_1^2 r_\alpha^2.$$

The first term in the summation can be estimated by, see Theorem 3.11 in [319],

$$\sum_{\alpha \in \mathcal{J}_1} \|u_\alpha\|_1^2 r_\alpha^2 = \sum_{|\alpha|>\mathsf{N}} \|u_\alpha\|^2 r_\alpha^2 \leq C^2(\mathcal{A}) \|f\|_{H^{-1}}^2 \sum_{|\alpha|>\mathsf{N}} (\sum_{k=1}^{\infty} C_k^2 q_k^2)^{|\alpha|}. \quad (11.4.7)$$

We write $\alpha \in \mathcal{J}_2$ into $\alpha = \alpha^{(1)} + \alpha^{(2)}$, where $\alpha^{(1)} \in \mathcal{J}_{\mathsf{N},\mathsf{n}}$ and all components in $\alpha^{(2)}$ where α_k, $k \leq \mathsf{n}$ are zeros. Let $|\alpha^{(1)}| = l$ and $|\alpha^{(2)}| = |\alpha| - l$, where l is a nonnegative integer.

$$\begin{aligned}\sum_{\alpha \in \mathcal{J}_2} \|u_\alpha\|^2 r_\alpha^2 &= \sum_{|\alpha|\geq 1,\, \alpha \in \mathcal{J}_2} \sum_{l=0}^{|\alpha|-1} \sum_{|\alpha^{(1)}|=l} \|u_\alpha\|^2 r_\alpha^2 \\ &\leq C^2(\mathcal{A}) \|f\|_{H^{-1}}^2 \sum_{|\alpha|\geq 1,\, \alpha \in \mathcal{J}_2} \sum_{l=0}^{|\alpha|-1} \sum_{|\alpha^{(1)}|=l} \\ &\qquad \frac{|\alpha|!}{\alpha^{(1)}!\alpha^{(2)}!} (\prod_k C_k^2 q_k^2)^{|\alpha_k|}, \end{aligned} \quad (11.4.8)$$

where we recall from [319] the estimate $\|u_\alpha\|^2 r_\alpha^2 \leq C^2(\mathcal{A}) \|f\|_{H^{-1}}^2 \frac{|\alpha|!}{\alpha^{(1)}!\alpha^{(2)}!} (\prod_k C_k^2 q_k^2)^{|\alpha_k|}$ for $\alpha \in \mathcal{J}_2$. Then by the multinomial expansion, the summation in (11.4.8) can be estimated by

$$\sum_{|\alpha|\geq 1,\,\alpha\in\mathcal{J}_2}\sum_{l=0}^{|\alpha|-1}\sum_{|\alpha^{(1)}|=l}\frac{|\alpha|!}{\alpha^{(1)}!\alpha^{(2)}!}(\prod_k C_k^2q_k^2)^{|\alpha_k|}$$

$$=\sum_{|\alpha|\geq 1,\,\alpha\in\mathcal{J}_2}\sum_{l=0}^{|\alpha|-1}\frac{|\alpha|!}{l!(|\alpha|-l)!}\sum_{|\alpha^{(1)}|=l}\frac{l!(|\alpha|-l)!}{\alpha^{(1)}!\alpha^{(2)}!}(\prod_k C_k^2q_k^2)^{|\alpha_k|}$$

$$\leq\sum_{|\alpha|=1}^{\infty}\sum_{l=0}^{|\alpha|-1}\frac{|\alpha|!}{l!(|\alpha|-l)!}(\hat{q}-\hat{q}_W)^i\hat{q}_W)^{n-i},$$

where $\hat{q}=\sum_k C_k^2q_k^2$ and $\hat{q}_W=\sum_{k>\mathsf{n}} C_k^2q_k^2$. From here and (11.4.8), it holds that

$$\sum_{\alpha\notin\mathcal{J}_{\mathsf{N},\mathsf{n}}}\|u_\alpha\|^2 r_\alpha^2=\sum_{|\alpha|>\mathsf{N}}\|u_\alpha\|^2 r_\alpha^2+\sum_{|\alpha|=1}^{\mathsf{N}}\sum_{l=0}^{|\alpha|-1}\|u_\alpha\|^2 r_\alpha^2$$
$$\leq C^2(\mathcal{A})\|f\|_{H^{-1}}^2\,(\sum_{|\alpha|=1}^{\mathsf{N}}\sum_{l=1}^{|\alpha|-1}\frac{|\alpha|!}{l!(|\alpha|-l)!}(\hat{q}-\hat{q}_W)^i(\hat{q}_W)^{n-i}+\sum_{|\alpha|>\mathsf{N}}\hat{q}^{|\alpha|}).$$

Now we claim that

$$\sum_{|\alpha|=1}^{\mathsf{N}}\sum_{l=1}^{|\alpha|-1}\frac{|\alpha|!}{l!(|\alpha|-l)!}(\hat{q}-\hat{q}_W)^l(\hat{q}_W)^{n-l}+\sum_{|\alpha|>\mathsf{N}}\hat{q}^{|\alpha|}\leq\frac{\hat{q}_W}{(1-\hat{q})^2}+\frac{(\hat{q}-\hat{q}_W)^{p+1}}{1-\hat{q}}.$$

In fact, by the binomial expansion

$$\sum_{|\alpha|=1}^{\mathsf{N}}\sum_{i=1}^{|\alpha|-1}\frac{|\alpha|!}{i!(|\alpha|-i)!}(\hat{q}-\hat{q}_W)^i(\hat{q}_W)^{n-i}+\sum_{|\alpha|>\mathsf{N}}\hat{q}^{|\alpha|}$$
$$\leq\sum_{|\alpha|=1}^{\mathsf{N}}(\hat{q}^{|\alpha|}-(\hat{q}-\hat{q}_W)^{|\alpha|})+\sum_{|\alpha|>\mathsf{N}}\hat{q}^{|\alpha|}$$
$$\leq\sum_{|\alpha|=1}^{\infty}(\hat{q}^{|\alpha|}-(\hat{q}-\hat{q}_W)^{|\alpha|})+\sum_{|\alpha|>\mathsf{N}}(\hat{q}-\hat{q}_W)^{|\alpha|}$$
$$\leq\frac{\hat{q}}{1-\hat{q}}-\frac{\hat{q}-\hat{q}_W}{1-\hat{q}+\hat{q}_W}+\frac{(\hat{q}-\hat{q}_W)^{p+1}}{1-\hat{q}+\hat{q}_W}$$
$$=\frac{\hat{q}_W}{(1-\hat{q}+\hat{q}_W)(1-\hat{q})}+\frac{(\hat{q}-\hat{q}_W)^{p+1}}{1-\hat{q}+\hat{q}_W}$$
$$\leq\frac{\hat{q}_W}{(1-\hat{q})^2}+\frac{(\hat{q}-\hat{q}_W)^{p+1}}{1-\hat{q}}.$$

Thus we arrive at (11.4.6).

Let us introduce the Ritz-Galerkin projection and its error estimate. The Ritz-Galerkin projection π_h is defined from $H_0^1(\mathcal{D})$ to S_h such that

$$A(\pi_h w - w, v_h) = 0, \quad \forall\, v_h \in S_h,\; w \in H_0^1(\mathcal{D}).$$

Denote by $\mathcal{T}_h$ the inverse of π_h in S_h, i.e., $\mathcal{T}_h g$ solves the deterministic problem $A^h(\pi_h w, v) = (g, v)$ with homogeneous Dirichlet boundary condition: $\pi_h w = \mathcal{T}_h g \in S_h$ and

$$A(w_h, v) = A^h(\pi_h w, v) = (g, v), \quad \forall\, v \in S_h. \tag{11.4.9}$$

For example, $\mathcal{A}u = -\sum_{i,j=1}^d D_i(D_j u) + a_0 u$, $a_0 > 0$. Then $\mathcal{A}$ is uniformly positive definite and

$$A^h(\pi_h u, v_h) = \sum_{i,j=1}^d (D_j \pi_h u, D_i v_h) + a_0(\pi_h u, v_h) = \sum_{i,j=1}^d (D_j w_h, D_i v_h) + a_0(w_h, v_h) = A(w_h, v_h).$$

The standard error estimate for π_h (associated with $\mathcal{A}$ and its bilinear form) is the following [446]:

$$\|\pi_h w - w\| + h\,\|\pi_h w - w\| \le h^s\,\|w\|_s, \quad \forall\, w \in H^s(\mathcal{D}) \cap H_0^1(\mathcal{D}), \quad 1 \le s \le r. \tag{11.4.10}$$

Lemma 11.4.4 (Error estimate of FEM error) *Assume the domain $\mathcal{D}$ is convex and open bounded with smooth boundary condition. Suppose that $f \in H^{m-1}(\mathcal{D})$, then $u \in R\mathbb{L}^2(\mathcal{F}, H^{m+1}(\mathcal{D}) \cap H_0^1(\mathcal{D}))$ solves (11.4.1). Suppose that $u_{\mathsf{N},\mathsf{n}}^h \in S_h$ is the FEA of (11.4.5). Under assumptions on S_h, we then have*

$$\sum_{\alpha \in \mathcal{J}_{\mathsf{N},\mathsf{n}}} \left\|u_\alpha - u_\alpha^h\right\|_1^2 r_\alpha^2 \le C\Big(\frac{1 - (\hat{q} - \hat{q}_W)^{\mathsf{N}+1}}{1 - (\hat{q} - \hat{q}_W)}\Big(\frac{1 - \bar{q}_{\mathsf{n}}^{\mathsf{N}+1}}{1 - \bar{q}_{\mathsf{n}}}\Big)^2 + \frac{1 - \tilde{q}_{\mathsf{n}}^{\mathsf{N}+1}}{1 - \tilde{q}_{\mathsf{n}}}\Big) h^{2m}\,\|f\|_{m-1}^2,$$

where C is a constant depending solely on m, A_1, A_2 and $\mathcal{D}$, $\hat{q} = \sum_{k \ge 1} C_k^2 q_k^2 < 1$ and $\hat{q}_W = \sum_{k > N} C_k^2 q_k^2$. All other constants are defined in Theorem 11.4.1.

Proof. Denote $e_\alpha = \pi_h u_\alpha - u_\alpha^h \in S_h$ and $\eta_\alpha = u_\alpha - \pi_h u_\alpha$, where π_h is the Ritz-Galerkin projection operator. From (11.4.3) and (11.4.5), one can get the following *error equation* for each $\alpha \in \mathcal{J}_{\mathsf{N},\mathsf{n}}$,

$$A^h(e_\alpha, v) = \sum_{k=1}^{\mathsf{n}} \sqrt{\alpha_k}(\mathcal{M}_k e_{\alpha - \varepsilon_k}, v) + \sum_{k=1}^{\mathsf{n}} \sqrt{\alpha_k}(\mathcal{M}_k \eta_{\alpha - \varepsilon_k}, v) \tag{11.4.11}$$

for all $v \in S_h$. This error equation can be rewritten as, by the definition of $\mathcal{T}_h$,

$$e_\alpha = \sum_{k=1}^{\mathsf{n}} \sqrt{\alpha_k}\,\mathcal{T}_h \mathcal{M}_k^h e_{\alpha - \varepsilon_k} + F_\alpha, \quad F_\alpha = \sum_{k=1}^{\mathsf{n}} \sqrt{\alpha_k}\,\mathcal{T}_h \mathcal{M}_k^h \eta_{\alpha - \varepsilon_k} \in S_h. \tag{11.4.12}$$

To address the dependence on the right-hand side and γ, we denote $e_\alpha(g;\gamma)$ the solution of (11.4.12), where $\gamma \in \mathcal{J}_{\mathsf{N},\mathsf{n}}$ and $F_\alpha = g\mathbf{1}_{\alpha=\gamma} \in H^1$. Noting that $e_\alpha(g;\gamma) = 0$ if $|\alpha| < |\gamma|$, we have

$$\sum_{\alpha\in\mathcal{J}_{\mathsf{N},\mathsf{n}}} \|e_\alpha(F_\gamma;\gamma)\|_1^2 r_\alpha^2 = \sum_{\alpha,\alpha+\gamma\in\mathcal{J}_{\mathsf{N},\mathsf{n}}} \|e_{\alpha+\gamma}(F_\gamma;\gamma)\|_1^2 r_{\alpha+\gamma}^2. \tag{11.4.13}$$

Define $\overline{e}_\alpha = e_\alpha(\alpha!)^{-1/2}$. By the linearity of the error equation (11.4.12), we have

$$\overline{e}_{\alpha+\gamma}(F_\gamma;\gamma) = \overline{e}_\alpha(F_\gamma(\gamma!)^{-1/2};(0)).$$

The term in the right-hand side $\overline{e}_\alpha(F_\gamma(\gamma!)^{-1/2};(0))$ can be estimated as follows. Following the arguments in the proof of Theorem 4.5 in [319], we have, for $|\alpha| = n$,

$$(\alpha!)^{1/2}\overline{e}_\alpha(F_\gamma(\gamma!)^{-1/2};(0)) = e_\alpha(F_\gamma(\gamma!)^{-1/2};(0)) = \frac{1}{\sqrt{\alpha!}}\sum_{\sigma\in\mathcal{P}_n} \mathcal{T}_h\mathcal{M}_{k_{\sigma_n}}\cdots \mathcal{T}_h\mathcal{M}_{k_{\sigma_1}}F_\gamma(\gamma!)^{-1/2}, \tag{11.4.14}$$

where $\mathcal{P}_n$ is the permutation group of the set $\{1,2,\cdots,n\}$. Thus by $\|\mathcal{T}_h\mathcal{M}_k v\|_1 \leq C_k\|v\|_1$,[1] we have, from (11.4.14),

$$\|\overline{e}_{\alpha+\gamma}(F_\gamma;\gamma)\|_1 \leq C_{\mathcal{A}}\frac{|\alpha|!}{\alpha!\sqrt{\gamma!}}\|F_\gamma\|_1\prod_k C_k^{\alpha_k}. \tag{11.4.15}$$

Thus we have, by the triangle inequality, (11.4.13) and (11.4.15),

$$\begin{aligned}
\Big(\sum_{\alpha\in\mathcal{J}_{\mathsf{N},\mathsf{n}}}\|e_\alpha\|_1^2 r_\alpha^2\Big)^{1/2} &\leq \sum_{\gamma\in\mathcal{J}_{\mathsf{N},\mathsf{n}}}\left(\sum_{\alpha\in\mathcal{J}_{\mathsf{N},\mathsf{n}}}\|e_\alpha(F_\gamma;\gamma)\|_1^2 r_\alpha^2\right)^{1/2}\\
&= \sum_{\gamma\in\mathcal{J}_{\mathsf{N},\mathsf{n}}}\left(\sum_{\alpha\in\mathcal{J}_{\mathsf{N},\mathsf{n}}}\|e_{\alpha+\gamma}(F_\gamma;\gamma)\|_1^2 r_{\alpha+\gamma}^2\right)^{1/2}\\
&\leq \sum_{\gamma\in\mathcal{J}_{\mathsf{N},\mathsf{n}}}\left(\sum_{\alpha\in\mathcal{J}_{\mathsf{N},\mathsf{n}}} C_{\mathcal{A}}^2\frac{(|\alpha|!)^2}{(\alpha!)^2\gamma!}\|F_\gamma\|_1^2\prod_k C_k^{2\alpha_k} r_{\alpha+\gamma}^2(\alpha+\gamma)!\right)^{1/2}\\
&= C_{\mathcal{A}}\sum_{\gamma\in\mathcal{J}_{\mathsf{N},\mathsf{n}}}\|F_\gamma\|_1 r_\gamma\left(\sum_{\alpha\in\mathcal{J}_{\mathsf{N},\mathsf{n}}}\frac{|\alpha|!\,|\gamma|!(\alpha+\gamma)!}{\alpha!\gamma!\ \ |\alpha+\gamma|!}[\frac{|\alpha|!}{\alpha!}\prod_k(C_k q_k)^{2\alpha_k}]\right)^{1/2},
\end{aligned}$$

where we recall the weights (11.4.2) in the last two steps. Then by the fact that $\frac{|\alpha|!|\gamma|!}{\alpha!\gamma!}\frac{(\alpha+\gamma)!}{|\alpha+\gamma|!} \leq 1$ [295, Lemma B.2]), we have

[1] The constant here is usually not the same as in the estimate $\|\mathcal{A}^{-1}\mathcal{M}_k\|_1 \leq C_k\|v\|_1$. Here we use the same constant for simplicity.

$$\left(\sum_{\alpha\in\mathcal{J}_{\mathsf{N},\mathsf{n}}}\|e_\alpha\|_1^2 r_\alpha^2\right)^{1/2} \le C_{\mathcal{A}}\sum_{\gamma\in\mathcal{J}_{\mathsf{N},\mathsf{n}}}\|F_\gamma\|_1 r_\gamma\left(\sum_{\alpha\in\mathcal{J}_{\mathsf{N},\mathsf{n}}}[\frac{|\alpha|!}{\alpha!}\prod_k (C_k q_k)^{2\alpha_k}]\right)^{1/2}$$

$$\le C_{\mathcal{A}}\sum_{\gamma\in\mathcal{J}_{\mathsf{N},\mathsf{n}}}\|F_\gamma\|_1 r_\gamma\left(\sum_{n=0}^{\mathsf{N}}(\sum_{k=1}^{\mathsf{n}} C_k^2 q_k^2)^n\right)^{1/2}$$

$$= C_{\mathcal{A}}\left(\frac{1-(\hat{q}-\hat{q}_W)^{\mathsf{N}+1}}{1-\hat{q}+\hat{q}_W}\right)^{1/2}\sum_{\gamma\in\mathcal{J}_{\mathsf{N},\mathsf{n}}}\|F_\gamma\|_1 r_\gamma, \quad (11.4.16)$$

where we denote $\sum_{k=1}^{\infty} C_k^2 q_k^2 = \hat{q}$ and $\sum_{k>N} C_k^2 q_k^2 = \hat{q}_W$.

It remains to estimate $\|F_\gamma\|_1$. Recalling (11.4.12) and $\|\mathcal{T}_h\mathcal{M}_k^h f\|_1 \le C_k\|f\|_1$, we have,

$$\|F_\gamma\|_1 = \left\|\sum_{k=1}^{\mathsf{n}}\sqrt{\gamma_k}\mathcal{T}_h\mathcal{M}_k^h\eta_{\gamma-\varepsilon_k}\right\|_1 \le \sum_{k=1}^{\mathsf{n}}\sqrt{\gamma_k}\left\|\mathcal{T}_h\mathcal{M}_k^h\eta_{\gamma-\varepsilon_k}\right\|_1$$

$$\le \sum_{k=1}^{\mathsf{n}}\sqrt{\gamma_k}C_k\|\eta_{\gamma-\varepsilon_k}\|_1 \le Ch^m\sum_{k=1}^{\mathsf{n}}\sqrt{\gamma_k}C_k\|u_{\gamma-\varepsilon_k}\|_{m+1}, \quad (11.4.17)$$

where we have applied the error estimate (11.4.10) for $\pi_h u - u$.

Assume that for any v in $H^{m+1}(\mathcal{D})$, it holds that

$$\|\mathcal{A}^{-1}\mathcal{M}_k v\|_{m+1} \le \tilde{C}_k\|v\|_{m+1}. \quad (11.4.18)$$

Similar to the proof of (11.4.15), there exists constants $C_{m,\mathcal{A}}$ such that

$$\|u_{\gamma-\varepsilon_k}\|_{m+1} \le C_{m,\mathcal{A}}\frac{|\gamma-\varepsilon_k|!}{\sqrt{(\gamma-\varepsilon_k)!}}\|f\|_{m-1}\prod_{j=1}^{\mathsf{n}}\tilde{C}_j^{(\gamma-\varepsilon_k)_j}. \quad (11.4.19)$$

From here and (11.4.17), we then have

$$\|F_\gamma\|_1 \le Ch^m\sum_{k=1}^{\mathsf{n}}\sqrt{\gamma_k}C_k\|u_{\gamma-\varepsilon_k}\|_{m+1} \le C_{m,\mathcal{A}}Ch^m\|f\|_{m-1}\frac{|\gamma|!}{\sqrt{\gamma!}}\prod_{j=1}^{\mathsf{n}}\bar{C}_j^{\gamma_j}\sum_{k=1}^{\mathsf{n}}\frac{\gamma_k}{|\gamma|}$$

$$= C_{m,\mathcal{A}}Ch^m\|f\|_{m-1}\frac{|\gamma|!}{\sqrt{\gamma!}}\prod_{j=1}^{\mathsf{n}}\bar{C}_j^{\gamma_j},$$

where $\bar{C}_k = \max(\tilde{C}_k, C_k)$. Hence by the multinomial expansion and (11.4.2), we have that

$$\begin{aligned}\sum_{\gamma\in\mathcal{J}_{\mathsf{N},\mathsf{n}}} \|F_\gamma\|_1 r_\gamma &\le Ch^m \|f\|_{m-1} \sum_{n=0}^{\mathsf{N}} \sum_{|\gamma|=n} \frac{\sqrt{|\gamma|!}}{\sqrt{\gamma!}} \prod_{j=1}^{\mathsf{n}} \bar{C}_j^{\gamma_j} q_j^{\gamma_j} \\ &\le Ch^m \|f\|_{m-1} \sum_{n=0}^{\mathsf{N}} \sum_{|\gamma|=n} \frac{|\gamma|!}{\gamma!} \prod_{j=1}^{\mathsf{n}} \bar{C}_j^{\gamma_j} q_j^{\gamma_j} \\ &\le Ch^m \|f\|_{m-1} \sum_{n=0}^{\mathsf{N}} (\sum_{j=1}^{\mathsf{n}} \bar{C}_j q_j)^n \\ &= Ch^m \|f\|_{m-1} \frac{1-\bar{q}_{\mathsf{n}}^{\mathsf{N}+1}}{1-\bar{q}_{\mathsf{n}}}, \quad (\bar{q}_{\mathsf{n}} = \sum_{j=1}^{\mathsf{n}} \bar{C}_j q_j < 1). \quad (11.4.20)\end{aligned}$$

Form (11.4.20), (11.4.16), and (11.4.19), we then conclude that

$$\begin{aligned}\sum_{\alpha\in\mathcal{J}_{\mathsf{N},\mathsf{n}}} \left\|u_\alpha - u_\alpha^h\right\|_1^2 r_\alpha^2 &\le 2 \sum_{\alpha\in\mathcal{J}_{\mathsf{N},\mathsf{n}}} \|e_\alpha\|_1^2 r_\alpha^2 + 2 \sum_{\alpha\in\mathcal{J}_{\mathsf{N},\mathsf{n}}} \|\eta_\alpha\|_1^2 r_\alpha^2 \\ &\le C \frac{1-(\hat{q}-\hat{q}_W)^{\mathsf{N}+1}}{1-(\hat{q}-\hat{q}_W)} (\frac{1-\bar{q}_{\mathsf{n}}^{\mathsf{N}+1}}{1-\bar{q}_{\mathsf{n}}})^2 h^{2m} \|f\|_{m-1}^2 \\ &\quad + Ch^{2m} \|f\|_{m-1}^2 \sum_{\alpha\in\mathcal{J}_{\mathsf{N},\mathsf{n}}} \frac{(|\alpha|!)^2}{\alpha!} \prod_{j=1}^{\mathsf{n}} \tilde{C}_j^{2\alpha} r_\alpha^2 \\ &\le Ch^{2m} \|f\|_{m-1}^2 (\frac{1-(\hat{q}-\hat{q}_W)^{\mathsf{N}+1}}{1-(\hat{q}-\hat{q}_W)} (\frac{1-\bar{q}_{\mathsf{n}}^{\mathsf{N}+1}}{1-\bar{q}_{\mathsf{n}}})^2 + \frac{1-\tilde{q}_{\mathsf{n}}^{\mathsf{N}+1}}{1-\tilde{q}_{\mathsf{n}}}),\end{aligned}$$

where $\tilde{q}_{\mathsf{n}} = \sum_{j=1}^{\mathsf{n}} \tilde{C}_j^2 q_j^2 < \bar{q}_{\mathsf{n}} < 1$ and we have used the following fact

$$\sum_{\alpha\in\mathcal{J}_{\mathsf{N},\mathsf{n}}} \frac{(|\alpha|!)^2}{\alpha!} \prod_{j=1}^{\mathsf{n}} \tilde{C}_j^{2\alpha} r_\alpha^2 = \sum_{\alpha\in\mathcal{J}_{\mathsf{N},\mathsf{n}}} \frac{|\alpha|!}{\alpha!} \prod_{j=1}^{\mathsf{n}} \tilde{C}_j^{2\alpha} q^{2\alpha} = \sum_{n=0}^{\mathsf{N}} (\sum_{j=1}^{\mathsf{n}} \tilde{C}_j^2 q_j^2)^n \le \frac{1-\tilde{q}_{\mathsf{n}}^{\mathsf{N}+1}}{1-\tilde{q}_{\mathsf{n}}}.$$

Remark 11.4.5 *Here we assume that the coefficients in $\mathcal{A}$ and $\mathcal{M}$ are sufficiently smooth. The operator $\mathcal{A}$ can be a second-order differential operator of general form but has to be positive definite.*

Remark 11.4.6 *In the proof, we assume for simplicity $A = A^h$ and $M = M^h$, i.e., we assume no extra errors introduced by numerical integration of the bilinear form A and M. The conclusion is also valid if these integration errors are considered and the convergence rate will not change as long as a high-order numerical integration method (higher than the convergence rate) is adopted. Strang's First Lemma (e.g., [80, Theorem 4.1.1]) sheds light on this issue: for each α,*

$$\begin{aligned}\left\|u_\alpha^h - u_\alpha\right\|_1 &\le C \inf_{v_h\in S_h} \left\{ \|v_h - u_\alpha\|_1 + \sup_{v_h\in S_h} \frac{\left|A(v_h,w_h) - A^h(v_h,w_h)\right|}{\|w_h\|_1} \right\} \\ &\quad + \sum_{k=1}^{\mathsf{n}} C_k \sqrt{\alpha_k} \left\|u_{\alpha-\varepsilon_k}^h - u_{\alpha-\varepsilon_k}\right\|_1.\end{aligned}$$

11.4.2 Numerical results

We consider a two-dimensional stochastic elliptic problem (11.4.1) in the following form

$$-\Delta u = \mathrm{div}(\nabla u) \diamond \dot{W} + 1, \quad x \in \mathcal{D} = [0,1] \times [0,1], \tag{11.4.21}$$

with homogeneous Dirichlet boundary conditions. Recall that $\dot{W}$ is a spatial white noise, $\dot{W} = \sum_{k=1}^{\infty} \phi_k(x)\xi_k$, where $\{\phi_k(x)\}$ is a CONS in $L^2(\mathcal{D})$ and ξ_k are mutually independent standard normal random variables. Here we take $\phi_k(x)$ as a proper reordering of the CONS with basis functions $m_l(x_1)m_n(x_2)$, where $\{m_l(x_1)\}$ is a CONS in $L^2([0,1])$, such that $l+n$ is increasing and l starts from 0. For example,

$$\phi_1(x) = m_0(x_1)m_0(x_2), \ \ \phi_2(x) = m_0(x_1)m_1(x_2), \ \ \phi_3(x) = m_1(x_1)m_0(x_2)$$
$$\phi_4(x) = m_0(x_1)m_2(x_2), \ \ \phi_5(x) = m_1(x_1)m_1(x_2), \ \ \phi_6(x) = m_0(x_1)m_2(x_2), \cdots$$

and we can deduce that

$$\phi_{19} = m_2(x_1)m_3(x_2), \quad \phi_{20} = m_1(x_1)m_4(x_2), \quad \phi_{21} = m_0(x_1)m_5(x_2).$$

In the computation we take $m_0(x_1) = 1$ and $m_l(x_1) = \sqrt{2}\cos(l\pi x_1)$ when $l \geq 1$. Equation (11.4.21) can be rewritten as

$$-\Delta u = \sum_{k=1}^{\infty} \mathrm{div}(\phi_k \nabla u) \diamond \xi_k + 1.$$

Let $u_{\mathsf{N},\mathsf{n}} = \sum_{n=0}^{\mathsf{N}} \sum_{|\alpha|=n} u_\alpha \xi_\alpha$ be the truncated WCE of the solution up to polynomial order N. Then the truncated propagator reads

$$-\Delta u_\alpha = \sum_{k=1}^{\mathsf{n}} \sqrt{\alpha_k} \mathrm{div}(\phi_k \nabla u_{\alpha-\varepsilon_k}) + f 1_{|\alpha|=0}, \quad \text{in } \mathcal{D}, \quad u_\alpha|_{\partial\mathcal{D}} = 0. \tag{11.4.22}$$

The weighted norm $\|u\|^2_{R\mathbb{L}^2(\mathcal{F};H^1(\mathcal{D}))}$ is computed by $\sum_{k=0}^{\infty} \sum_{|\alpha|=k} \|u_\alpha\|^2_{H^1} r_\alpha^2$, where

$$r_\alpha = \frac{q^\alpha}{|\alpha|!}, \quad q_k = \frac{1}{(1+k)C_k}, \quad C_k = \|\phi_k\|_\infty.$$

Each equation in the propagator (11.4.22) is solved with a spectral element method with 10 by 10 elements and 6 by 6 nodes (Gauss-Legendre-Lobatto nodes in both x_1 and x_2-directions) in each elements. In Figure 11.2, we plot the errors with respect to N with $\mathsf{n} = 21$, which are computed by

$$\left\| u_h^{\mathsf{N},\mathsf{n}} - u_h^{\mathsf{N}-1,\mathsf{n}} \right\|_{R\mathbb{L}^2(\mathcal{F};H^1(\mathcal{D}))} = \Big(\sum_{|\alpha|=\mathsf{N}} \|u_\alpha\|^2_{H^1} r_\alpha^2 \Big)^{1/2}.$$

We observe that a spectral convergence in N, the order of WCE, can be obtained, in agreement with our error estimate in Theorem 11.4.1.

Fig. 11.2. Weighted-H^1 errors in the polynomial orders N of WCE for Equation (11.4.22). A spectral element method with uniform partition (10×10 elements) with 6×6 Gauss-Legendre-Lobatto nodes in each elements. The number of random variable used is $\mathsf{n} = 21$.

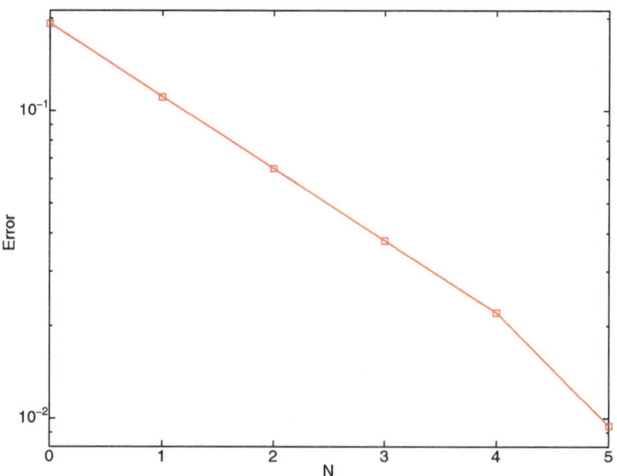

11.5 Application of Wick-Malliavin approximation to nonlinear SPDEs

In Chapter 11.3, we showed both numerically and theoretically that the accuracy of Wick-Malliavin approximation to elliptic equation with lognormal coefficient (11.3.8) is decreasing with the noise intensity, which is similar to perturbation methods. In this section, we show that the zeroth order Wick-Malliavin approximation leads to the same deterministic systems as those derived from perturbation methods for SPDEs with quadratic nonlinearity driven by Gaussian random fields, such as stochastic Burgers equation and stochastic Navier-Stokes equation.

Consider the following stochastic Burgers equation

$$\partial_t v + v\partial_x v = \nu\partial_x^2 v + \sigma\cos(x)\sum_{k=1}^{\infty}\lambda_k m_k(t)\xi_k, \quad (t,x)\in(0,T]\times(0,2\pi). \tag{11.5.1}$$

with deterministic initial condition $v(0,x)$ and periodic boundary conditions. Here $\{m_k(t)\}$ is CONS in $L^2([0,T])$, ξ_k's are mutually independent standard Gaussian random variables and λ_k's are real numbers. When $\lambda_k = 1$ for all $k\geq 1$, then $\sum_{k=1}^{\infty}\lambda_k m_k(t)\xi_k$ is the white noise, see Chapter 2.2.

Consider the WCE for the stochastic Burgers equation (11.5.1). Suppose that $v = \sum_{\alpha\in\mathcal{J}_{\mathsf{N},\mathsf{n}}} v_\alpha\xi_\alpha$. By (11.2.7) and Lemma 11.2.1, we know that, see also [225, 348],

$$v^2 = \sum_{\gamma\in\mathcal{J}}\sum_{(0)\leq\beta\leq\alpha} B(\alpha,\beta,\gamma)v_{\alpha-\beta+\gamma}v_{\beta+\gamma}\xi_\alpha,$$

where $B(\alpha,\beta,\gamma)$ is defined in (11.2.9). Then we can readily obtain the propagator for (11.5.1):

$$\partial_t v_\alpha + \frac{1}{2}\sum_{\gamma\in\mathcal{J}}\sum_{(0)\le\beta\le\alpha} B(\alpha,\beta,\gamma)\partial_x(u_{\alpha-\beta+\gamma}u_{\beta+\gamma}) = \nu\partial_x^2 u_\alpha + \sigma(x)\sum_{k=1}^{\infty} 1_{\{\alpha_j=\delta_{j,k}\}}\lambda_k m_k(t), \tag{11.5.2}$$

where for fixed k, $1_{\{\alpha_j=\delta_{j,k}\}} = 1$ and otherwise is equal to 0. The initial conditions for u_α are

$$u_\alpha(0,x) = \delta_{|\alpha|=0}u_0(x). \tag{11.5.3}$$

Again, we are solving a truncated propagator: for $\alpha\in\mathcal{J}_{\mathsf{N},\mathsf{n}}$,

$$\partial_t v_\alpha + \sum_{\gamma\in\mathcal{J}_{\mathsf{N},\mathsf{n}}}\sum_{(0)\le\beta\le\alpha}\frac{1}{2}B(\alpha,\beta,\gamma)\partial_x(v_{\alpha-\beta+\gamma}v_{\beta+\gamma}) = \nu\partial_x^2 u_\alpha + \sigma(x)\sum_{k=1}^{\infty} 1_{\{\alpha_j=\delta_{j,k}\}}m_k(t) \tag{11.5.4}$$

with the initial condition (11.5.3). By the Wick-Malliavin approximation, the truncation propagator can be approximated by

$$\partial_t u_\alpha + \sum_{\gamma\in\mathcal{J}_{\mathsf{Q},\mathsf{n}}}\sum_{(0)\le\beta\le\alpha}\frac{1}{2}B(\alpha,\beta,\gamma)\partial_x(u_{\alpha-\beta+\gamma}u_{\beta+\gamma}) = \nu\partial_x^2 u_\alpha + \sigma(x)\sum_{k=1}^{\infty} 1_{\{\alpha_j=\delta_{j,k}\}}\lambda_k m_k(t) \tag{11.5.5}$$

with the initial condition (11.5.3). Here we used u_α to represent an approximation of v_α as these two are generally different. In particular, when $\mathsf{Q}=0$, we have

$$\partial_t u_\alpha + \sum_{(0)\le\beta\le\alpha}\frac{1}{2}\partial_x(u_{\alpha-\beta}u_\beta) = \nu\partial_x^2 u_\alpha + \sigma(x)\sum_{k=1}^{\infty} 1_{\{\alpha_j=\delta_{j,k}\}}\lambda_k m_k(t). \tag{11.5.6}$$

When $\mathsf{Q}\ge\mathsf{N}$, we have from (11.2.11) that the truncated propagator (11.5.5) coincides with (11.5.4).

Now, we apply a perturbation method to solve (11.5.1): first write the solution in a power series expansion:

$$u = \tilde{u}_{(0)} + \sum_{|\alpha|=1}^{\infty}\tilde{u}_\alpha\prod_k \xi_k^{\alpha_k}$$

and then plug this expansion into (11.5.1) and compare the coefficient of $\prod_\alpha \xi_k^{\alpha_k}$ to obtain an equation that $\tilde{u}_\alpha$ satisfies:

$$\partial_t \tilde{u}_\alpha + \sum_{(0)\le\beta\le\alpha}\frac{1}{2}\partial_x(\tilde{u}_{\alpha-\beta}\tilde{u}_\beta) = \nu\partial_x^2 \tilde{u}_\alpha + \sigma(x)\sum_{k=1}^{\infty} 1_{\{\alpha_j=\delta_{j,k}\}}\lambda_k m_k(t). \tag{11.5.7}$$

We observe that $\tilde{u}_\alpha$ satisfies the exact same equation as (11.5.6) and also the same initial condition (11.5.3). A similar observation is made in [462] when a one-dimensional Burgers equation is considered with a linear combination of several Gaussian random variables as additive noise.

Remark 11.5.1 *Here we illustrated the idea for SPDEs with temporal white noise instead of spatial white noise or spatial-temporal white noise. For SPDEs with temporal white noise, the regularity can be low in time but can be high in space. For SPDEs with spatial white noise, the regularity in space can be low as we have seen in Chapter 10. Thus we need to use more terms in Wiener chaos expansion and Wick-Malliavin approximation.*

11.6 Wick-Malliavin approximation: extensions for non-Gaussian white noise

Let $\{m_k(x)\}_{k\geq 1}$ be a CONS in $L^2(\mathcal{D})$, $\mathcal{D}\subseteq \mathbb{R}^n$. Then we can define the white noise as

$$\dot{\mathfrak{N}}(x) = \sum_{k=1}^{\infty} m_k(x)\xi_k, \tag{11.6.1}$$

where ξ_k are mutually *uncorrelated* random variables obeying the same distribution with mean zero and variance one.

For example, when $\mathcal{D} = ([0,T])$ and $\{m_k(t)\}_{k\geq 1}$ is a CONS on $L^2\left([0,T]\right)$, the stochastic process $\dot{\mathfrak{N}}$ is exactly the Gaussian white noise if ξ_k are mutually independent standard Gaussian random variables. As in Chapter 2.2, the process $\int_0^t \dot{\mathfrak{N}}(s)\,ds$ is Brownian motion:

$$\mathfrak{N}(t) =: \int_0^t \dot{\mathfrak{N}}(s)\,ds = \sum_k \int_0^t m_k\,(s)\,ds\xi_k, \quad 0\leq t\leq T. \tag{11.6.2}$$

It is exactly the formulation of Brownian motion using an orthogonal expansion.

Even if ξ_k are not standard Gaussian random variables, we still have the following properties:

- $\mathfrak{N}(t)$ has uncorrelated increments;
- $\mathbb{E}[\mathfrak{N}(t)] = 0$;
- $\mathbb{E}[\mathfrak{N}(t)\mathfrak{N}(s)] = t\wedge s$. In particular, when $t=s$, $\mathbb{E}[(\mathfrak{N}(t))^2] = t$.

We can also define stochastic integrals in Ito's sense: for any continuous deterministic $f(t)$ on $[0,T]$:

$$\int_0^t f(s)d\mathfrak{N}(s) = \lim_{|\pi_n|\to 0}\sum_{i=1}^{n} f\left(t_i\right)\left(\mathfrak{N}_{t_{i+1}} - \mathfrak{N}_{t_i}\right) \text{ in } \mathbb{L}^2,$$

where $\Pi_n = \{t_i = t_i^n, 0\leq i\leq n\}$ is a partition of $[0,T]$.

Example 11.6.1 (Non-Gaussian white noise) *Let ξ_k be i.i.d. and $\mathbb{P}(\xi_k = \pm 1) = 1/2$. Then the process $\mathfrak{N}_t$ in (11.6.2) is not Gaussian since its characteristic function is for each $n>1$, $\theta\in\mathbb{R}$,*

$$\mathbb{E}[\exp\left(i\theta\sum_{k=1}^{n}\int_0^t m_k(s)ds\xi_k\right)] = \prod_{k=1}^{n}\cos\left(\theta\int_0^t m_k(s)ds\right).$$

Let ξ_k be uniform and i.i.d. on $[-\sqrt{3},\sqrt{3}]$. The stochastic process (11.6.2) *is non-Gaussian as*

$$\mathbb{E}[\exp\left(i\theta\sum_{k=1}^{n}\int_0^t m_k(s)ds\xi_k\right)] = \prod_{k=1}^{n}\frac{\sin\left(\theta\int_0^t m_k(s)ds\right)}{\left(\theta\int_0^t m_k(s)ds\right)}.$$

Recall that the characteristic function $\mathbb{E}[e^{i\theta X}]$ determines uniquely a distribution of the random variable X. Thus, to show that a process (random variable) is not Gaussian, it is equivalent to show that its characteristic function is not the same as the characteristic function of a Gaussian process (random variable). For i.i.d. standard Gaussian random variables ξ_k, the characteristic function is

$$\mathbb{E}[\exp\left(i\theta\sum_{k=1}^{n}\int_0^t m_k(s)ds\xi_k\right)] = \prod_{k=1}^{n}\exp\left(-\frac{\theta^2\int_0^t m_k(s)ds}{2}\right).$$

For many different distributions, we can derive generalized polynomial chaos using orthogonal polynomials, which are listed in Table 11.1. More generally, the whole family of Askey scheme of orthogonal polynomials can be used to construct polynomials chaos for various distributions, see, e.g., [488].

Table 11.1. Commonly used distributions (measures) and corresponding orthogonal polynomials.

Distribution	Orthogonal polynomials	Support	Alias
Gaussian	Hermite polynomials	$\mathbb{R}$	Wiener chaos (Hermite chaos)
Uniform	Legendre polynomials	$[a,b]$	Legendre chaos
Beta	Jacobi polynomials	$[a,b]$	Jacobi chaos
Gamma	Laguerre polynomials	$[0,\infty)$	Laguerre chaos
Poisson	Charlier polynomial	$\{0,1,2,\ldots\}$	Charlier chaos
Binomial	Krawtchouk polynomial	$\{0,1,2,\ldots,N\}$	Krawtchouk chaos

Assume that ξ obeys some distribution in Table 11.1 and P_k be the k-th order orthogonal polynomial corresponding to that measure. Assume also that

$$\int_{\mathcal{D}} P_k(\alpha\xi+\beta)P_l(\alpha\xi+\beta)\,d\mu = \delta_{l,k}\sqrt{k!},$$

where $\mathcal{D}$ is the support of the random variable ξ and α, β are constants such that $\alpha\xi+\beta$ are well defined in the domain of P_k's (the support $\mathcal{D}$). Here μ is the corresponding distribution (measure) of ξ. Then we define that

$$\mathfrak{N}_k(\xi) = P_k(\alpha\xi + \beta), \quad k \geq 0. \tag{11.6.3}$$

Then by the completeness of these polynomials in $\mathbb{L}^2(\mu)$, the set $\{\mathfrak{N}_k\}_{k=1}^{\infty}$ is a complete orthogonal basis of $\mathbb{L}^2(\Omega, \sigma(\xi), \mathbb{P})$.

Example 11.6.2 *Assume that ξ obeys a uniform distribution on $[-\sqrt{3}, \sqrt{3}]$. The orthogonal basis consists of Legendre polynomials, which can de defined by the Rodrigues formula*

$$L_k(x) = \frac{(-1)^k}{2^k k!} \frac{d^k}{dx^k} \left[(1 - x^2)^k\right]. \tag{11.6.4}$$

While Legendre polynomials L_k's are defined on $[-1, 1]$, we need the polynomials $L_k(\frac{x}{\sqrt{3}})$ instead ($\alpha = \frac{1}{\sqrt{3}}$, $\beta = 0$ in (11.6.3)). Moreover, we can check from the definition of the Legendre polynomials and integration by parts that for any $k, l \geq 0$,

$$\frac{1}{2\sqrt{3}} \int_{-\sqrt{3}}^{\sqrt{3}} L_k(\frac{x}{\sqrt{3}}) L_k(\frac{x}{\sqrt{3}}) dx = \frac{1}{2} \int_{-1}^{1} L_k(y) L_k(y) dy = \frac{1}{(2k+1)} \delta_{k,l}. \tag{11.6.5}$$

Define

$$\mathfrak{N}_k(\xi) = \sqrt{k!(2k+1)} L_k(\frac{\xi}{\sqrt{3}}). \tag{11.6.6}$$

Then the set $\{\mathfrak{N}_k(\xi)\}_{k=1}^{\infty}$ is a complete orthogonal basis of $\mathbb{L}^2(\Omega, \sigma(\xi), \mathbb{P})$.

For a multi-index $\alpha \in \mathcal{J}$, we define the polynomial in a product fashion:

$$\mathfrak{N}_\alpha = \prod_k \mathfrak{N}_{\alpha_k}(\xi_k). \tag{11.6.7}$$

Note that $\mathbb{E}[\mathfrak{N}_\alpha^2] = \alpha!$. The set $\{\mathfrak{N}_\alpha, \alpha \in J\}$ is a complete orthogonal basis in $\mathbb{L}^2(\Omega, \sigma(\xi_k, k \geq 1), \mathbb{P})$.

Similar to the Cameron-Martin theorem (Theorem 2.3.6), we can use orthogonal expansions to represent square-integrable stochastic processes.

Theorem 11.6.3 (Generalized polynomial chaos expansion) *Let $\mathcal{F} = \sigma$
$(\xi_k,\ k \geq 1)$. Then $\{\mathfrak{N}_\alpha\}_{\alpha \in \mathcal{J}}$ from (11.6.7) is a complete orthogonal system of $\mathbb{L}^2$
$(\Omega, \mathcal{F}, \mathbb{P})$: for each $\eta \in \mathbb{L}^2(\Omega, \mathcal{F}, \mathbb{P})$,*

$$\eta = \sum_\alpha \eta_\alpha \mathfrak{N}_\alpha, \quad \eta_\alpha = \frac{\mathbb{E}[\eta \mathfrak{N}_\alpha]}{\alpha!}, \ \mathbb{E}[(\mathfrak{N}_\alpha)^2] = \alpha!.$$

Moreover, $\mathbb{E}[\eta^2] = \sum_\alpha \eta_\alpha^2 \alpha! < \infty$.

The orthogonal expansion in the theorem is called a generalized polynomial chaos expansion, cf. Wiener (Hermite polynomial) chaos expansion in Chapter 2.5.3. The basis $\{\mathfrak{N}_\alpha\}$ is called a generalized Cameron-Martin basis.

Now we define the Wick product for the generalized Cameron-Martin basis and then for square-integrable stochastic processes.

$$\mathfrak{N}_\alpha \diamond \mathfrak{N}_\beta = \mathfrak{N}_{\alpha+\beta}, 1 \diamond \mathfrak{N}_\alpha = \mathfrak{N}_\alpha, \quad \alpha, \beta \in \mathcal{J}.$$

For $u = \sum_\alpha u_\alpha \mathfrak{N}_\alpha, v = \sum_\alpha v_\alpha \mathfrak{N}_\alpha$, where u_α, $v_\alpha \in \mathbb{R}$

$$u \diamond v = \sum_\alpha \sum_{\beta \leq \alpha} u_\beta v_{\alpha-\beta} \mathfrak{N}_\alpha.$$

The following example shows that the propagators of Wick-nonlinear equations are formally the same regardless of the distribution of random variables or stochastic processes, see, e.g., [349].

Example 11.6.4 *Consider the equation*

$$\mathcal{A}u - u^{\diamond 3} + \sum_{k=1}^{\infty} \mathcal{M}_k u \diamond \xi_n = f.$$

Here $u^{\diamond 3}$ denotes $u \diamond u \diamond u$. The operators $\mathcal{A}$ and $\mathcal{M}_k$ are the same with those in Chapter 11.4.

By Theorem 11.6.3, we can seek a chaos solution: $u = \sum_{\alpha \in \mathcal{J}} u_\alpha \mathfrak{N}_\alpha$. By the definition of Wick product,

$$u^{\diamond 3} = u \diamond u \diamond u = \sum_{\alpha,\beta,\gamma \in \mathcal{J}} u_\alpha u_\beta u_\gamma \mathfrak{N}_{\alpha+\beta+\gamma} = \sum_{\theta \in \mathcal{J}} \Big(\sum_{\alpha+\beta+\gamma=\theta} u_\alpha u_\beta u_\gamma \Big) \mathfrak{N}_\theta. \tag{11.6.8}$$

Then by the orthogonality of generalized polynomial chaos expansion, we obtain the propagator (the coefficients of chaos expansion):

$$\mathcal{A}u_\theta + \sum_{\alpha+\beta+\gamma=\theta} u_\alpha u_\beta u_\gamma + \sum_{k=1}^{\infty} \mathcal{M}_k u_{\theta-\epsilon_k} = f_\theta, \quad f_\theta = \frac{\mathbb{E}[f \mathfrak{N}_\theta]}{\theta!}, \theta \in \mathcal{J}. \tag{11.6.9}$$

The propagator (11.6.9) holds for any random variable ξ_k's as long as they have distributions corresponding to polynomial chaos listed in Table 11.1. For example, ξ_k's can be all Gaussian random variables or ξ_{2k}'s are Gaussian random variables while ξ_{2k-1}'s have uniform distributions.

For equations with polynomial nonlinearity, we can use the Wick-Malliavin expansion to obtain propagators. Let us define the Malliavin derivative of the generalized Cameron-Martin basis and then that of square-integrable stochastic processes.

$$\mathfrak{D}\mathfrak{N}_\alpha = \sum_{|\gamma|=1, \gamma \leq \alpha} \frac{\alpha!}{(\alpha-p)!} \mathfrak{N}_{\alpha-\gamma} = \sum_{\beta \in \mathcal{J}} \sum_{|\gamma|=1, \beta+\gamma=\alpha} \frac{(\gamma+\beta)!}{\beta!} \mathfrak{N}_\beta.$$

For $u = \sum_\alpha u_\alpha \mathfrak{N}_\alpha \in \mathbb{L}^2$,

$$\mathfrak{D}_\gamma u = \sum_{|\alpha|\geq 1} \alpha_k u_\alpha \mathfrak{N}_{\alpha(k)} = \sum_{\alpha\geq\gamma} \frac{\alpha!}{(\alpha-\gamma)!} u_\alpha \mathfrak{N}_{\alpha-\gamma}, \quad |\gamma| = 1,$$

$$\mathfrak{D} u = \sum_{|\gamma|=1} \mathfrak{D}_\gamma u = \sum_{\alpha\geq\gamma} \sum_{|\gamma|=1} \frac{\alpha!}{(\alpha-\gamma)!} u_\alpha \mathfrak{N}_{\alpha-\gamma} = \sum_{\alpha\in\mathcal{J}} \sum_{|\gamma|=1} \frac{(\alpha+\gamma)!}{\alpha!} u_{\alpha+\gamma} \mathfrak{N}_\alpha.$$

Here $\alpha(k) = \alpha - \epsilon_k \geq 0$ and $\epsilon_k \in \mathcal{J}$ with one at k-th element and $|\epsilon_k| = 1$. Higher-order Malliavin derivatives can be defined as follows:

$$\mathfrak{D}_\gamma^n u = \sum_{\alpha\geq\gamma} \frac{\alpha!}{(\alpha-\gamma)!} u_\alpha \mathfrak{N}_{\alpha-\gamma}, \quad |\gamma| = 1,$$

$$\mathfrak{D}^n u = \sum_{|\gamma|=n} \mathfrak{D}_\gamma^n u = \sum_{|\gamma|=n} \sum_{\alpha\geq\gamma} \frac{\alpha!}{(\alpha-\gamma)!\gamma!} u_\alpha \mathfrak{N}_{\alpha-\gamma} = \sum_{\alpha\in\mathcal{J}} \sum_{|\gamma|=n} \frac{(\alpha+\gamma)!}{\alpha!\gamma!} u_{\alpha+\gamma} \mathfrak{N}_\alpha.$$

In general, the Mikulevicius-Rozovsky formula (11.2.5) is not valid for general polynomial chaos other than Hermite polynomial chaos. However, it is still possible to define Wick-Malliavin approximation due to the polynomial nature of generalized polynomial chaos. In fact, the idea of Mikulevicius-Rozovsky formula is based on the linearization coefficient problem of Hermite polynomials. For univariate Hermite polynomials, the linearization coefficient problem is to find the coefficient $a_{k,n,m}$ such that

$$H_n(x)H_m(x) = \sum_{k=0}^{m+n} c_{k,n,m} H_k(x) = \sum_{q=0}^{m\wedge n} b_{n,m,q} H_{m+m-2q}, \quad b_{n,m,q} = B(n,m,q) \frac{\sqrt{n!m!}}{\sqrt{(n+m-2q)!}}. \tag{11.6.10}$$

The Wick-product keeps the highest order polynomial by letting $q = 0$. The Wick-Malliavin approximation keeps all higher-order polynomials by letting $0 \leq q \leq Q$ where $0 \leq Q \leq n \wedge m$. Similarly, the linearization coefficient problem for general orthogonal polynomials is

$$P_n(x)P_m(x) = \sum_{k=0}^{m+n} a_{k,n,m} P_k(x) = \sum_{0\leq 2q\leq m+n} \tilde{b}_{n,m,q} P_{m+n-2q}(x). \tag{11.6.11}$$

Similar to the idea of Wick-Malliavin approximation, we define the following approximation

$$P_n(x)P_m(x) \approx P_n(x) \diamond_{\mathsf{Q}} P_m(x) = \sum_{0\leq 2q\leq \mathsf{Q}} \tilde{b}_{n,m,q} P_{m+n-2q}(x), \quad \mathsf{Q} \leq m+n.$$

Thus for (11.6.3) and (11.6.7), the Wick-Malliavin approximation is defined by

$$\mathfrak{N}_\alpha \diamond_{\mathsf{Q}} \mathfrak{N}_\beta = \sum_{0\leq 2|\gamma|\leq \mathsf{Q}} \tilde{b}_{\alpha,\beta,\gamma} \mathfrak{N}_{\alpha+\beta-\gamma}(x), \quad \mathsf{Q} \leq |\alpha+\beta|. \tag{11.6.12}$$

Assume that $u = \sum_{\alpha\in\mathcal{J}} u_\alpha \mathfrak{N}_\alpha$ and $v = \sum_{\alpha\in\mathcal{J}} v_\alpha \mathfrak{N}_\alpha$. Then

$$u \diamond_{\mathsf{Q}} v = \sum_{\alpha,\beta\in\mathcal{J}} u_\alpha v_\beta \mathfrak{N}_\alpha \diamond_{\mathsf{Q}} \mathfrak{N}_\beta = \sum_{\alpha,\beta\in\mathcal{J}} u_\alpha v_\beta \sum_{0\le 2|\gamma|\le \mathsf{Q}(\alpha,\beta)} \tilde{b}_{\alpha,\beta,\gamma} \mathfrak{N}_{\alpha+\beta-\gamma}(x), \mathsf{Q}(\alpha,\beta) \le |\alpha+\beta|. \quad (11.6.13)$$

11.6.1 Numerical results

Consider the following Burgers equation with additive noise:

$$\partial_t u + u\partial_x u = \nu\partial_x^2 u + \sigma/\mathsf{n}\sum_{k=1}^{\mathsf{n}} \cos(2k\pi x)\cos(2k\pi t)\xi_k, \quad x \in (0, 2\pi), \quad (11.6.14)$$

with deterministic initial condition $u_0(x) = 1 + \sin(2x)$ and periodic boundary conditions. We assume that ξ_k's are all standard Gaussian random variables or uniform random variables on $[-1, 1]$. The propagator of the Burgers equation is

$$\partial_t u_\alpha + \frac{1}{2}\sum_{q\in\mathcal{J}_{\mathsf{N},\mathsf{n}}}\sum_{\beta\le\alpha} \mathbb{C}(\alpha,\beta,q)\partial_x(u_{\alpha-\beta+q}u_{\beta+q}) = \nu\partial_x^2 u_\alpha + \sigma\sum_{k=1}^{\mathsf{n}} 1_{\{\alpha_j=\delta_{j,k},|\alpha|=1\}} m_k(t), \quad (11.6.15)$$

where for fixed k, $1_{\{\alpha_j=\delta_{j,k},|\alpha|=1\}} = 1$ and otherwise is equal to 0. The initial condition for u_α are

$$u_\alpha(0, x) = \delta_{|\alpha|=0} u_0(x). \quad (11.6.16)$$

If ξ_k's are standard Gaussian random variables, $\mathbb{C}(\alpha,\beta,q) = \Phi(\alpha,\beta,q)$. If ξ_k's are normalized uniform random variables over $[-1, 1]$, $\mathbb{C}(\alpha,\beta,q) = \Psi(\alpha,\beta,q)$. For the Legendre polynomial chaos expansion, we have for $u, v \in \mathbb{L}^2$,

$$uv = u \diamond v + \sum_{Q=1}^{\infty}\sum_{\alpha\in\mathcal{J}}\sum_{\kappa\in\mathcal{J},|\kappa|=Q}\sum_{(0)\le\beta\le\alpha} u_{\beta+\kappa}v_{\alpha-\beta+\kappa}\Psi(\alpha,\beta,\kappa)\frac{\mathfrak{N}_\alpha}{\sqrt{\alpha!}},$$

where $\mathfrak{N}_\alpha$ is the Legendre polynomial chaos (11.6.6)–(11.6.7) and

$$\Psi(\alpha,\beta,q) = \frac{A_q A_\beta A_{\alpha-\beta}}{A_{\alpha+q}} \frac{\sqrt{2\alpha+1}}{2(\alpha+q)+1}\sqrt{(2(\beta+q)+1)(2(\alpha-\beta+q)+1)}. \quad (11.6.17)$$

Here $A_q = \frac{\Gamma(q+1/2)}{q!\Gamma(1/2)}$ and $\Gamma(\cdot)$ is the Gamma function.

Let us first check the convergence in σ. In Figure 11.3, we plot the error in the second and fourth moments of the Wick-Malliavin approximation for (11.6.15) with Gaussian noise. Again, we don't have any exact solution and choose a reference solution from a stochastic collocation method with a very fine resolution in random space. We observe that the error is decreasing faster with larger Q up to certain σ. We can actually observe a further decrease in errors with smaller σ when we increase the resolution in physical space and time (numerical results are not shown here). In other words, in the current plot, the further decease trend in σ for large Q is dominated by the space-time discretization error. We have a very similar observation in Figure 11.4 for (11.6.15) with uniform noise.

Fig. 11.3. Relative errors in moments of Wick-Malliavin approximation for (11.6.15) with different level Q and a single Gaussian random variable: $T = 1$ and $\mathsf{n} = 1$. Left: Error in the second moment; Right: Error in the fourth moment.

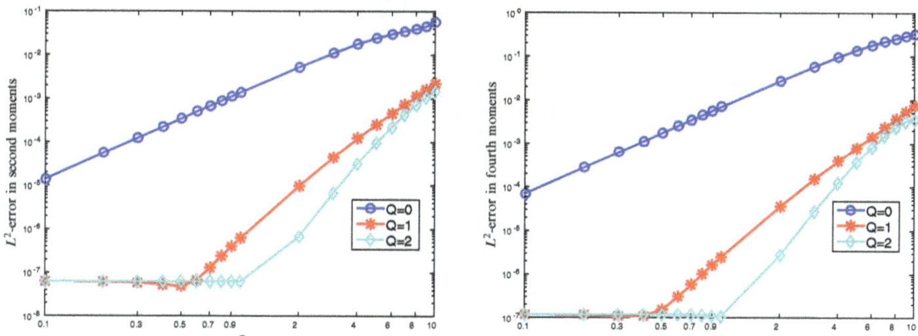

Now we present the computational error behavior with time where $\mathsf{n} = 4$ and we use $\mathsf{N} = 8$ for (11.6.15) with Gaussian noise and (11.6.15) with uniform noise. Here we obtain the reference solution by computing the solution with polynomial chaos methods without the Wick-Malliavin approximation. For

Fig. 11.4. Relative errors in moments of Wick-Malliavin approximation for (11.6.15) with different level Q and a single normalized uniform random variable: $T = 1$ and $\mathsf{n} = 1$. Left: Error in the second moment; Right: Error in the fourth moment.

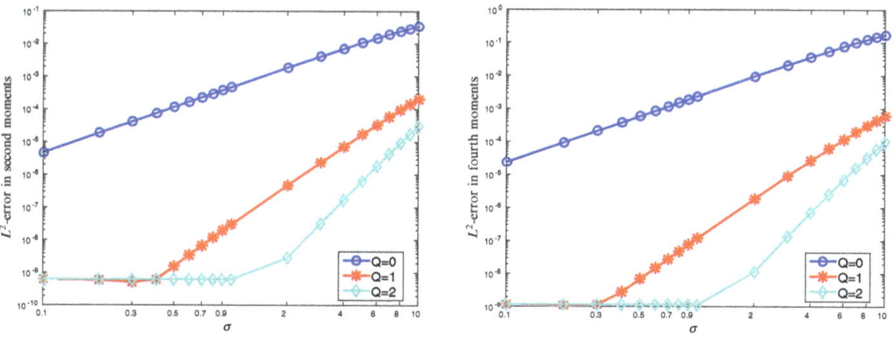

this problem, we observe that errors in the second moment and the fourth moment at $t = 0.9$ are changing slowly with time, see Figure 11.5 for Gaussian noise and Figure 11.6 for uniform noise. The error behaviors are similar but the errors from the uniform noise is smaller. This effect can be explained as follows: the variance of the uniform random variable (1/3) is smaller than that of the standard Gaussian. Also, it is interesting to observe that when $\mathsf{Q} = 3$, the errors of second moments are actually decreasing with time. One possible reason is that lower level Wick-Malliavin approximations may lead to large accumulation errors from the Wick-Malliavin approximation while for high level approximation (level 3 in this case) can allow small accumulation errors from the Wick-Malliavin approximation such that the total error is dominated by space-time discretization errors.

Fig. 11.5. Relative errors in moments of Wick-Malliavin approximation for (11.6.15) with Gaussian noise using different level Q: $\sigma = 10$ and $\mathsf{n} = 4$.

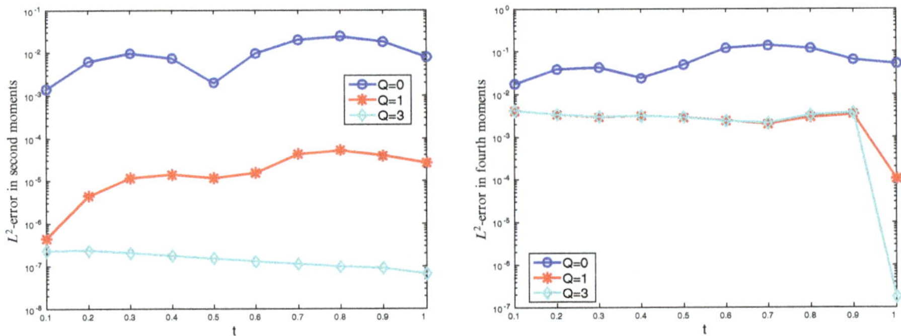

Fig. 11.6. Relative errors in moments of Wick-Malliavin approximation for (11.6.15) with uniform noise using with different level Q: $\sigma = 10$ and $\mathsf{n} = 4$.

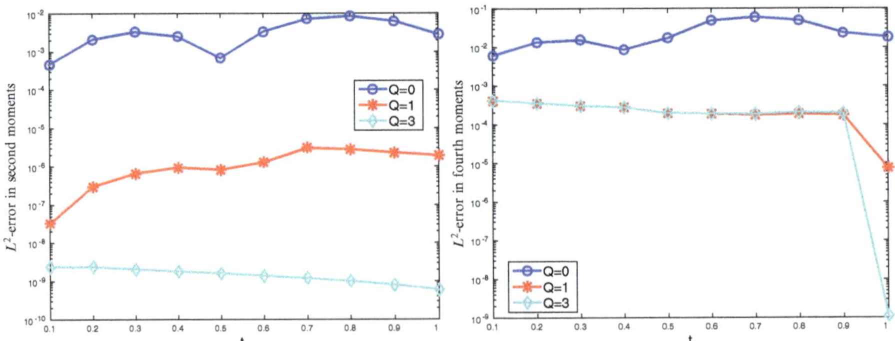

11.6.2 Malliavin derivatives for Poisson noises

The Malliavin derivatives of random variables depend on the nature of associated orthogonal polynomials. For Poisson noises, the associated measure of a Poisson random variable is $\sum_{k\in S} \frac{e^{-\lambda}\lambda^k}{k!}\delta(x-k)$ (the distribution is $\mathbb{P}(x=k) = \frac{e^{-\lambda}\lambda^k}{k!}$ and the mean is λ), where $S = \{0,1,2,\ldots\}$. The corresponding orthogonal polynomials are Charlier polynomials, the generating function of which is

$$\sum_{n=0}^{\infty} \frac{C_n(x;\lambda)}{n!} t^n = e^{-\lambda t}(1+t)^x. \tag{11.6.18}$$

The Charlier polynomials satisfy the following orthogonality:

$$\sum_{k\in S} \frac{e^{-\lambda}\lambda^k}{k!} C_m(k;\lambda)C_n(k;\lambda) = n!\lambda^n \delta_{m,n}. \tag{11.6.19}$$

Similar to Jacobi and Hermite polynomials, the Charlier polynomials can be computed via a recurrence relation

$$\begin{aligned} C_0(x;\lambda) &= 1, \quad C_1(x;\lambda) = x-\lambda, \\ C_{n+1}(x;\lambda) &= (x-\lambda-n)C_n(x;\lambda) - \lambda n C_{n-1}(x;\lambda). \end{aligned}$$

The first few Charlier polynomials are

$$\begin{aligned} C_0(x;\lambda) &= 1, \quad C_1(x;\lambda) = x-\lambda, \\ C_2(x;\lambda) &= x^2 = 2\lambda x - x + \lambda^2, \\ C_3(x;\lambda) &= x^3 - 3\lambda x^2 - 3x^2 + 3\lambda^2 x + 3\lambda x + 2x - \lambda^3. \end{aligned}$$

In general, the Charlier polynomial can be written as

$$C_n(x;\lambda) = \sum_{k=0}^{n} \binom{n}{k} (-1)^{n-k} x(x-1)\cdots(x-k+1). \tag{11.6.20}$$

Consider the product of two Charlier polynomials, which can be represented by linear combinations of Charlier polynomials. Moreover, by (11.6.18), we have

$$C_n(k;\lambda)C_m(k;\lambda) = \sum_{l\le m+n}\sum_{p\ge 0} \frac{m!n!\lambda^p}{p!(p-m+l)!(p-n+l)!(m+n-l-2p)!} C_l(k;\lambda)$$

where $p-m+l,\ p-n+l, m+n-l-2p \ge 0$. Letting $l = m+n-2q$, we then have

$$C_n(k;\lambda)C_m(k;\lambda) = \sum_{2q=0}^{m+n} b_{m,n,q} C_{m+n-2q}(k;\lambda), \tag{11.6.21}$$

where

$$b_{m,n,q} = \sum_{0\le p\le q,\, p+m\ge 2q,\, p+n\ge 2q} \frac{m!n!\lambda^p}{p!(p+m-2q)!(p+n-2q)!(2q-2p)!}.$$

Now we define the Charlier polynomial chaos and the Wick-Malliavin approximation of a product of two square-integrable random variables. Suppose that $u = \sum_{m=0}^{\infty} u_m C_m(\xi;\lambda) \in \mathbb{L}^2$ and $v = \sum_{n=0}^{\infty} v_n C_n(\xi;\lambda) \in \mathbb{L}^2$. When $uv \in \mathbb{L}^2$,

$$uv = \sum_{m,n=0}^{\infty} u_m v_n C_m(\xi;\lambda) C_n(\xi;\lambda) = \sum_{m,n=0}^{\infty} u_m v_n \sum_{2q=0}^{m+n} a_{q,m,n} C_{m+n-2q}(\xi;\lambda). \tag{11.6.22}$$

Define the Wick product as

$$C_m(\xi;\lambda) \diamond C_n(\xi;\lambda) = a_{0,m,n} C_{m+n}(\xi;\lambda) = C_{m+n}(\xi;\lambda).$$

As before, the idea is to simplify the form of the product in computation. Based on similar ideas, we can define the Wick-Malliavin product as follows:

$$C_n(\xi;\lambda) \diamond_{\mathsf{Q}} C_m(\xi;\lambda) = \sum_{2q=0}^{\mathsf{Q}} a_{q,m,n} C_{m+n-2q}(\xi;\lambda), \quad \mathsf{Q} = 0,1,2,\ldots. \tag{11.6.23}$$

When $\mathsf{Q} = 0$,

$$C_n(\xi;\lambda) \diamond_0 C_m(\xi;\lambda) = a_{0,m,n} C_{m+n}(\xi;\lambda) = C_n(\xi;\lambda) \diamond C_m(\xi;\lambda). \tag{11.6.24}$$

When $\mathsf{Q} \ge m+n$, then

$$C_n(\xi;\lambda) \diamond_{\mathsf{Q}} C_m(\xi;\lambda) = C_n(\xi;\lambda) C_m(\xi;\lambda). \tag{11.6.25}$$

We can interpret the operator $\diamond_{\mathsf{Q}}$ as the Wick-Malliavin approximation as Q goes to $m+n$ from below. With the approximation (11.6.23), we can define

$$u \diamond_{\mathsf{Q}} v = \sum_{m,n=0}^{\infty} u_m v_n C_m(\xi;\lambda) \diamond_{\mathsf{Q}} C_n(\xi;\lambda) = \sum_{m,n=0}^{\infty} u_m v_n \sum_{2q=0}^{\mathsf{Q}} a_{q,m,n} C_{m+n-2q}(\xi;\lambda). \tag{11.6.26}$$

When $\mathsf{Q} = \infty$, $u \diamond_{\mathsf{Q}} v$ is exactly uv if every term is well defined.

For multiple Poisson random variables ($\xi_k' s$ $k \ge 2$), the Charlier polynomial chaos basis can be defined as

$$\mathfrak{C}_\alpha = \prod_k C_{\alpha_k}(\xi_k;\lambda_k), \quad \alpha = (\alpha_1, \alpha_2, \ldots) \in \mathcal{J}.$$

The Wick-Malliavin product for these elements in the polynomial chaos basis is

$$\mathfrak{C}_\alpha \diamond_{\mathsf{Q}} \mathfrak{C}_\beta = \prod_k \left(C_{\alpha_k}(\xi_k;\lambda_k) \diamond_{q_k} C_{\beta_k}(\xi_k;\lambda_k) \right), \quad \mathsf{Q} = (q_1, q_2, \ldots) \in \mathcal{J}, \text{ and } \alpha, \beta \in \mathcal{J}. \tag{11.6.27}$$

This definition of Wick-Malliavin product is a bit different from that in (11.6.12) where Q is a scalar. When Q = (0), the Wick-Malliavin product becomes the Wick product.

$$\mathfrak{C}_\alpha \diamond \mathfrak{C}_\beta = \mathfrak{C}_{\alpha+\beta}. \tag{11.6.28}$$

For two square-integrable stochastic processes, u, v, we define

$$u \diamond_{\mathsf{Q}} v = \sum_{\alpha \in \mathcal{J}} u_\alpha \mathfrak{C}_\alpha \diamond_{\mathsf{Q}} \sum_{\beta \in \mathcal{J}} v_\beta \mathfrak{C}_\beta = \sum_{\alpha,\beta \in \mathcal{J}} u_\alpha v_\beta \left(\mathfrak{C}_\alpha \diamond_{\mathsf{Q}} \mathfrak{C}_\beta\right), \tag{11.6.29}$$

when uv is also square integrable.

Example 11.6.5 *Let ξ_k's be all Poisson random variables with mean λ. Consider the following Burgers equation*

$$\partial_t u + u\partial_x u = \nu\partial_x^2 u + \sigma \sum_{k=1}^{\mathsf{n}} m_k(t,x) C_1(\xi_k), \quad x \in (-\pi, \pi), \tag{11.6.30}$$

with a deterministic initial condition and periodic boundary conditions. Here $m_k(t,x)$'s are some real-valued function in t and x. A Wick-Malliavin approximation of (11.6.30)

$$\partial_t v + v \diamond_{\mathsf{Q}} \partial_x v = \nu\partial_x^2 v + \sigma \sum_{k=1}^{\mathsf{n}} m_k(t,x) C_1(\xi_k), \quad x \in (-\pi, \pi), \tag{11.6.31}$$

where $\mathsf{Q} = (q_1, q_2, \ldots, q_{\mathsf{n}}) \in \mathcal{J}_{\mathsf{N},\mathsf{n}}$.

The propagator of the Burgers equation (11.6.30) reads, for any $\alpha \in \mathcal{J}_{\mathsf{n}}$,

$$\partial_t u_\alpha + \sum_{\gamma,\beta \in \mathcal{J}} \sum_{\beta+\gamma=\alpha+2\theta,\, \theta \in \mathcal{J}} b_{\beta,\gamma,\theta} \partial_x u_\beta u_\gamma = \nu\partial_x^2 u_\alpha + \sigma \sum_{k=1}^{\mathsf{n}} 1_{\{\alpha_j=\delta_{j,k}, |\alpha|=1\}} m_k(t) \tag{11.6.32}$$

where for fixed k, $1_{\{\alpha_j=\delta_{j,k},|\alpha|=1\}} = 1$ and otherwise is equal to 0 and $b_{\alpha,\beta,\gamma}$ is the multivariate version of (11.6.21). The initial condition for u_α are

$$u_\alpha(0,x) = \delta_{\{|\alpha|=0\}} u_0(x). \tag{11.6.33}$$

In computation, we are only interested in a truncation of the propagator (with a bit abuse of notation): for $\alpha \in \mathcal{J}_{\mathsf{N},\mathsf{n}}$,

$$\begin{aligned} \partial_t u_\alpha + \sum_{\gamma,\beta \in \mathcal{J}_{\mathsf{N},\mathsf{n}}} \sum_{\beta+\gamma=\alpha+2\theta,\, \theta \in \mathcal{J}_{\mathsf{N},\mathsf{n}}} & b_{\beta,\gamma,\theta} \partial_x u_\beta u_\gamma \\ &= \nu\partial_x^2 u_\alpha + \sigma \sum_{k=1}^{\mathsf{n}} 1_{\{\alpha_j=\delta_{j,k}, |\alpha|=1\}} m_k(t). \end{aligned} \tag{11.6.34}$$

A Wick-Malliavin approximation is then, for $\alpha \in \mathcal{J}_{\mathsf{N},\mathsf{n}}$,

$$\partial_t u_\alpha + \sum_{\gamma,\beta \in \mathcal{J}_{\mathsf{N},\mathsf{n}}} \sum_{\substack{\beta+\gamma=\alpha+2\theta \\ \theta \in \mathcal{J}_{\mathsf{N},\mathsf{n}},\, \theta \leq \mathsf{Q}}} b_{\beta,\gamma,\theta} \partial_x u_\beta u_\gamma = \nu\partial_x^2 u_\alpha + \sigma \sum_{k=1}^{\mathsf{n}} 1_{\{\alpha_j=\delta_{j,k}, |\alpha|=1\}} m_k(t). \tag{11.6.35}$$

Numerical results

Consider the Burgers equation (11.6.30) with the deterministic initial condition $u_0(x) = 1 - \sin(x)$ and periodic boundary conditions. More numerical examples can be found in [509].

We are seeking a numerical solution of the form

$$u_{\mathsf{N},\mathsf{n}}^{\mathsf{Q}}(t,x) = \sum_{\alpha \in \mathcal{J}_{\mathsf{N},\mathsf{n}}} u_\alpha(t,x)\mathfrak{C}_\alpha,$$

where $u_\alpha(t,x)$ satisfies the Wick-Malliavin approximation (11.6.35). To solve the propagator, we use the following time discretization

$$u_\alpha^{n+1} - \nu\tfrac{\delta t}{2}\partial_x^2 u_\alpha^{n+1} = u_\alpha^n + \nu\tfrac{\delta t}{2}\partial_x^2 u_\alpha^n + \sum_{\gamma,\beta\in\mathcal{J}_{\mathsf{N},\mathsf{n}}} \sum_{\substack{\beta+\gamma=\alpha+2\theta \\ \theta\in\mathcal{J}_{\mathsf{N},\mathsf{n}},\, \theta\le \mathsf{Q}}} b_{\beta,\gamma,\theta}\partial_x u_\beta^n u_\gamma^n$$
$$+\sigma \sum_{k=1}^{\mathsf{n}} 1_{\{\alpha_j=\delta_{j,k},|\alpha|=1\}} m_k((t_n+t_{n+1})/2). \qquad (11.6.36)$$

In space, we use Fourier collocation method as in Chapter 7.4.

We compute the following error in the second moments

$$\rho_2(t) = \frac{\left\|\mathbb{E}[u_{\mathsf{N},\mathsf{Q}}^2] - \mathbb{E}[u_{\mathrm{ref}}^2]\right\|_{l^2}}{\|\mathbb{E}[u_{\mathrm{ref}}^2]\|_{l^2}}. \qquad (11.6.37)$$

Here the reference solution u_{ref} is computed with taking $\xi = k$, $k = 0,1,2,\ldots,K$ where $\mathbb{P}(\xi = K) = e^{-\lambda K}\lambda^K/K! > 10^{-16}$ and $\mathbb{P}(\xi = K+1) < 10^{-16}$. In computation, 100 Fourier collocation points are used and the time step size is specified in each figure.

For simplicity, we consider a single random variable where $\mathsf{n} = 1$, $m_1(x,t) = 1$, and $\lambda = 1$. We investigate the accuracy of the Wick-Malliavin approximation when Q varies. We take Charlier polynomial chaos up to six order ($\mathsf{N} = 6$). We observe in Figure 11.7 that the accuracy decreases with increasing Q.

We then check the effect of Q for various polynomial chaos order N. In Figure 11.8, we observe that for $\mathsf{N} = 1,2,3,4$, the accuracy can be improved when $\mathsf{Q} \le \mathsf{N} - 1$ but the improvement in accuracy is less significant when $\mathsf{Q} > \mathsf{N} - 1$, see also [509]. Moreover, when $\mathsf{Q} = \mathsf{N}$, the Wick-Malliavin approximation (11.6.35) leads to the same propagator as in (11.6.34).

An adaptive Wick-Malliavin approximation method. Now let us describe a simple but effective adaptive method proposed in [509]. The goal is to keep the relative error $\rho_2(t)$ at time grid no more than ϵ, say, 10^{-10}. The N-adaptivity refers to the refinement in N to keep $\rho_2(t) \le \epsilon$. In other words, when $\rho_2(t) > \epsilon$, we need to increase the number N. The Q-adaptivity refers to the refinement in Q when $\rho_2(t) > \epsilon$ if $\mathsf{Q} \le \mathsf{N}$. In Figure 11.9, we fix either N or Q to keep the relative error $\rho_2(t)$ no larger than ϵ.

11.7 Summary and bibliographic notes

For elliptic equations with spatial noise, we considered two cases of the coefficients: lognormal coefficient and spatial white noise. The use of Wick product leads to significant reduction of the computational cost by sparsifying the resulting linear system of deterministic elliptic equations.

Fig. 11.7. Different levels of Wick-Malliavin approximation of the Burger equations using (11.6.31): the relative errors $\rho_2(t)$ versus t. Here $\nu = 1/2$ and $\sigma = 0.1$. The polynomial chaos order $\mathsf{N} = 6$ and the time step size is $\delta t = 2 \times 10^{-4}$. These figures are adapted from [509].

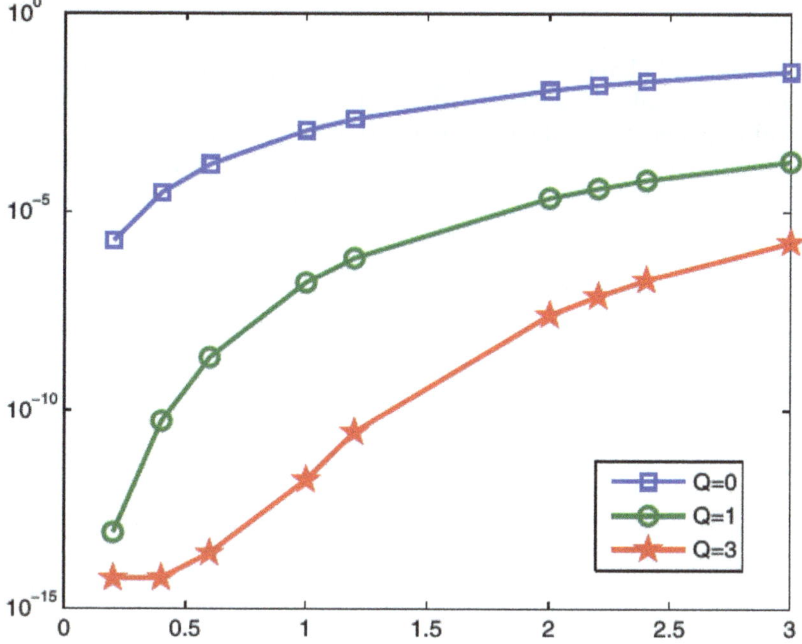

- For elliptic equations with lognormal coefficients, the Wick-Malliavin approximation was used in the framework of WCE to sparsify the resulting linear systems. The Wick-Malliavin approximation can be seen as high-order perturbation methods in terms of noise intensity, see Theorem 11.3.4. Numerical results show that the Wick-Malliavin approximation can work well even when the magnitude of noise is relatively large.
- For elliptic equations with spatial-white-noise coefficients, solutions lie in a weighted stochastic Sobolev space. The error estimate of finite element approximation and WCE are derived in Theorem 11.4.1. It is shown that in a proper stochastic Sobolev space, the finite element error will not be

polluted in the linear system of deterministic elliptic equations (propagator), i.e., the finite element approximation error can be small even if there is a large number of equations in the propagator.
- The Wick-Malliavin approximation are applied to nonlinear equations as well. We showed that the zeroth order Wick-Malliavin approximation leads to the same system of PDEs with that resulted from stochastic perturbation methods, see Chapter 11.5.
- The Wick-Malliavin approximation is generalized for non-Gaussian white noise. One-dimensional Burgers equations driven by non-Gaussian white noises are considered and numerical results are presented using the corresponding generalized polynomial chaos. An adaptive method with varying polynomial chaos orders and levels of the Wick-Malliavin approximation is presented.

The Wick-Malliavin approximation provides a systematic way to reduce the computational cost of generalized polynomial chaos methods, especially for nonlinear equations. The approximation can be thought as a high-order perturbation analysis method. However, it is not clear that under what conditions the Wick-Malliavin approximation errors will not blow up with the truncation parameters.

Bibliographic notes. In different applications, the following form of $a(x,\omega)$ has been used in (11.1.1):

- $a(x,\omega)$ is a bounded stochastic field, i.e., for a.e. ω, $a(x,\omega)$ is uniformly bounded in x ($a(x,\omega$ is uniformly bounded in both x and ω.), see, e.g., in [12].
- $a(x,\omega)$ is a lognormal field, i.e., $\ln(a(x,\omega))$ is a Gaussian process, see, e.g., [73, 124, 140, 141, 220, 462, 469].
- $a(x,\omega)$ is a Gaussian field, see, e.g., [319, 469]. In this case, the elliptic problem is not well posed in the classical sense and usually is accompanied by the Wick product where the solution lies in some weighted space along stochastic direction, see, e.g., [319, 469].

When $a(x,\omega)$ is uniformly positive, bounded, and sufficiently regular, the well-posedness and some finite element analysis has been established, see, e.g., [12] for some rigorous error analysis under the root assumption of finite dimensional noise. See also, e.g., [11] where deterministic integration methods are used (SCM), [107] (Monte Carlo methods), [1, 25] (multilevel Monte Carlo methods), [281] (quasi-Monte Carlo methods), and [282] (multilevel quasi-Monte Carlo methods).

For lognormal diffusion, the well-posedness question has been considered in, e.g., [73, 74, 141, 220, 320]. In the random space, several integration methods have been considered, see, e.g., [220, 320] (WCE), [124](SCM) [75, 166, 440] (Monte Carlo and multilevel Monte Carlo methods), [165] (quasi-Monte Carlo methods), and [164] (multilevel quasi-Monte Carlo methods).

Fig. 11.8. Different levels of Wick-Malliavin approximation of the Burgers equations using (11.6.31): the relative errors $\rho_2(0.5)$ versus Q. Here $\nu = 1$ and $\sigma = 1$. The time step size is $\delta t = 5 \times 10^{-4}$. These figures are adapted from [509].

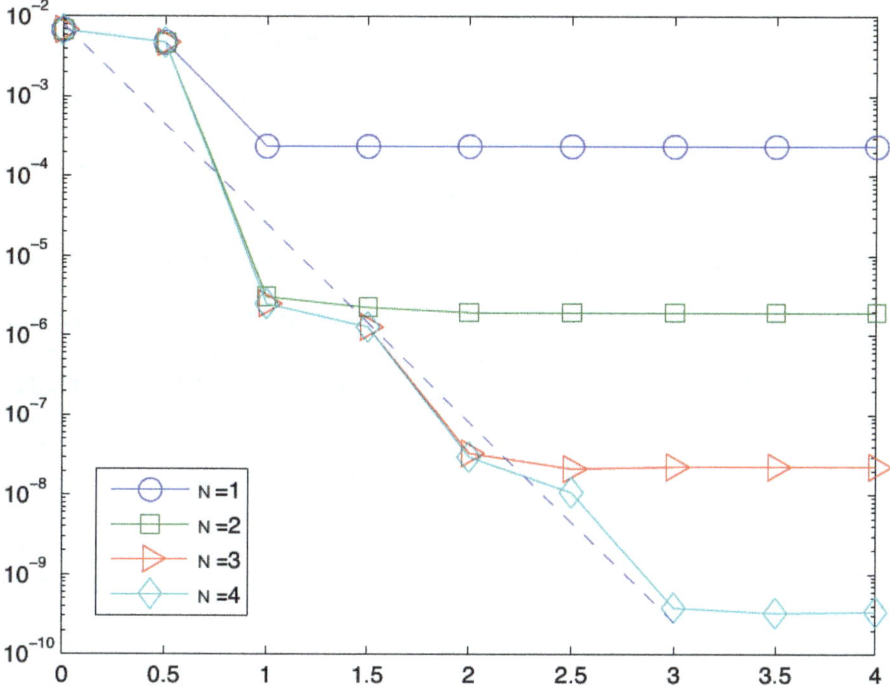

Fig. 11.9. The Charlier polynomial chaos method for the Burgers equation with N- or Q-adaptivity: the relative error $\rho_2(t)$ versus t. Here $\nu = 1/2$ and time step $\delta t = 2 \times 10^{-4}$. Left: Q-adaptivity with the polynomial chaos order N = 6 and $\sigma = 1$; Right. N-adaptivity with the level of Wick-Malliavin approximation Q = 3 and $\sigma = 0.1$. These figures are adapted from [509].

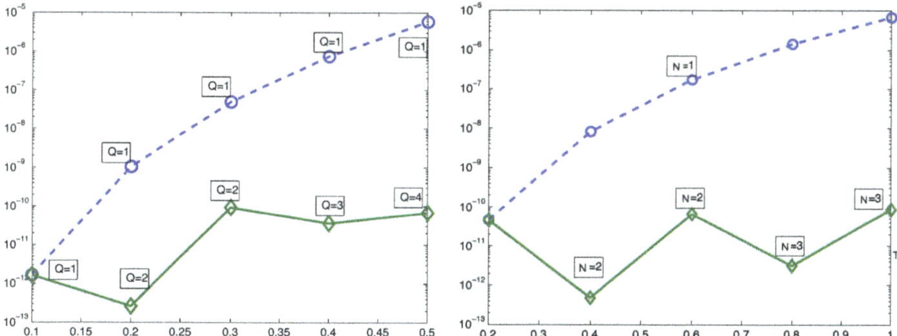

Another Wick-type model for elliptic equations with lognormal coefficient is proposed in [466, 467]:

$$-\mathrm{div}\big((a^{-1})^{\diamond(-1)}\big) \diamond \nabla u) = f.$$

This is also a second-order approximation (in the intensity of noise) of the model
(11.1.1) and is numerically demonstrated to be a more accurate approximation than (11.1.2).

Karhunen–Loève expansion. A convergence study of the Karhunen–Loève expansion has been presented numerically in [231] and theoretically in [135], where error analysis of truncating Karhunen–Loève expansion is provided with finite element methods for obtaining numerically $\phi_k(x)$, the elements of CONS.

Interpretation of WCE Solutions to SPDEs. With Wiener chaos methods, Rozovsky and his colleagues construct solutions in weighted Wiener chaos spaces for SPDEs, especially for linear equations, see [315, 319, 345, 347, 348, 384], etc. Note that the solutions are not always in $L^2(\mathbb{F})$, even for a simple wave equation with additive white noise in two dimensions. Thus, weighted spaces should be carefully introduced. Rozovsky and his colleagues provide a systematic approach for the Wiener chaos solution. By comparing it with the approach based on white noise theory, more appropriate weighted spaces are carefully chosen for different problems, offering more flexible solution spaces than the framework in [223].

For Wiener chaos method for white noise SPDEs, a lot of work has been done for linear equations, such as [34, 327, 330, 332, 333, 443, 469]. Convergence analysis can be found in [34, 62, 63, 469], etc. However, the error analysis is far from what is practically demanded, e.g., for nonlinear equations and how to balance the errors from deterministic solvers and truncation in random space.

Convergence rate of WCE for elliptic equations with lognormal coefficients. The key argument for the convergence of WCE is to estimate the derivatives with respect to parameters (random variables). There are two approaches for estimating these derivatives. The first one is to use multivariate Taylor's expansion to obtain the regularity estimation of the elliptic equations that the multivariate derivatives of solutions satisfy, see, e.g., [73, 74, 220]. Once these derivatives are estimated, the convergence rate of WCE can be found [220]. The second approach is to directly estimate the WCE coefficients of the solution, see, e.g., [140, 141], where some weighted spaces in random space are used.

The Mikulevicius-Rozovsky formula (11.2.5) was mentioned in [229, Theorem 4.10] but has been forgotten until the formula is derived in [346] with a simple proof for square-integrable random fields.

The Wick-Malliavin approximation has also been applied to some SPDEs with quadratic nonlinearity, driven by some Gaussian random fields [462] or driven by discrete processes [509]. A general framework for Wick-Malliavin approximation for random fields with given distribution has been developed in [349]. However, no rigorous analysis of the Wick-Malliavin approximation for these problems is available.

11.8 Suggested practice

Exercise 11.8.1 *Show that the formula* (11.2.11) *holds.*

Exercise 11.8.2 *Show that the Legendre polynomials defined by* (11.6.4) *satisfy the following relations*

$$\int_{-1}^{1} L_l(x)L_k(x)\,dx = \frac{2}{2k+1}\delta_{k,l},$$

and the three-term recurrence relation

$$(k+1)L_{k+1}(x) = (2k+1)xL_k(x) - kL_{k-1}(x),$$

where $L_0(x) = 1$, $L_1(x) = 1$. *Apply Legendre polynomial chaos* (11.6.6) *and the explicit fourth-order Runge-Kutta to solve the following linear model*

$$\frac{dy}{dt} = -\xi y, \quad t \in [0,5],$$

where ξ *obeys the uniform distribution on* $[0,1]$.

Exercise 11.8.3 *Apply the multistage method as in Exercise 6.6.2 to solve the linear model in the last exercise over the time interval* $[0,100]$.

Exercise 11.8.4 *Derive a* Q*-level Wick-Malliavin approximation and write down the corresponding propagator for the following equation*

$$\partial_t u - \partial_x^2 u = u - u^2 + \sum_{k=1}^{m} \cos(kx)\xi_k, \quad x \in (0,2\pi),$$

under the following conditions

- $m = 1$, ξ_1 *is a standard Gaussian random variable;*
- $m = 2$, ξ_k*'s (*$k = 1,2$*) are i.i.d. standard Gaussian random variables;*
- $m = 1$, ξ_1 *obeys the uniform distribution on* $[0,1]$*;*
- $m = 2$, ξ_k*'s (*$k = 1,2$*) obey the uniform distribution on* $[0,1]$ *and are i.i.d..*

12

Epilogue

Stochastic partial differential equations usually have solutions of low regularity due to the nature of infinite dimensional rough noises. The low regularity results in an enormous amount of computational time spent on Monte Carlo simulations. Despite of the simplicity of Monte Carlo methods, the slow convergence of Monte Carlo methods is the main bottleneck in computing numerical solutions to SPDEs. Although substantial improvements in Monte Carlo methods have been made in recent years, it is still desirable to have further accelerated sampling techniques. Depending on the specific problem, the integration in random space can be made effective using different methods such as quasi-Monte Carlo methods, Wiener chaos methods, and stochastic collocation methods.

In addition to methods of integration in random space, it is of great importance to understand the underlying SPDEs. While numerical methods are usually discussed for general equations, it is appreciated for numerical SPDEs that a specialized numerical method can be applied to solve a small class of SPDEs. For example, for linear equations with deterministic coefficients, we can make full use of linearity as done in Chapters 6 and 7.

In this work, we apply the Wong-Zakai approximation to stochastic differential equations with white noise. Our focus is to present how to use deterministic integration methods in random space, particularly Wiener chaos methods and stochastic collocation methods.

12.1 A review of this work

In Part I, we consider numerical methods for stochastic ordinary differential equations (SDEs). For stochastic differential equations with constant time delay, we derive three schemes from the Wong-Zakai approximation using

Z. Zhang, G.E. Karniadakis, *Numerical Methods for Stochastic Partial Differential Equations with White Noise*,
Applied Mathematical Sciences 196, DOI 10.1007/978-3-319-57511-7_12

the spectral approximation, the predictor-corrector scheme, the mid-point scheme, and the Milstein scheme. For stochastic ordinary differential equations with or without delay, we observe that the convergence order of numerical schemes via the Wong-Zakai approximation is not determined by the Wong-Zakai approximation but depends on further discretization in time. For example, under the assumption of Lipschitz continuous coefficients, the Wong-Zakai approximation itself is of order half in the mean square sense; however, the Milstein scheme based on the Wong-Zakai approximation (called Milstein-like scheme in Chapter 4) is of order one.

In practice, the coefficients of SDEs are not Lipschitz continuous or not even of linear growth. We consider stochastic differential equations with non-Lipschitz continuous coefficients both in drift and diffusion. Under a one-sided Lipschitz condition on coefficients, we present a fundamental limit theorem, i.e., a relationship between the local truncation error and global error in the mean-square sense for numerical schemes for nonlinear stochastic differential equations. We present an explicit balanced scheme so that we can efficiently integrate stochastic differential equations with super-linearly growing coefficients over a finite time interval.

In Part II, we consider Wiener chaos and stochastic collocation methods for linear advection-diffusion-reaction equations with multiplicative noises.

We present a recursive multistage Wiener chaos expansion method (WCE) and a recursive multistage stochastic collocation method (SCM) for longer time integration of linear stochastic advection-reaction-diffusion equations with finite dimensional noises. To compute the first two moments of the solution with such a recursive multistage procedure, we first compute the covariance matrix of the solution at different physical points at a time step and then recursively compute the covariance matrix of the solution at the next time step using the covariance matrix at the previous time step. We continue this process before we reach the final integration time.

We compare the recursive multistage WCE with methods of characteristics plus a standard Monte Carlo sampling strategy and show that the multistage WCE is more efficient than the standard Monte Carlo methods if high accuracy in the first two moments is desired.

We also compare WCE and SCM in conjunction with the recursive multistage procedure. Although WCE theoretically exhibits higher-order convergence than SCM, we show that both methods are comparable in performance, depending on the underlying problem. The computational cost is proportional to the fourth power of the number of nodes or modes employed in physical space but the cost can be reduced to the second power if we make full use of the sparsity of the solution.

For SCM, we also discuss a benchmark problem for stochastic nonlinear conservation laws–a stochastic piston problem in one-dimensional physical space. The problem of a moving piston (with the piston velocity being a Brownian motion) into a tube is modeled with stochastic Euler equations driven by white noise. After splitting the stochastic Euler equations into two parts (by Lie-Trotter splitting), we truncate the Brownian motion with its spectral expansion and applied SCM to obtain variances of the shock locations at different time instants. The conclusion is that SCM is efficient for a short time simulation and quasi-Monte Carlo methods are more efficient for a relatively longer time simulation.

We also illustrate the efficiency of SCM with the Euler scheme in time through a linear stochastic ordinary differential equations: error estimates show that SCM using a sparse grid of Smolyak type is efficient for short time integration and for small magnitudes of noises.

Our conclusion on integration methods in random space is as follows. WCE and SCM are efficient for longer time integration of linear problems using our recursive approach and for a small number of noises within short time simulation. However, if time increases, we have already employed many random variables, either from increments of Brownian motion or from the modes of spectral truncation of Brownian motion. Hence, deterministic integration methods are not efficient anymore since their computational cost grows exponentially with the number of random variables. We then have to use randomized sampling strategies, such as Monte Carlo methods or randomized quasi-Monte Carlo methods, possibly together with variance reduction methods to reduce the statistical errors.

For both WCE and SCM, we apply the Wong-Zakai approximation using a spectral approximation of Brownian motion. However, we use different stochastic products for WCE and SCM because of computational efficiency. In practice, WCE is associated with the Ito-Wick product, which yields a weakly coupled system of PDEs for linear equations. SCM is associated with the Stratonovich product, which yields a decoupled system of PDEs. These different formulations lead to different numerical performance but both methods are comparable in performance for linear problems.

In Part III, we consider elliptic equations with additive noise and multiplicative noise.

Using a spectral approximation of Brownian motion we discuss a semilinear equation with additive spatial white noise. We find that for problems in two or three dimensions in physical space, we cannot expect better convergence from the spectral approximation of Brownian motion than that from piecewise linear approximation. However, we may expect high-order convergence when we have solutions of high regularity. For example, for elliptic equations with additive noise in one-dimensional physical space or even higher-order equations in two- or three-dimensional physical space, we can expect high regularity and benefit from the spectral truncation of Brownian motion.

For elliptic equations with multiplicative noise, we consider WCE for lognormal coefficient as well as spatial white noise as coefficient. In the former case, we use the Wick-Malliavin approximation to reduce the computational cost. It is shown that the Wick-Malliavin approximation can be a higher-order perturbation even when the noise intensity is relatively large. In the latter case (spatial white noise as coefficient), the solution lies in some weighted stochastic Sobolev spaces. Though the WCE can lead to a weakly coupled linear system of deterministic equations which is of great convenience in computation, it is not clear what the physical meaning of these numerical solutions is as no bounded second-order moments of these solutions exist.

For those who are interested in general numerical methods for SPDEs, we have included a brief review in Chapter 3.

12.2 Some open problems

What is the most important question to ask in numerical SPDEs? In general, the answer depends on what you want from SPDEs. Are you seeking the average behavior of solutions to SPDEs, e.g., mean and covariance of solutions or some behavior along trajectories or the probability distribution of solutions? This question immediately leads to different treatment in numerical methods and senses of convergence of numerical methods. Currently, the main focus is on mean-square convergence and weak convergence (in moments or functionals of solutions). In applications, where the interest is the probability distribution of solutions, the current methodology can be very inefficient using Monte Carlo methods or its variants, or deterministic integration methods.

For *stochastic partial differential equations with space-time noise*, deterministic integration methods in random space are too expensive as many random variables should be employed to truncate space-time noise. Monte Carlo methods and associated variance reduction methods including multilevel Monte Carlo method could potentially be applied to resolve this issue. Further, some *model reduction* methods could be applied to reduce the heavy computational load, e.g., some homogenization for multiscale stochastic partial differential equations [119].

For *long-time integration of nonlinear stochastic differential equations using deterministic integration methods in random space*, dimensionality in random space is still the essential difficulty: the number of random variables grows linearly and the number of Wiener chaos modes or stochastic collocation points grows exponentially. For linear equations solved with the recursive multistage WCE or SCM, we will have fast increasing computational cost for some statistics of the solutions other than the first two moments, e.g., the computational cost for third-order moments will be proportional to the sixth power of nodes or modes employed in physical space. For nonlinear equations, the recursive multistage approach fails as nonlinear equations usually have strong dependence on the initial condition and the superposition principle does not work anymore.

To lift the curse of dimensionality, we have to suppress the history data and restart from time to time to keep low dimensionality in random space (and thus low computational cost). To suppress the history data, we should employ some reduction methods, such as functional analysis of variance, see, e.g., [133, 171, 504], to reduce the number of used random variables before integrating over the next time interval.

Appendices

A

Basics of probability

A.1 Probability space

Definition A.1.1 (probability measure) *A probability measure $\mathbb{P}$ on a measurable space $(\Omega, \mathcal{F})$ is a function from $\mathcal{F}$ to $[0,1]$ such that*

- $\mathbb{P}(\emptyset) = 0$, *and* $\mathbb{P}(\Omega) = 1$;
- *If* $\{A_n\}_{n\geq 1} \in \mathcal{F}$ *and* $A_i \cap A_j \neq \emptyset$ *if* $i \neq j$, *then* $\mathbb{P}(\cup_{n=1}^{\infty} A_n) = \sum_{n=1}^{\infty} \mathbb{P}(A_n)$.

Definition A.1.2 (probability space) *A triple $(\Omega, \mathcal{F}, \mathbb{P})$ is called a probability space if*

- Ω *is a sample space which is a collection of all samples;*
- $\mathcal{F}$ *is a σ-algebra*[1] *on Ω;*
- $\mathbb{P}$ *is a probability measure on $(\Omega, \mathcal{F})$.*

Definition A.1.3 (complete probability space) *A probability space $(\Omega, \mathcal{F}, \mathbb{P})$ is said to be a* ***complete probability space*** *if for all $B \in \mathcal{F}$ with $\mathbb{P}(B) = 0$ and all $A \subseteq B$ one has $A \in \mathcal{F}$.*

A.1.1 Random variable

Denote $\sigma(\mathcal{D}) = \cap\{\mathcal{H} | \mathcal{H} \text{ is a } \sigma - \text{algebra of } \Omega,\ \mathcal{D} \subseteq \mathcal{H}\}$. We call $\sigma(\mathcal{D})$ a σ-algebra generated by $\mathcal{D}$.

Definition A.1.4 ($\mathcal{F}$-measurable) *If $(\Omega, \mathcal{F}, \mathbb{P})$ is a given probability space, then a function $Y : \Omega \to \mathbb{R}^n$ is called $\mathcal{F}$-measurable if $Y^{-1}(U) = \{\omega \in \Omega | Y(\omega) \in U\} \in \mathcal{F}$ holds for all open sets $U \in \mathbb{R}^n$ (or, equivalently, for all Borel sets $U \in \mathbb{R}^n$.)*

[1] A σ-algebra on a set X is a collection of subsets of X that includes the empty subset and is closed under complement and under countable unions.

Z. Zhang, G.E. Karniadakis, *Numerical Methods for Stochastic Partial Differential Equations with White Noise*,
Applied Mathematical Sciences 196, DOI 10.1007/978-3-319-57511-7

If $X: \Omega \to \mathbb{R}^n$ is a function, then $\sigma(X)$ is the smallest σ-algebra on Ω containing all the sets $X^{-1}(U)$ for all open sets U in $\mathbb{R}^n$.

Definition A.1.5 (Random variable) *Suppose that $(\Omega, \mathcal{F}, \mathbb{P})$ is a given complete probability space. A* random variable *X is an $\mathcal{F}$-measurable function $X: \Omega \to \mathbb{R}^n$.*

Theorem A.1.6 (Doob-Dynkin theorem) *If $XY: \Omega \to \mathbb{R}^n$ are two given functions the Y is $\sigma(X)$-measurable* if and only if *there exists a Borel measurable function $g: \mathbb{R}^n \to \mathbb{R}^n$ and $Y = g(X)$.*

Every random variable induces a probability measure μ_X (*distribution of* X) on $\mathbb{R}^n$:

$$\mu_X(B) = \mathbb{P}(X^{-1}(B)).$$

If $\int_\Omega |X(\omega)|\, d\mathbb{P}(\omega) < \infty$, the expectation of X w.r.t $\mathbb{P}$ is defined by

$$\mathbb{E}[X] = \int_\Omega X(\omega)\, d\mathbb{P}(\omega) = \int_{\mathbb{R}^n} x\, d\mu_X(x).$$

For continuous random variables,

$$\mathbb{E}[X] = \int_0^\infty \mathbb{P}(X > \lambda)\, d\lambda, \quad \text{for } X \geq 0.$$

For discrete random variables,

$$\mathbb{E}[X] = \sum_{n=0}^\infty \mathbb{P}(X \geq n), \text{ for } X \geq 0.$$

The p-th moment of X is defined as (if the integrals are well defined)

$$\mathbb{E}[X^p] = \int_\Omega X^p\, d\mathbb{P}(\omega) = \int_{\mathbb{R}^n} x^p\, d\mu_X(x).$$

The centered moments are defined by $\mathbb{E}[|X - \mathbb{E}[X]|^p]$, $p = 1, 2, \ldots$. When $p = 2$, the centered moment is also called the variance.

A.2 Conditional expectation

Given a probability space $(\Omega, \mathcal{F}, \mathbb{P})$, a sub-$\sigma$-algebra $\mathcal{G} \subseteq \mathcal{F}$ and an integrable random variable X (or $\mathbb{E}[|X|] < \infty$).

Definition A.2.1 (Conditional expectation) *The conditional expectation of X given $\mathcal{G}$, denoted by $\mathbb{E}[X|\mathcal{G}]$, is a random variable Y such that*

- *Y is $\mathcal{G}$-measurable;*

- $\int_A Y\, dP = \int_A X\, dP$ *for all* $A \in \mathcal{G}$.

Example A.2.2 *Let* $\Omega = [0,1)$, $\mathcal{F} = \mathcal{B}([0,1))$, *and let* $\mathbb{P}$ *be Lebesgue measure. Then a random variable is simply a Borel measurable function* $X : [0,1) \to \mathbb{R}$. *Let* $\mathcal{G} = \{\emptyset, [0, \frac{1}{2}), [\frac{1}{2}, 1), [0,1)\}$. *Then*

$$\mathbb{E}[X|\mathcal{G}](x) = \begin{cases} 2\int_0^{\frac{1}{2}} X(y)\, dy, & x \in [0, \frac{1}{2}) \\ 2\int_{\frac{1}{2}}^1 X(y)\, dy, & x \in [\frac{1}{2}, 1), \end{cases}$$

In other words, the conditional expectation simply performs a "partial average" over the partition of the underlying probability space.

Example A.2.3 *Let* $\{\Lambda_i\}_{i=1}^N$ *be a disjoint partition of* Ω:

$$\cup_{i=1}^N \Lambda_i = \Omega, \quad \Lambda_i \cap \Lambda_j = \emptyset \ (i \neq j).$$

Suppose that $\mathbb{P}(\Lambda_i) > 0$ *for* $i = 1, 2, \ldots, N$ *and* $\mathcal{G} = \sigma(\Lambda_1, \cdots, \Lambda_n)$. *Then a version of the conditional expectation for an integrable random variable* X *is*

$$\mathbb{E}[X|\mathcal{G}](\omega) = \sum_{i=1}^N 1_{\Lambda_i}(\omega) \frac{\mathbb{E}[1_{\Lambda_i} X]}{\mathbb{P}(\Lambda_i)}.$$

When $\omega \in \Lambda_j$ $(j = 1, 2, \cdots, N)$, *then*

$$\mathbb{E}[X|\mathcal{G}](\omega) = \frac{\mathbb{E}[1_{\Lambda_j} X]}{\mathbb{P}(\Lambda_j)}.$$

A.2.1 Properties of conditional expectation

- (Linearity) $\mathbb{E}[aX + bY|\mathcal{G}] = a\mathbb{E}[X|\mathcal{G}] + b\mathbb{E}[Y|\mathcal{G}]$.
- $\mathbb{E}[\mathbb{E}[X|\mathcal{G}]] = \mathbb{E}[X]$, for any $\mathcal{G}$.
- If X is $\mathcal{G}$-measurable, then $\mathbb{E}[X|\mathcal{G}] = X$.
- If X is independent of $\mathcal{G}$, then $\mathbb{E}[X|\mathcal{G}] = \mathbb{E}[X]$.

  If $\mathcal{H}$ is independent of $\sigma(\sigma(X), \mathcal{G})$, then $\mathbb{E}[X|\sigma(\mathcal{H}, \mathcal{G})] = \mathbb{E}[X|\mathcal{G}]$.
- ("Taking out what is known") If $\mathbb{E}[XY]$ is well defined and Y is $\mathcal{G}$-measurable, then $\mathbb{E}[XY|\mathcal{G}] = Y\mathbb{E}[X|\mathcal{G}]$.
- (Tower property) If $\mathbb{E}[X]$ is well defined (or simply $\mathbb{E}[|X|] < \infty$), $\mathcal{H} \subseteq \mathcal{G}$, $\mathbb{E}[\mathbb{E}[X|\mathcal{G}]|\mathcal{H}] = \mathbb{E}[X|\mathcal{H}] = \mathbb{E}[\mathbb{E}[X|\mathcal{H}]|\mathcal{G}]$.
- (conditional Jensen's inequality) If ϕ is convex, $\mathbb{E}[|\phi(X)|] < \infty$, then

$$\phi(\mathbb{E}[X|\mathcal{G}]) \leq \mathbb{E}[\phi(X)|\mathcal{G}].$$

Theorem A.2.4 *Consider a sequence of random variables $\{X_n\}$ on a probability space $(\Omega, \mathcal{F}, \mathbb{P})$. Let $\mathcal{G} \subseteq \mathcal{F}$ be a σ-algebra. Then the following results hold.*

- *(conditional Monotone Convergence Theorem). If $X_n \geq 0$ and is increasing with a limit X, then $\mathbb{E}[X_n|\mathcal{G}]$ is increasing and*

$$\lim_{n\to\infty} \mathbb{E}[X_n|\mathcal{G}] = \mathbb{E}[X|\mathcal{G}] \quad a.s..$$

- *(conditional Fatou Lemma). If $X_n \geq 0$, then*

$$\liminf_{n\to\infty} \mathbb{E}[X_n|\mathcal{G}] \geq \mathbb{E}[\liminf_{n\to\infty} X_n|\mathcal{G}] \ a.s..$$

- *(conditional Dominated Convergence Theorem). If $|X_n| \leq Y$ and $\mathbb{E}[Y] < \infty$ and $X_n \to X$ a.s., then*

$$\mathbb{E}[|X_n - X|\,|\mathcal{G}] \to 0 \ a.s., \ and \ \lim_{n\to\infty} \mathbb{E}[X_n|\mathcal{G}] = \mathbb{E}[X|\mathcal{G}] \ a.s..$$

Theorem A.2.5 (Best estimator/predictor) *Let Z be a Y-measurable random variable and X is square-integrable $\mathbb{E}[X^2] < \infty$. Then*

$$\mathbb{E}[|X - \mathbb{E}[X|Y]|^2] \leq \mathbb{E}[|X - Z|^2].$$

A.2.2 Filtration and Martingales

On a probability space $(\Omega, \mathcal{F}, \mathbb{P})$, a **filtration** refers to an increasing sequence of σ-algebra:

$$\mathcal{F}_0 \subseteq \mathcal{F}_1 \subseteq \mathcal{F}_2 \subseteq \cdots \subseteq \mathcal{F}_n \subseteq \cdots .$$

A **natural filtration** (w.r.t. X) is the smallest σ-algebra that contains information of X. It is generated by X and $\mathcal{F}_n^X = \sigma(X_1, \ldots, X_n)$ with $\mathcal{F}_0^X = \{\emptyset, \Omega\}$. If $\lim_{n\to\infty} \mathcal{F}_n \subseteq \mathcal{F}$. Then we call $(\Omega, \mathcal{F}, \{\mathcal{F}_n\}_{n\geq 1}\, \mathbb{P})$ a **filtered probability space**. A stochastic process $\{X_n\}$ on a filtered probability space is an **adapted process** if X_n is $\mathcal{F}_n$-measurable for each n.

Definition A.2.6 (martingale) *The process and filtration $\{(X_n, \mathcal{F}_n)\}$ is called a martingale if for each n*

- *X_n is $\mathcal{F}_n$-measurable.*
- $\mathbb{E}[|X_n|] < \infty$.
- $\mathbb{E}[X_{n+1}|\mathcal{F}_n] = X_n$.

A submartingale is defined by replacing the third condition with ($\mathbb{E}[X_{n+1}|\mathcal{F}_n] \geq X_n$). A supermartingale is defined if the third condition is replaced with $\mathbb{E}[X_{n+1}|\mathcal{F}_n] \leq X_n$.

A.3 Continuous time stochastic process

Definition A.3.1 *Let $(\Omega, \mathcal{F}, \mathbb{P})$ be a probability space and let $T \subseteq \mathbb{R}$ be time. A collection of random variables X_t, $t \in T$ with values in $\mathbb{R}$ is called a* ***stochastic process.***

If X_t takes values in $S = \mathbb{R}^d$, it is called a vector-valued stochastic process (often abbreviated as stochastic process).

If the time T can be a discrete subset of $\mathbb{R}$, then X_t is called a discrete time stochastic process.

If time is an interval, $\mathbb{R}^+$ or $\mathbb{R}$, it is called a stochastic process with continuous time. For any fixed $\omega \in \Omega$, one can regard $X_t(\omega)$ as a function of t (called a sample function of the stochastic process).

Definition A.3.2 *A stochastic process is measurable if $X : \Omega \otimes T \mapsto S$ is measurable with respect to the product σ-algebra $\mathcal{F} \otimes \mathcal{B}(T)$.*

Definition A.3.3 (filtration) *A family of sub-σ-algebras $\mathcal{F}_t \subseteq \mathcal{F}$ indexed by $t \in [0, \infty)$ is called a filtration if it is increasing $\mathcal{F}_s \subseteq \mathcal{F}_t$ when $0 \leq s \leq t < \infty$.*

A stochastic process X is adapted to the filtration $\{\mathcal{F}_t\}_{t\in[0,\infty)}$ if X_t is $\mathcal{F}_t$-measurable for every $t \in [0, \infty)$.

We assume that the filtration $\{\mathcal{F}_t\}_{t\in[0,\infty)}$ satisfies the so-called *usual conditions*, i.e.,

- $\mathcal{F}_0$ contains all the $\mathbb{P}$-negligible sets (hence so does every $\mathcal{F}_t$),
- The filtration $\{\mathcal{F}_t\}_{t\in[0,\infty)}$ is right-continuous, i.e., $\mathcal{F}_t = \mathcal{F}_{t+} =: \cap_{s>t}\mathcal{F}_s = \cap_{n=1}^{\infty}\mathcal{F}_{t+\frac{1}{n}}$.

Let $X(t, \omega) \in T \times \Omega \to \mathbb{R}$ be a stochastic process. For a fixed $\omega_0 \in \Omega$, we call $X(t, \omega_0)$ a realization (a sample) of the stochastic process $X(t)$.

We say a stochastic process is mean-square continuous if

$$\lim_{\varepsilon\to 0} \mathbb{E}[|X(t+\varepsilon) - X(t)|^2] = 0.$$

B

Semi-analytical methods for SPDEs

Here we recall some semi-analytical methods of obtaining solutions of SPDEs, especially stochastic transformation methods and integrating factor methods that can transform SPDEs into deterministic PDEs. For integrating factor methods, we refer to Chapter 3.

Consider the following stochastic Burgers equation on $(0,T] \times (0,1)$:

$$\partial_t u + u\partial_x u = \mu \partial_x^2 u + \sigma(t,x)\dot{W}(t), \quad u(0,x) = u_0(x),\, u(t,0) = u(t,1). \tag{B.0.1}$$

Here $W(t)$ is a standard Brownian motion and $\mu > 0$. If $\sigma(t,x)$ depends on t, it can be readily checked that the solution is

$$u(t,x) = v(t, Y(t,x)) + \int_0^t \sigma(s)W(s)\, ds, \tag{B.0.2}$$

where $Y(t,x) = x - \int_0^t \sigma(s)W(s)\, ds$ and $v(t,x)$ satisfies the deterministic Burgers equation

$$\partial_t v + vv_x = \mu \partial_x^2, \quad v(0,x) = u_0(x),\, v(t,0) = v(t,1). \tag{B.0.3}$$

The solution to the following Stratonovich Burgers equation on $(0,T] \times (0,1)$

$$\partial_t u + (u + \sigma \dot{W}(t)) \circ \partial_x u = \mu \partial_x^2 u, \quad u(0,x) = u_0(x),\, u(t,0) = u(t,1) \tag{B.0.4}$$

is given by

$$u(t,x) = v(t, x - \sigma W(t)), \tag{B.0.5}$$

where $v(t,x)$ satisfies Equation (B.0.3).

Since we usually don't have analytical solutions to Equation (B.0.3), we first find a numerical solution for v in Equation (B.0.3) and then obtain the solution to Equation (B.0.1) using (B.0.2) and the solution to (B.0.4)

Z. Zhang, G.E. Karniadakis, *Numerical Methods for Stochastic Partial Differential Equations with White Noise*,
Applied Mathematical Sciences 196, DOI 10.1007/978-3-319-57511-7

using (B.0.5). For the above stochastic Burgers equation, the methodology exploits (stochastic) analytical transforms and numerical methods and is thus called semi-analytical. This approach is very efficient when periodic boundary conditions are imposed since no SPDEs are solved.

Here are more equations that can be transformed to deterministic ones. The stochastic Korteweg-de Vries (KdV) equation

$$\partial_t u + u\partial_x u + \mu\partial_x^3 u + \gamma u = f(t,\omega) \tag{B.0.6}$$

can be transformed to the following KdV equation on $(0,T]$

$$\partial_t v + v\partial_x v + \mu\partial_x^3 v + \gamma v = 0, \tag{B.0.7}$$

if we let $u(t,x) = v(t, x - \int_0^t g(s,\omega)\,ds) + g(s,\omega)$, where $g(s,\omega) = e^{-\gamma t}\int_0^t e^{\gamma s}$ $f(s,\omega)\,ds$. Here $f(t,\omega)$ can be very rough, e.g., white noise. The Navier-Stokes equation with additive random forcing also can be transformed into a deterministic one. Through the substitutions $u(t,x) = U(t, x - \int_0^t g(s,\omega)\,ds)$ and $p = P(t, x - \int_0^t g(s,\omega)\,ds)$, $g(s,\omega) = \int_0^t f(s,\omega)\,ds$, we obtain from

$$\partial_t u + u\nabla u = \mu\Delta u - \nabla p + f(t,\omega). \tag{B.0.8}$$

that

$$\partial_t U + U\nabla U = \mu\Delta U - \nabla P. \tag{B.0.9}$$

Though the boundary conditions for (B.0.8) may be different from those for (B.0.9), the initial conditions are the same. Moreover, if we are given periodic boundary conditions for (B.0.8), we can still apply periodic boundary conditions to (B.0.9).

For multiplicative noise, we can apply similar techniques if the noise is only time-dependent. Consider for example the advection-diffusion equation

$$\partial_t u + f(t,\omega)\partial_x u = \mu\partial_x^2 u.\ \mu \geq 0.$$

Let $u(t,x,\omega) = v(t, x - \int_0^t f(s,\omega)\,ds)$. Then we have

$$\partial_t v = \mu\partial_x^2 v.$$

Here again the noise $f(t,\omega)$ can be rough, e.g., white noise. However, if $f(t,\omega)$ is white noise, we have to interpret the product $f(t,\omega)\partial_x u$ using Stratonovich product: $f(t,\omega)\circ\partial_x u$ as in (B.0.4).

C

Gauss quadrature

We recall some basic facts about Gauss quadrature, which is used in the construction of sparse grid collocation methods.

C.1 Gauss quadrature

Definition C.1.1 *Suppose that*

$$I(f) = \int_a^b f(x)dx \approx I_n(f) = \sum_{k=0}^{n} A_k f(x_k),$$

When $I_n(f)$ has a polynomial exactness $(2n+1)$, we will call it Gauss-Legendre quadrature rule, and the corresponding points x_k, $k = 0, 1\cdots, n$ are called Gauss-Legendre points.

$I(f) \approx I_n(f)$ has a polynomial exactness $(2n+1)$ if and only if

$$\int_a^b x^i dx = \sum_{k=0}^{n} A_k x_k^i, \quad i = 0, 1, \cdots, 2n+1.$$

Example C.1.2 *Find A_0, A_1 and x_0, x_1 such that the following rule is a Gauss quadrature.*

$$\int_{-1}^{1} f(x)dx \approx A_0 f(x_0) + A_1 f(x_1).$$

Solution. When $n = 1$, we need a polynomial exactness $2 + 1 = 3$. Thus

Z. Zhang, G.E. Karniadakis, *Numerical Methods for Stochastic Partial Differential Equations with White Noise*, Applied Mathematical Sciences 196, DOI 10.1007/978-3-319-57511-7

$$f(x) = 1, \quad A_0 + A_1 = \int_{-1}^{1} 1dx = 2,$$

$$f(x) = x, \quad A_0x_0 + A_1x_1 = \int_{-1}^{1} xdx = 0,$$

$$f(x) = x^2, \quad A_0x_0^2 + A_1x_1^2 = \int_{-1}^{1} x^2dx = \frac{2}{3},$$

$$f(x) = x^3, \quad A_0x_0^3 + A_1x_1^3 = \int_{-1}^{1} x^3dx = 0.$$

We then obtain $A_0 = A_1 = 1$, $x_0 = -\frac{1}{\sqrt{3}}, x_1 = \frac{1}{\sqrt{3}}$. So the desired Gauss quadrature rule on $[-1, 1]$ is

$$\int_{-1}^{1} f(x)dx \approx f\left(-\frac{1}{\sqrt{3}}\right) + f\left(\frac{1}{\sqrt{3}}\right).$$

$$I(g) = \int_{-1}^{1} g(t)dt \approx \sum_{k=0}^{n} A_k g(t_k),$$

The zeros of $P_{n+1}(t)$ are the Gauss quadrature points and the Gauss quadrature weights are

$$A_k = \int_{-1}^{1} \prod_{\substack{j=0 \\ j \neq k}}^{n} \frac{t - t_j}{t_k - t_j} dt, \quad k = 0, 1, \cdots, n.$$

When $n = 0$, $t_0 = 0, A_0 = 2$,

$$\int_{-1}^{1} g(t)dt \approx 2g(0).$$

When $n = 1$, $t_0 = -\frac{1}{\sqrt{3}}, t_1 = \frac{1}{\sqrt{3}}, A_0 = 1, A_1 = 1$,

$$\int_{-1}^{1} g(t)dt \approx g\left(-\frac{1}{\sqrt{3}}\right) + g\left(\frac{1}{\sqrt{3}}\right).$$

When $n = 2$, $t_0 = -\sqrt{\frac{3}{5}}, t_1 = 0, t_2 = \sqrt{\frac{3}{5}}$, $A_0 = \frac{5}{9}$, $A_1 = \frac{8}{9}$, $A_2 = \frac{5}{9}$.

$$\int_{-1}^{1} g(t)dt \approx \frac{5}{9}g\Big(-\sqrt{\frac{3}{5}}\Big) + \frac{8}{9}g(0) + \frac{5}{9}g\Big(\sqrt{\frac{3}{5}}\Big).$$

Remark C.1.3 (Gauss quadrature rule on $[a,b]$) *Using the linear transformation $x = \frac{a+b}{2} + \frac{b-a}{2}t$, we have*

$$I(f) = \int_a^b f(x)dx = \int_{-1}^1 \frac{b-a}{2} f\left(\frac{a+b}{2} + \frac{b-a}{2}t\right) dt.$$

By Gauss quadrature rule on $[-1,1]$, we have

$$I_n(f) = \sum_{k=0}^{n} \frac{b-a}{2} A_k f\left(\frac{a+b}{2} + \frac{b-a}{2}t_k\right).$$

Quadrature rules for integration with weights. Consider the following integration

$$I(f) = \int_a^b \rho(x)f(x)dx, \quad \rho(x) \in C(a,b),$$

where $f(x)$ has enough number of derivatives over $[a,b]$ (smooth enough). The weight function $\rho(x)$ satisfies the following conditions:

1. $x \in (a,b)$, $\rho(x) \geq 0$;
2. $\int_a^b \rho(x)dx > 0$;
3. $k = 0,1,2,\ldots$, $\int_a^b x^k \rho(x)dx$ is well defined.

Recall the definition of polynomial exactness for the quadrature rule:

$$\int_a^b \rho(x)f(x)dx \approx \sum_{k=0}^{n} A_k f(x_k). \tag{C.1.1}$$

We say the quadrature rule (C.1.1) has polynomial exactness m when (C.1.1) is exact for $f(x) = 1, x, x^2, \cdots, x^m$ but (C.1.1) is not exact for $f(x) = x^{m+1}$. We call a quadrature rule Gauss quadrature rule when the polynomial exactness is $(2n+1)$.

Example C.1.4 *Consider the following integral with weights $I(f) = \int_0^1 \frac{f(x)}{\sqrt{x}}$ dx*

$$I(f) \approx Af\left(\frac{1}{5}\right) + Bf(1).$$

Find A, B to make the polynomial exactness as high as possible.

Solution. Here we have two unknowns. We can ask for at least polynomial exactness 1. When $f(x) = 1$, $I(f) = \int_0^1 \frac{1}{\sqrt{x}}dx = 2 = A + B$. When $f(x) = x$, $I(f) = \int_0^1 \frac{x}{\sqrt{x}}dx = \frac{2}{3} = \frac{1}{5}A + B$.

$$\begin{aligned} A + B &= 2 \\ \frac{1}{5}A + B &= \frac{2}{3} \end{aligned}$$

This gives $A = \frac{5}{3}, B = \frac{1}{3}$. The quadrature rule is

$$I(f) \approx \frac{5}{3} f\left(\frac{1}{5}\right) + \frac{1}{3} f(1).$$

When $f(x) = x^2$, $I(f) = \int_0^1 \frac{x^2}{\sqrt{x}} dx = \frac{2}{5}$. The quadrature rule gives the same value: $\frac{5}{3}\left(\frac{1}{5}\right)^2 + \frac{1}{3}(1)^2 = \frac{2}{5}$.
When $f(x) = x^3$, $I(f) = \int_0^1 \frac{x^3}{\sqrt{x}} dx = \frac{2}{7}$, but the quadrature rule gives a different value $\frac{5}{3}\left(\frac{1}{5}\right)^3 + \frac{1}{3}(1)^3 = \frac{26}{75}$. The polynomial exactness is 2.

One fundamental theorem for Gauss quadrature is that the Gauss quadrature points are exactly the $n+1$ zeros of the $n+1$-th order orthogonal polynomial (with respect to the weight $\rho(x)$). For example, when $\rho = 1$, the orthogonal polynomials are Legendre polynomials. The Gauss quadrature points are exactly the zeros of $n+1$-th order Legendre polynomial. When $\rho(x) = (1-x)^\alpha (1+x)^\beta$ $(\alpha, \beta > -1)$, the corresponding orthogonal polynomial is called the Jacobi polynomial. The quadrature is then called Gauss-Jacobi quadrature. When $\rho(x) = \exp(-x^2/2)$, the corresponding orthogonal polynomial is called the Hermite polynomial and the quadrature is called Gauss-Hermite quadrature.

Gauss-Lobatto quadrature

Definition C.1.5 *Suppose that a positive function $\rho(x)$ satisfies the conditions in* (C.1).

$$I(f) = \int_a^b f(x)\rho(x)\, dx \approx I_n(f) = \sum_{k=0}^{n} A_k f(x_k), \qquad x_0 = a,\, x_n = b.$$

When $I_n(f)$ has a polynomial exactness $(2n-1)$, we will call it Gauss-Lobatto quadrature rule, and the corresponding points x_k, $k = 0, 1 \cdots, n$ are called Gauss-Lobatto quadrature points.

When $\rho(x) = 1$, the Gauss-Lobatto quadrature is called Gauss-Legendre-Lobatto quadrature. The Gauss-Lobatto quadrature points are the zeros of $(1-x^2)\partial_x P_n(x)$, where $P_n(x)$ is the n-th order Legendre polynomial.

C.2 Gauss-Hermite quadrature

Let $\psi(y)$, $y \in \mathbb{R}$, be a smooth function and $Q_n\psi$ be a Gauss-Hermite quadrature applied to ψ, i.e.

$$\begin{aligned} \mathbb{E}[\psi(\xi)] &= \frac{1}{(2\pi)^{1/2}} \int_{\mathbb{R}} \psi(y) \exp(-y^2/2)\, dy \qquad \text{(C.2.1)} \\ &\approx Q_n\psi := \sum_{i=1}^{n} \psi(\mathsf{y}_i)\mathsf{w}_i, \end{aligned}$$

where ξ is a standard Gaussian random variable, $\mathsf{y}_i = \mathsf{y}_i^{(n)}$, $i = 1, 2, \ldots, n$, are roots of the nth Hermite polynomial

$$H_n(y) = (-1)^n \exp(y^2/2) \frac{d^n}{dy^n} \exp(-y^2/2)$$

and the associated weights $\mathsf{w}_i = \mathsf{w}_i^{(n)}$ are given by

$$\mathsf{w}_i = \frac{n!}{n^2 \left[H_{n-1}(\mathsf{y}_i)\right]^2}. \tag{C.2.2}$$

For instance,

$$\begin{aligned}
&n = 1: \ \mathsf{w}_1 = 1, \quad \mathsf{y}_1 = 0; \qquad\qquad (C.2.3)\\
&n = 2: \ \mathsf{w}_1 = \mathsf{w}_2 = 1/2, \ \ \mathsf{y}_1 = -1, \ \mathsf{y}_2 = 1;\\
&n = 3: \ \mathsf{w}_1 = \mathsf{w}_3 = 1/6, \ \mathsf{w}_2 = 2/3, \ \ \mathsf{y}_1 = -\sqrt{3}, \ \mathsf{y}_3 = -\sqrt{3}, \ \mathsf{y}_2 = 0.
\end{aligned}$$

The quadrature $Q_n\psi$ is exactly equal to $\mathbb{E}[\psi(\xi)]$ for polynomials $\psi(y)$ of order $\leq 2n - 1$.

D

Some useful inequalities and lemmas

We give a summary of basic inequalities and lemmas that we use in the book.

Cauchy-Schwarz inequality (aka. Cauchy inequality). If $f, g \in L^2(D)$, then $fg \in L^1(D)$ and

$$\int_D fg \leq (\int_D f^2)^{1/2} \int_D g^2)^{1/2}.$$

Hölder inequality. If $\frac{1}{p} + \frac{1}{q} = 1$, $p, q \geq 1$ and $f \in L^p(D)$ and $g \in L^q(D)$, then $fg \in L^1(D)$ and

$$\int_D fg \leq (\int_D f^p)^{1/p} (\int_D g^q)^{1/q}.$$

Young inequality. If $\frac{1}{p} + \frac{1}{q} = 1$, $p, q > 0$, then for all a, $b \geq 0$, we have

$$ab \leq \frac{a^p}{p} + \frac{b^q}{q}.$$

Gronwall inequality (a.k.a. Gronwall's lemma, Gronwall-Bellman inequality) Assume that $u(t)$, $k(t)$, $\varphi(t) \geq 0$ are continuous on $t \in [t_0, T]$ and

$$u(t) \leq \varphi(t) + \int_{t_0}^t k(s) u(s)\, ds, \text{ for all } t \in [t_0, T].$$

Then $u(t)$, $t \in [t_0, T]$ is bounded by

$$u(t) \leq \varphi(t) + \int_{t_0}^t k(s) \varphi(s) \exp\left(\int_s^t k(\theta)\, \mathrm{d}\theta \right) ds.$$

Z. Zhang, G.E. Karniadakis, *Numerical Methods for Stochastic Partial Differential Equations with White Noise*, Applied Mathematical Sciences 196, DOI 10.1007/978-3-319-57511-7

If $\varphi(t)$ is a constant, then

$$u(t) \leq \varphi \exp\left(\int_{t_0}^{t} k(s)\,\mathrm{d}s\right).$$

Nonlinear Gronwall inequality. Assume that $u(t)\,k(t) \geq 0$ are continuous on $t \in [t_0, T]$ and

$$u^2(t) \leq M + \int_{t_0}^{t} k(s)u(s)\,ds,\ M \geq 0, \text{ for all } t \in [t_0, T].$$

Then $u(t)$, $t \in [t_0, T]$ is bounded by

$$u(t) \leq \sqrt{M} + 2\int_{t_0}^{t} k(s)\,ds.$$

Discrete Gronwall inequality. Assume that u_n, K_n, and k_n are nonnegative sequences and

$$u_n \leq K_n + \sum_{j=0}^{n-1} k_j u_j, \quad n \geq 1.$$

Then it holds that for $n \geq 0$,

$$u_n \leq K_n + \sum_{j=0}^{n-1} k_j K_j \prod_{j<i<n} (1 + k_i).$$

Poincare inequality. Let $p \geq 1$ and Ω is a bounded subset in $\mathbb{R}^d$. There then exists a constant C depending only on Ω and p such that for any $u \in W_0^{1,p}$

$$\|u\|_{L^p(\Omega)} \leq C\,\|\nabla u\|_{L^p(\Omega)}\,.$$

Here the constant C is called the Poincare constant.

Littlewood-Paley inequality [301]. Suppose that $f(x) = \sum_{i=1}^{\infty} s_k m_k(x)$ exists in $L^2([a,b])$, a and b are finite. If $f \in L^p([a,b])$, $1 < p < \infty$, then there exist constants $L > 0$ and $M > 0$ such that

$$L\,\|f\|_{L^p} \leq \left\|\left(\sum_{i=1}^{\infty} s_k^2 m_k^2(x)\right)^{1/2}\right\|_{L^p} \leq M\,\|f\|_{L^p}\,.$$

The function within the norm in the middle term is called the Littlewood-Paley function.

Markov inequality. Assume that ϕ is a monotonically increasing function from the nonnegative reals to the nonnegative reals. If X is a random variable and $\mathbb{E}[\phi(X)] < \infty$, $c > 0$, and $\phi(c) > 0$, then

$$\mathbb{P}(|X| \geq c) \leq \frac{\mathbb{E}[\phi(|X|)]}{\phi(c)}.$$

Chebyshev inequality. Let X be a random variable with $|\mathbb{E}[X]| \leq \infty$ and $\mathbb{V}\text{ar}[X] = \sigma^2 < \infty$. Then for any real number $c > 0$,

$$\mathbb{P}(|X - \mu| \geq c\sigma) \leq \frac{1}{c^2}.$$

Jensen inequality. If X is a random variable, ϕ is a convex function and $\mathbb{E}[\phi(X)] < \infty$, then

$$\phi(\mathbb{E}[X]) \leq \mathbb{E}[\phi(X)].$$

An example of convex functions in this book is $\phi(x) = x^p$, $p \geq 1$.

Central limit theorem. X_i are i.i.d. (independent and identically distributed) and $\mathbb{E}[X_i^2] < \infty$ and also $\mu = \mathbb{E}[X_1]$ and $\sigma^2 = \text{Var}(X_1)$. $S_n = \sum_{i=1}^n X_i$. Then

$$P\left\{a \leq \frac{S_n - n\mu}{\sigma\sqrt{n}} \leq b\right\} \to \int_a^b \frac{1}{\sqrt{2\pi}} e^{-\frac{x^2}{2}}\, dx, \text{ as } n \to \infty.$$

Borel-Cantelli Lemma. Let $\{A_n\}$ be a sequence of events in a probability space. If $\sum_{n=1}^{\infty} \mathbb{P}(A_n) < \infty$, then

$$\mathbb{P}\left(\limsup_{n\to\infty} A_n\right) = 0. \quad \text{Recall that } \limsup_{n\to\infty} A_n = \bigcap_{n=1}^{\infty} \bigcup_{k\geq n}^{\infty} A_k.$$

Burkholder-Davis-Gundy inequality. For any $1 \leq p < \infty$, there exist constants $c_p, C_p > 0$ such that for all (local) martingales X with $X_0 = 0$ and stopping times τ, the following inequality holds:

$$c_p\mathbb{E}[[X]_\tau^{p/2}] \leq \mathbb{E}[(X_\tau^*)^p] \leq C_p\mathbb{E}\left[[X]_\tau^{p/2}\right].$$

Here $X_t^* = \sup_{s\leq t} |X_s|$ is the maximum process X_t and $[X]$ is the quadratic variation of X. Furthermore, for continuous (local) martingales, this statement holds for all $0 < p < \infty$.

Fubini theorem. This is called Fubini-Tonelli theorem but often called Fubini theorem.

Consider two σ-finite measure spaces $(X, \mathcal{E}, \mu)$ and $(Y, \mathcal{F}, \nu)$, and the product measure space $(X \times Y, \mathcal{E} \otimes \mathcal{F}, \pi)$. If f is a measurable function on the product measure space such that any one of the three integrals

$$\int_{X\times Y} f(x,y)\pi(dxdy), \quad \int_X \left[\int_Y f(x,y)\nu(dy)\right] \mu(dx), \quad \int_Y \left[\int_X f(x,y)\mu(dx)\right] \nu(dy)$$

is finite. Then

$$\int_{X\times Y} f(x,y)\pi(dxdy) = \int_X \left[\int_Y f(x,y)\nu(dy)\right] \mu(dx) = \int_Y \left[\int_X f(x,y)\mu(dx)\right] \nu(dy).$$

Recall that a measure defined on a σ-algebra of subsets of a set X is called σ-finite if X is the countable union of measurable sets with finite measure

E

Computation of convergence rate

Suppose that g_n is a good approximation of f, say, $\|f - g_n\| \sim Cn^{-r}$ ($\|f - g_n\|$ is proportional to Cn^{-r}), where $r > 0$ and C does not depend on n. To determine the convergence rate of approximation methods, we can use the following formula

$$\frac{\log(\|f - g_{n_2}\| \,/\, \|f - g_{n_1}\|)}{\log(n_2/n_1)}.$$

Denote by $E_n = \|f - g_n\|$. Suppose that $E_n \sim Cn^{-r}$. We then have

$$\frac{E_{n_2}}{E_{n_1}} \sim \left(\frac{n_2}{n_1}\right)^{-r}.$$

Taking the logarithm over both sides leads to the formula above.

When f is not known, we can replace f with some f_N obtained with a numerical method where N is sufficiently large so that $f - f_N$ is much smaller than $f_N - g_n$

$$\|f - g_n\| = \|(f - f_N) + f_N - g_n\| \approx \|f_N - g_n\|.$$

We call this f_N as a *reference solution* and measure the convergence rate by

$$\frac{\log(\|f_N - g_{n_2}\| \,/\, \|f_N - g_{n_1}\|)}{\log(n_2/n_1)}, \quad n_1,\, n_2 \ll N.$$

Z. Zhang, G.E. Karniadakis, *Numerical Methods for Stochastic Partial Differential Equations with White Noise*,
Applied Mathematical Sciences 196, DOI 10.1007/978-3-319-57511-7

References

1. A. Abdulle, A. Barth, C. Schwab, Multilevel Monte Carlo methods for stochastic elliptic multiscale PDEs. Multiscale Model. Simul. **11**, 1033–1070 (2013)
2. M. Abramowitz, I.A. Stegun (eds.), *Handbook of Mathematical Functions, with Formulas, Graphs, and Mathematical Tables* (Dover, Mineola, 1972). 10th printing, with corrections
3. P. Acquistapace, B. Terreni, An approach to Ito linear equations in Hilbert spaces by approximation of white noise with coloured noise. Stoch. Anal. Appl. **2**, 131–186 (1984)
4. I.A. Adamu, G.J. Lord, Numerical approximation of multiplicative SPDEs. Int. J. Comput. Math. **89**, 2603–2621 (2012)
5. A. Alabert, I. Gyöngy, On numerical approximation of stochastic Burgers' equation, in *From Stochastic Calculus to Mathematical Finance* (Springer, Berlin, 2006), pp. 1–15
6. E.J. Allen, S.J. Novosel, Z. Zhang, Finite element and difference approximation of some linear stochastic partial differential equations. Stoch. Stoch. Rep. **64**, 117–142 (1998)
7. V.V. Anh, W. Grecksch, A. Wadewitz, A splitting method for a stochastic Goursat problem. Stoch. Anal. Appl. **17**, 315–326 (1999)
8. L. Arnold, *Stochastic Differential Equations: Theory and Applications* (Wiley-Interscience, New York, 1974)
9. A. Ashyralyev, M. Akat, An approximation of stochastic hyperbolic equations: case with Wiener process. Math. Methods Appl. Sci. **36**, 1095–1106 (2013)
10. R. Askey, J.A. Wilson, *Some Basic Hypergeometric Orthogonal Polynomials that Generalize Jacobi Polynomials* (American Mathematical Society, Providence, 1985)

Z. Zhang, G.E. Karniadakis, *Numerical Methods for Stochastic Partial Differential Equations with White Noise*,
Applied Mathematical Sciences 196, DOI 10.1007/978-3-319-57511-7

11. I. Babuska, F. Nobile, R. Tempone, A stochastic collocation method for elliptic partial differential equations with random input data. SIAM J. Numer. Anal. **45**, 1005–1034 (2007)
12. I. Babuska, R. Tempone, G.E. Zouraris, Galerkin finite element approximations of stochastic elliptic partial differential equations. SIAM J. Numer. Anal. **42**, 800–825 (2004)
13. J. Bäck, F. Nobile, L. Tamellini, R. Tempone, Stochastic spectral Galerkin and collocation methods for PDEs with random coefficients: a numerical comparison, in *Spectral and High Order Methods for Partial Differential Equations* (Springer, Berlin, Heidelberg, 2011), pp. 43–62
14. C.T.H. Baker, E. Buckwar, Numerical analysis of explicit one-step methods for stochastic delay differential equations. LMS J. Comput. Math. **3**, 315–335 (2000)
15. V. Bally, Approximation for the solutions of stochastic differential equations. I. L^p-convergence. Stoch. Stoch. Rep. **28**, 209–246 (1989)
16. V. Bally, Approximation for the solutions of stochastic differential equations. II. Strong convergence. Stoch. Stoch. Rep. **28**, 357–385 (1989)
17. V. Bally, Approximation for the solutions of stochastic differential equations. III. Jointly weak convergence. Stoch. Stoch. Rep. **30**, 171–191 (1990)
18. V. Bally, A. Millet, M. Sanz-Solé, Approximation and support theorem in Hölder norm for parabolic stochastic partial differential equations. Ann. Probab. **23**, 178–222 (1995)
19. X. Bardina, M. Jolis, L. Quer-Sardanyons, Weak convergence for the stochastic heat equation driven by Gaussian white noise. Electron. J. Probab. **15**(39), 1267–1295 (2010)
20. X. Bardina, I. Nourdin, C. Rovira, S. Tindel, Weak approximation of a fractional SDE. Stoch. Process. Appl. **120**, 39–65 (2010)
21. A. Barth, A. Lang, Milstein approximation for advection-diffusion equations driven by multiplicative noncontinuous martingale noises. Appl. Math. Optim. **66**, 387–413 (2012)
22. A. Barth, A. Lang, Simulation of stochastic partial differential equations using finite element methods. Stochastics **84**, 217–231 (2012)
23. A. Barth, A. Lang, L^p and almost sure convergence of a Milstein scheme for stochastic partial differential equations. Stoch. Process. Appl. **123**, 1563–1587 (2013)
24. A. Barth, A. Lang, C. Schwab, Multilevel Monte Carlo method for parabolic stochastic partial differential equations. BIT Numer. Math. **53**, 3–27 (2013)
25. A. Barth, C. Schwab, N. Zollinger, Multi-level Monte Carlo finite element method for elliptic PDEs with stochastic coefficients. Numer. Math. **119**, 123–161 (2011)
26. M. Barton-Smith, A. Debussche, L. Di Menza, Numerical study of two-dimensional stochastic NLS equations. Numer. Methods Partial Differ. Equ. **21**, 810–842 (2005)

27. C. Bauzet, On a time-splitting method for a scalar conservation law with a multiplicative stochastic perturbation and numerical experiments. J. Evol. Equ. **14**, 333–356 (2014)
28. C. Bayer, P.K. Friz, Cubature on Wiener space: pathwise convergence. Appl. Math. Optim. **67**, 261–278 (2013)
29. S. Becker, A. Jentzen, P.E. Kloeden, An exponential Wagner-Platen type scheme for SPDEs. SIAM J. Numer. Anal. **54**, 2389–2426 (2016)
30. G. Ben Arous, M. Grădinaru, M. Ledoux, Hölder norms and the support theorem for diffusions. Ann. Inst. H. Poincaré Probab. Stat. **30**, 415–436 (1994)
31. A. Bensoussan, R. Glowinski, A. Răşcanu, Approximation of the Zakai equation by the splitting up method. SIAM J. Control Optim. **28**, 1420–1431 (1990)
32. A. Bensoussan, R. Glowinski, A. Răşcanu, Approximation of the Zakai equation by the splitting up method. SIAM J. Control Optim. **28**, 1420–1431 (1990)
33. A. Bensoussan, R. Glowinski, A. Răşcanu, Approximation of some stochastic differential equations by the splitting up method. Appl. Math. Optim. **25**, 81–106 (1992)
34. F.E. Benth, J. Gjerde, Convergence rates for finite element approximations of stochastic partial differential equations. Stoch. Stoch. Rep. **63**, 313–326 (1998)
35. M. Bieri, C. Schwab, Sparse high order FEM for elliptic SPDEs. Comput. Methods Appl. Mech. Eng. **198**, 1149–1170 (2009)
36. D. Blömker, Amplitude equations for stochastic partial differential equations, in *Interdisciplinary Mathematical Sciences*, vol. 3 (World Scientific Publishing Co. Pte. Ltd., Hackensack, NJ, 2007)
37. D. Blömker, A. Jentzen, Galerkin approximations for the stochastic Burgers equation. SIAM J. Numer. Anal. **51**, 694–715 (2013)
38. D. Blömker, M. Kamrani, S.M. Hosseini, Full discretization of the stochastic Burgers equation with correlated noise. IMA J. Numer. Anal. **33**(3), 825–848 (2013)
39. F. Bonizzoni, F. Nobile, Perturbation analysis for the Darcy problem with log-normal permeability. SIAM/ASA J. Uncertain. Quant. **2**, 223–244 (2014)
40. N. Bou-Rabee, E. Vanden-Eijnden, A patch that imparts unconditional stability to explicit integrators for Langevin-like equations. J. Comput. Phys. **231**, 2565–2580 (2012)
41. H. Breckner, Approximation of the solution of the stochastic Navier-Stokes equation. Optimization **49**, 15–38 (2001)
42. N. Bruti-Liberati, E. Platen, Strong predictor-corrector Euler methods for stochastic differential equations. Stoch. Dyn. **8**, 561–581 (2008)
43. Z. Brzeźniak, M. Capiński, F. Flandoli, Stochastic partial differential equations and turbulence. Math. Models Methods Appl. Sci. **1**, 41–59 (1991)

44. Z. Brzeźniak, E. Carelli, A. Prohl, Finite-element-based discretizations of the incompressible Navier-Stokes equations with multiplicative random forcing. IMA J. Numer. Anal. **33**, 771–824 (2013)
45. Z. Brzeźniak, A. Carroll, Approximations of the Wong-Zakai type for stochastic differential equations in M-type 2 Banach spaces with applications to loop spaces, in *Séminaire de Probabilités XXXVII* (Springer, Berlin, 2003), pp. 251–289
46. Z. Brzeźniak, F. Flandoli, Almost sure approximation of Wong-Zakai type for stochastic partial differential equations. Stoch. Process. Appl. **55**, 329–358 (1995)
47. Z. Brzeźniak, A. Millet, On the splitting method for some complex-valued quasilinear evolution equations, in *Stochastic Analysis and Related Topics*, ed. by L. Decreusefond, J. Najim (Springer, Berlin, 2012), pp. 57–90
48. R. Buckdahn, É. Pardoux, Monotonicity methods for white noise driven quasi-linear SPDEs, in *Diffusion Processes and Related Problems in Analysis*, Evanston, IL, 1989, vol. I (Birkhäuser, Boston, 1990), pp. 219–233
49. E. Buckwar, T. Sickenberger, A comparative linear mean-square stability analysis of Maruyama- and Milstein-type methods. Math. Comput. Simul. **81**, 1110–1127 (2011)
50. E. Buckwar, R. Winkler, Multistep methods for SDEs and their application to problems with small noise. SIAM J. Numer. Anal. **44**, 779–803 (2006)
51. E. Buckwar, R. Winkler, Multi-step Maruyama methods for stochastic delay differential equations. Stoch. Anal. Appl. **25**, 933–959 (2007)
52. A. Budhiraja, L. Chen, C. Lee, A survey of numerical methods for nonlinear filtering problems. Phys. D: Nonlinear Phenom. **230**, 27–36 (2007)
53. A. Budhiraja, G. Kallianpur, Approximations to the solution of the Zakai equation using multiple Wiener and Stratonovich integral expansions. Stoch. Stoch. Rep. **56**, 271–315 (1996)
54. A. Budhiraja, G. Kallianpur, The Feynman-Stratonovich semigroup and Stratonovich integral expansions in nonlinear filtering. Appl. Math. Optim. **35**, 91–116 (1997)
55. A. Budhiraja, G. Kallianpur, Two results on multiple Stratonovich integrals. Stat. Sinica **7**, 907–922 (1997)
56. R.E. Caflisch, Monte Carlo and quasi-Monte Carlo methods. Acta Numer. **7**, 1–49 (1998)
57. R.H. Cameron, W.T. Martin, The orthogonal development of non-linear functionals in series of Fourier-Hermite functionals. Ann. Math. (2) **48**, 385–392 (1947)
58. C. Canuto, M.Y. Hussaini, A. Quarteroni, T.A. Zang, *Spectral Methods* (Springer, Berlin, 2006)

59. W. Cao, Z. Zhang, Simulations of two-step Maruyama methods for nonlinear stochastic delay differential equations. Adv. Appl. Math. Mech. **4**, 821–832 (2012)
60. W. Cao, Z. Zhang, On exponential mean-square stability of two-step Maruyama methods for stochastic delay differential equations. J. Comput. Appl. Math. **245**, 182–193 (2013)
61. W. Cao, Z. Zhang, G.E. Karniadakis, Numerical methods for stochastic delay differential equations via Wong-Zakai approximation. SIAM J. Sci. Comput. **37**(1), A295–A318 (2015)
62. Y. Cao, On convergence rate of Wiener-Ito expansion for generalized random variables. Stochastics **78**, 179–187 (2006)
63. Y. Cao, Z. Chen, M. Gunzburger, Error analysis of finite element approximations of the stochastic Stokes equations. Adv. Comput. Math. **33**, 215–230 (2010)
64. Y. Cao, H. Yang, L. Yin, Finite element methods for semilinear elliptic stochastic partial differential equations. Numer. Math. **106**, 181–198 (2007)
65. Y. Cao, L. Yin, Spectral Galerkin method for stochastic wave equations driven by space-time white noise. Commun. Pure Appl. Anal. **6**, 607–617 (2007)
66. Y. Cao, L. Yin, Spectral method for nonlinear stochastic partial differential equations of elliptic type. Numer. Math. Theory Methods Appl. **4**, 38–52 (2011)
67. Y. Cao, R. Zhang, K. Zhang, Finite element and discontinuous Galerkin method for stochastic Helmholtz equation in two- and three-dimensions. J. Comput. Math. **26**, 702–715 (2008)
68. Y. Cao, R. Zhang, K. Zhang, Finite element method and discontinuous Galerkin method for stochastic scattering problem of Helmholtz type in $\mathbb{R}^d$ $(d = 2, 3)$. Potential Anal. **28**, 301–319 (2008)
69. E. Carelli, A. Prohl, Rates of convergence for discretizations of the stochastic incompressible Navier-Stokes equations. SIAM J. Numer. Anal. **50**, 2467–2496 (2012)
70. T. Cass, C. Litterer, On the error estimate for cubature on Wiener space. Proc. Edinb. Math. Soc. (2) **57**, 377–391 (2014)
71. H.D. Ceniceros, G.O. Mohler, A practical splitting method for stiff SDEs with applications to problems with small noise. Multiscale Model. Simul. **6**, 212–227 (2007)
72. M. Chaleyat-Maurel, D. Michel, A Stroock Varadhan support theorem in nonlinear filtering theory. Probab. Theory Relat. Fields **84**, 119–139 (1990)
73. J. Charrier, Strong and weak error estimates for elliptic partial differential equations with random coefficients. SIAM J. Numer. Anal. **50**, 216–246 (2012)

74. J. Charrier, A. Debussche, Weak truncation error estimates for elliptic PDEs with lognormal coefficients. Stoch. PDE: Anal. Comp. **1**, 63–93 (2013)
75. J. Charrier, R. Scheichl, A.L. Teckentrup, Finite element error analysis of elliptic PDEs with random coefficients and its application to multilevel Monte Carlo methods. SIAM J. Numer. Anal. **51**, 322–352 (2013)
76. G.-Q. Chen, Q. Ding, K.H. Karlsen, On nonlinear stochastic balance laws. Arch. Ration. Mech. Anal. **204**, 707–743 (2012)
77. H. Cho, D. Venturi, G.E. Karniadakis, Statistical analysis and simulation of random shocks in stochastic Burgers equation. Proc. R. Soc. Lond. Ser. A Math. Phys. Eng. Sci. **470** (2014), 20140080, 21
78. P. L. Chow, J.-L. Jiang, J.-L. Menaldi, Pathwise convergence of approximate solutions to Zakai's equation in a bounded domain, in *Stochastic Partial Differential Equations and Applications*, Trento, 1990 (Longman Scientific & Technical, Harlow, 1992), pp. 111–123
79. I. Chueshov, A. Millet, Stochastic two-dimensional hydrodynamical systems: Wong-Zakai approximation and support theorem. Stoch. Anal. Appl. **29**, 570–611 (2011)
80. P.G. Ciarlet, *The Finite Element Method for Elliptic Problems* (SIAM, Philadelphia, PA, 2002)
81. Z. Ciesielski, Hölder conditions for realizations of Gaussian processes. Trans. Am. Math. Soc. **99**, 403–413 (1961)
82. K.A. Cliffe, M.B. Giles, R. Scheichl, A.L. Teckentrup, Multilevel Monte Carlo methods and applications to elliptic PDEs with random coefficients. Comput. Vis. Sci. **14**, 3–15 (2011)
83. F. Comte, V. Genon-Catalot, Y. Rozenholc, Penalized nonparametric mean square estimation of the coefficients of diffusion processes. Bernoulli **13**, 514–543 (2007)
84. R. Courant, K.O. Friedrichs, *Supersonic Flow and Shock Waves* (Interscience Publishers, Inc., New York, 1948)
85. S. Cox, J. van Neerven, Convergence rates of the splitting scheme for parabolic linear stochastic Cauchy problems. SIAM J. Numer. Anal. **48**, 428–451 (2010)
86. S. Cox, J. van Neerven, Pathwise Hölder convergence of the implicit-linear Euler scheme for semi-linear SPDEs with multiplicative noise. Numer. Math. **125**, 259–345 (2013)
87. D. Crisan, Exact rates of convergence for a branching particle approximation to the solution of the Zakai equation. Ann. Probab. **31**, 693–718 (2003)
88. D. Crisan, Particle approximations for a class of stochastic partial differential equations. Appl. Math. Optim. **54**, 293–314 (2006)
89. D. Crisan, J. Gaines, T. Lyons, Convergence of a branching particle method to the solution of the Zakai equation. SIAM J. Appl. Math. **58**, 1568–1590 (1998)

90. D. Crisan, T. Lyons, A particle approximation of the solution of the Kushner-Stratonovitch equation. Probab. Theory Relat. Fields **115**, 549–578 (1999)
91. D. Crisan, J. Xiong, Numerical solutions for a class of SPDEs over bounded domains, in *Conference Oxford sur les méthodes de Monte Carlo séquentielles* (EDP Sciences, Les Ulis, 2007), pp. 121–125
92. G. Da Prato, *Kolmogorov Equations for Stochastic PDEs* (Birkhäuser, Basel, 2004)
93. G. Da Prato, A. Debussche, R. Temam, Stochastic Burgers' equation. NoDEA Nonlinear Differ. Equ. Appl. **1**, 389–402 (1994)
94. G. Da Prato, J. Zabczyk, *Stochastic Equations in Infinite Dimensions* (Cambridge University Press, Cambridge, 1992)
95. A.M. Davie, J.G. Gaines, Convergence of numerical schemes for the solution of parabolic stochastic partial differential equations. Math. Comp. **70**, 121–134 (2001)
96. P.J. Davis, P. Rabinowitz, *Methods of Numerical Integration.* Computer Science and Applied Mathematics, 2nd edn. (Academic, Orlando, FL, 1984)
97. D.A. Dawson, Stochastic evolution equations. Math. Biosci. **15**, 287–316 (1972)
98. D.A. Dawson, E.A. Perkins, Measure-valued processes and renormalization of branching particle systems, in *Stochastic Partial Differential Equations: Six Perspectives*, ed. by R.A. Carmona, B. Rozovskii (eds.) (AMS, Providence, RI, 1999), pp. 45–106
99. A. de Bouard, A. Debussche, A stochastic nonlinear Schrödinger equation with multiplicative noise. Commun. Math. Phys. **205**, 161–181 (1999)
100. A. De Bouard, A. Debussche, A semi-discrete scheme for the stochastic nonlinear Schrödinger equation. Numer. Math. **96**, 733–770 (2004)
101. A. de Bouard, A. Debussche, Weak and strong order of convergence of a semidiscrete scheme for the stochastic nonlinear Schrödinger equation. Appl. Math. Optim. **54**, 369–399 (2006)
102. A. de Bouard, A. Debussche, Random modulation of solitons for the stochastic Korteweg-de Vries equation. Ann. Inst. H. Poincaré Anal. Non Linéaire **24**, 251–278 (2007)
103. A. de Bouard, A. Debussche, The nonlinear Schrödinger equation with white noise dispersion. J. Funct. Anal. **259**, 1300–1321 (2010)
104. A. de Bouard, A. Debussche, L. Di Menza, Theoretical and numerical aspects of stochastic nonlinear Schrödinger equations. Monte Carlo Methods Appl. **7**, 55–63 (2001)
105. A. de Bouard, A. Debussche, Y. Tsutsumi, White noise driven Korteweg-de Vries equation. J. Funct. Anal. **169**, 532–558 (1999)
106. A. de Bouard, A. Debussche, On the stochastic Korteweg-de Vries equation. J. Funct. Anal. **154**, 215–251 (1998)

107. M.K. Deb, I.M. Babuska, J.T. Oden, Solution of stochastic partial differential equations using Galerkin finite element techniques. Comput. Methods Appl. Mech. Eng. **190**, 6359–6372 (2001)
108. A. Debussche, The 2D-Navier-Stokes equations perturbed by a delta correlated noise, in *Probabilistic Methods in Fluids* (World Scientific Publishers, River Edge, NJ, 2003), pp. 115–129
109. A. Debussche, Weak approximation of stochastic partial differential equations: the nonlinear case. Math. Comp. **80**, 89–117 (2011)
110. A. Debussche, J. Printems, Numerical simulation of the stochastic Korteweg-de Vries equation. Phys. D **134**, 200–226 (1999)
111. A. Debussche, J. Printems, Convergence of a semi-discrete scheme for the stochastic Korteweg-de Vries equation. Discrete Contin. Dyn. Syst. Ser. B **6**, 761–781 (2006)
112. A. Debussche, J. Printems, Weak order for the discretization of the stochastic heat equation. Math. Comp. **78**, 845–863 (2009)
113. A. Debussche, J. Vovelle, Scalar conservation laws with stochastic forcing. J. Funct. Anal. **259**, 1014–1042 (2010)
114. A. Deya, M. Jolis, L. Quer-Sardanyons, The Stratonovich heat equation: a continuity result and weak approximations. Electron. J. Probab. **18**(3), 34 (2013)
115. J. Dick, F.Y. Kuo, I.H. Sloan, High-dimensional integration: the quasi-Monte Carlo way. Acta Numer. **22**, 133–288 (2013)
116. P. Dörsek, Semigroup splitting and cubature approximations for the stochastic Navier-Stokes equations. SIAM J. Numer. Anal. **50**, 729–746 (2012)
117. H. Doss, Liens entre équations différentielles stochastiques et ordinaires. C. R. Acad. Sci. Paris Sér. A-B **283**, Ai, A939–A942 (1976)
118. Q. Du, T. Zhang, Numerical approximation of some linear stochastic partial differential equations driven by special additive noises. SIAM J. Numer. Anal. **40**, 1421–1445 (2002)
119. J. Duan, W. Wang, *Effective Dynamics of Stochastic Partial Differential Equations* (Elsevier, Amsterdam, 2014)
120. Y. Duan, X. Yang, On the convergence of a full discretization scheme for the stochastic Navier-Stokes equations. J. Comput. Anal. Appl. **13**, 485–498 (2011)
121. Y. Duan, X. Yang, The finite element method of a Euler scheme for stochastic Navier-Stokes equations involving the turbulent component. Int. J. Numer. Anal. Model. **10**, 727–744 (2013)
122. M.A. El-Tawil, A.-H.A. El-Shikhipy, Approximations for some statistical moments of the solution process of stochastic Navier-Stokes equation using WHEP technique. Appl. Math. Inf. Sci. **6**, 1095–1100 (2012)
123. H.C. Elman, C.W. Miller, E.T. Phipps, R.S. Tuminaro, Assessment of collocation and Galerkin approaches to linear diffusion equations with random data. Int. J. Uncertain. Quant. **1**, 19–33 (2011)

124. O. Ernst, B. Sprungk, Stochastic collocation for elliptic PDEs with random data: the lognormal case, in *Sparse Grids and Applications - Munich 2012*, ed. by J. Garcke, D. Pflüger (Springer International Publishing, Cham, 2014), pp. 29–53
125. L.C. Evans, *Partial Differential Equations* (AMS, Providence, RI, 1998)
126. J. Feng, D. Nualart, Stochastic scalar conservation laws. J. Funct. Anal. **255**, 313–373 (2008)
127. G. Ferreyra, A Wong-Zakai-type theorem for certain discontinuous semimartingales. J. Theor. Probab. **2**, 313–323 (1989)
128. F. Flandoli, V.M. Tortorelli, Time discretization of Ornstein-Uhlenbeck equations and stochastic Navier-Stokes equations with a generalized noise. Stoch. Stoch. Rep. **55**, 141–165 (1995)
129. W. Fleming, Distributed parameter stochastic systems in population biology, in *Control Theory, Numerical Methods and Computer Systems Modelling*, ed. by A. Bensoussan, J. Lions (Springer, Berlin, 1975), pp. 179–191
130. P. Florchinger, F. Le Gland, Time-discretization of the Zakai equation for diffusion processes observed in correlated noise, in *Analysis and Optimization of Systems (Antibes, 1990)* (Springer, Berlin, 1990), pp. 228–237
131. P. Florchinger, F. Le Gland, Time-discretization of the Zakai equation for diffusion processes observed in correlated noise. Stoch. Stoch. Rep. **35**, 233–256 (1991)
132. P. Florchinger, F. Le Gland, Particle approximation for first order stochastic partial differential equations, in *Applied Stochastic Analysis*, New Brunswick, NJ, 1991 (Springer, Berlin, 1992), pp. 121–133
133. J. Foo, G.E. Karniadakis, Multi-element probabilistic collocation method in high dimensions. J. Comput. Phys. **229**, 1536–1557 (2010)
134. J.-P. Fouque, J. Garnier, G. Papanicolaou, K. Sølna, *Wave Propagation and Time Reversal in Randomly Layered Media*. Stochastic Modelling and Applied Probability, vol. 56 (Springer, New York, 2007)
135. P. Frauenfelder, C. Schwab, R.A. Todor, Finite elements for elliptic problems with stochastic coefficients. Comput. Methods Appl. Mech. Eng. **194**, 205–228 (2005)
136. M.I. Freidlin, Random perturbations of reaction-diffusion equations: the quasideterministic approximation. Trans. Am. Math. Soc. **305**, 665–697 (1988)
137. P. Friz, H. Oberhauser, Rough path limits of the Wong-Zakai type with a modified drift term. J. Funct. Anal. **256**, 3236–3256 (2009)
138. P. Friz, H. Oberhauser, On the splitting-up method for rough (partial) differential equations. J. Differ. Equ. **251**, 316–338 (2011)
139. J.G. Gaines, Numerical experiments with S(P)DE's, in *Stochastic Partial Differential Equations (Edinburgh, 1994)* (Cambridge University Press, Cambridge, 1995), pp. 55–71

140. J. Galvis, M. Sarkis, Approximating infinity-dimensional stochastic Darcy's equations without uniform ellipticity. SIAM J. Numer. Anal. **47**, 3624–3651 (2009)
141. J. Galvis, M. Sarkis, Regularity results for the ordinary product stochastic pressure equation. SIAM J. Math. Anal. **44**, 2637–2665 (2012)
142. A. Ganguly, Wong-Zakai type convergence in infinite dimensions. Electron. J. Probab. **18**(31), 34 (2013)
143. J. Garcia, Convergence of stochastic integrals and SDE's associated to approximations of the Gaussian white noise. Adv. Appl. Stat. **10**, 155–177 (2008)
144. M. Gardner, White and brown music, fractal curves and one-over-f fluctuations. Sci. Am. **238**, 16–27 (1978)
145. L. Gawarecki, V. Mandrekar, *Stochastic Differential Equations in Infinite Dimensions with Applications to Stochastic Partial Differential Equations*. Probability and its Applications (New York) (Springer, Heidelberg, 2011)
146. K. Gawędzki, M. Vergassola, Phase transition in the passive scalar advection. Phys. D **138**, 63–90 (2000)
147. M. Geissert, M. Kovács, S. Larsson, Rate of weak convergence of the finite element method for the stochastic heat equation with additive noise. BIT Numer. Math. **49**, 343–356 (2009)
148. A. Genz, B.D. Keister, Fully symmetric interpolatory rules for multiple integrals over infinite regions with Gaussian weight. J. Comput. Appl. Math. **71**, 299–309 (1996)
149. M. Gerencsér, I. Gyöngy, Finite difference schemes for stochastic partial differential equations in Sobolev spaces. Appl. Math. Optim. **72**, 77–100 (2015)
150. A. Germani, L. Jetto, M. Piccioni, Galerkin approximation for optimal linear filtering of infinite-dimensional linear systems. SIAM J. Control Optim. **26**, 1287–1305 (1988)
151. A. Germani, M. Piccioni, A Galerkin approximation for the Zakai equation, in *System Modelling and Optimization (Copenhagen, 1983)* (Springer, Berlin, 1984), pp. 415–423
152. A. Germani, M. Piccioni, Semidiscretization of stochastic partial differential equations on $\mathbf{R}^d$ by a finite-element technique. Stochastics **23**, 131–148 (1988)
153. T. Gerstner, Sparse Grid Quadrature Methods for Computational Finance, PhD thesis, University of Bonn, Habilitation, 2007
154. T. Gerstner, M. Griebel, Numerical integration using sparse grids. Numer. Algorithms **18**, 209–232 (1998)
155. R.G. Ghanem, P.D. Spanos, *Stochastic Finite Elements: A Spectral Approach* (Springer, New York, 1991)
156. M.B. Giles, Multilevel Monte Carlo path simulation. Oper. Res. **56**, 607–617 (2008)
157. M.B. Giles, Multilevel Monte Carlo methods, in *Monte Carlo and Quasi-Monte Carlo Methods 2012* (Springer, Berlin, 2013)

158. M.B. Giles, C. Reisinger, Stochastic finite differences and multilevel Monte Carlo for a class of SPDEs in finance. SIAM J. Financ. Math. **3**, 572–592 (2012)
159. H. Gilsing, T. Shardlow, SDELab: a package for solving stochastic differential equations in MATLAB. J. Comput. Appl. Math. **205**, 1002–1018 (2007)
160. C.J. Gittelson, J. Könnö, C. Schwab, R. Stenberg, The multi-level Monte Carlo finite element method for a stochastic Brinkman problem. Numer. Math. **125**, 347–386 (2013)
161. E. Gobet, G. Pagès, H. Pham, J. Printems, Discretization and simulation of the Zakai equation. SIAM J. Numer. Anal. **44**, 2505–2538 (2006)
162. N.Y. Goncharuk, P. Kotelenez, Fractional step method for stochastic evolution equations. Stoch. Process. Appl. **73**, 1–45 (1998)
163. D. Gottlieb, S.A. Orszag, *Numerical Analysis of Spectral Methods: Theory and Applications* (SIAM, Philadelphia, PA, 1977)
164. I. Graham, F. Kuo, J. Nichols, R. Scheichl, C. Schwab, I. Sloan, Quasi-Monte Carlo finite element methods for elliptic PDEs with lognormal random coefficients. Numer. Math. **131**, 329–368 (2015)
165. I.G. Graham, F.Y. Kuo, D. Nuyens, R. Scheichl, I.H. Sloan, Quasi-Monte Carlo methods for elliptic PDEs with random coefficients and applications. J. Comput. Phys. **230**, 3668–3694 (2011)
166. I.G. Graham, R. Scheichl, E. Ullmann, Mixed finite element analysis of lognormal diffusion and multilevel Monte Carlo methods. Stoch. PDE: Anal. Comp. **4**, 41–75 (2016)
167. W. Grecksch, P.E. Kloeden, Time-discretised Galerkin approximations of parabolic stochastic PDEs. Bull. Aust. Math. Soc. **54**, 79–85 (1996)
168. W. Grecksch, H. Lisei, Approximation of stochastic nonlinear equations of Schrödinger type by the splitting method. Stoch. Anal. Appl. **31**, 314–335 (2013)
169. W. Grecksch, B. Schmalfuß, Approximation of the stochastic Navier-Stokes equation. Mat. Apl. Comput. **15**, 227–239 (1996)
170. W. Grecksch, C. Tudor, *Stochastic Evolution Equations. A Hilbert Space Approach.* Mathematical Research, vol. 85 (Akademie, Berlin, 1995)
171. M. Griebel, Sparse grids and related approximation schemes for higher dimensional problems, in *Foundations of Computational Mathematics, Santander 2005* (Cambridge University Press, Cambridge, 2006), pp. 106–161
172. M. Griebel, M. Holtz, Dimension-wise integration of high-dimensional functions with applications to finance. J. Complex. **26**, 455–489 (2010)
173. M. Grigoriu, Control of time delay linear systems with Gaussian white noise. PrEM **12**, 89–96 (1997)
174. M. Grigoriu, *Stochastic Systems: Uncertainty Quantification and Propagation* (Springer, London, 2012)

175. C. Gugg, H. Kielhöfer, M. Niggemann, On the approximation of the stochastic Burgers equation. Commun. Math. Phys. **230**, 181–199 (2002)
176. Q. Guo, W. Xie, T. Mitsui, Convergence and stability of the split-step θ-Milstein method for stochastic delay Hopfield neural networks. Abstr. Appl. Anal. 12 pp. (2013), no. 169214
177. S.J. Guo, On the mollifier approximation for solutions of stochastic differential equations. J. Math. Kyoto Univ. **22**, 243–254 (1982)
178. B. Gustafsson, H.-O. Kreiss, J. Oliger, *Time Dependent Problems and Difference Methods* (Wiley, New York, 1995)
179. I. Gyöngy, On the approximation of stochastic differential equations. Stochastics **23**, 331–352 (1988)
180. I. Gyöngy, On the approximation of stochastic partial differential equations. I. Stochastics **25**, 59–85 (1988)
181. I. Gyöngy, On the approximation of stochastic partial differential equations. II. Stoch. Stoch. Rep. **26**, 129–164 (1989)
182. I. Gyöngy, The stability of stochastic partial differential equations and applications. I. Stoch. Stoch. Rep. **27**, 129–150 (1989)
183. I. Gyöngy, The stability of stochastic partial differential equations. II. Stoch. Stoch. Rep. **27**, 189–233 (1989)
184. I. Gyöngy, The approximation of stochastic partial differential equations and applications in nonlinear filtering. Comput. Math. Appl. **19**, 47–63 (1990)
185. I. Gyöngy, On the support of the solutions of stochastic differential equations. Teor. Veroyatnost. i Primenen. **39**, 649–653 (1994)
186. I. Gyöngy, Lattice approximations for stochastic quasi-linear parabolic partial differential equations driven by space-time white noise. I. Potential Anal. **9**, 1–25 (1998)
187. I. Gyöngy, A note on Euler's approximations. Potential Anal. **8**, 205–216 (1998)
188. I. Gyöngy, Lattice approximations for stochastic quasi-linear parabolic partial differential equations driven by space-time white noise. II. Potential Anal. **11**, 1–37 (1999)
189. I. Gyöngy, Approximations of stochastic partial differential equations, in *Stochastic Partial Differential Equations and Applications (Trento, 2002)* (Dekker, New York, 2002), pp. 287–307
190. I. Gyöngy, N. Krylov, On the rate of convergence of splitting-up approximations for SPDEs, in *Stochastic Inequalities and Applications* (Birkhäuser, Basel, 2003), pp. 301–321
191. I. Gyöngy, N. Krylov, On the splitting-up method and stochastic partial differential equations. Ann. Probab. **31**, 564–591 (2003)
192. I. Gyöngy, N. Krylov, An accelerated splitting-up method for parabolic equations. SIAM J. Math. Anal. **37**, 1070–1097 (2005)

193. I. Gyöngy, N. Krylov, Accelerated finite difference schemes for linear stochastic partial differential equations in the whole space. SIAM J. Math. Anal. **42**, 2275–2296 (2010)
194. I. Gyöngy, T. Martínez, On numerical solution of stochastic partial differential equations of elliptic type. Stochastics **78**, 213–231 (2006)
195. I. Gyöngy, G. Michaletzky, On Wong-Zakai approximations with δ-martingales. Proc. R. Soc. Lond. Ser. A Math. Phys. Eng. Sci. **460**, 309–324 (2004)
196. I. Gyöngy, D. Nualart, Implicit scheme for quasi-linear parabolic partial differential equations perturbed by space-time white noise. Stoch. Process. Appl. **58**, 57–72 (1995)
197. I. Gyöngy, D. Nualart, Implicit scheme for stochastic parabolic partial differential equations driven by space-time white noise. Potential Anal. **7**, 725–757 (1997)
198. I. Gyöngy, D. Nualart, M. Sanz-Solé, Approximation and support theorems in modulus spaces. Probab. Theory Relat. Fields **101**, 495–509 (1995)
199. I. Gyöngy, T. Pröhle, On the approximation of stochastic differential equation and on Stroock-Varadhan's support theorem. Comput. Math. Appl. **19**, 65–70 (1990)
200. I. Gyöngy, A. Shmatkov, Rate of convergence of Wong-Zakai approximations for stochastic partial differential equations. Appl. Math. Optim. **54**, 315–341 (2006)
201. I. Gyöngy, P.R. Stinga, Rate of convergence of Wong-Zakai approximations for stochastic partial differential equations, in *Seminar on Stochastic Analysis, Random Fields and Applications VII*, ed. by R.C. Dalang, M. Dozzi, F. Russo (Springer, Basel, 2013), pp. 95–130
202. L.G. Gyurkó, T.J. Lyons, Efficient and practical implementations of cubature on Wiener space, in *Stochastic Analysis 2010* (Springer, Heidelberg, 2011), pp. 73–111
203. M. Hairer, J. Maas, A spatial version of the Itô-Stratonovich correction. Ann. Probab. **40**, 1675–1714 (2012)
204. M. Hairer, M.D. Ryser, H. Weber, Triviality of the 2D stochastic Allen-Cahn equation. Electron. J. Probab. **17**, 1–14 (2012)
205. M. Hairer, J. Voss, Approximations to the stochastic Burgers equation. J. Nonlinear Sci. **21**, 897–920 (2011)
206. E.J. Hall, Accelerated spatial approximations for time discretized stochastic partial differential equations. SIAM J. Math. Anal. **44**, 3162–3185 (2012)
207. E.J. Hall, Higher order spatial approximations for degenerate parabolic stochastic partial differential equations. SIAM J. Math. Anal. **45**, 2071–2098 (2013)
208. R.Z. Has′minskiĭ, *Stochastic Stability of Differential Equations* (Sijthoff & Noordhoff, Alphen, 1980)

209. E. Hausenblas, Numerical analysis of semilinear stochastic evolution equations in Banach spaces. J. Comput. Appl. Math. **147**, 485–516 (2002)
210. E. Hausenblas, Approximation for semilinear stochastic evolution equations. Potential Anal. **18**, 141–186 (2003)
211. E. Hausenblas, Weak approximation for semilinear stochastic evolution equations, in *Stochastic Analysis and Related Topics VIII* (Birkhäuser, Basel, 2003), pp. 111–128
212. E. Hausenblas, Wong-Zakai type approximation of SPDEs of Lévy noise. Acta Appl. Math. **98**, 99–134 (2007)
213. E. Hausenblas, Weak approximation of the stochastic wave equation. J. Comput. Appl. Math. **235**, 33–58 (2010)
214. J. He, Numerical analysis for stochastic age-dependent population equations with diffusion, in *Advances in Electronic Commerce, Web Application and Communication*, ed. by D. Jin, S. Lin (Springer, Berlin, 2012), pp. 37–43
215. R.L. Herman, A. Rose, Numerical realizations of solutions of the stochastic KdV equation. Math. Comput. Simul. **80**, 164–172 (2009)
216. J.S. Hesthaven, S. Gottlieb, D. Gottlieb, *Spectral Methods for Time-Dependent Problems*. Cambridge Monographs on Applied and Computational Mathematics, vol. 21 (Cambridge University Press, Cambridge, 2007)
217. D.J. Higham, An algorithmic introduction to numerical simulation of stochastic differential equations. SIAM Rev. **43**, 525–546 (2001)
218. D.J. Higham, X. Mao, A.M. Stuart, Strong convergence of Euler-type methods for nonlinear stochastic differential equations. SIAM J. Numer. Anal. **40**, 1041–1063 (2002)
219. D.J. Higham, X. Mao, L. Szpruch, Convergence, non-negativity and stability of a new Milstein scheme with applications to finance. Discrete Contin. Dyn. Syst. Ser. B **18**, 2083–2100 (2013)
220. V.H. Hoang, C. Schwab, N-term Wiener chaos approximation rates for elliptic PDEs with lognormal Gaussian random inputs. Math. Models Methods Appl. Sci. **24**, 797–826 (2014)
221. D.G. Hobson, L.C.G. Rogers, Complete models with stochastic volatility. Math. Financ. **8**, 27–48 (1998)
222. N. Hofmann, T. Müller-Gronbach, A modified Milstein scheme for approximation of stochastic delay differential equations with constant time lag. J. Comput. Appl. Math. **197**, 89–121 (2006)
223. H. Holden, B. Øksendal, J. Uboe, T. Zhang, Stochastic Partial Differential Equations (Birkhäuser, Boston, MA, 1996)
224. H. Holden, N.H. Risebro, Conservation laws with a random source. Appl. Math. Optim. **36**, 229–241 (1997)
225. T.Y. Hou, W. Luo, B. Rozovskii, H.-M. Zhou, Wiener chaos expansions and numerical solutions of randomly forced equations of fluid mechanics. J. Comput. Phys. **216**, 687–706 (2006)

226. Y. Hu, Semi-implicit Euler-Maruyama scheme for stiff stochastic equations, in *Stochastic Analysis and Related Topics* (Birkhäuser, Boston, MA, 1996), pp. 183–202
227. Y. Hu, G. Kallianpur, J. Xiong, An approximation for the Zakai equation. Appl. Math. Optim. **45**, 23–44 (2002)
228. Y. Hu, S.-E.A. Mohammed, F. Yan, Discrete-time approximations of stochastic delay equations: the Milstein scheme. Ann. Probab. **32**, 265–314 (2004)
229. Y.-Z. Hu, J.-A. Yan, Wick calculus for nonlinear Gaussian functionals. Acta Math. Appl. Sin. Engl. Ser. **25**, 399–414 (2009)
230. C. Huang, S. Gan, D. Wang, Delay-dependent stability analysis of numerical methods for stochastic delay differential equations. J. Comput. Appl. Math. **236**, 3514–3527 (2012)
231. S.P. Huang, S.T. Quek, K.K. Phoon, Convergence study of the truncated Karhunen-Loeve expansion for simulation of stochastic processes. Int. J. Numer. Methods Eng. **52**, 1029–1043 (2001)
232. M. Hutzenthaler, A. Jentzen, On a perturbation theory and on strong convergence rates for stochastic ordinary and partial differential equations with non-globally monotone coefficients. ArXiv (2014). `https://arxiv.org/abs/1401.0295`
233. M. Hutzenthaler, A. Jentzen, Numerical approximation of stochastic differential equations with non-globally Lipschitz continuous coefficients. Mem. Am. Math. Soc. **236**, 1 (2015)
234. M. Hutzenthaler, A. Jentzen, P.E. Kloeden, Strong and weak divergence in finite time of Euler's method for stochastic differential equations with non-globally Lipschitz continuous coefficients. Proc. R. Soc. A **467**, 1563–1576 (2011)
235. M. Hutzenthaler, A. Jentzen, P.E. Kloeden, Strong convergence of an explicit numerical method for SDEs with nonglobally Lipschitz continuous coefficients. Ann. Appl. Probab. **22**, 1611–1641 (2012)
236. M. Hutzenthaler, A. Jentzen, P.E. Kloeden, Divergence of the multilevel Monte Carlo Euler method for nonlinear stochastic differential equations. Ann. Appl. Probab. **23**, 1913–1966 (2013)
237. M. Hutzenthaler, A. Jentzen, X. Wang, Exponential integrability properties of numerical approximation processes for nonlinear stochastic differential equations. Math. Comp. (2017). `https://doi.org/10.1090/mcom/3146`
238. S.M. Iacus, *Simulation and Inference for Stochastic Differential Equations*. Springer Series in Statistics (Springer, New York, 2008). With R examples
239. N. Ikeda, S. Nakao, Y. Yamato, A class of approximations of Brownian motion. Publ. Res. Inst. Math. Sci. **13**, 285–300 (1977/1978)
240. N. Ikeda, S. Taniguchi, The Itô-Nisio theorem, quadratic Wiener functionals, and 1-solitons. Stoch. Process. Appl. **120**, 605–621 (2010)

241. N. Ikeda, S. Watanabe, *Stochastic Differential Equations and Diffusion Processes* (North-Holland Publishing Co., Amsterdam, 1981)
242. K. Ito, Approximation of the Zakai equation for nonlinear filtering. SIAM J. Control Optim. **34**, 620–634 (1996)
243. K. Itô, M. Nisio, On the convergence of sums of independent Banach space valued random variables. Osaka J. Math. **5**, 35–48 (1968)
244. K. Ito, B. Rozovskii, Approximation of the Kushner equation for nonlinear filtering. SIAM J. Control Optim. **38**, 893–915 (2000)
245. M. Jardak, C.-H. Su, G.E. Karniadakis, Spectral polynomial chaos solutions of the stochastic advection equation. J. Sci. Comput. **17**, 319–338 (2002)
246. A. Jentzen, Pathwise numerical approximation of SPDEs with additive noise under non-global Lipschitz coefficients. Potential Anal. **31**, 375–404 (2009)
247. A. Jentzen, Taylor expansions of solutions of stochastic partial differential equations. Discrete Contin. Dyn. Syst. Ser. B **14**, 515–557 (2010)
248. A. Jentzen, Higher order pathwise numerical approximations of SPDEs with additive noise. SIAM J. Numer. Anal. **49**, 642–667 (2011)
249. A. Jentzen, P.E. Kloeden, The numerical approximation of stochastic partial differential equations. Milan J. Math. **77**, 205–244 (2009)
250. A. Jentzen, P.E. Kloeden, Overcoming the order barrier in the numerical approximation of stochastic partial differential equations with additive space-time noise. Proc. R. Soc. Lond. Ser. A Math. Phys. Eng. Sci. **465**, 649–667 (2009)
251. A. Jentzen, P.E. Kloeden, *Taylor Approximations for Stochastic Partial Differential Equations* (SIAM, Philadelphia, PA, 2011)
252. A. Jentzen, M. Röckner, A Milstein scheme for SPDEs. Found. Comput. Math. **15**, 313–362 (2015)
253. G.-S. Jiang, C.-W. Shu, Efficient implementation of weighted ENO schemes. J. Comput. Phys. **126**, 202–228 (1996)
254. E. Kalpinelli, N. Frangos, A. Yannacopoulos, Numerical methods for hyperbolic SPDEs: a Wiener Chaos approach. Stoch PDE: Anal Comp. **1**, 606–633 (2013)
255. I. Karatzas, S.E. Shreve, *Brownian Motion and Stochastic Calculus*, 2nd edn. (Springer, New York, 1991)
256. M.A. Katsoulakis, G.T. Kossioris, O. Lakkis, Noise regularization and computations for the 1-dimensional stochastic Allen-Cahn problem. Interfaces Free Bound. **9**, 1–30 (2007)
257. P.E. Kloeden, A. Jentzen, Pathwise convergent higher order numerical schemes for random ordinary differential equations. Proc. R. Soc. Lond. Ser. A Math. Phys. Eng. Sci. **463**, 2929–2944 (2007)
258. P.E. Kloeden, G.J. Lord, A. Neuenkirch, T. Shardlow, The exponential integrator scheme for stochastic partial differential equations: pathwise error bounds. J. Comput. Appl. Math. **235**, 1245–1260 (2011)

259. P.E. Kloeden, E. Platen, *Numerical Solution of Stochastic Differential Equations* (Springer, Berlin, 1992)
260. P.E. Kloeden, T. Shardlow, The Milstein scheme for stochastic delay differential equations without using anticipative calculus. Stoch. Anal. Appl. **30**, 181–202 (2012)
261. P.E. Kloeden, S. Shott, Linear-implicit strong schemes for Itô-Galerkin approximations of stochastic PDEs. J. Appl. Math. Stoch. Anal. **14**, 47–53 (2001)
262. L. Kocis, W.J. Whiten, Computational investigations of low-discrepancy sequences. ACM Trans. Math. Softw. **23**, 266–294 (1997)
263. F. Konecny, On Wong-Zakai approximation of stochastic differential equations. J. Multivar. Anal. **13**, 605–611 (1983)
264. R. Korn, E. Korn, G. Kroisandt, *Monte Carlo Methods and Models in Finance and Insurance* (CRC Press, Boca Raton, FL, 2010)
265. G.T. Kossioris, G.E. Zouraris, Fully-discrete finite element approximations for a fourth-order linear stochastic parabolic equation with additive space-time white noise. M2AN Math. Model. Numer. Anal. **44**, 289–322 (2010)
266. G.T. Kossioris, G.E. Zouraris, Finite element approximations for a linear Cahn-Hilliard-Cook equation driven by the space derivative of a space-time white noise. Discrete Contin. Dyn. Syst. Ser. B **18**, 1845–1872 (2013)
267. G.T. Kossioris, G.E. Zouraris, Finite element approximations for a linear fourth-order parabolic SPDE in two and three space dimensions with additive space-time white noise. Appl. Numer. Math. **67**, 243–261 (2013)
268. P. Kotelenez, *Stochastic Ordinary and Stochastic Partial Differential Equations* (Springer, New York, 2008)
269. M. Kovács, S. Larsson, F. Lindgren, Weak convergence of finite element approximations of linear stochastic evolution equations with additive noise. BIT Numer. Math. **52**, 85–108 (2012)
270. M. Kovács, S. Larsson, F. Lindgren, Weak convergence of finite element approximations of linear stochastic evolution equations with additive noise II. Fully discrete schemes. BIT Numer. Math. **53**, 497–525 (2013)
271. M. Kovács, S. Larsson, A. Mesforush, Finite element approximation of the Cahn-Hilliard-Cook equation. SIAM J. Numer. Anal. **49**, 2407–2429 (2011)
272. M. Kovács, S. Larsson, F. Saedpanah, Finite element approximation of the linear stochastic wave equation with additive noise. SIAM J. Numer. Anal. **48**, 408–427 (2010)
273. R.H. Kraichnan, Small-scale structure of a scalar field convected by turbulence. Phys. Fluids **11**, 945–953 (1968)
274. I. Kröker, C. Rohde, Finite volume schemes for hyperbolic balance laws with multiplicative noise. Appl. Numer. Math. **62**, 441–456 (2012)

275. R. Kruse, Consistency and stability of a Milstein-Galerkin finite element scheme for semilinear SPDE. Stoch. Partial Differ. Equ. Anal. Comput. **2**, 471–516 (2014)
276. R. Kruse, Optimal error estimates of Galerkin finite element methods for stochastic partial differential equations with multiplicative noise. IMA J. Numer. Anal. **34**, 217–251 (2014)
277. R. Kruse, *Strong and Weak Approximation of Semilinear Stochastic Evolution Equations* (Springer, Cham, 2014)
278. N.V. Krylov, *Introduction to the Theory of Diffusion Processes* (AMS, Providence, RI, 1995)
279. U. Küchler, E. Platen, Strong discrete time approximation of stochastic differential equations with time delay. Math. Comput. Simul. **54**, 189–205 (2000)
280. H. Kunita, Stochastic partial differential equations connected with nonlinear filtering, in *Nonlinear Filtering and Stochastic Control (Cortona, 1981)* (Springer, Berlin, 1982), pp. 100–169
281. F.Y. Kuo, C. Schwab, I.H. Sloan, Quasi-Monte Carlo finite element methods for a class of elliptic partial differential equations with random coefficients. SIAM J. Numer. Anal. **50**, 3351–3374 (2012)
282. F.Y. Kuo, C. Schwab, I.H. Sloan, Multi-level quasi-Monte Carlo finite element methods for a class of elliptic PDEs with random coefficients. Found. Comput. Math. **15**, 411–449 (2015)
283. T.G. Kurtz, P. Protter, Weak limit theorems for stochastic integrals and stochastic differential equations. Ann. Probab. **19**, 1035–1070 (1991)
284. T.G. Kurtz, P. Protter, Wong-Zakai corrections, random evolutions, and simulation schemes for SDEs, in *Stochastic Analysis* (Academic, Boston, MA, 1991), pp. 331–346
285. T.G. Kurtz, J. Xiong, Numerical solutions for a class of SPDEs with application to filtering, in *Stochastics in Finite and Infinite Dimensions.* Trends in Mathematics (Birkhäuser, Boston, MA, 2001), pp. 233–258
286. H.J. Kushner, On the differential equations satisfied by conditional probability densities of Markov processes, with applications. J. Soc. Indust. Appl. Math. Ser. A Control **2**, 106–119 (1964)
287. D.F. Kuznetsov, Strong approximation of multiple Ito and Stratonovich stochastic integrals: multiple Fourier series approach, St. Petersburg State Polytechnic University, St. Petersburg, Russian edition, 2011
288. A. Lang, A Lax equivalence theorem for stochastic differential equations. J. Comput. Appl. Math. **234**, 3387–3396 (2010)
289. A. Lang, Almost sure convergence of a Galerkin approximation for SPDEs of Zakai type driven by square integrable martingales. J. Comput. Appl. Math. **236**, 1724–1732 (2012)
290. A. Lang, P.-L. Chow, J. Potthoff, Almost sure convergence for a semidiscrete Milstein scheme for SPDEs of Zakai type. Stochastics **82**, 315–326 (2010)

291. S. Larsson, A. Mesforush, Finite-element approximation of the linearized Cahn-Hilliard-Cook equation. IMA J. Numer. Anal. **31**, 1315–1333 (2011)
292. M.P. Lazarev, P. Prasad, S.K. Sing, An approximate solution of one-dimensional piston problem. Z. Angew. Math. Phys. **46**, 752–771 (1995)
293. F. Le Gland, Splitting-up approximation for SPDEs and SDEs with application to nonlinear filtering, in *Stochastic Partial Differential Equations and Their Applications* (Springer, Berlin, 1992), pp. 177–187
294. O.P. Le Maitre, O.M. Knio, *Spectral Methods for Uncertainty Quantification* (Springer, Dordrecht, 2010)
295. C.Y. Lee, B. Rozovskii, A stochastic finite element method for stochastic parabolic equations driven by purely spatial noise. Commun. Stoch. Anal. 4 (2010), 271–297
296. R.J. LeVeque, *Numerical Methods for Conservation Laws.* Lectures in Mathematics ETH Zürich (Birkhäuser, Basel, 1990)
297. H.W. Liepmann, A. Roshko, *Elements of Gasdynamics* (Wiley, New York, 1957)
298. G. Lin, C.H. Su, G.E. Karniadakis, The stochastic piston problem. Proc. Natl. Acad. Sci. U.S.A. **101**, 15840–15845 (2004)
299. F. Lindner, R. Schilling, Weak order for the discretization of the stochastic heat equation driven by impulsive noise. Potential Anal. **38**, 345–379 (2013)
300. C. Litterer, T. Lyons, High order recombination and an application to cubature on Wiener space. Ann. Appl. Probab. **22**, 1301–1327 (2012)
301. J.E. Littlewood, R.E.A.C. Paley, Theorems on Fourier series and power series (II). Proc. Lond. Math. Soc. **S2-42**, 52–89 (1937)
302. D. Liu, Convergence of the spectral method for stochastic Ginzburg-Landau equation driven by space-time white noise. Commun. Math. Sci. **1**, 361–375 (2003)
303. H. Liu, On spectral approximations of stochastic partial differential equations driven by Poisson noise, PhD thesis, University of Southern California, 2007
304. J. Liu, A mass-preserving splitting scheme for the stochastic Schrödinger equation with multiplicative noise. IMA J. Numer. Anal. **33**, 1469–1479 (2013)
305. J. Liu, Order of convergence of splitting schemes for both deterministic and stochastic nonlinear Schrödinger equations. SIAM J. Numer. Anal. **51**, 1911–1932 (2013)
306. M. Liu, W. Cao, Z. Fan, Convergence and stability of the semi-implicit Euler method for a linear stochastic differential delay equation. J. Comput. Appl. Math. **170**, 255–268 (2004)
307. J.A. Londoño, A.M. Ramirez, Numerical performance of some Wong-Zakai type approximations for stochastic differential equations, Technical report, Department of Mathematics, National University of Colombia, Bogota, Colombia, 2006

308. G.J. Lord, C.E. Powell, T. Shardlow, *An Introduction to Computational Stochastic PDEs* (Cambridge University Press, Cambridge, 2014)
309. G.J. Lord, J. Rougemont, A numerical scheme for stochastic PDEs with Gevrey regularity. IMA J. Numer. Anal. **24**, 587–604 (2004)
310. G.J. Lord, T. Shardlow, Postprocessing for stochastic parabolic partial differential equations. SIAM J. Numer. Anal. **45**, 870–889 (2007)
311. G.J. Lord, A. Tambue, A modified semi-implicit Euler-Maruyama scheme for finite element discretization of SPDEs. ArXiv (2010). `https://arxiv.org/abs/1004.1998`
312. G.J. Lord, A. Tambue, Stochastic exponential integrators for the finite element discretization of SPDEs for multiplicative and additive noise. IMA J. Numer. Anal. **33**, 515–543 (2013)
313. G.J. Lord, V. Thümmler, Computing stochastic traveling waves. SIAM J. Sci. Comput. **34**, B24–B43 (2012)
314. S.V. Lototskiĭ, B.L. Rozovskiĭ, The passive scalar equation in a turbulent incompressible Gaussian velocity field. Uspekhi Mat. Nauk **59**, 105–120 (2004)
315. S. Lototsky, R. Mikulevicius, B.L. Rozovskii, Nonlinear filtering revisited: a spectral approach. SIAM J. Control Optim. **35**, 435–461 (1997)
316. S. Lototsky, B. Rozovskii, Stochastic differential equations: a Wiener chaos approach, in *From Stochastic Calculus to Mathematical Finance* (Springer, Berlin, 2006), pp. 433–506
317. S.V. Lototsky, Wiener chaos and nonlinear filtering. Appl. Math. Optim. **54**, 265–291 (2006)
318. S.V. Lototsky, B.L. Rozovskii, Wiener chaos solutions of linear stochastic evolution equations. Ann. Probab. **34**, 638–662 (2006)
319. S.V. Lototsky, B.L. Rozovskii, Stochastic partial differential equations driven by purely spatial noise. SIAM J. Math. Anal. **41**, 1295–1322 (2009)
320. S.V. Lototsky, B.L. Rozovskii, X. Wan, Elliptic equations of higher stochastic order. M2AN Math. Model. Numer. Anal. **44**, 1135–1153 (2010)
321. S.V. Lototsky, K. Stemmann, Solving SPDEs driven by colored noise: a chaos approach. Quart. Appl. Math. **66**, 499–520 (2008)
322. T. Lyons, N. Victoir, Cubature on Wiener space. Proc. R. Soc. Lond. Ser. A Math. Phys. Eng. Sci. **460**, 169–198 (2004)
323. V. Mackevičius, On Ikeda-Nakao-Yamato type approximations. Litovsk. Mat. Sb. **30**, 752–757 (1990)
324. V. Mackevičius, On approximation of stochastic differential equations with coefficients depending on the past. Liet. Mat. Rink. **32**, 285–298 (1992)
325. V. Mackyavichyus, Symmetric stochastic differential equations with nonsmooth coefficients. Mat. Sb. (N.S.) **116**(158), 585–592, 608 (1981)

326. P. Malliavin, Stochastic calculus of variation and hypoelliptic operators, in *Proceedings of the International Symposium SDE*, Kyoto 1976, ed. by K. Ito, (Springer, Kinokuniya, 1978), pp. 195–263
327. H. Manouzi, A finite element approximation of linear stochastic PDEs driven by multiplicative white noise. Int. J. Comput. Math. **85**, 527–546 (2008)
328. H. Manouzi, Numerical solutions of SPDEs involving white noise. Int. J. Tomogr. Stat. **10**, 96–108 (2008)
329. H. Manouzi, SPDEs driven by additive and multiplicative white noises: a numerical study. Math. Comput. Simul. **81**, 2234–2243 (2011)
330. H. Manouzi, M. Seaid, Solving Wick-stochastic water waves using a Galerkin finite element method. Math. Comput. Simul. **79**, 3523–3533 (2009)
331. H. Manouzi, M. Seaid, M. Zahri, Wick-stochastic finite element solution of reaction-diffusion problems. J. Comput. Appl. Math. **203**, 516–532 (2007)
332. H. Manouzi, T.G. Theting, Mixed finite element approximation for the stochastic pressure equation of Wick type. IMA J. Numer. Anal. **24**, 605–634 (2004)
333. H. Manouzi, T.G. Theting, Numerical analysis of the stochastic Stokes equations of Wick type. Numer. Methods Partial Differ. Equ. **23**, 73–92 (2007)
334. X. Mao, *Stochastic Differential Equations and Their Applications* (Horwood Publishing Limited, Chichester, 1997)
335. X. Mao, L. Szpruch, Strong convergence and stability of implicit numerical methods for stochastic differential equations with non-globally Lipschitz continuous coefficients. J. Comput. Appl. Math. **238**, 14–28 (2013)
336. X. Mao, L. Szpruch, Strong convergence rates for backward Euler-Maruyama method for non-linear dissipative-type stochastic differential equations with super-linear diffusion coefficients. Stochastics **85**, 144–171 (2013)
337. T. Martínez, M. Sanz-Solé, A lattice scheme for stochastic partial differential equations of elliptic type in dimension $d \geq 4$. Appl. Math. Optim. **54**, 343–368 (2006)
338. R. Marty, On a splitting scheme for the nonlinear Schrödinger equation in a random medium. Commun. Math. Sci. **4**, 679–705 (2006)
339. G. Mastroianni, G. Monegato, Error estimates for Gauss-Laguerre and Gauss-Hermite quadrature formulas, in *Approximation and Computation* (Birkhäuser, Boston, 1994), pp. 421–434
340. J. Matoušek, On the L^2-discrepancy for anchored boxes. J. Complex. **14**, 527–556 (1998)
341. J.C. Mattingly, A.M. Stuart, D.J. Higham, Ergodicity for SDEs and approximations: locally Lipschitz vector fields and degenerate noise. Stoch. Process. Appl. **101**, 185–232 (2002)

342. E.J. McShane, Stochastic differential equations and models of random processes, in *Proceedings of the Sixth Berkeley Symposium on Mathematical Statistics and Probability (Univ. California, Berkeley, Calif., 1970/1971), Vol. III: Probability theory, Berkeley, Calif., 1972* (University of California Press, Berkeley, 1972), pp. 263–294
343. W.C. Meecham, D.-T. Jeng, Use of the Wiener-Hermite expansion for nearly normal turbulence. J. Fluid Mech. **32**, 225–249 (1968)
344. R. Mikulevicius, B. Rozovskii, Separation of observations and parameters in nonlinear filtering, in *Proceedings of the 32nd IEEE Conference on Decision and Control*, vol. 2 (1993), pp. 1564–1569
345. R. Mikulevicius, B. Rozovskii, Linear parabolic stochastic PDEs and Wiener chaos. SIAM J. Math. Anal. **29**, 452–480 (1998)
346. R. Mikulevicius, B. Rozovskii, On unbiased stochastic Navier-Stokes equations. Probab. Theory Relat. Fields **54**, 787–834 (2012)
347. R. Mikulevicius, B.L. Rozovskii, Fourier–Hermite expansions for nonlinear filtering. Theory Probab. Appl. **44**, 606–612 (2000)
348. R. Mikulevicius, B.L. Rozovskii, Stochastic Navier-Stokes equations for turbulent flows. SIAM J. Math. Anal. **35**, 1250–1310 (2004)
349. R. Mikulevicius, B.L. Rozovskii, On distribution free Skorokhod–Malliavin calculus. Stoch. PDE: Anal. Comp. **4**, 319–360 (2016)
350. A. Millet, P.-L. Morien, On implicit and explicit discretization schemes for parabolic SPDEs in any dimension. Stoch. Process. Appl. **115**, 1073–1106 (2005)
351. A. Millet, M. Sanz-Solé, A simple proof of the support theorem for diffusion processes, in *Séminaire de Probabilités, XXVIII*. Lecture Notes in Mathematics, vol. 1583 (Springer, Berlin, 1994), pp. 36–48
352. A. Millet, M. Sanz-Solé, Approximation and support theorem for a wave equation in two space dimensions. Bernoulli **6**, 887–915 (2000)
353. G.N. Mil′shteĭn, A theorem on the order of convergence of mean-square approximations of solutions of systems of stochastic differential equations. Teor. Veroyatnost. i Primenen. **32**, 809–811 (1987)
354. G.N. Milstein, *Numerical Integration of Stochastic Differential Equations* (Kluwer Academic Publishers Group, Dordrecht, 1995)
355. G.N. Milstein, E. Platen, H. Schurz, Balanced implicit methods for stiff stochastic systems. SIAM J. Numer. Anal. **35**, 1010–1019 (1998)
356. G.N. Milstein, Y.M. Repin, M.V. Tretyakov, Numerical methods for stochastic systems preserving symplectic structure. SIAM J. Numer. Anal. **40**, 1583–1604 (2002)
357. G.N. Milstein, M.V. Tretyakov, Numerical methods in the weak sense for stochastic differential equations with small noise. SIAM J. Numer. Anal. **34**, 2142–2167 (1997)
358. G.N. Milstein, M.V. Tretyakov, *Stochastic Numerics for Mathematical Physics* (Springer, Berlin, 2004)

359. G.N. Milstein, M.V. Tretyakov, Numerical integration of stochastic differential equations with nonglobally Lipschitz coefficients. SIAM J. Numer. Anal. **43**, 1139–1154 (2005)
360. G.N. Milstein, M.V. Tretyakov, Computing ergodic limits for Langevin equations. Phys. D **229**, 81–95 (2007)
361. G.N. Milstein, M.V. Tretyakov, Solving parabolic stochastic partial differential equations via averaging over characteristics. Math. Comp. **78**, 2075–2106 (2009)
362. J. Ming, M. Gunzburger, Efficient numerical methods for stochastic partial differential equations through transformation to equations driven by correlated noise. Int. J. Uncertain. Quant. **3**, 321–339 (2013)
363. S. Mishra, C. Schwab, Sparse tensor multi-level Monte Carlo finite volume methods for hyperbolic conservation laws with random initial data. Math. Comp. **81**, 1979–2018 (2012)
364. S. Mishra, C. Schwab, J. Šukys, Multi-level Monte Carlo finite volume methods for nonlinear systems of conservation laws in multi-dimensions. J. Comput. Phys. **231**, 3365–3388 (2012)
365. S. Mishra, C. Schwab, J. Šukys, Multilevel Monte Carlo finite volume methods for shallow water equations with uncertain topography in multi-dimensions. SIAM J. Sci. Comput. **34**, B761–B784 (2012)
366. Y.S. Mishura, G.M. Shevchenko, Approximation schemes for stochastic differential equations in a Hilbert space. Teor. Veroyatn. Primen. **51**, 476–495 (2006)
367. K. Mohamed, M. Seaid, M. Zahri, A finite volume method for scalar conservation laws with stochastic time-space dependent flux functions. J. Comput. Appl. Math. **237**, 614–632 (2013)
368. S.E.A. Mohammed, *Stochastic Functional Differential Equations* (Pitman Publishing Ltd., Boston, MA, 1984)
369. C.M. Mora, Numerical solution of conservative finite-dimensional stochastic Schrödinger equations. Ann. Appl. Probab. **15**, 2144–2171 (2005)
370. M. Motamed, F. Nobile, R. Tempone, A stochastic collocation method for the second order wave equation with a discontinuous random speed. Numer. Math. 1–44 (2012)
371. T. Müller-Gronbach, K. Ritter, An implicit Euler scheme with nonuniform time discretization for heat equations with multiplicative noise. BIT Numer. Math. **47**, 393–418 (2007)
372. T. Müller-Gronbach, K. Ritter, Lower bounds and nonuniform time discretization for approximation of stochastic heat equations. Found. Comput. Math. **7**, 135–181 (2007)
373. T. Muller-Gronbach, K. Ritter, L. Yaroslavtseva, Derandomization of the Euler scheme for scalar stochastic differential equations. J. Complex. **28**, 139–153 (2012)

374. D. Mumford, The dawning of the age of stochasticity. Atti Accad. Naz. Lincei Cl. Sci. Fis. Mat. Natur. Rend. Lincei (9) Mat. Appl. (2000), pp. 107–125. Mathematics towards the third millennium (Rome, 1999)
375. A. Neuenkirch, L. Szpruch, First order strong approximations of scalar SDEs defined in a domain. Numer. Math. **128**, 103–136 (2014)
376. H. Niederreiter, *Random Number Generation and Quasi-Monte Carlo Methods* (SIAM, Philadelphia, PA, 1992)
377. M. Ninomiya, Application of the Kusuoka approximation with a tree-based branching algorithm to the pricing of interest-rate derivatives under the HJM model. LMS J. Comput. Math. **13**, 208–221 (2010)
378. F. Nobile, R. Tempone, Analysis and implementation issues for the numerical approximation of parabolic equations with random coefficients. Int. J. Numer. Methods Eng. **80**, 979–1006 (2009)
379. F. Nobile, R. Tempone, C.G. Webster, An anisotropic sparse grid stochastic collocation method for partial differential equations with random input data. SIAM J. Numer. Anal. **46**, 2411–2442 (2008)
380. F. Nobile, R. Tempone, C.G. Webster, A sparse grid stochastic collocation method for partial differential equations with random input data. SIAM J. Numer. Anal. **46**, 2309–2345 (2008)
381. E. Novak, K. Ritter, Simple cubature formulas with high polynomial exactness. Constr. Approx. **15**, 499–522 (1999)
382. A. Nowak, A Wong-Zakai type theorem for stochastic systems of Burgers equations. Panamer. Math. J. **16**, 1–25 (2006)
383. A. Nowak, K. Twardowska, On support and invariance theorems for a stochastic system of Burgers equations. Demonstratio Math. **39**, 691–710 (2006)
384. D. Nualart, B. Rozovskii, Weighted stochastic Sobolev spaces and bilinear SPDEs driven by space-time white noise. J. Funct. Anal. **149**, 200–225 (1997)
385. D. Nualart, M. Zakai, On the relation between the Stratonovich and Ogawa integrals. Ann. Probab. **17**, 1536–1540 (1989)
386. S. Ogawa, Quelques propriétés de l'intégrale stochastique du type noncausal. Jpn. J. Appl. Math. **1**, 405–416 (1984)
387. S. Ogawa, On a deterministic approach to the numerical solution of the SDE. Math. Comput. Simul. **55**, 209–214 (2001)
388. B. Øksendal, *Stochastic Differential Equations*. Universitext, 6th edn. (Springer, Berlin, 2003). An introduction with applications
389. G. Pagès, H. Pham, Optimal quantization methods for nonlinear filtering with discrete-time observations. Bernoulli **11**, 893–932 (2005)
390. G. Pagès, J. Printems, Optimal quadratic quantization for numerics: the Gaussian case. Monte Carlo Methods Appl. **9**, 135–165 (2003)
391. R. Paley, N. Wiener, *Fourier Transforms in the Complex Domain* (AMS Colloquium Publications, New York, 1934)
392. M.D. Paola, A. Pirrotta, Time delay induced effects on control of linear systems under random excitation. PrEM **16**, 43–51 (2001)

393. K. Petras, SmolPack: a software for Smolyak quadrature with Clenshaw-Curtis basis-sequence (2003). `http://people.sc.fsu.edu/~jburkardt/c_src/smolpack/smolpack.html`
394. P. Pettersson, G. Iaccarino, J. Nordström, A stochastic Galerkin method for the Euler equations with Roe variable transformation. J. Comput. Phys. **257**, 481–500 (2014)
395. P. Pettersson, J. Nordström, G. Iaccarino, Boundary procedures for the time-dependent Burger's equation under uncertainty. Acta Math. Sci. Ser. B Engl. Ed. **30**, 539–550 (2010)
396. J. Picard, Approximation of nonlinear filtering problems and order of convergence, in *Filtering and Control of Random Processes (Paris, 1983)* (Springer, Berlin, 1984), pp. 219–236
397. J. Picard, Approximation of stochastic differential equations and application of the stochastic calculus of variations to the rate of convergence, in *Stochastic Analysis and Related Topics (Silivri, 1986)* (Springer, Berlin, 1988), pp. 267–287
398. F. Pizzi, Stochastic collocation and option pricing, Master's thesis, Politecnico di Milano, 2012
399. C. Prevot, M. Rockner, *A Concise Course on Stochastic Partial Differential Equations* (Springer, Berlin, 2007)
400. J. Printems, On the discretization in time of parabolic stochastic partial differential equations. M2AN Math. Model. Numer. Anal. **35**, 1055–1078 (2001)
401. P. Protter, Approximations of solutions of stochastic differential equations driven by semimartingales. Ann. Probab. **13**, 716–743 (1985)
402. R. Qi, X. Yang, Weak convergence of finite element method for stochastic elastic equation driven by additive noise. J. Sci. Comput. **56**, 450–470 (2013)
403. L. Quer-Sardanyons, M. Sanz-Solé, Space semi-discretisations for a stochastic wave equation. Potential Anal. **24**, 303–332 (2006)
404. C. Reisinger, Mean-square stability and error analysis of implicit time-stepping schemes for linear parabolic SPDEs with multiplicative Wiener noise in the first derivative. Int. J. Comput. Math. **89**, 2562–2575 (2012)
405. A.J. Roberts, A step towards holistic discretisation of stochastic partial differential equations. ANZIAM J. **45**, C1–C15 (2003/04)
406. C. Roth, A combination of finite difference and Wong-Zakai methods for hyperbolic stochastic partial differential equations. Stoch. Anal. Appl. **24**, 221–240 (2006)
407. C. Roth, Weak approximations of solutions of a first order hyperbolic stochastic partial differential equation. Monte Carlo Methods Appl. **13**, 117–133 (2007)
408. B.L. Rozovskiĭ, *Stochastic Evolution Systems* (Kluwer, Dordecht, 1990)
409. G. Rozza, D.B.P. Huynh, A.T. Patera, Reduced basis approximation and a posteriori error estimation for affinely parametrized elliptic coer-

cive partial differential equations: application to transport and continuum mechanics. Arch. Comput. Methods Eng. **15**, 229–275 (2008)
410. S. Sabanis, Euler approximations with varying coefficients: the case of superlinearly growing diffusion coefficients. Ann. Appl. Probab. **26**, 2083–2105 (2016)
411. S. Sabanis, A note on tamed Euler approximations. Electron. Commun. Probab. **18**, 1–10 (2013), no. 47
412. M. Sango, Splitting-up scheme for nonlinear stochastic hyperbolic equations. Forum Math. **25**, 931–965 (2013)
413. B. Saussereau, I.L. Stoica, Scalar conservation laws with fractional stochastic forcing: existence, uniqueness and invariant measure. Stoch. Process. Appl. **122**, 1456–1486 (2012)
414. B. Schmalfuss, On approximation of the stochastic Navier-Stokes equations. Wiss. Z. Tech. Hochsch. Leuna-Merseburg **27**, 605–612 (1985)
415. W. Schoutens, *Stochastic Processes and Orthogonal Polynomials*. Lecture Notes in Statistics, vol. 146 (Springer, New York, 2000)
416. H. Schurz, Numerical analysis of stochastic differential equations without tears, in *Handbook of Stochastic Analysis and Applications* (Dekker, New York, 2002), pp. 237–359
417. C. Schwab, *p- and hp-Finite Element Methods* (The Clarendon Press Oxford University Press, New York, 1998)
418. C. Schwab, C.J. Gittelson, Sparse tensor discretizations of high-dimensional parametric and stochastic PDEs. Acta Numer. **20**, 291–467 (2011)
419. C. Schwab, R.A. Todor, Karhunen-Loeve approximation of random fields by generalized fast multipole methods. J. Comput. Phys. **217**, 100–122 (2006)
420. T. Shardlow, Numerical methods for stochastic parabolic PDEs. Numer. Funct. Anal. Optim. **20**, 121–145 (1999)
421. T. Shardlow, Weak convergence of a numerical method for a stochastic heat equation. BIT Numer. Math. **43**, 179–193 (2003)
422. G. Shevchenko, Rate of convergence of discrete approximations of solutions to stochastic differential equations in a Hilbert space. Theory Probab. Math. Stat. **69**, 187–199 (2004)
423. I.H. Sloan, S. Joe, *Lattice Methods for Multiple Integration* (Oxford University Press, New York, 1994)
424. I.H. Sloan, H. Woźniakowski, When are quasi-Monte Carlo algorithms efficient for high-dimensional integrals? J. Complex. **14**, 1–33 (1998)
425. S.A. Smolyak, Quadrature and interpolation formulas for tensor products of certain classes of functions. Sov. Math. Dokl. **4**, 240–243 (1963)
426. A.R. Soheili, M.B. Niasar, M. Arezoomandan, Approximation of stochastic parabolic differential equations with two different finite difference schemes. Bull. Iran. Math. Soc. **37**, 61–83 (2011)

427. C. Soize, R. Ghanem, Physical systems with random uncertainties: chaos representations with arbitrary probability measure. SIAM J. Sci. Comput. **26**, 395–410 (2004)
428. V.N. Stanciulescu, M.V. Tretyakov, Numerical solution of the Dirichlet problem for linear parabolic SPDEs based on averaging over characteristics, in *Stochastic Analysis 2010* (Springer, Heidelberg, 2011), pp. 191–212
429. M. Stoyanov, M. Gunzburger, J. Burkardt, Pink noise, $1/f^{\alpha}$ noise, and their effect on solutions of differential equations. Int. J. Uncertain. Quant. **1**, 257–278 (2011)
430. R.L. Stratonovič, A new form of representing stochastic integrals and equations. Vestnik Moskov. Univ. Ser. I Mat. Meh. **1964**, 3–12 (1964)
431. R.L. Stratonovich, A new representation for stochastic integrals and equations. SIAM J. Control **4**, 362–371 (1966)
432. D.W. Stroock, S.R.S. Varadhan, On the support of diffusion processes with applications to the strong maximum principle, in *Proceedings of the Sixth Berkeley Symposium on Mathematical Statistics and Probability*, Berkeley, CA, vol. III (Univ. California Press, Berkeley, 1972), pp. 333–359
433. M. Sun, R. Glowinski, Pathwise approximation and simulation for the Zakai filtering equation through operator splitting. Calcolo **30**, 219–239 (1993)
434. H.J. Sussmann, On the gap between deterministic and stochastic ordinary differential equations. Ann. Probab. **6**, 19–41 (1978)
435. H.J. Sussmann, Limits of the Wong-Zakai type with a modified drift term, in *Stochastic Analysis* (Academic, Boston, MA, 1991), pp. 475–493
436. L. Szpruch, V-stable tamed Euler schemes. ArXiv (2013). `https://arxiv.org/abs/1310.0785`
437. D. Talay, Stochastic Hamiltonian systems: exponential convergence to the invariant measure, and discretization by the implicit Euler scheme. Markov Process. Relat. Fields **8**, 163–198 (2002)
438. D.M. Tartakovsky, M. Dentz, P.C. Lichtner, Probability density functions for advective-reactive transport with uncertain reaction rates. Water Resour. Res. **45** (2009)
439. M. Tatang, G. McRae, Direct treatment of uncertainty in models of reaction and transport, tech. rep., Department of Chemical Engineering, MIT, 1994
440. A. Teckentrup, R. Scheichl, M. Giles, E. Ullmann, Further analysis of multilevel Monte Carlo methods for elliptic PDEs with random coefficients. Numer. Math. **125**, 569–600 (2013)
441. A.L. Teckentrup, Multilevel Monte Carlo methods for highly heterogeneous media, in *Proceedings of the Winter Simulation Conference, Winter Simulation Conference* (2012), pp. 32:1–32:12

442. G. Tessitore, J. Zabczyk, Wong-Zakai approximations of stochastic evolution equations. J. Evol. Equ. **6**, 621–655 (2006)
443. T.G. Theting, Solving Wick-stochastic boundary value problems using a finite element method. Stoch. Stoch. Rep. **70**, 241–270 (2000)
444. T.G. Theting, Solving parabolic Wick-stochastic boundary value problems using a finite element method. Stoch. Stoch. Rep. **75**, 49–77 (2003)
445. T.G. Theting, Numerical solution of Wick-stochastic partial differential equations, in *Proceedings of the International Conference on Stochastic Analysis and Applications* (Kluwer Academic Publishers, Dordrecht, 2004), pp. 303–349
446. V. Thomée, *Galerkin Finite Element Methods for Parabolic Problems*, 2nd edn. (Springer, Berlin, 2006)
447. L.N. Trefethen, *Spectral Methods in Matlab* (SIAM, Philadelphia, PA, 2000)
448. M.V. Tretyakov, Z. Zhang, A fundamental mean-square convergence theorem for SDEs with locally Lipschitz coefficients and its applications. SIAM J. Numer. Anal. **51**, 3135–3162 (2013)
449. L.S. Tsimring, A. Pikovsky, Noise-induced dynamics in bistable systems with delay. Phys. Rev. Lett. **87**, 250602 (2001)
450. C. Tudor, Wong-Zakai type approximations for stochastic differential equations driven by a fractional Brownian motion. Z. Anal. Anwend. **28**, 165–182 (2009)
451. C. Tudor, M. Tudor, On approximation of solutions for stochastic delay equations. Stud. Cerc. Mat. **39**, 265–274 (1987)
452. M. Tudor, Approximation schemes for stochastic equations with hereditary argument. Stud. Cerc. Mat. **44**, 73–85 (1992)
453. K. Twardowska, On the approximation theorem of the Wong-Zakai type for the functional stochastic differential equations. Probab. Math. Stat. **12**, 319–334 (1991)
454. K. Twardowska, An extension of the Wong-Zakai theorem for stochastic evolution equations in Hilbert spaces. Stoch. Anal. Appl. **10**, 471–500 (1992)
455. K. Twardowska, An approximation theorem of Wong-Zakai type for nonlinear stochastic partial differential equations. Stoch. Anal. Appl. **13**, 601–626 (1995)
456. K. Twardowska, An approximation theorem of Wong-Zakai type for stochastic Navier-Stokes equations. Rend. Sem. Mat. Univ. Padova **96**, 15–36 (1996)
457. K. Twardowska, Wong-Zakai approximations for stochastic differential equations. Acta Appl. Math. **43**, 317–359 (1996)
458. K. Twardowska, T. Marnik, M. Pasławaska-Południak, Approximation of the Zakai equation in a nonlinear filtering problem with delay. Int. J. Appl. Math. Comput. Sci. **13**, 151–160 (2003)
459. G. Vage, Variational methods for PDEs applied to stochastic partial differential equations. Math. Scand. **82**, 113–137 (1998)

460. K. Vasilakos, A. Beuter, Effects of noise on a delayed visual feedback system. J. Theor. Biol. **165**, 389–407 (1993)
461. D. Venturi, D. Tartakovsky, A. Tartakovsky, G.E. Karniadakis, exact PDF equations and closure approximations for advective-reactive transport. J. Comput. Phys. **243**, 323–343 (2013)
462. D. Venturi, X. Wan, R. Mikulevicius, B.L. Rozovskii, G.E. Karniadakis, Wick-Malliavin approximation to nonlinear stochastic partial differential equations: analysis and simulations. Proc. R. Soc. Edinb. Sect. A. **469** (2013)
463. J.B. Walsh, A stochastic model of neural response. Adv. Appl. Probab. **13**, 231–281 (1981)
464. J.B. Walsh, Finite element methods for parabolic stochastic PDE's. Potential Anal. **23**, 1–43 (2005)
465. J.B. Walsh, On numerical solutions of the stochastic wave equation. Ill. J. Math. **50**, 991–1018 (2006)
466. X. Wan, A note on stochastic elliptic models. Comput. Methods Appl. Mech. Eng. **199**, 2987–2995 (2010)
467. X. Wan, A discussion on two stochastic elliptic modeling strategies. Commun. Comput. Phys. **11**, 775–796 (2012)
468. X. Wan, G. Karniadakis, Beyond Wiener-Askey expansions: handling arbitrary PDFs. J. Sci. Comput. **27**, 455–464 (2006)
469. X. Wan, B. Rozovskii, G.E. Karniadakis, A stochastic modeling methodology based on weighted Wiener chaos and Malliavin calculus. Proc. Natl. Acad. Sci. U.S.A. **106**, 14189–14194 (2009)
470. X. Wan, B.L. Rozovskii, The Wick–Malliavin approximation of elliptic problems with log-normal random coefficients. SIAM J. Sci. Comput. **35**, A2370–A2392 (2013)
471. P. Wang, D.M. Tartakovsky, Uncertainty quantification in kinematic-wave models. J. Comput. Phys. **231**, 7868–7880 (2012)
472. X. Wang, S. Gan, B-convergence of split-step one-leg theta methods for stochastic differential equations. J. Appl. Math. Comput. **38**, 489–503 (2012)
473. X. Wang, S. Gan, A Runge-Kutta type scheme for nonlinear stochastic partial differential equations with multiplicative trace class noise. Numer. Algorithms **62**, 193–223 (2013)
474. X. Wang, S. Gan, The tamed Milstein method for commutative stochastic differential equations with non-globally Lipschitz continuous coefficients. J. Differ. Equ. Appl. **19**, 466–490 (2013)
475. X. Wang, S. Gan, Weak convergence analysis of the linear implicit Euler method for semilinear stochastic partial differential equations with additive noise. J. Math. Anal. Appl. **398**, 151–169 (2013)
476. X. Wang, S. Gan, J. Tang, Higher order strong approximations of semilinear stochastic wave equation with additive space-time white noise. SIAM J. Sci. Comput. **36**, A2611–A2632 (2015)

477. G.W. Wasilkowski, H. Woźniakowski, Explicit cost bounds of algorithms for multivariate tensor product problems. J. Complex. **11**, 1–56 (1995)
478. G.B. Whitham, *Linear and Nonlinear Waves* (Wiley, New York, 1974)
479. N. Wiener, The homogeneous chaos. Am. J. Math. **60**, 897–936 (1938)
480. M. Wiktorsson, Joint characteristic function and simultaneous simulation of iterated Itô integrals for multiple independent Brownian motions. Ann. Appl. Probab. **11**, 470–487 (2001)
481. E. Wong, M. Zakai, On the convergence of ordinary integrals to stochastic integrals. Ann. Math. Stat. **36**, 1560–1564 (1965)
482. E. Wong, M. Zakai, On the relation between ordinary and stochastic differential equations. Int. J. Eng. Sci. **3**, 213–229 (1965)
483. F. Wu, X. Mao, L. Szpruch, Almost sure exponential stability of numerical solutions for stochastic delay differential equations. Numer. Math. **115**, 681–697 (2010)
484. D. Xiu, Fast numerical methods for stochastic computations: a review. Commun. Comput. Phys. **5**, 242–272 (2009)
485. D. Xiu, *Numerical Methods for Stochastic Computations: A Spectral Method Approach* (Princeton University Press, Princeton, 2010)
486. D. Xiu, J. Hesthaven, High-order collocation methods for differential equations with random inputs. SIAM J. Sci. Comput. **27**, 1118–1139 (2005)
487. D. Xiu, G.E. Karniadakis, Modeling uncertainty in steady state diffusion problems via generalized polynomial chaos. Comput. Methods Appl. Mech. Eng. **191**, 4927–4948 (2002)
488. D. Xiu, G.E. Karniadakis, The Wiener-Askey polynomial chaos for stochastic differential equations. SIAM J. Sci. Comput. **24**, 619–644 (2002)
489. J. Xu, J. Li, Sparse Wiener chaos approximations of Zakai equation for nonlinear filtering, in *Proceedings of the 21st Annual International Conference on Chinese Control and Decision Conference, CCDC'09*, Piscataway, NJ (IEEE Press, New York, 2009), pp. 910–913
490. Y. Yan, Semidiscrete Galerkin approximation for a linear stochastic parabolic partial differential equation driven by an additive noise. BIT Numer. Math. **44**, 829–847 (2004)
491. Y. Yan, Galerkin finite element methods for stochastic parabolic partial differential equations. SIAM J. Numer. Anal. **43**, 1363–1384 (2005)
492. X. Yang, Y. Duan, Y. Guo, A posteriori error estimates for finite element approximation of unsteady incompressible stochastic Navier-Stokes equations. SIAM J. Numer. Anal. **48**, 1579–1600 (2010)
493. X. Yang, W. Wang, Y. Duan, The approximation of a Crank-Nicolson scheme for the stochastic Navier-Stokes equations. J. Comput. Appl. Math. **225**, 31–43 (2009)

494. R.-M. Yao, L.-J. Bo, Discontinuous Galerkin method for elliptic stochastic partial differential equations on two and three dimensional spaces. Sci. China Ser. A **50**, 1661–1672 (2007)
495. H. Yoo, Semi-discretization of stochastic partial differential equations on $\mathbf{R}^1$ by a finite-difference method. Math. Comp. **69**, 653–666 (2000)
496. N. Yoshida, Stochastic shear thickening fluids: strong convergence of the Galerkin approximation and the energy equality. Ann. Appl. Probab. **22**, 1215–1242 (2012)
497. K. Yosida, *Functional Analysis*, vol. 11 (Springer, Berlin, 1995), p. 14. Reprint of the sixth (1980) edition
498. M. Zahri, M. Seaid, H. Manouzi, M. El-Amrani, Wiener-Itô chaos expansions and finite-volume solution of stochastic advection equations, in *Finite Volumes for Complex Applications IV* (ISTE, London, 2005), pp. 525–538
499. M. Zakai, On the optimal filtering of diffusion processes. Z. Wahrscheinlichkeitstheorie und Verw. Gebiete **11**, 230–243 (1969)
500. G. Zhang, M. Gunzburger, Error analysis of a stochastic collocation method for parabolic partial differential equations with random input data. SIAM J. Numer. Anal. **50**, 1922–1940 (2012)
501. H. Zhang, S. Gan, L. Hu, The split-step backward Euler method for linear stochastic delay differential equations. J. Comput. Appl. Math. **225**, 558–568 (2009)
502. L. Zhang, Q.M. Zhang, Convergence of numerical solutions for the stochastic Navier-Stokes equation. Math. Appl. (Wuhan) **21**, 504–509 (2008)
503. Z. Zhang, H. Ma, Order-preserving strong schemes for SDEs with locally lipschitz coefficients. Appl. Numer. Math. **112**, 1–16 (2017)
504. Z. Zhang, M. Choi, G.E. Karniadakis, Error estimates for the ANOVA method with polynomial chaos interpolation: tensor product functions. SIAM J. Sci. Comput. **34**, A1165–A1186 (2012)
505. Z. Zhang, B. Rozovskii, M.V. Tretyakov, G.E. Karniadakis, A multistage Wiener chaos expansion method for stochastic advection-diffusion-reaction equations. SIAM J. Sci. Comput. **34**, A914–A936 (2012)
506. Z. Zhang, M.V. Tretyakov, B. Rozovskii, G.E. Karniadakis, A recursive sparse grid collocation method for differential equations with white noise. SIAM J. Sci. Comput. **36**, A1652–A1677 (2014)
507. Z. Zhang, M.V. Tretyakov, B. Rozovskii, G.E. Karniadakis, Wiener chaos vs stochastic collocation methods for linear advection-diffusion equations with multiplicative white noise. SIAM J. Numer. Anal. **53**, 153–183 (2015)
508. Z. Zhang, X. Yang, G. Lin, G.E. Karniadakis, Numerical solution of the Stratonovich- and Ito-Euler equations: application to the stochastic piston problem. J. Comput. Phys. **236**, 15–27 (2013)

509. M. Zheng, B. Rozovsky, G.E. Karniadakis, Adaptive Wick-Malliavin approximation to nonlinear SPDEs with discrete random variables. SIAM J. Sci. Comput. **37**, A1872–A1890 (2015)
510. B. Zhibaĭtis, V. Matskyavichyus, Gaussian approximations of Brownian motion in a stochastic integral. Liet. Mat. Rink. **33**, 508–526 (1993)
511. B. Žibaitis, Mollifier approximation of Brownian motion in stochastic integral. Litovsk. Mat. Sb. **30**, 717–727 (1990)
512. X. Zong, F. Wu, C. Huang, Convergence and stability of the semi-tamed Euler scheme for stochastic differential equations with non-Lipschitz continuous coefficients. Appl. Math. Comput. **228**, 240–250 (2014)

Index

Z. Zhang, G.E. Karniadakis, *Numerical Methods for Stochastic Partial Differential Equations with White Noise*,
Applied Mathematical Sciences 196, DOI 10.1007/978-3-319-57511-7

GPSR Compliance
The European Union's (EU) General Product Safety Regulation (GPSR) is a set of rules that requires consumer products to be safe and our obligations to ensure this.

If you have any concerns about our products, you can contact us on

ProductSafety@springernature.com

In case Publisher is established outside the EU, the EU authorized representative is:

Springer Nature Customer Service Center GmbH
Europaplatz 3
69115 Heidelberg, Germany

www.ingramcontent.com/pod-product-compliance
Ingram Content Group UK Ltd.
Pitfield, Milton Keynes, MK11 3LW, UK
UKHW021008290726
14059UKWH00001BA/29
* 9 7 8 3 3 1 9 5 7 5 1 0 0 *